Sustainable Landscape Planning and Natural Resources Management

IEREK Interdisciplinary Series for Sustainable Development

Sustainable Landscape Planning and Natural Resources Management (SLNR) is a peer-reviewed book series that presents groundbreaking research through the intersection of two main areas that of which are 'Sustainable Landscape Planning' and 'Natural Resources Management'. The series initially covers a wide range of academic disciplines in sustainability, including landscape ecology, green urbanism, urban regeneration, land restoration and renaturation, energy management, and nature conservation. Moreover, it also incorporates the latest developments and research in engineering topics that contribute to more sustainable cities, such as Geosciences, infrastructure, civil engineering, geology, mining, water, and natural resources conservation and management.

By showcasing methods for discovering, managing, and employing natural resources, the series aims to provide approaches that enable the responsible use of these resources. Ultimately, the SLNR series is committed to advancing the United Nations' Sustainable Development Goals for 2030.

The wide range of topics covered in this series are primarily related to sustainable development, involving durable infrastructure projects, civil engineering, urban regeneration, the built, sustainability, climate change mitigation, energy management, geosciences, geology, and the future of cities, all of which depend on the availability of natural resources and effective wealth management. The series also features conceptual and empirical contributions from various interrelated fields of earth science, with an emphasis on the era of new discovery. It focuses on providing practical solutions to ensure sustainable development and provides new case studies that offer concrete examples of how to manage natural resources effectively.

The series encompasses the best research papers from various international conferences organized by IEREK conferences, also accepts proposals from other international conferences, including interdisciplinary contributions. It exhibits innovative approaches and highlights how they can best support both economic and sustainable development through the responsible use of natural resources and global, local, and individual activities for the welfare of all societies on Earth.

The SLNR series is a valuable source for experts in government, international organizations, consultancies, industry, business sectors, and academics. Published in cooperation with IEREK, the Springer SLNR series displays important new research in various sectors, delivering valuable insights and practical solutions to the complex challenges of sustainable development for the readers.

Kartika Sharma
Editor

Bio-prospecting of Novel Microbial Bioactive Compounds for Sustainable Development

Management of Natural Resources Through Microbial Conversion into Valuable Products

Springer

Editor
Kartika Sharma
Department of Biological Science
College of Basic Science
Bikaner, Rajasthan, India

ISSN 2948-1910 ISSN 2948-1929 (electronic)
Sustainable Landscape Planning and Natural Resources Management
IEREK Interdisciplinary Series for Sustainable Development
ISBN 978-3-031-82177-6 ISBN 978-3-031-82178-3 (eBook)
https://doi.org/10.1007/978-3-031-82178-3

This Springer imprint is published by the registered company Springer Nature Switzerland AG
The registered company address is: Gewerbestrasse 11, 6330 Cham, Switzerland

Preface

Microbes have been a valuable source of bioactive natural chemicals for over several years. Some of these compounds have been turned into medications to treat immune-system-related disorders, cancer, and infections. Novel compounds produced by microbes were traditionally found using traditional bioprospecting, which involved isolating possible producers and screening extracts from them using a range of bioassays. As a result of the discovery of the majority of the naturally occurring compounds identified by this method, the pipeline for novel medications derived from metabolites produced by microbes began to close.

The book *Bio-prospecting of Novel Microbial Bioactive Compounds for Sustainable Development: Management of Natural Resources Through Microbial Conversion into Valuable Products* covers recent advancements in the production of bioactive compounds, including therapeutics, pharmaceutical, biopesticides, enzymes, biosensors, biofertilizers, nutraceutical products. It offers a thorough understanding of how to use microbes or bioactive substances derived from microbes to create sustainable solutions. Additionally, the use of microbe-based goods in producing greener technology and controlling environmental pollution is highlighted. The book promotes the commercialization of bioactive substances obtained from microorganisms and their use in environmental protection research, agriculture, and the biochemical industries. The book also provides insightful information about the principles of microbial prospecting. Researchers, academics, and undergraduate and graduate students working in the field of microbial bioprospection of bioactive chemicals for conversion into useful products might use this as a reference book.

Bikaner, India

Kartika Sharma

Acknowledgment

First and foremost, I want to express my gratitude and praise to God, the Almighty, for providing me with the knowledge, opportunity, and numerous blessings that have allowed me to complete the book project. I appreciate the contributions made by each of the authors of the different chapters. From the first outlines to the development of the whole chapters and the subsequent revisions based on reviewer feedback, it had been a somewhat drawn-out process. I truly appreciate that the author agreed to go through this procedure. I also thank the members of our review panels for their work and expertise, many of which had to be completed on short notice. I am grateful to everyone at Springer Nature, particularly to Mr. Yogesh Padmanaban and Dr. Christian Witschel, and IEREK, particularly to Dr. Mourad Amer and Mr. Mayar Ehab with whom I had correspondence, for their guidance and assistance in the writing of this book.

Kartika Sharma

Contents

About the Editor

Dr. Kartika Sharma is Assistant Professor, Department of Biological Science, College of Basic Science, Bikaner, Rajasthan, India. Dr. Sharma completed her doctoral degree in Environmental Biology from Songkla University, Songkhla, Thailand, in 2023. Her specialization is bioactive compounds characterization and Heavy metal remediation. She is also working on the remediation of soil and water pollutants using recent technology like nanoparticles, biochar, etc. She has published more than 21 research articles in leading international and journals having high impact factors. She has received many awards and fellowships from different government organizations. She is involved in teaching and research and her area of research is plant and microbial biotechnology. She is an active reviewer for journals, including 3 Biotech, Chemosphere, Science of Total Environment, Gene, Journal of Hazardous Materials, PLoS One, etc.

Role of Microorganisms in Biosurfactant Production from Agricultural and Industrial Wastes

Asma Abid, Nour Elhouda Mekhadmi, Randa Mlik, and Kamilia Bireche

Abstract

Biosurfactants, characterized by their amphiphilic nature, are synthesized by a diverse array of microorganisms. These microorganisms possess the enzymatic capabilities to decompose complex organic molecules present in agricultural and industrial wastes. Through sophisticated fermentation and biodegradation processes, these substrates are converted into valuable biosurfactants. Agricultural residues, including crop remains, food processing by-products, and industrial effluents, serve as abundant feedstocks for microbial biosurfactant production. The resulting biosurfactants exhibit desirable properties such as biodegradability and low toxicity, making them highly suitable for various industrial applications. However, optimizing the production process remains challenging, necessitating careful consideration of factors such as microbial strain selection, fermentation conditions, and downstream processing. Continuous research is imperative to deepen our understanding of microbial metabolism and enhance bioprocess optimization, thereby promoting greater efficiency and sustainability in the production of biosurfactants from agricultural and industrial wastes. This chapter highlights the crucial role of microorganisms in converting agricultural and industrial wastes into biosurfactants, emphasizing the need for continued research to overcome production challenges and promote sustainable practices.

Keywords

Biosurfactants · Microorganisms · Agricultural wastes · Industrial wastes

1 Introduction

Surfactants are amphiphilic substances that lower the overall energy of a system by displacing high-energy bulk molecules at a contact. They possess a hydrophobic segment that has minimal attraction to the surrounding medium and a hydrophilic group that is drawn to the surrounding medium (Fig. 1). Surfactants have been employed in industrial applications for their sticky, flocculating, wetting, foaming, demulsifying, and penetrating properties (Abdel-Rahem, 2024). Biosurfactants, also known as tension-active biomolecules, are a type of surfactants biologically produced by various microorganisms. These include bacteria, filamentous fungi, yeast, and even some animals and plants (Ceresa et al., 2023). These compounds' amphiphilic characteristics and low molecular weight point to their genesis in microbes (Uzoigwe et al., 2015). To put it functionally, biosurfactants lower surface and interfacial tension by concentrating at water/air, oil/air, or oil/water interfaces. Although they have long been linked to hydrocarbon breakdown, new studies have revealed their broader functions in microbial physiology. Antibacterial and anticancer effects are among the therapeutic potentials of biosurfactants, which microbes make to alter cellular or environmental properties; these

A. Abid (✉) · K. Bireche
Laboratory of Valorization and Promotion of Saharan Resources (VPRS), Faculty of Mathematics and Matter Sciences, University of Ouargla, Ouargla, Algeria
e-mail: abid.asma@univ-ouargla.dz

N. E. Mekhadmi
Department of Biology, SNV Faculty, University of Chahid Hamma Lakhdar, El-Oued, Algeria

N. E. Mekhadmi
Laboratory of Biodiversity and Applications of Biotechnology in the Agriculture Field, University of Chahid Hamma Lakhdar, El-Oued, Algeria

R. Mlik
National Institute of Agronomic Research of Algeria, Adrar, Algeria

K. Sharma (ed.), *Bio-prospecting of Novel Microbial Bioactive Compounds for Sustainable Development*, Sustainable Landscape Planning and Natural Resources Management, https://doi.org/10.1007/978-3-031-82178-3_1

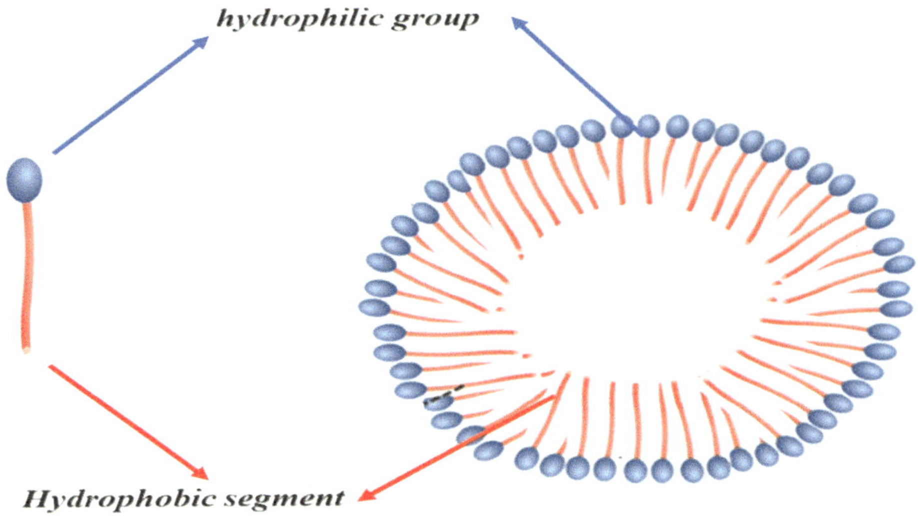

Fig. 1 The surfactant forms (hydrophobic segment and hydrophilic group)

biosurfactants are essential for microbial processes such as biofilm creation, cell motility, and fruiting body formation (Sharma et al., 2021).

With climate change and the increasing global population, pursuing innovative, efficient, and sustainable solutions is crucial. Biosurfactants from various microbes possess unique benefits compared to synthetic surfactants. These advantages include mild manufacturing conditions, adaptability, improved biodegradability, and minimized damage to living cells. Biosurfactants are in high demand due to their capacity to lower surface tension, stabilize emulsions, and biodegrade. They are widely used in diverse industries such as food, agriculture, and environmental restoration (Gayathiri et al., 2022). Furthermore, a wide variety of biosurfactants have the potential to be commercially produced and widely used in the pharmaceutical, cosmetics, and food industries. These biosurfactants are considered less environmentally harmful and have lower toxicity levels. Environmental microorganisms, especially in the rhizosphere and with plants, produce biosurfactants that aid mobility, communication, and biofilm formation. Thus, they profoundly affect plant-microorganism interactions. Biosurfactants eradicate plant diseases, boost beneficial microbial nutrient availability, and remediate agricultural soil (Das & Rao, 2024). According to Domínguez Rivera et al. (2019), biosurfactants made from agro-industrial wastes have advantages such as lower production costs, renewable materials, large-scale substrate production, product preservation, non-toxicity to microorganisms, and environmentally friendly constituents (Domínguez Rivera et al., 2019).

Biosurfactants' effectiveness and eco-friendliness make them highly sought after for use in biomedical and pharmaceutical applications. Due to their antioxidant, antibacterial, and anti-inflammatory qualities, they are appropriate for wound healing, immunomodulation, and medicine administration. However, the exorbitant manufacturing and purifying expenses hinder the extensive use of these chemicals. To address this issue, it is necessary to develop genetically modified strains that can create certain biosurfactants. Regulatory compliance and toxicity testing are required to ensure these strains' safe use. Although there are challenges, the engagement of key players in the market bodes well for biosurfactants in these industries (Ceresa et al., 2023).

The agro-industry generates lots of garbage with different qualities. The global agricultural wastes surpass 2 billion tonnes. Thus, exploring ways to exploit agro-industrial wastes better to maximize industry benefits is important. Inadequate handling of these waste materials negatively impacts the environment and the economic well-being of numerous countries. The current imperative is to develop effective and environmentally friendly waste management strategies. This involves two key aspects: firstly, the efficient transformation of wastes into valuable products and byproducts at reasonable treatment expenses, and secondly, evaluating the impact of these practices on soil quality and productivity (Singh et al., 2021). The literature discusses different types of wastes and byproducts resulting from agro-industrial processes utilized in biosurfactant production. These include wastes from oil processing, starch, sugar industry, fruits and vegetables, distilleries, animal fat, and petroleum byproducts. The utilization of industrial wastes offers several benefits, including cost reduction in production, increased availability of affordable and renewable substrates, large-scale substrate production, unchanged fundamental functional properties of the product, non-harmful nature of the product to microorganisms, and the eco-friendly and safe composition of all product components (Banat et al., 2014).

Historically, the predominant surfactants employed in industrial settings were derived from petrochemical sources.

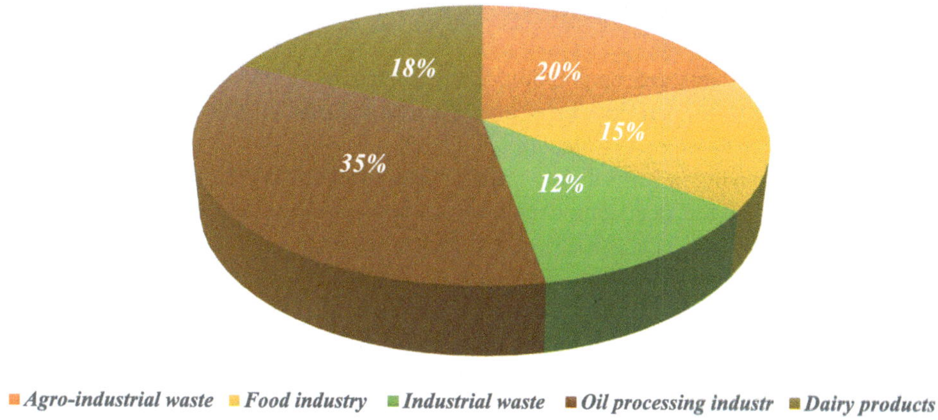

Fig. 2 The Contribution of different waste materials in biosurfactant production

However, these surfactants exhibit only partial biodegradability and toxicity towards living creatures, contributing to various environmental problems. In addition, the production of these items exhausts the Earth's finite petrochemical resources (Uzoigwe et al., 2015). The global output of synthetic surfactants surpasses 15 million tons annually (Fung et al., 2023), catering to diverse industrial sectors, including food processing, pharmaceuticals, cosmetics, detergent manufacturing, environmental cleanup (bioremediation), and enhanced oil recovery. In 2010, Reznik et al., (2010) published a study stating that the worldwide output of artificial surfactants reached 13 million tons in 2008. This production increased by 2% in 2009, resulting in annual revenue of US$2,433 billion. It was anticipated that the market would experience a consistent growth rate of 28% from 2009 to 2012, followed by an approximate increase of 3.5–4%. Nevertheless, numerous research studies have indicated that biosurfactants possess multiple advantages over petrochemical-based or synthetic surfactants. They demonstrate outstanding surface activity and emulsification characteristics while possessing little toxicity and enhanced biodegradability. They have high efficacy at low concentrations and can function well under various environmental circumstances, including pH, temperature, salinity, alkalinity, and acidity. Additional significant attributes of these biomolecules encompass enhanced environmental compatibility, reduced critical micelle concentration, heightened selectivity, specific activity, and the capacity to be synthesized from renewable, cost-effective resources (Rawat et al., 2020).

There has been a significant rise in the concern for environmental safety in recent years. The management of waste materials originating from diverse sources poses a formidable challenge. Utilizing garbage as a substrate for biosurfactant manufacturing offers a promising way to address wastes disposal issues (Yañez-Ocampo et al., 2017). Scientists have employed various waste products in varying amounts, as depicted in Fig. 2, to create biosurfactants for specific purposes. The distribution of waste sources is as follows: agro-industrial wastes account for 20%, the food industry contributes 15%, industrial wastes make up 12%, the oil processing industry is responsible for 35%, and dairy products contribute 18% (Jimoh & Lin, 2019).

In light of this, the chapter focuses on the considerable contribution that microorganisms make in the process of synthesizing biosurfactants from waste materials that are produced by industrial and agricultural processes.

2 Microbial Production of Biosurfactant

The dual nature of the biosurfactants, with both hydrophilic (water-loving) and hydrophobic (water-hating) regions, allows them to reduce surface tension at the interface between water and oil, making them valuable tools in various applications. Here's a closer look at the major types of biosurfactants produced by microbes. Different bacteria, yeast, and fungi species, especially those with filaments, create biosurfactants in varying quantities. They have roughly 60 distinct congeners and homologs. The primary category of microbes responsible for producing biosurfactants comprises *Pseudomonas* species, which were initially identified as producers of these secondary metabolites. *Acinetobacter sp., Arthrobacter sp., Bacillus sp.* and *Candida sp.* are more microbes that generate biosurfactants. It is commonly recognized that *Pseudomonas* species can create a wide range of bioactive metabolites on land and in the ocean (Abacha et al., 2016). Bhatnagar and Kim (2012) revealed that these unique bacteria generate around 800 bioactive compounds, the majority of which are antibiotic agents and others with a variety of other characteristics (Bhatnagar & Kim, 2012).

2.1 Bacterial Biosurfactant

Bacterial biosurfactants are amphiphilic compounds produced by various bacteria species like *Pseudomonas*, *Bacillus*, *Acinetobacter*, *Stenotrophomonas*, and *Burkholderia*, playing a crucial role in soil bioremediation by enhancing the degradation of hydrocarbons and heavy metals (Fardami et al., 2022). These biosurfactants exhibit excellent surface activity, emulsification properties, and biodegradability, making them environmentally friendly alternatives to chemical surfactants (Ingle, 2023). They are structurally diverse, including glycolipids, lipopeptides, fatty acids, and neutral lipids, with different functions and applications across industries like agriculture, petroleum, cosmetics, and medicine (Fatima et al., 2022). *Pseudomonas species*, known for producing rhamnolipids, have been studied for their effectiveness in cleaning applications and inhibiting pathogenic fungi like *Fusarium oxysporum*, showcasing the multifunctional properties of bacterial biosurfactants. Gaining a comprehensive understanding of the physiology, genetics, and biochemistry of strains that produce biosurfactants is crucial for improving production techniques and minimizing costs in many applications.

2.2 Biosurfactants Produced by Yeast

Yeast biosurfactants are popular due to their eco-friendliness and industrial applications. Yeasts like *Yarrowia lipolytica* and JAF-11 are effective biosurfactant makers, lowering surface tension and functioning as emulsifiers (Kim et al., 2023). These biosurfactants are attractive to pharmaceutical, food, cosmetic, and petrochemical industries because of their high biodegradability, low toxicity, and structural variety (Drumond de Souza et al., 2023). Screening, recovery, purification, and characterization methods are advancing to improve yeast-produced biosurfactant design, synthesis, and industrial use. Cleanly manufacturing biosurfactants using yeasts and low-cost substrates like agro-industrial byproducts should lower production costs and boost yields.

2.3 Biosurfactants Produced by Fungi

Fungal biosurfactants are used in pesticides, medicines, industrial materials, and environmental bioremediation (Drumond de Souza et al., 2023). These amphiphilic compounds with hydrophobic and hydrophilic moieties are biodegradable, low-toxic, and antibacterial, making them potential substitutes for chemically synthesized surfactants. Fungi, especially Candida species, produce biosurfactants with antibacterial, antifungal, antiviral, antiadhesion, and anticancer activities, making them useful in pharmaceutical and medicinal applications (Stainsby et al., 2022). Actinobacteria create biosurfactants with biomedical and environmental implications, particularly mycolic acid-containing *Rhodococcus*, *Corynebacterium*, *Dietzia*, *Gordonia*, and *Tsukamurella* (Galieva et al., 2023). Biosurfactants from *Bacillus* subtilis H1 can inhibit fungal plant diseases, suggesting biocontrol in agriculture (Fardami et al., 2022). Overall, fungal-produced biosurfactants are a valuable, ecologically benign resource with many industrial applications.

3 Classification of Biosurfactants

Biosurfactants are often categorized primarily by their chemical composition and microbial origin, in contrast to chemically synthesized surfactants, which are typically classified based on the type of their polar groups. In general, biosurfactants are divided into two groups: high and low-molecular-weight molecules. Glycolipids, lipopeptides, and phospholipids are examples of low-mass surfactants; polymeric and particulate surfactants are examples of high-mass surfactants, Table 1 summarizes the classification and the producing organism (Shoeb et al., 2013).

4 Biosurfactant Production Process

Carbon and nitrogen sources are essential factors in the production of these biosurfactants and greatly influence their cost of production, of which considerable efforts have been made to use sub-agro-industrial products and renewable resources as substrates in the production process. Studies have shown that using inexpensive substrates, such as raw materials or wastes, significantly affects biosurfactant production costs (Rawat et al., 2020). The different substrates from the wastes were used for the production of biosurfactants, especially the agricultural and industrial wastes. Figure 3 depicts the biosurfactant production process.

5 Factor Affecting on the Biosurfactant Production

The producer strain and the culture conditions impact the biosurfactant's composition and emulsifying activity. Therefore, the kind of polymer formed and the amount of biosurfactant produced are determined by various factors, including the carbon and nitrogen sources, their respective C:N ratios, dietary constraints, and chemical and physical

Table 1 Biosurfactants classification and the producing organism

	Class of Biosurfactant		Producing organism
Low molecular mass	Glycolipids	Rhamnoselipids	*Pseudomonas sp.*
		Trehalolipids	*Mycobacterium tuberculosis, Rhodococcus erythropolis*
	Lipopeptides	Corynomycolates	*Arthrobacter sp.*
		Surfactin Iturin Fengucin	*Bacillus licheniformis, Bacillus subtilis*
	Phospholipids and fatty acid	Spiculisporic acid Phosphatidylethanolamine Corynomycolic acid	*Penicillium spiculisporum, Acinetobacter sp., Rhodococcus erythropolis, Corynebacterium lepus*
High molecular mass	Polymeric	Emulsan Liposan Mannoprotein	*Acinetobacter calcoaceticus RAG-1* *Candida lipolytica* *Saccharomyces cerevisiae*
	Particulate surfactants	Vesicles Whole cells	*Acinetobacter ssp.* *Corynebacteria*

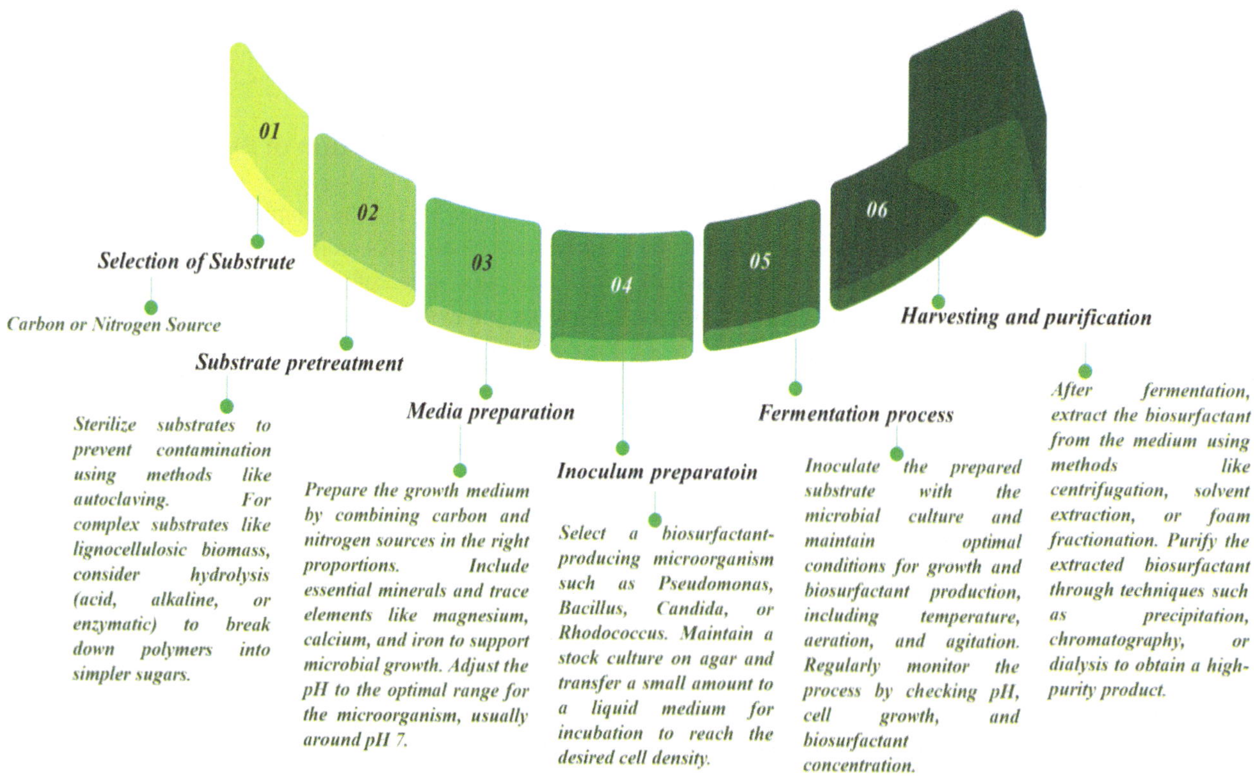

Fig. 3 Graphical depiction of biosurfactant production

characteristics, including pH, temperature, aeration, and divalent cations (2023b; Aslam et al., 2023a).

5.1 Carbon Sources

The nature of the carbon substrate influences and affects the quantity and quality of biosurfactant synthesis. We have already discussed the various carbon sources and how they affect the production of BSs. Several researchers have proposed using agro-industrial wastes as the primary source of carbohydrates to reduce production costs. According to reports, crude oil, diesel, glucose, sucrose, and glycerol are good sources of carbon substrate for the synthesis of biosurfactants (Jimoh & Lin, 2019).

5.2 Nitrogen Sources

Since protein and enzyme synthesis depend on nitrogen, it is critical for microbial development in the biosurfactant culture medium. Ammonium salts such as ammonium nitrate and ammonium sulphate, urea, peptones, meat and malt extract, yeast extract, and so on are examples of nitrogen sources. *Arthrobacter paraffineusprefers* ammonium salts and urea as nitrogen sources for biosurfactant synthesis, while *P. aeruginosa* produces the most significant amounts of surfactant when nitrate supplies nitrogen (Tripathi et al., 2019).

5.3 Environmental Factors

These significantly impact the yield and properties of the biosurfactant that is produced. It is always required to optimize the bioprocess to obtain significant quantities of biosurfactants since variations in temperature, pH, aeration, or agitation speed might alter the final product:

5.3.1 PH Factor

The growth of microbes and the fermentation of biosurfactants necessitate an optimal pH (Chen et al., 2018).

5.3.2 Temperature Factor

Temperature is a crucial environmental factor in the generation of biosurfactants. The development of each microorganism is linked to a specific temperature range, ideally between 25 and 30 °C. Both bacterial growth and carbohydrate fermentation require an optimal and carefully controlled temperature (Tripathi et al., 2019).

5.3.3 Agitation and Aeration Factor

In the case of the fermentation process, agitation is a crucial component since it affects a variety of processes, including the medium's viscosity, the growth of microorganisms' cells, and the distribution of heat, nutrients, and oxygen. Aeration, on the other hand, provides the oxygen gas required for fermentation and cell growth. It also lessens the amount of exhaust gases produced during the procedure. Adamczak and colleagues saw the highest production of BSs investigating how *C. antarctica* produces BSs. 50% of dissolved oxygen and a 1 vvm airflow rate were set as the parameters for the fermentation process. Because of the high foam formation 44, the increase in airflow rate suggested lower production of BSs. Wei and his colleagues 45 found that raising the agitation rate from 50 to 200 rpm boosted rhamnolipid production by about 80% and raised the rate of cell growth from 0.22 to 0.72 $h-^{1}$. They also found that microbial growth and rhamnolipid synthesis are positively impacted by increased dissolved oxygen levels (Wei et al., 2005).

6 Characterization of Biosurfactants

Various methods can be employed to discover biosurfactants, including haemolysis, axisymmetric drop shape analysis, rapid drop collapsing, colorimetry, and thin layer chromatography (Mouafo et al., 2023). Screening tests like surface tension measurement, emulsification index, oil displacement, drop collapse, and haemolysis are commonly used to detect biosurfactant activity. Techniques such as the hemolysis test, parafilm M test, lipase production test, CTAB agar plate method, and emulsification assay are routinely utilized for screening biosurfactant-producing strains (Alvionita et al., 2023). These diverse methods play a crucial role in identifying and studying the properties of biosurfactants.

The chemical characterization of purified biosurfactants involves various techniques to determine their structure and composition. The chemical nature of biosurfactants produced by diverse microbes has been studied using LC–MS, GCMS, FTIR, NMR, and thin-layer chromatography. (Mouafo et al., 2023). For example, LC–MS and GC–MS have identified glycolipids, whereas FTIR and NMR have confirmed biosurfactant chemical structures. Biosurfactants with lipopeptides have been analysed using thin-layer chromatography. Successful screening approaches such as oil spreading assay, blood agar hemolysis, emulsification assay, and foaming activity have identified biosurfactant-producing microorganisms in distinct strains. Biosurfactants are used in pharmaceuticals, detergents, and agriculture. Therefore, these methods help comprehend their molecular composition, functional groups, and structural features (Chauhan et al., 2022).

7 Properties of Biosurfactants

Microbial surfactants are characterized by their ability to travel across surfaces, withstand changes in pH, temperature, and ionic properties, degrade naturally, have minimal toxicity, possess emulsifying and demulsifying capabilities, and exhibit antimicrobial activity (Chandran & Das, 2010).

7.1 Surface and Interface Activity

Surfactants help decrease surface strain and interfacial pressure. *Pseudomonas aeruginosa* secretes rhamnolipids, which reduce the surface tension of water to 26 mN m^{-1} and the interfacial tension of water and hexadecane to less than 1 mN $m-^{1}$ (Chakrabarti, 2012). Biosurfactants have greater potency and efficacy and considerably lower Critical Micelle Concentration than chemical surfactants. A smaller quantity of biosurfactant is needed to achieve the highest possible decrease in surface tension (Chakrabarti, 2012).

7.2 Temperature and pH Tolerance

The utilization of extremophiles for the synthesis of biosurfactants has received considerable interest in recent years, owing to its significant potential for use in commercial settings. Several biosurfactants resist natural variables, such as temperature and pH, while maintaining their surface activity. The biosurfactant synthesized by *Arthrobacter protophormiae* exhibited stability throughout a wide range of temperatures (30–100 °C) and pH levels (2–12). Given the severe temperature, pH, and pressure conditions involved in industrial operations, it is essential to discover new microbial products that can perform well under these circumstances (Onyekonwu & Ogolo, 2010).

7.3 Biodegradability Property

Microbial-derived chemicals are more readily biodegradable than synthetic surfactants, which makes them well-suited for natural applications such as bioremediation and biosorption (Das & Mukherjee, 2007).

7.4 Low Toxicity Property

Although less research exists on the toxicity of biosurfactants, they are usually regarded as safe and suitable for use in the pharmaceutical, cosmetic, and food industries (Guerra-Santos et al., 1984).

7.5 Emulsion Framing and Emulsion

Biosurfactants function as substances that can either emulsify or de-emulsify. An emulsion is a mixture consisting of two fluids that do not mix well together. One fluid is scattered as droplets in the other fluid. The diameter of these droplets is usually larger than 0.1 mm. Emulsions can be classified into two types: water-in-oil or oil-in-water emulsions. These substances have low stability, but this can be improved by adding substances like biosurfactants. With the addition of these additives, the substances can be kept as stable emulsions for a considerable duration, ranging from months to years (Hu & Ju, 2001).

8 Agro-Industerial Wastes as Substrate Used for Biosurfactant Production

Agricultural wastes play a crucial role in impacting biosurfactant production efficiency by serving as low-cost substrates for microbial biosurfactant production, thus enhancing the economic viability of biosurfactant manufacturing. Utilizing agricultural wastes such as corncobs, sunflower stalks, cassava, palm kernel, and sawdust as substrates in submerged fermentation processes can lead to the production of high added-value biosurfactants with promising tenso-active and emulsifying properties (Sharon, 2023). The use of waste materials for biosurfactant production not only reduces environmental pollution but also contributes to cost-effective manufacturing processes, addressing the high production costs associated with biosurfactants and enhancing their commercialization potential. By optimizing production processes through the utilization of environment-friendly low-cost substrates and efficient product recovery methods, biosurfactant production efficiency can be significantly improved, making biosurfactants more competitive in the market. As Novel Methodologies and Approaches in Biosurfactant Production Research.

Environmental *Bacillus sp.* strains utilize potato waste water as a biosurfactant substrate, resulting in a cost-effective manufacturing process (Kavyarathna et al., 2023). This innovative method of biosurfactant manufacture utilizes inexpensive food wastes as substrates to mitigate environmental degradation and decrease manufacturing expenses effectively. The technique employs naturally occurring *Bacillus sp.* strains to enhance the effectiveness of biosurfactant synthesis from potato waste water. This approach utilizes inexpensive waste substrates to generate biosurfactants at a cost-effective rate, facilitating the production of sustainable and environmentally friendly food sector products.

The research conducted by Santos et al. (2023) explores the possible use of hemicellulosic liquors derived from corncobs and sunflower stalks as substrates for biosurfactant production. By employing submerged fermentation with *Bacillus* subtilis, we successfully generated biosurfactants from these agro-industrial wastes. Following the alkaline processing of hemicellulose-rich hydrolysates, we conducted tests on glucose and liquor concentrations to enhance the synthesis of biosurfactants. The sunflower stalk hemicellulose liquor demonstrated the maximum cell concentration and surface tension reduction, while the corncob liquor exhibited the highest emulsification index. These results were quite encouraging. 2.5% glucose, 20% hemicellulose liquor, and 1% mineral salts are the most favourable conditions for biosurfactant production. These results underscore the potential of these inexpensive substrates to be employed in the development of beneficial and valuable biosurfactants (Santos et al., 2023).

Factorial Designs (FD) and Response Surface Methodology (RSM) are statistical optimisation techniques employed to improve biosurfactant production by utilising refuse substrates. This approach leads to cost savings and increased market competitiveness. FD and RSM were extensively employed in this work to enhance the efficiency of biosurfactant production. FDs such as the Plackett–Burman Design and Taguchi design were used to evaluate the impacts and dependencies of factors. At the same time, RSM was employed to identify the best compositions of medium and culture conditions. Implementing Central Composite Design and Box-Behnken Design facilitated the optimisation of biosurfactant production in RSM by reducing the number of required testing runs. The integration of Artificial Neural Network demonstrated the ability of artificial intelligence to predict biosurfactant production, suggesting potential advancements in the future's cost-effective and efficient manufacturing of biosurfactants (Sharon, 2023).

9 Environmental and Economic Benefits of Biosurfactants

9.1 Environmental Benefits of Biosurfactants

Microbial surfactants, an example of a bio-economic product, have seen a growth in their utilization due to the recent push for ecologically friendly technologies, this is because of the environmental concerns raised over chemical surfactants. According to Georgiou et al. (1992), these biological molecules are seen as a sustainable and environmentally advantageous alternative to conventional surfactants manufactured using chemicals and fossil fuels (Georgiou et al., 1992). Thus, biosurfactants are also environmentally acceptable when used in waste treatment and bioremediation sectors. Biosurfactants are classified into four categories based on the chemical structure of the hydrophobic component: (1) glycolipid type, (2) fatty acid type, (3) lipopeptide type, and (4) polymer type. Because of their ability to reduce surface tension, stabilize emulsions, promote foam, and degrade biologically, there is much interest in using them instead of chemical surfactants (Gayathiri et al., 2022). In general, biosurfactants exhibit better qualities than chemically synthesized surfactants, making their usage in the oil recovery process safer for the environment. Their ability to endure harsh environments, their simplicity of culturing, their high-scale production capabilities, their eco-friendly nature, and their diverse nature render them sufficiently effective in the implementation of various domains, including microbial degradation (Gayathiri et al., 2022).

Biosurfactants are multifaceted substances that are found to be extensively used in biodegradation and bioremediation processes to eliminate harmful pollutants. They are also employed in the dairy, pharmaceutical, and other industries. Biosurfactants are still of interest due to their many applications and advantages. In agriculture, these compounds are employed to eradicate plant diseases and increase beneficial plant microbes' bioavailability of nutrients. Using biosurfactants for agricultural soil remediation can significantly enhance the quality of agricultural soil (Purwasena et al., 2019). Some of the ideas discussed are using sustainable by-products as substrates, cutting down on wastes, and eventually recycling treated wastes. This has led to increased exposure to these microbial products during production. Increased biodegradability and low toxicity are provided by whole surfactants, which mitigate the negative impacts of synthetic surfactants (Gayathiri et al., 2022).

Biosurfactants exhibit better qualities than chemically synthesized surfactants, making their usage in the oil recovery process safer for the environment. They are sufficiently effective in the implementation of different domains, including microbial degradation, due to their endurance in harsh environments, ease of culturing, high-scale production, eco-friendly nature, and diverse nature (Gayathiri et al., 2022). Biosurfactants are multifaceted substances that are found to be extensively used in biodegradation and bioremediation processes to eliminate harmful pollutants. They are also employed in the dairy, pharmaceutical, and other industries (Dadrasnia & Ismail, 2015). Biosurfactants are still of interest due to their many applications and advantages. In agriculture, these compounds are employed to eradicate plant diseases and increase beneficial plant microbes' bioavailability of nutrients. Using biosurfactants for agricultural soil remediation can significantly enhance the quality of agricultural soil (Purwasena et al., 2019). Some of the ideas discussed are using sustainable by-products as substrates, cutting down on wastes, and eventually recycling treated wastes. This has led to increased exposure to these microbial products during production. Increased biodegradability and low toxicity are provided by whole surfactants, which mitigate the negative impacts of synthetic surfactants (Gayathiri et al., 2022).

9.2 Economic Benefits of Biosurfactants

Large-scale production for most microbial surface-active agents has not reached a satisfactory economic level due to low yields. The raw resources they consume are essential in increasing the financial viability of bacteria-producing biosurfactants. For example, using waste products helps to maximize the production of the biosurfactant (over other metabolites) while lowering the cost of that resource (Bognolo, 1999). Optimizing reactor design is necessary for commercially successful biosurfactant production since it minimizes material loss and guarantees maximal biosurfactant extraction. It can produce biosurfactants more affordably and competitively than synthetic surfactants by using agro-industrial wastes from plants and animals to make biosurfactants. To obtain a more commercially viable production process, researchers have introduced an alternative approach using solid-state fermentation (Kapadia Sanket & Yagnik, 2013). Several studies have been conducted that will be important for developing economically efficient industrial-scale biotechnological processes in the future (Banat et al., 2014).

10 Applications of Biosurfactants

10.1 Environmental Applications

Figure 4 illustrates the various ways in which biosurfactants are used in environmental applications. Due to their

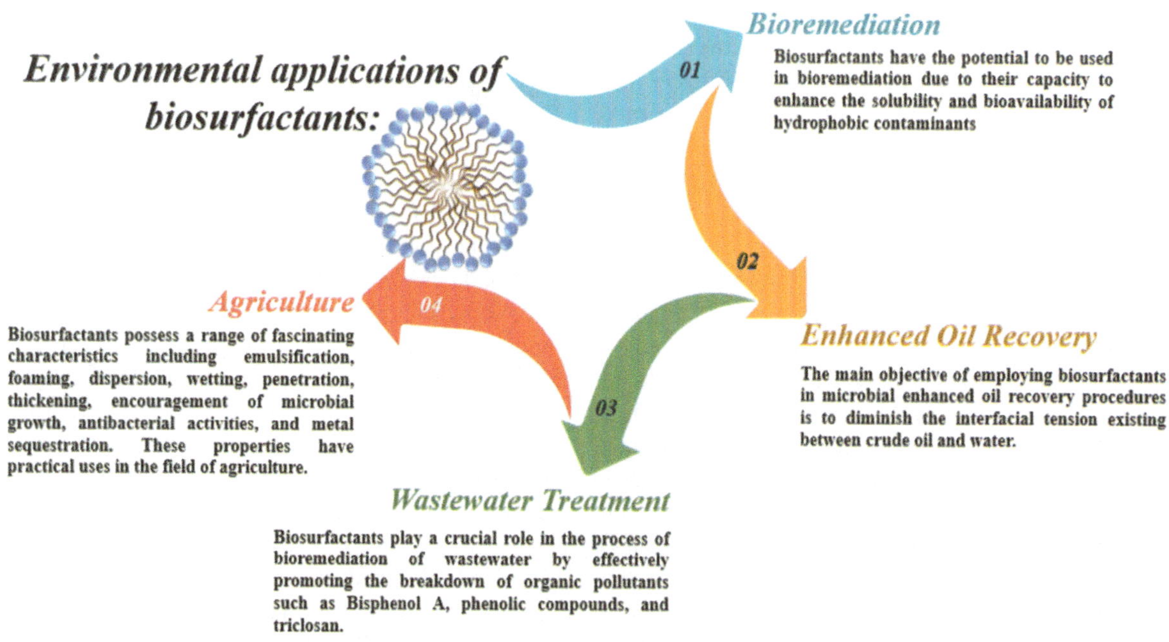

Fig. 4 Environmental applications of biosurfactant

biodegradable and non-toxic properties, these substances are highly suitable for a wide range of environmental uses; these encompass:

10.1.1 Bioremediation

Biosurfactants can improve hydrophobic pollutants' solubility and bioavailability, making them useful in bioremediation. Biosurfactants generate micelles that attach to heavy metal ions. Biosurfactants' metal binding and metal selectivity help transport and retrieve pollutants (Mishra et al., 2021). Biosurfactants improve mass transfer efficiency during biodegradation, making them essential in bioremediation. Biosurfactants can enhance, impede, or have no effect on mass transfer in microbial cells. The outcome depends on the bacteria's physiological traits and substrate features. Microorganisms can dissolve, pseudo-solubilize in biosurfactant micelles, or adhere to nonaqueous phases to access substrates. Biosurfactant micelles increase pollutant solubility, boosting soil desorption and biodegradation. This is demonstrated with *Pseudomonas* alcaligenes PA-10 and polyaromatic hydrocarbons. Increased micelles may weaken the contaminant by reducing mass transfer. Biosurfactants increase hydrophobic contaminants' pseudo-solubility, making them easier for microbes to digest and improve soil cleaning. This procedure can be in situ or ex situ. In situ methods are cheaper but slower and less uniform due to site heterogeneity. In contrast, ex-situ methods involve soil excavation and transfer. Research shows that biosurfactants remove organic contaminants and toxic metals from soils. Due to their lower environmental impact, they may be better than synthetic surfactants (Singh et al., 2009).

10.1.2 Enhanced Oil Recovery

Biosurfactants in microbial-enhanced oil recovery reduce crude oil–water interfacial tension. Flood recovery processes become more efficient, allowing them to obtain more oil. This requires identifying, developing, and synthesizing cost-effective biosurfactants. Biosurfactants with high surface activity, low CMC, and temperature and pH tolerance are ideal. They must dissolve quickly and form stable emulsions (Walter et al., 2010). Biosurfactants are used in microbial-enhanced oil recovery in two ways: (1) direct injection into the reservoir or (2) activating indigenous bacteria in the oil reservoir or injecting specialized microbes to produce biosurfactants on-site. Biosurfactants are sometimes mixed with metal ions. These additives promote polar interactions between ions and biosurfactant molecules, improving their efficacy. Lab trials using sand-packed columns have revealed that biosurfactants in Microbial Enhanced Oil Recovery may recover 95% of oil. This shows biosurfactants can boost oil recovery (Nikolova & Gutierrez, 2020).

10.1.3 Waste Water Treatment

Biosurfactants, especially rhamnolipids, help bioremediate waste water by breaking down organic contaminants such as Bisphenol A, phenolic compounds, and triclosan. Clearance rates reach 90%. These compounds change heavy metals like zinc and copper and lead to better dissolve and help remove them from polluted places. Rhamnolipids reduce surface tension and improve emulsification, enhancing oil–water separation. This is crucial in industrial or oil disasters (Liu et al., 2018). Rhamnolipids also boost enzymatic activity and pollutant accessibility, accelerating breakdown. Research shows they can boost laccase catalytic activity, reducing Bisphenol A by 65% in industrial waste water. They also modify enzyme structures to help decompose waste water pollutants, including phenolic compounds and petroleum hydrocarbons (PAHs and TPH). *Stenotrophomonas sp.* reduces pollutant concentrations by 70–90% using rhamnolipids (Patel & Patel, 2020). Personal care products contain triclosan, which rhamnolipids break down. Di-rhamnolipids break down 94% in water–sediment systems. Overall, rhamnolipids and other biosurfactants are a sustainable and adaptive waste water treatment system pollution solution. This highlights their environmental remediation potential and ability to reduce pollutants and safeguard human health (Malkapuram et al., 2021).

10.1.4 Agriculture Wastes Treatment

Biosurfactants have remarkable properties such as emulsification, foaming, dispersion, wetting, penetration, thickening, microbial growth, antibacterial activity, and metal sequestration. These qualities are helpful in agriculture. According to Thavasi et al. (2014), these compounds control plant diseases due to their antibacterial capabilities. Research has demonstrated that biosurfactants help fight plant diseases and improve resistance (Thavasi et al., 2014). This shows their potential to enhance agricultural pesticides and plant disease resistance:

Disease Control

Studies have shown that plants' biosurfactants can control fungi and other infections. Rhamnolipids and Sophorolipids lyse oomycete plant-pathogenic fungi zoospores. These fungi cause hydroponic plant diseases. Zoospore membranes rupture with biosurfactants, causing lysis and motility inhibition. This stops the disease from spreading. *Pseudomonas* and *Serratia strains* that produce biosurfactants have also been used on plant surfaces to limit fungal disease growth and spread (Thavasi et al., 2014).

Induced Systemic Resistance (ISR)

Biosurfactants can evoke amazement beyond their antibacterial properties. Rhamnolipids have been shown to boost plant systemic resistance. These chemicals cause plants to produce pathogenesis-related proteins and other defence mechanisms. This innate immune response, called ISR, helps plants resist several illnesses, lowering the requirement for pesticides (Salwan et al., 2023).

10.2 Therapeutic and Pharmaceutical Potential Applications

Medical and pharmaceutical applications for biosurfactants demonstrate their therapeutic potential. Commercial antibiotics contain powerful antibacterial and antifungal biosurfactants such as lipopeptides and glycolipids. They effectively prevent and stop biofilm buildup, which is crucial for reducing hospital device and surface infections (Sharma et al., 2019). Biosurfactants also disrupt microorganism cell membranes and biofilms, which may reduce doses and adverse effects while battling resistant strains. Since they affect the architecture of various enveloped viruses, they may be antiviral medications. Biological compounds like biosurfactants help wounds heal by encouraging cell development and tissue regeneration. This reduces infections and speeds healing (Balakrishnan et al., 2023). Some biosurfactants can also slow tumour cell proliferation, making cancer cells more susceptible to chemotherapy. Their immunomodulatory effects assist in treating immune-mediated illnesses and improve vaccine formulations, showing their potential to promote health and fight disease (Sajid et al., 2020).

10.3 Food Industry Applications

Biosurfactants are becoming more popular in food processing due to their natural origin and versatility. As emulsifying and stabilizing agents, they improve the stability of food emulsions such as dressings and sauces, protecting product quality and shelf life. Biosurfactants improve salad dressing consistency and thermal resistance compared to traditional emulsifiers. They also reduce calorie content and increase texture, volume, and quality in baked goods like bread and cookies. Despite limited coverage, biosurfactants may improve ice cream texture, creaminess, and physical properties during processing. These chemicals' antibacterial and anti-adhesive properties restrict microbe development and reduce their attachment to food surfaces, extending food shelf life. In dairy processing, this is beneficial. Biosurfactants contain antioxidants that limit food oxidation, improve product shelf life, and reduce the risk of oxidative stress-related disorders. Biosurfactants reduce synthetic chemicals, making them popular in the food business. This adoption enhances food quality, safety, and sustainability (Ribeiro et al., 2020).

10.4 Cosmetic Industry Applications

Biosurfactants like Sophorolipids, Rhamnolipids, and Mannosyl erythritol lipids are used more in cosmetics due to their natural and eco-friendly properties. Sophorolipids' hygroscopic qualities make them useful in moisturizers and emollients. They contain moisturizers, lipsticks, eye shadows, antibacterial acne treatments, and deodorants. Rhamnolipids, which produce less skin irritation, are utilized in insect repellents, acne treatments, antidandruff shampoos, and anti-wrinkle products. They are antimicrobial in cosmetics and personal hygiene. Pseudozyma yeasts produce mannosylerythritol, which has good industrial and biological qualities. These ingredients hydrate dehydrated skin, restore damaged hair, and fight ageing in skin care products. Their ceramide-like properties help maintain the skin's protective barrier. These biosurfactants indicate a shift towards eco-friendly cosmetics (Fracchia et al., 2014).

11 Challenges and Future Directions in Biosurfactant Research

Low yield, high manufacturing costs, and poor fermentation hinder biosurfactant commercialization. Future research could address these issues:

11.1 Improving Biosurfactant Yields

The biosurfactant-synthesis microbial strains used nowadays often have low yield and productivity. Scientists are employing metabolic and cellular engineering to improve biosurfactant synthesis by improving metabolic processes. Examples include enhancing biosynthetic gene expression and using molecular methods to boost microbial output. Intelligent methods, including artificial intelligence and statistical optimization, have improved culture medium composition to increase yields (Dhanya, 2021).

11.2 Enhancement of Process Economics

Biosurfactant manufacturing costs depend on raw material costs and downstream processing efficiency. Use cheap, sustainable resources and agricultural-industrial byproducts to cut costs. Downstream processing technologies like chemical-free recovery reduce production costs and environmental damage (Aslam et al., 2023b).

11.3 Advancements in the Fermentation Process

Are required for maximum biosurfactant production. Various fermentation methods, such as batch, fed-batch, and continuous fermentation, have been studied. Immobilized or resting cells and solid-state fermentation can boost production and streamline manufacturing. Notably, bioprocess coupling has shown promise in reducing manufacturing costs by synthesizing enzymes and biosurfactants (Beuker et al., 2014).

12 Conclusion

This chapter specifically examines the crucial role that microorganisms have in the conversion of agricultural and industrial waste materials into valuable biosurfactants. The objective of this chapter is to provide a comprehensive summary of the procedure. Biosurfactants refer to compounds that possess both hydrophilic and hydrophobic characteristics. A diverse range of bacteria, yeast, and fungus use complex fermentation and biodegradation processes to produce biosurfactants. These compounds provide many significant advantages, such as their ability to undergo natural degradation, their few adverse effects, and their remarkable reactivity with surfaces. Consequently, their skills are highly sought after in the food business, pharmaceutical manufacturing, cosmetic development, and environmental pollution cleanup. The use of waste materials generated by agro-industrial operations has the potential to reduce the production costs of biosurfactants. Consequently, the minimization of environmental pollutants leads to an enhancement in both the economic and ecological sustainability of the process. Nevertheless, obstacles have to be surmounted in order to enhance the efficiency of microbial strains, optimize fermentation conditions, and enhance downstream processing. Hence, it is essential to carry out continuous research to enhance the effectiveness and scalability of the biosurfactant manufacturing process. Such measures will promote environmentally conscious actions and protect the natural ecosystem.

References

Abacha, Y., Sanderson, P., & Rahman, P. (2016). Microbial process development for fermentation based biosurfactant production. In *The handbook of microbial bioresources*.

Abdel-Rahem, R. A. (2024). A mini review on synthetic and biosurfactants: Origin and structure, physicochemical properties, applications, cost of production, toxicity, biodegradability and environmental effects. *Tenside Surfactants Detergents, 61*(1), 105–114.

Alvionita, M., et al. (2023). Isolation of biosurfactant producing bacteria from Mangrove area. In *AIP conference proceedings*. AIP Publishing.

Aslam, R., et al. (2023a). Biosurfactants: types, sources, and production. In *Advancements in biosurfactants research* (pp. 3–24). Springer

Aslam, R., et al. (2023b). *Advancements in biosurfactants research*. Springer.

Balakrishnan, S., et al. (2023). Biosurfactants: An antiviral perspective. In *Multifunctional microbial biosurfactants* (pp. 431–454). Springer.

Banat, I. M., et al. (2014). Cost effective technologies and renewable substrates for biosurfactants' production. *Frontiers in Microbiology, 5*, 122316.

Beuker, J., Syldatk, C., & Hausmann, R. (2014) Bioreactors for the production of biosurfactants. *Biosurfactants: Production and Utilization*, 117–128.

Bhatnagar, I., & Kim, S.-K. (2012). Pharmacologically prospective antibiotic agents and their sources: A marine microbial perspective. *Environmental Toxicology and Pharmacology, 34*(3), 631–643.

Bognolo, G. (1999). Biosurfactants as emulsifying agents for hydrocarbons. *Colloids and Surfaces a: Physicochemical and Engineering Aspects, 152*(1–2), 41–52.

Ceresa, C., et al. (2023). Harnessing the potential of biosurfactants for biomedical and pharmaceutical applications. *Pharmaceutics, 15*(8), 2156.

Chakrabarti, S. (2012). *Bacterial biosurfactant: Characterization, antimicrobial and metal remediation properties*.

Chandran, P., & Das, N. (2010). Biosurfactant production and diesel oil degradation by yeast species Trichosporon asahii isolated from petroleum hydrocarbon contaminated soil. *International Journal of Engineering, Science and Technology, 2*(12), 6942–6953.

Chauhan, V., Dhiman, V. K., & Kanwar, S. S. (2022). Purification and characterization of a novel bacterial Lipopeptide (s) biosurfactant and determining its antimicrobial and cytotoxic properties. *Process Biochemistry, 120*, 114–125.

Chen, C., et al. (2018). Optimization and characterization of biosurfactant production from kitchen waste oil using Pseudomonas aeruginosa. 25, 14934–14943.

Dadrasnia, A., & Ismail, S. (2015). Biosurfactant production by Bacillus salmalaya for lubricating oil solubilization and biodegradation. *International Journal of Environmental Research and Public Health, 12*(8), 9848–9863.

Das, K., & Mukherjee, A. K. (2007). Crude petroleum-oil biodegradation efficiency of Bacillus subtilis and Pseudomonas aeruginosa strains isolated from a petroleum-oil contaminated soil from North-East India. *Bioresource Technology, 98*(7), 1339–1345.

Das, S., & Rao, K. B. (2024). A comprehensive review of biosurfactant production and its uses in the pharmaceutical industry. *Archives of Microbiology, 206*(2), 60.

Dhanya, M. (2021). Biosurfactant-enhanced bioremediation of petroleum hydrocarbons: Potential issues, challenges, and future prospects. *Bioremediation for Environmental Sustainability*, 215–250.

Domínguez Rivera, Á., Martínez Urbina, M. Á., & López y López, V. E. (2019). Advances on research in the use of agro-industrial waste in biosurfactant production. *World Journal of Microbiology and Biotechnology, 35*(10), 155.

Drumond de Souza, L. M., et al. (2023). Fungal biosurfactants: Applications in agriculture and environmental bioremediation processes. 243–254.

Fardami, A. Y., et al. (2022). A review on biosurfactant properties, production and producing microorganisms. *10*(1), 5–12.

Fatima, S., et al. (2022) Bioremediation of contaminated soils by bacterial biosurfactants. In *Advances in remediation techniques for polluted soils and groundwater* (pp. 67–85). Elsevier.

Fracchia, L., et al. (2014). Industrial applications of biosurfactants. *Biosurfactants: Production and Utilization—Processes, Technologies, and Economics, 159*, 245–260.

Fung, K. C., et al. (2023). From wastewater treatment plants to the oceans: a review on synthetic chemical surfactants (SCSs) and perspectives on marine-safe biosurfactants. *15*(14), 11436.

Galieva, G., et al. (2023) Biosurfactants produced by the novel strain Bacillus subtilis H1 as an efficient tool to biocontrol fungal diseases of tomato and wheat. In *E3S web of conferences*. EDP Sciences.

Gayathiri, E., et al. (2022). Biosurfactants: Potential and eco-friendly material for sustainable agriculture and environmental safety—A review. *Agronomy, 12*(3), 662.

Georgiou, G., Lin, S.-C., & Sharma, M. M. (1992). Surface–active compounds from microorganisms. *Bio/technology, 10*(1), 60–65.

Guerra-Santos, L., Käppeli, O., & Fiechter, A. (1984). Pseudomonas aeruginosa biosurfactant production in continuous culture with glucose as carbon source. *Applied and Environmental Microbiology, 48*(2), 301–305.

Hu, Y., & Ju, L.-K. (2001). Purification of lactonic sophorolipids by crystallization. *Journal of Biotechnology, 87*(3), 263–272.

Ingle, A. P., et al. (2023). Biosurfactants in nanotechnology: Recent advances and applications. 173–194.

Jimoh, A. A., & Lin, J. (2019). Biosurfactant: A new frontier for greener technology and environmental sustainability. *Ecotoxicology and Environmental Safety, 184*, 109607.

Kapadia Sanket, G., & Yagnik, B. (2013). Current trend and potential for microbial biosurfactants. *Asian Journal of Experimental Biological Sciences, 4*(1), 1–8.

Kavyarathna, V., et al. (2023). Applications of biosurfactants produced from food-related waste as substrate and their role in renewable resource utilization. In *Applications of next generation biosurfactants in the food sector* (pp. 403–447). Elsevier.

Kim, J.-S., et al. (2023). Production of a new biosurfactant by a new yeast species isolated from Prunus mume Sieb. *et Zucc, 33*(8), 1023.

Liu, Y., et al. (2018). Effect of surfactants on the interaction of phenol with laccase: Molecular docking and molecular dynamics simulation studies. *Journal of Hazardous Materials, 357*, 10–18.

Malkapuram, S. T., et al. (2021). A review on recent advances in the application of biosurfactants in wastewater treatment. *Sustainable Energy Technologies and Assessments, 48*, 101576.

Mishra, S., et al. (2021). Biosurfactant is a powerful tool for the bioremediation of heavy metals from contaminated soils. *Journal of Hazardous Materials, 418*, 126253.

Mouafo, H. T., et al. (2023). Methods of purification and characterization of biosurfactants: An overview. *Journal of Advances in Biology Biotechnology, 26*(5), 35–53.

Nikolova, C., & Gutierrez, T. (2020). Use of microorganisms in the recovery of oil from recalcitrant oil reservoirs: Current state of knowledge, technological advances and future perspectives. *10*, 2996.

Onyekonwu, M. O., & Ogolo, N. A. (2010). Investigating the use of nanoparticles in enhancing oil recovery. In *SPE Nigeria annual international conference and exhibition*. SPE.

Patel, K., & Patel, M. (2020). Improving bioremediation process of petroleum wastewater using biosurfactants producing Stenotrophomonas sp. S1VKR-26 and assessment of phytotoxicity. *Bioresource Technology, 315*, 123861.

Purwasena, I. A., et al. (2019). Stability test of biosurfactant produced by Bacillus licheniformis DS1 using experimental design and its application for MEOR. *Journal of Petroleum Science and Engineering, 183*, 106383.

Rawat, G., Dhasmana, A., & Kumar, V. J. E. S. (2020). Biosurfactants: the next generation biomolecules for diverse applications. *3*. 353–369.

Reznik, G. O., et al. (2010). Use of sustainable chemistry to produce an acyl amino acid surfactant. *Applied Microbiology and Biotechnology, 86*, 1387–1397.

Ribeiro, B. G., Guerra, J. M., & Sarubbo, L. A. (2020). Biosurfactants: Production and application prospects in the food industry. *Biotechnology Progress, 36*(5), e3030.

Sajid, M., et al. (2020). Biosurfactants: Potential applications as immunomodulator drugs. *Immunology Letters, 223*, 71–77.

Salwan, R., et al. (2023). Insights into plant beneficial microorganism-triggered induced systemic resistance. *7*, 100140.

Santos, B. L. P., et al. (2023). Use of agro-industrial waste for biosurfactant production: a comparative study of hemicellulosic liquors from corncobs and sunflower stalks. *Sustainability, 15*(8).

Sharma, D., Misba, L., & Khan, A. U. (2019). Antibiotics versus biofilm: An emerging battleground in microbial communities. *Antimicrobial Resistance and Infection Control, 8*, 1–10.

Sharma, J., Sundar, D., & Srivastava, P. (2021). Biosurfactants: potential agents for controlling cellular communication, motility, and antagonism. *8*, 727070.

Sharon, A. B., et al. (2023). Statistical optimization strategies on waste substrates for solving high-cost challenges in biosurfactants production: a review. In *iop conference series: earth and environmental science*. IOP Publishing.

Shoeb, E., et al. (2013). Classification and industrial applications of biosurfactants. *Academic Research International, 4*(3), 243.

Singh, R., et al. (2021). Utilisation of agro-industrial waste for sustainable green production: A review. *Environmental Sustainability, 4*(4), 619–636.

Singh, A., Kuhad, R. C., & Ward, O. P. (2009). *Advances in applied bioremediation*. Springer.

Stainsby, F. M., Hodar, J., & Vaughan, H. (2022) Biosurfactant production by mycolic acid-containing actinobacteria. In *Actinobacteria-diversity, applications and medical aspects*. IntechOpen.

Thavasi, R., Marchant, R., & Banat, I. M. (2014). 15 Biosurfactant applications in agriculture. *Biosurfactants: Production and Utilization—Processes, Technologies, and Economics, 159*, 313.

Tripathi, L., et al. (2019). Biosynthesis of rhamnolipid by a Marinobacter species expands the paradigm of biosurfactant synthesis to a new genus of the marine microflora. *Microbial Cell Factories, 18*, 1–12.

Uzoigwe, C., Ennis, C. J., Rahman, P. K. (2015). Production of biosurfactants using eco-friendly microorganisms. *Environmental Sustainability: Role of Green Technologies*, 185–204.

Walter, V., Syldatk, C., & Hausmann, R. (2010). Screening concepts for the isolation of biosurfactant producing microorganisms. *Biosurfactants*, 1–13.

Wei, Y.-H., Chou, C.-L., & Chang, J.-S. (2005). Rhamnolipid production by indigenous Pseudomonas aeruginosa J4 originating from petrochemical wastewater. *Biochemical Engineering Journal, 27*(2), 146–154.

Yañez-Ocampo, G., et al. (2017). Utilization of agroindustrial waste for biosurfactant production by native bacteria from chiapas. *Open Agriculture, 2*(1), 341–349.

Microbial Bio Plastic Production: Progress and Prospects

Julie Dogra Bandral, Skarma Choton, and Upma Dutta

Abstract

One of the main problems with the environment is contamination with plastic endangers wildlife species and people in the food chain. Unlike traditional plastics made from petroleum, which are non-renewable and pollute the environment, bio plastics are made from renewable biomass sources. From an ecologically friendly and sustainable standpoint, bio plastic is an alternative to traditional plastics due to its biological compatibility and biodegradable properties. Using microorganisms to manufacture biodegradable plastics is a novel and sustainable process known as microbial manufacturing of bio plastics. In addition to lowering the release of greenhouse gases and reducing reliance on fossil fuels, microbial generation of bio plastics addresses the problem of plastic waste as well as pollution. It has been used in packaging, bio-engineering and biomedical fields. PHA and its nanocomposite copolymers have been applied to the development of tissue, bone scaffolds, implants for surgery and drug delivery for the repair of wounds and blood vessels. Despite the great market potential of PHAs, their high production costs prevent their prominent use; therefore, cheaper carbon sources, like industrial and agricultural bio-wastes, should be used to minimize the cost of producing PHAs. In addition, optimizing the conditions for development in hosts and using appropriate waste management techniques could result in an increase in the production of bio plastics.

Keywords

Bio plastic · Biodegradability · Biocompatibility · Bio-wastes · Nanocomposite

1 Introduction

Plastic garbage production is currently one of the biggest ecological problems and is growing daily. Since petroleum-based plastics do not degrade, the rapid growth in the number of people and the resulting rise in plastic pollution have had a severe negative impact on the natural flora and fauna and the chances of their survival. According to Chia et al. (2020), the world's plastic output reached 359 million metric tonnes in 2018. The PHA industry was assessed to be worth US$73.6 million in 2018 and is expected to grow to US$93.5 million by 2021 (Khatami et al., 2021). Nowadays, non-renewable petroleum products derived from fossil fuels like coal, natural gas and oil are used to make most plastics. The amount of plastic trash generated globally is only 18% re-used, 24% ignited, and 58% disposed of or discharged into the natural world, where it accumulates and persists for an extremely long time. The amount of plastic garbage produced worldwide is growing daily and it may take more than a century for plastics made from petroleum that are scattered throughout the environment—including the oceans—to mineralize). As a result, a remedy must be found to address this issue.

The accumulation of plastic leads to pollution and jeopardizes ecosystems as well as the human food web. The harm to the environment caused by plastic is long-lasting because it can take hundreds or thousands of centuries for it to break down. Researchers are focused on creating biodegradable polymers from sources of renewable energy

J. D. Bandral (✉) · S. Choton
Division of Post Harvest Management, Faculty of Horticulture and Forestry, Sher-E-Kashmir University of Agricultural Sciences and Technology of Jammu (J&K), Jammu, India
e-mail: jdbandralpht@gmail.com

U. Dutta
Division of Microbiology, Faculty of Basic Sciences, Sher-E-Kashmir University of Agricultural Sciences and Technology of Jammu (J&K), Jammu, India

K. Sharma (ed.), *Bio-prospecting of Novel Microbial Bioactive Compounds for Sustainable Development*, Sustainable Landscape Planning and Natural Resources Management, https://doi.org/10.1007/978-3-031-82178-3_2

because of the ongoing ecological threat posed by petroleum plastics' persistence in the environment. The latest generation of goods, materials and technology has been greatly aided by the growing understanding of green science and the preservation of the environment. Researchers have recently developed bio plastics, which are generated from living things, to combat non-biodegradable plastic waste.

Bio plastics, which are good for the environment, biodegradable and renewable, refers to plastic created from living things including germs, plants and wildlife. It can replace synthetic plastics for different uses. The polymer known as polyhydroxyalkanoates (PHA) is used to make bio plastics. Under stressful conditions, such as when they lack inorganic nutrients like phosphate, oxygen, nitrogen, etc., bacteria such as *Ralstonia eutropha*, *Cupriavidus necator*, *Pseudomonas* spp., *Comamonas* spp., *Alcaligenes latus* and other species synthesize PHA in the cytoplasmic region of these bacteria. PHAs are polyesters composed of (R)-3-hydroxyalkanoic acids, which share many of the thermoplastic properties of petrochemical-based plastics and are non-toxic, recyclable and biocompatible. Microbial procedures for fermentation are used to synthesize it, while microbe cell lysis is used to extract it. Poly-3-hydroxybutyrate (PHB), the most studied PHA polymer, is brittle and highly crystalline, similar to polypropylene as among other characteristics. Bio plastics can be classified as either biodegradable or non-biodegradable as Fig. 1 illustrates. Materials that break down over time include cellulose, starch, polyhydroxyalkanoates (PHA), polylactic acid (PLA) and so on. Non-biodegradable substances include bio-polyethylene terephthalate (bio-PET), polyol-polyurethane and bio-polyethylene (Narancic et al., 2020).

2 Bio Plastics Production Through Microbes

Using microorganisms to manufacture biodegradable plastics is a novel and efficient process known as microbial manufacturing of bio plastics (Fig. 2). Unlike traditional plastics derived from petroleum, which are non-renewable and contribute to environmental contamination, biological plastics are a kind of plastic made from naturally occurring biomass (Bhatt, 2023). Microbe cell factories are microorganisms that can produce chemicals that have commercial value. Production hosts for bio plastics can be either naturally occurring or genetically modified. Crude or polished kinds of biomass are used as substrates for the synthesis of bio polymers, which can be moulded into a variety of products. There are two types of fermentative routes available for producing bio plastics: direct (one-step process) and indirect (two steps). The former method, which employs microbial fermentation to directly transform biomass into bio plastics, is recommended over the latter. However, because of the unavailability of organisms with a recognized metabolic system that produces the necessary polymer, this approach typically justifies the use of metabolic engineering. In the indirect technique, fermentation produces bio-based monomers that are used to make plastics using chemical or enzymatic processes. These techniques were created in light of the demand for a bio-based substitute for the monomeric building blocks that are currently being synthesized using petrochemical processes.

The process for microbial synthesis of bio-based plastics can be broadly explained through the following steps:

(1) Biochemical screening or development of metabolically engineered strain for production host identification;

Fig. 1 Different categories of plastics and bio plastics (Moodley & Trois, 2022)

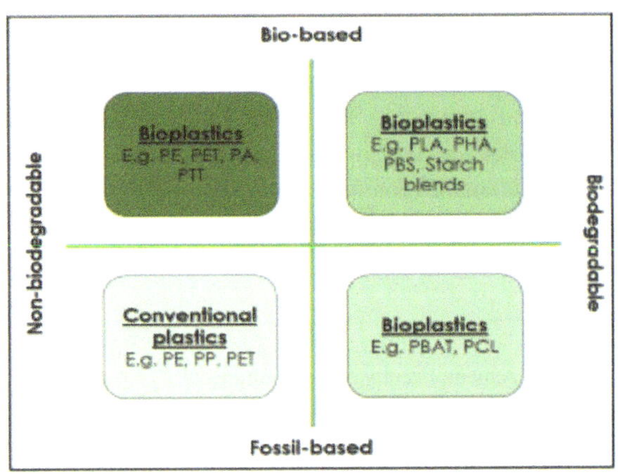

PE, polyethylene; PET, polyethylene terephthalate; PA, polyamide; PTT, polytrimethylene terephthalate; PLA, polylactic acid; PHA, polyhydroxyalkanoate; PBS, polybutylene succinate; PP, polypropylene; PBAT, polybutyleneadipate terephthalate; PBL, plastic barrier laminate

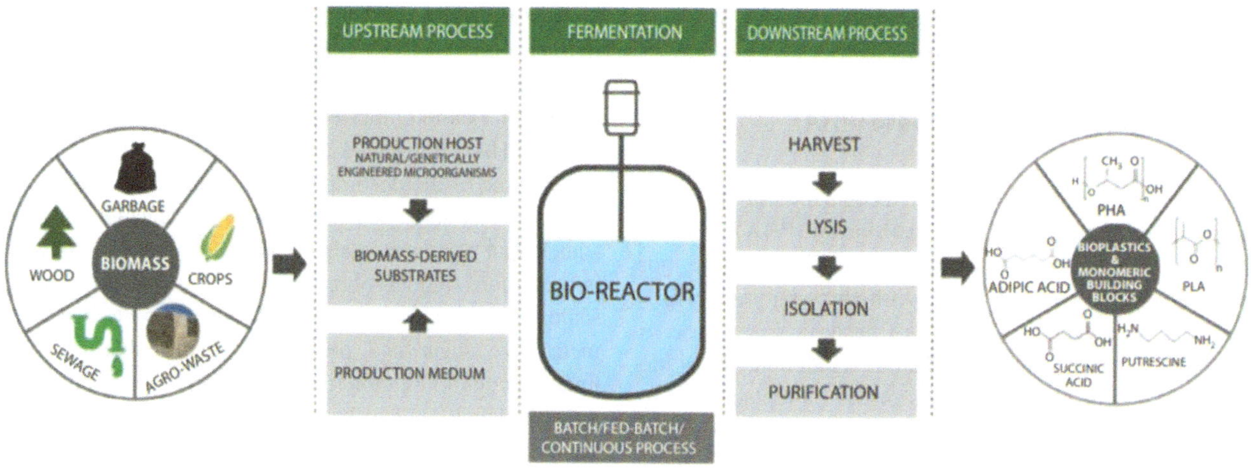

Fig. 2 Microbial production of bio plastic (Francis & Parayil, 2023)

(2) Bio-process creation on a pilot scale using appropriate substrate and production medium;

(3) Optimization of fermentation conditions to lower production costs and boost yield;

(4) Expansion of production procedure to satisfy market expectations;

(5) Creation/using effective procedures for selective biopolymer recovery.

3 Microorganisms Producing PHA and Its Co-Polymers

Many different types of microbes, including engineered and indigenous strains, are being used to produce PHA and its copolymers on an industrial scale. PHA synthesis is known to occur in a variety of prokaryotic microbes, such as photosynthetic bacteria andhalophiles. PHA is produced using a variety of bacterial species, including *Cupriavidusnecator, Alcaligeneslatus, Escherischia coli, Azobacter vinelandii, Bacillus megaterium* and *Pseudomonas oleovorans* (Kumar et al., 2014). Agro-industrial residues were used as a carbon source by bacterial species such as *Cupriavidusnecator, Alcaligenes eutrophus, Ralstoniaeutropha, Aeromonas hydrophila, Alcaligeneslatus, Pseudomonas putida* and *Wautersia eutropha* to produce PHA. *Acinetobacter junii* was found to manufacture 94.3% PHB by means of utilizing rice milling by-products as a substrate. The majority of prokaryotes, including halophiles and thermophiles, are said to be good PHA producers and are stored as intracellular particles in cells of microbes.

Botryococcus braunii, Aulosirafer tilissima, Nostoc muscorum, Microalgae consortium, Synechococcus elongates, Spirulina sp., Synechococcus subsalsus and *Synechocystis salina* are among the algae that produce PHA. In particular, *Nostoc muscorum* yields PHA (69% yield) when there is a phosphorus shortage (Pratap & Krishnan, 2023). The most promising microalgae are chlorella and spirulina species, which have emerged as abundant sources for the manufacturing of bio plastics. Microalgae have the capacity to create large amounts of biomass, accumulate large amounts of lipids and do not rely on sources of food as a substrate. Since growing micro-algae in a controlled growth system involves greater yields at a much lower cost, bio plastics from microalgae offers a more sustainable method and supports the bio economy. Additionally, it has been studied that mutant eukaryotic yeast genera such *Yarrowiali polytica, Rhodotoru laminuta* variant RY4, and *Saccharomyces cerevisiae* may make PHA. Recently there have been numerous reports on the manufacture of PHA using inexpensive sources of carbon by native or wild-type yeast organisms, such as *Candida tropicalis* BPU1 and *Wickerhamomyces anomalus* VIT-NN01. A new study found that using banana peels and chicken feather hydrolysis product, the marine yeast *Pichia kudriavzevii* VIT-NN02, derived from seaweeds, can create 79% of PHBV.

These investigations have shown that, by using inexpensive raw materials as carbon materials, both halophilic and non-halophilic yeasts actively contributed to the formation of PHA. Due to their special adaptability obtained by living in such natural habitats (rhizosphere, mangroves, seaweeds, estuarine sediment), halophilic yeasts can be developed as an effective producer of PHAs. To generate PHA and comprehend the mechanism behind it, additional eukaryotic wild-type yeast strains must be discovered (Ojha & Das, 2021).

4 Recent Developments in Microbial Bio Plastics Production

4.1 Polyhydroxyalkanoate (PHA)

Naturally occurring biopolymer recognized for its biodegradable properties polyhydroxyalkanoate (PHA) is mostly found inside bacteria as an energy source. One of the most prevalent forms of PHA is poly-3-hydroxybutyrate (PHB), which is produced by a variety of bacteria. From an environmentally benign and sustainable standpoint, these biologically produced plastic goods provide a replacement for traditional plastics due to their biocompatibility and ability to degrade. PHA is produced in both anaerobic and aerobic settings from a variety of Gram-positive and Gram-negative bacteria (Choi et al., 2020). *Bacillus megaterium* was found to produce PHA for the first time. It builds up inside the cells as 0.2–0.5 μm diameter, unable to dissolve in granules that act as an energy storage substance. While many bacterial communities do not require nutrient limits to create PHA, some PHA-producing bacteria exhibit improved polymer manufacturing in the presence of abundant carbon sources and nutritional restrictions (e.g., nitrogen, magnesium, oxygen, and phosphorus. The (R)-hydroxyalkanoic acid (HA) monomers, which fluctuate in the amount of alkyl side chain (R) and carbons, are the building blocks of PHA polymers and are in charge of producing a wide range of PHAs (Vu et al., 2020). Depending on the PHA manufacturer, the molecular weight of PHA might range from 50 to 100 kDa. PHAs' overall framework is depicted in Fig. 3.

On the basis of the length of their chains, PHAs can be categorized into three primary groups: long-chain length (lcl), medium-chain length (mcl) and short-chain length (scl). According to Kumar et al. (2014), PHAs can be homo polymers (made up of equal monomers) or heteropolymers (made up of distinct monomers) depending on how their monomeric structure is determined. Bacteria including *Bacillus megaterium, Burkholderia cepacia, Cupriavidus necator* and others can make scl-PHAs, which are defined as having three to five atoms of carbon in the framework. Six to fourteen carbon polymer chains, such as poly(3-hydroxyhexanoate) P(3HHx) and poly (3-hydroxyoctanoate) P (3HO) are known as Mcl-PHAs. Additionally, mcl-PHAs can be produced by certain bacteria, including *Pseudomonas aeruginosa, P. oleovorans* and *P. corrugata.*

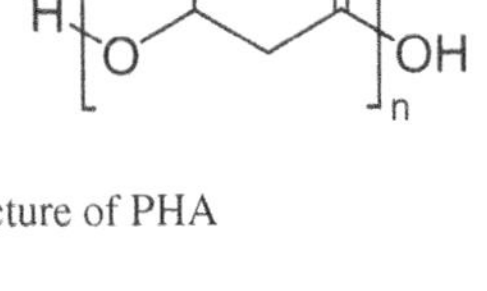

Fig. 3 General structure of PHA

According to Muneer et al. (2020), Lcl-PHAs are produced in trace amounts by microbes and include more than 14 carbon atoms. The groups' polymer physical characteristics differ; for example, mcl-PHAs have visco-elastic qualities, but scl-PHAs are rigid. Linear polyhydroxy alkanoates are an outcome of the decomposition of sugar and fats by bacteria. PHA is an increasingly ductile and less elastic substance than plastic; it is also biodegradable, which is why it is used more frequently in the medical industry. On a large scale, industrial environments are ideal for the bacteria to metabolize sugar and produce purified polyester, which involves combining more than 150 types of monomers.

Although PHA can be made from a variety of materials, corn, sucrose, sugar, and vegetable oils are the main sources of PHA in the food processing sector. Nevertheless, there are drawbacks to using these materials, such as rivalry from the food industry for substrate and high manufacturing costs that prevent these substrates from being widely recognized in the market (Yanez et al., 2020). For the synthesis of PHA, researchers are currently attempting to use less expensive substrates such as waste from the oil industry, waste from the dairy industry, waste feedstocks (WF), municipal wastes, lingo cellulosic feedstocks (LF) and waste from bio-diesel (Bhatia et al., 2021). Figure 4 illustrates how fermentation, which follows microbe inoculation onto a substrate like wheat, maize or rice, is the first step in the creation of PHA.

Under specific fermentation situations, bacteria use PHA biopolymers as storage molecules to collect carbon and energy within their cells. Following fermentation, the biomass (cell) is harvested via a process called centrifugation; filtration or sedimentation and the biomass, enzyme-covered PHA granules and protein structures are recovered. Several techniques are used in the process extraction step, including biological (freeze-drying), chemical (1,2-dichloroethane, sodium hypochlorite, chloroform/methanol, methylene chloride, sodium dodecyl sulphate and chloropropane) and mechanical (bead milling, high-pressure homogenization, ultrasonication). Nevertheless, combination approaches are mostly used to lower the quantity, price and negative impacts of chemicals. The isolated PHA is further refined in accordance with the specifications for the finished product through washing, grinding and polishing (Fig. 4).

Li and Wilkins (2020) have reported on certain recent advancements in feedstocks, strains, and similar process advancements related to PHA production. The prior treatment and hydrolysis procedures are used in conjunction with LF to produce PHA because they increase the availability of carbon sources for microbial fermentation and PHA synthesis. When compared to plastics made from synthetic materials, PHA is more expensive to produce on a big

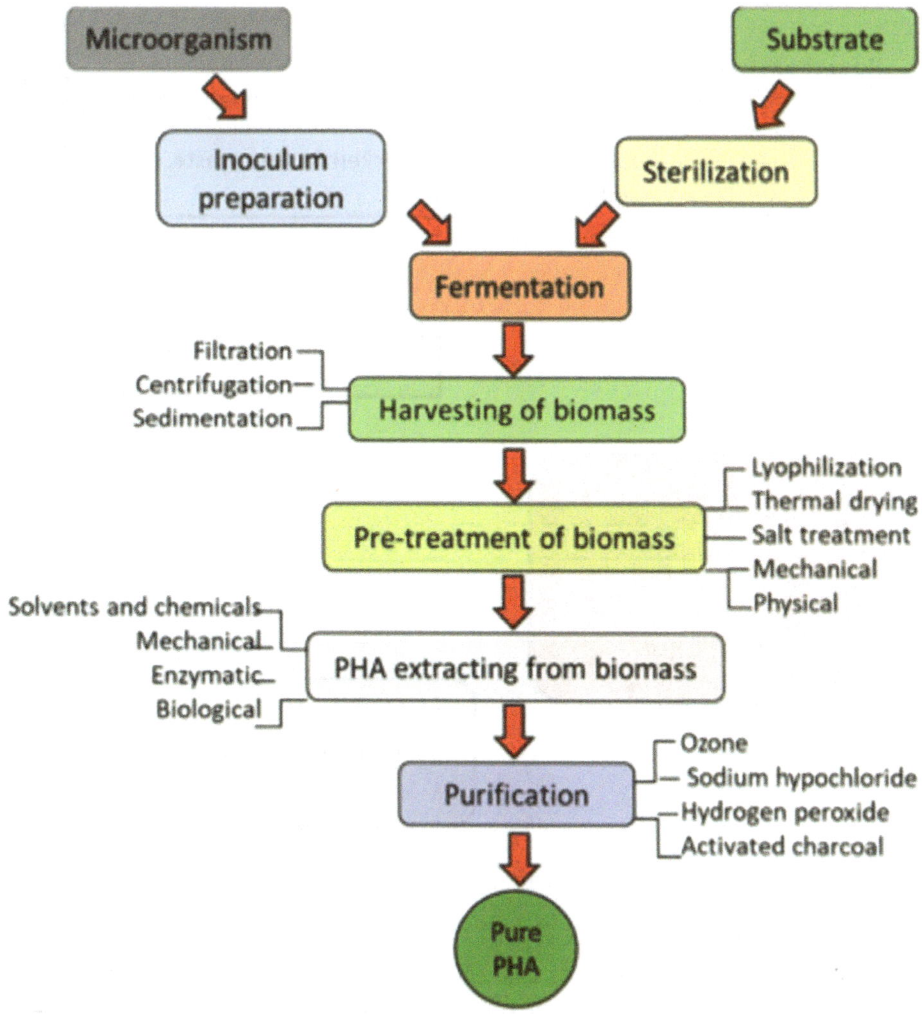

Fig. 4 Basic steps for production of PHA (Bezirhan et al., 2021)

scale due to fermentation procedures and carbon sources. Despite the huge commercial potential of PHAs, their high production costs prevent them from being widely used. However, the utilization of biological waste as inexpensive feed has helped to produce PHAs economically. Therefore, locating less costly sources of carbon, including industry and waste from agriculture, reduces the cost of making PHA and is a financially feasible choice (Guimaraes et al., 2022).

4.1.1 Use of Household Waste

Around the world, the amount of garbage generated by households has skyrocketed and this comprises leftovers from restaurants, cafeterias, kitchens, and the processing industry. It generally consists of proteins, lipids, and carbohydrates-polysaccharides, which include organic acids, polymers, cellulose, hemicelluloses, starch and inorganic molecules-as well as bread goods, dairy products, rice, fruit and vegetable wastes and meats (Li et al., 2021). PHAs are synthesized by bacteria using sugars that are produced from food waste through fermentation caused by bacteria. The right pre-processing technique is crucial depending on the bacterial strain and fermentation broth employed, as pre-processing facilitates breakdown and makes food waste accessible for enzyme breakdown in order to increase PHA synthesis. There have been reports of bacterial strains synthesizing PHA and its copolymers by employing whey as their carbon substrate, including *Hydrogenophaga pseudoflava* DSM1023, *Burkholderia sacchari*, *Caulobacter segnis* DSM 29,236, *Thermos thermophilus* HB8, *Sinorhizobium meliloti* and *Bacillus megaterium*. According to reports, sugar molasses is an accessible and environmentally friendly supply of sucrose for the efficient synthesis of PHA (Ojha & Das, 2018). The microbial synthesis of PHA/PHB from various home garbage is shown in Fig. 5.

4.1.2 Use of Fruits and Vegetable Wastes

Vegetable and fruit scraps are excellent sources of carbon for PHA synthesis. The peels of bananas have recently been utilized to create PHB by *Zobellela tiwanensis* strain DD5,

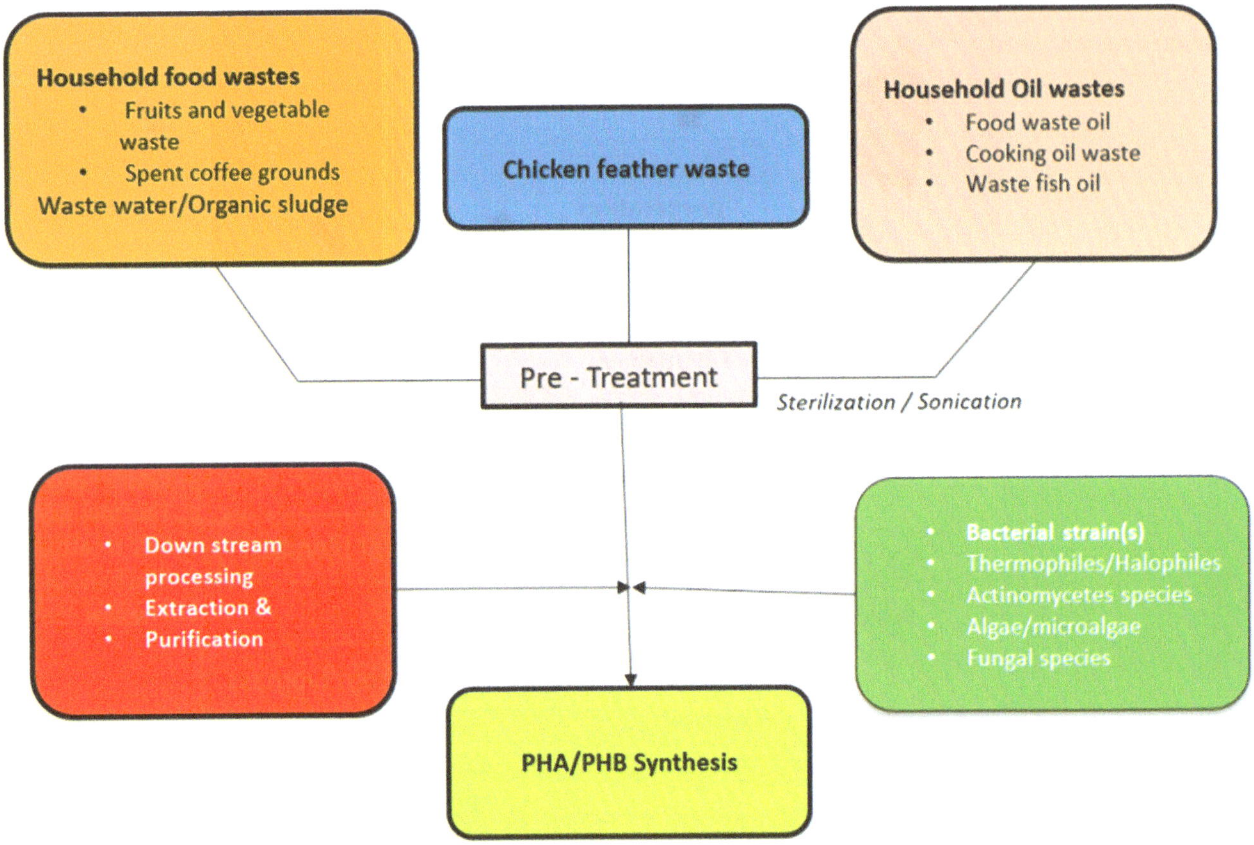

Fig. 5 Microbial production of PHA/PHB from different household wastes (Pratap & Krishnan, 2023)

and PHA by *Pichia kudriavzevii,* wherein banana peels were used as a carbon source and chicken feather hydrolysis products as a nitrogen source. Date waste was used by *H. mediterranei* DSM1411 to create PHBV (Alsafadi et al., 2020). Using pomegranate peels, *Bacillus halotolerans* produced 83% of PHA, while *Cupriavidus necator* generated 71% (Rayasamet al., 2020).

4.1.3 Use of Spent Coffee Grounds (SCGs)

The synthesis of PHA, biopolymer precursors and polymer composite manufacture has been documented with valorized spent coffee grounds (SCGs), which can also be used as an inexpensive carbon source for PHA production (Bomfim et al., 2022). After chemical pre-treatment, halophile strain *Halomonas* produces PHA from SCG hydrolysates (Kovalciket al., 2018), while *B. cepacia* uses SCG hydrolysates to create PHA

4.1.4 Use of Municipal Organic Solid Waste (OFMSW)

Municipal organic solid waste (OFMSW) is another source from which PHA is produced. *Burkholderia sp.* F24 was used to manufacture PHA from sugarcane bagasse and *B. cereus* was used to produce PHBV from wheat starch wastewater (WSW) (Sinaei et al., 2021). It has been observed that *Bacillus spp*. Was utilized to manufacture PHB from sugar cane bagasse, banana peel, maize cob, and teff straw. It has also been reported that waste from pineapples is used to make PHB using *C. necator* A-04, while *Burkholderia cepacian* USM produced PHA from rice husk by alkali treatment followed by enzyme hydrolysis.

4.1.5 Use of Chicken Feather

There is evidence of the bacterial use of chicken feathers as a substrate to create PHBs. Under ideal circumstances, exploiting chicken feather waste as the substrate (15.96 g/L), at 37 °C temperature, and 6.85 pH, *Pseudomonas aeruginosa* (NCBI ID MF18889) was isolated from the chicken (feather) waste which produced PHB amounting to 4.8 g/L (Murugan et al., 2021).

4.1.6 Use of Oils

PHA can also be made from a variety of oils, including canola oil, palm oil, palm olein, coconut oil, cooking oil, soy oil and olive oil distillate (Alves et al., 2023). Using canola oil as a substrate, *Cupriavidus necator* ATCC 17,699 was shown to generate mcl-PHA (90%). Fish oil leftovers as a carbon source were also used to study the formation of PHA

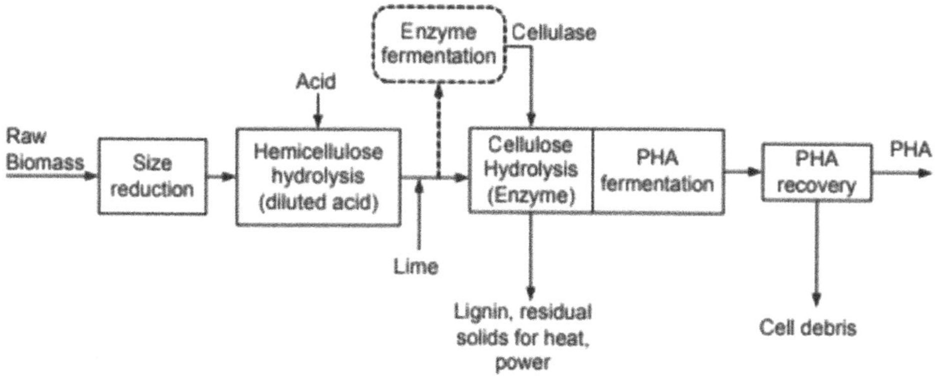

Fig. 6 PHA fermentation using lignocellulosic biomass (Yu, 2007)

through a two-stage process that included culture selection and accumulation of intracellular components. *P. resinovorans* from leftover frying oil was used in the process of fermentation to generate the mclPHA, whereas *C. necator* produced PHB. Food waste oil can be used by *Pseudomonas alcaligenes* to make PHA (Pan et al., 2021). Additionally, leftover cooking or fish oil is utilized as a low-cost carbon source to produce PHB utilizing *Cupriavidus necator* H16. It has been suggested that leftover animal fats from the meat processing and food processing sectors could provide an inexpensive carbon source for the synthesis of PHA. PHA polymer is produced by *Pseudomonas resinovorans* using un-hydrolyzed tallow; 15% of the entire biomass contains PHA (Pagliano et al., 2017).

4.1.7 Use of Plastics Wastes

Different pyrolysis products (oils and waxes) from various waste plastic types (polyethene terephthalate, polystyrene and polyolefins) have been investigated for their potential to produce polyhydroxy alkanoates (PHA) through the use of hydrocarbon-degrading bacterial strains as their main carbon substrate. Fluorescent *Pseudomonads, P. oleovorans* and *P. aeruginosa* have all been reported to produce MCL-PHAs utilizing n-alkanoic acids, hexadecane and octane, respectively (Fernandez et al., 2005). *Pseudomonas sp.* was fermented using styrene, terephthalic acid and polyethene wax in order to produce MCL-PHAs.

4.1.8 Use of Lignocellulosic Wastes

It has been claimed that the most powerful and environmentally friendly way for producing PHA is the lingo cellulosic feedstocks, which are the wealthiest and practical carbon sources. They contain 20–50% hemicelluloses, 40–50% cellulose, and 20–30% lignin. This technique totally prevents the decline of food crops. Utilizing lingo-cellulosic biomass hydrolysates (LBHs), such as pine biomass hydrolysate (PBH), barley biomass hydrolysate (BBH) and miscanthus biomass hydrolysate (MBH) as efficient carbon substrates, the growth conditions of *R. eutropha* 5119 were most recently optimized to generate PHA (Bhatia et al., 2019). Figure 6 illustrates a schematic procedure flow sheet for PHA production on lingo-cellulosic biomass that combines the simultaneous cellulose hydrolysis and PHA fermentation with biomass pre-treatment.

4.1.9 PHA Nanocomposites

Because of their adaptable qualities, scientific curiosity in the incorporation of nanostructured materials into PHA polymers has recently increased. PHA nanocomposites are created when nanostructured substances (>100 nm) are combined with PHA using a variety of fabrication techniques. In order to improve the mechanical and thermal barrier properties of PHA nano-composites, PHAs can be combined with various other polymers (nanofibers) or filled with nano-fillers such as organically modified montmorillonite (OMMT), nano-crystals, nano-particles, bioactive glass and multi-walled carbon nanotubes (MWCNTs) (Ojha & Das, 2021).

The manufacture of nanoparticles made of polymers (PNPs) in the shape of nano-capsules or nano-spheres is another strategy that has been proposed. PNPs, which can be produced in a variety of ways, including mini-emulsification and nano-precipitation, have been shown to be the most environmentally safe nano-carriers for bioactive substances intended to target sick cells with little adverse effects (Ojha & Das, 2021). PHA and its copolymers have been found to be a great, smart and active packaging biomaterial that can be a viable substitute for non-biodegradable packaging materials when blended with additional bio polymers like chitosan and cellulose that have been impregnated with antimicrobial nanoparticles.

4.2 Polyhydroxybutyrate (PHB)

PHB is the main secondary polymer in the PHA class. Its characteristics are comparable to those of the synthetic polymer polypropylene (PP), which makes it a useful material for the production of bioplastics that can take the place of plastic derived from petroleum. The characteristics of PHB

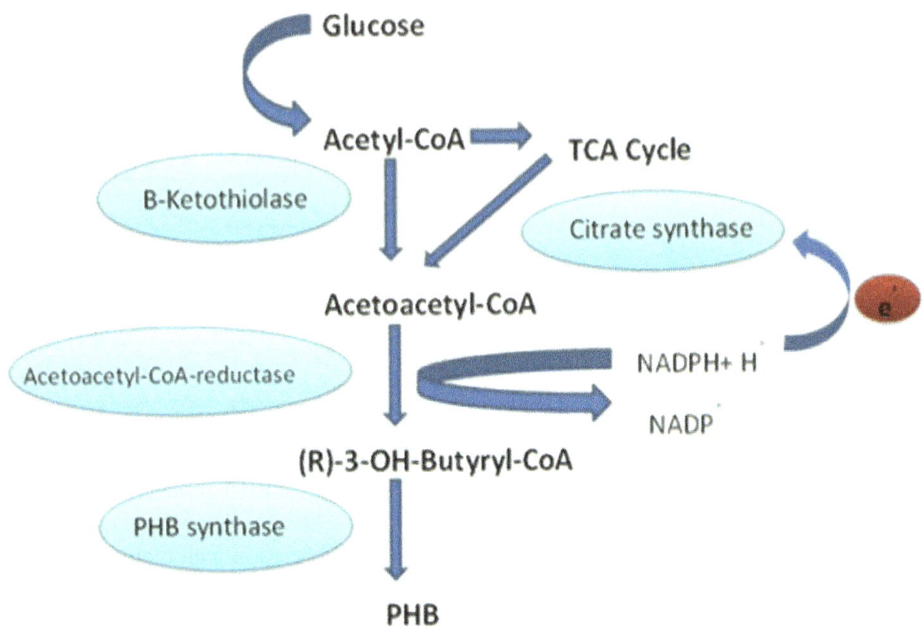

Fig. 7 PHB synthesis by bacteria (Sharma, 2019)

are influenced by the makeup of the carbon sources that the microorganisms employ to manufacture it (Zhang et al., 2022). Figure 7 shows the steps that acetyl coenzyme-A takes to finish its action, which produces PHB. Acetyl-CoA acetyltransferase's action converts two acetyl-CoAs to acetoacetyl-CoA in the initial step. Acetoacetyl-CoA reductase helps transform acetoacetyl-CoA into 3-D-hydroxybutyryl-CoA in the second step, while PHB catalyses the final polymerization step in the final step. Processing of microbes that use the raw substrate in the form of maize starch, glucose, or wastewater is the basis for the manufacture of the biopolymer poly-3-hydroxybutyrate (PHB). Every year, bacteria produce more poly-3-hydroxybutyrate (PHB), which has characteristics that are extremely comparable to those of petro-plastic polypropylene. PHB isolation is first based on physical properties at an industrial scale, and it subsequently develops into an apparent film with a high melting point of more than 130 °C (Chen & Patel, 2012). Many PHB copolymers, such as poly(3-hydroxybutyrate-co-3-hydroxyvalerate), P(3HB-co-3HV)), are now recognized with distinct structures and characteristics. These copolymers can be made by a variety of bacteria. This copolymer is more flexible and suitable for wider applications.

When compared with plastic derived from petroleum, PHB manufacture experiences a substantial obstacle in the form of high costs of production. Other inexpensive sources of PHB include dairy waste, agro-residues, maize steep liquor and molasses (Awasthi et al., 2021). Several methods, including the creation of an effective bacterial strain, fermentation and optimization of the recovery procedure have been employed to lower PHB manufacturing costs. PHB typically develops in bacteria when there is a high carbon content and low quantity of nitrogen, phosphorus and oxygen. Terpolymers are now being investigated as a substitute for copolymers with several secondary monomers in their condensation, such as P (3HB-co-3HV-co-3HHx) and P (3HB-co-3HV-co-4HB) generated by bacteria. When ammonium sulphate is provided as a nitrogen source and mannitol is provided as carbon at an acidity level of 7, the optimized strains of *Rhizobium elti* E1 and *Pseudomonas stutzeri* E114 increase the output of PHBs. A temperature of 30 °C and an incubation duration of 48 h were the ideal conditions for producing PHB with the maximum yield. In order to produce PHB, *Alcaligenes sp.* was optimized in batch culture utilizing cane molasses and urea as nutritional sources.

Agricultural materials such as steep corn liquor and cane molasses, which serve as carbon and nitrogen sources, are the most affordable sources for the synthesis of PHBs, which are environmentally friendly polyesters. PHB, thermoplastic polyester in intracellular carbon, is used by bacteria and fungi, which then build up inside the cell as inclusion structures. Due to their superior physical qualities over PHB, PHA co-polymers such as poly (hydroxybutyrate, hydroxyvalerate, or P(3HB-co-3HV), PHBV) are produced rather than homo-polymers polyhydroxybutyrate (PHB) (Montemurro et al., 2022).

5 Applications of Bio Plastic

Advanced microbial polymers have several uses in the fields of biomedical sciences, packaging, and bioengineering. PHA and PHB have been thoroughly investigated for use in industry and medicine. Figure 8 provides a schematic

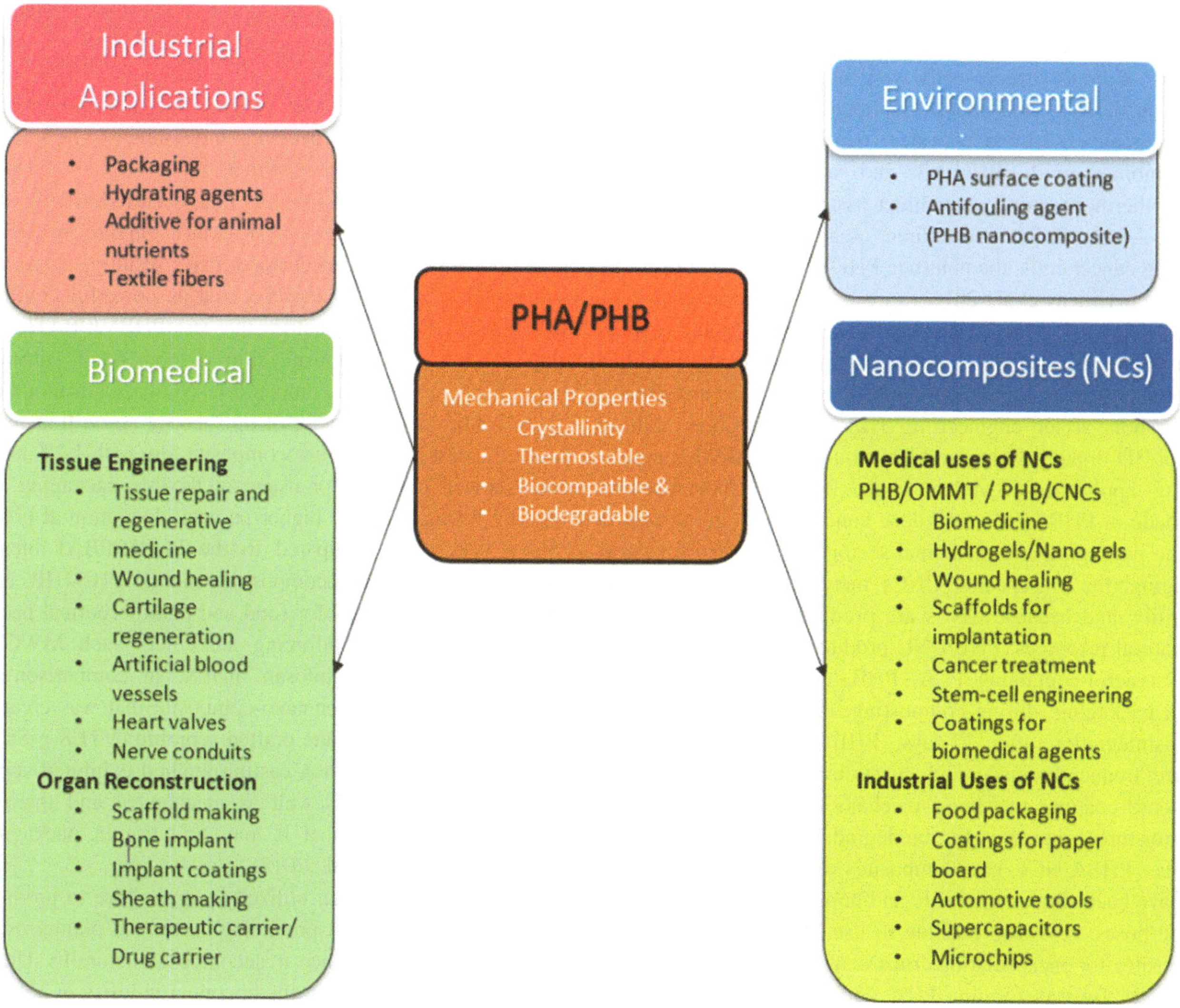

Fig. 8 Applications of bio plastics (PHA/PHB) and its nanocomposites (Pratap & Krishnan, 2023)

representation of the applications of PHBs, PHAs and their nano-composites. PHAs are mostly utilized in the creation of bio-plastics, which are an excellent alternative to traditional polymers in food packaging. For substances that come into proximity to food, the copolymers PHBV and poly((R)-3-hydroxybutyrate-co-(R)-3-hydroxyhexanoate can be employed. PHB and PHB composites are currently produced by a number of businesses for use in the manufacturing of packaging and plastic (Koller & Mukherjee, 2022). PHA finds use in biomedical fields such as tissue scaffolding, drug delivery systems, bone implants, nano-coatings, and prosthetic organ development. Because it lacks a lipopolysaccharide coating, PHA produced from Gram-positive bacterial strains is less immunogenic and therefore more therapeutically biocompatible. PHB, poly(3-hydroxybutyrate-co-3-hydroxyvalerate-co-3-hydroxyhexanoate), PHBV, and PHBHHx are commonly utilized as bio-plastic biomaterials. PHBHHx has also been employed as a medicine carrier for immune suppressants to manage autoimmune illnesses. In addition to being compatible with blood tissue, PHBV composite materials can be utilized in tissue engineering to facilitate the adherence, attachment, and proliferation of fibroblasts (Karbowniczek et al., 2021). Biocompatible surgical mesh, scaffolds, and sutures are made from PHB fibers.

PHA nanocarriers are employed in medication delivery and cancer treatment. Research interest in PHA nanoparticles (PHANPs) for applications in medicine has increased due to their non-toxicity, biocompatibility, and inherent ability to be broken down by non-specific esterases and lipases. In order to decrease the amplification of folate receptor α in breast cancer cells, linkers are attached to the functional moieties of PHANPs with different folate functional groups. PHANPs have functional groups of fluorescent tags linked to them in order to treat the area that is affected, do fluorescence imaging, and track both living and dead cancer cells. During the cancer cell recovery process, PHANPs with fluorescent tags affixed to the MT3 line of breast cancer cells were utilized as a nano-tracer. Drug delivery and cancer treatment use poly (3-hydroxyoctanoate-co-10-undecanoate) nanoparticles (PHOUNPs) combined with PHO-N3 and azide-terminated PHA. Furthermore, using an indirect assay for drug delivery on the A549 cell line and a direct assay for treating the A459 lung cancer cells, the nontoxic PHOUNPs had a noteworthy effect (Kilicay et al., 2016).

Because of their porosity, non-immunogenic nature, and biocompatibility, PHB/OMMT (organically modified montmorillonite) nanocomposites have been investigated for bone tissue engineering. They have also been utilized for 3D organ printing and lab-on-a-chip scaffold engineering applications (Mohan et al., 2021). Nanocomposites made of PHB/CNCs (cellulose nanocrystals) are utilized in the packaging and medicinal sectors. When used for packaging, the presence of CNCs increases their biodegradability, and because CNCs are produced using inexpensive natural resources, PHB/CNC production costs are reduced. Carvacrol-containing PLA, PHB, and CNCs are utilized in packaging; during composting, they also show superior disintegration after 17 days. PHB/CNC nano-composites are frequently made with plasticizers for usage as paper board coatings that boost mechanical strength and provide moisture protection for biodegradable packaging materials. PHB/CNCs nanocomposites during in vitro studies have been shown to be safe and non-toxic for use in human purposes. It is also feasible to use PHB/CNCs nano-composites for encapsulating drugs or fertilizers (Usurelu et al., 2022). By transforming PHA into PHA bio nano-composites, improved or exceptionally well multifunctional PHA nano-composites with antibacterial properties have been created, expanding its applicability in delicate situations. As of right now, PHAs that have been nano-mixed or nano-impregnated have improved thermal barrier, antibacterial and mechanical qualities. PHA bio-nano-composites, which have several functions, are anticipated to make a substantial contribution to the food packaging market in the near future (Ojha & Das, 2020). Additionally, PHB composites have been investigated for use in surgical and medication administration applications.

PHA nano-composites as an intelligent and active packaging substance: Biogenic nanoparticles have been extensively studied as active strengthening, smart green and environment-friendly nanofiller agents. A smart enriched PHBV-CS biopolymer supplemented with zinc oxide-silver nano-composites (ZnO–Ag NCs) was examined for the extended longevity of food packaging medium by Ojha and Das (2020). ZnO–Ag NCs function as a nanoactive agent to prolong food's shelf life and prevent decomposition due to impurities. According to them, the PHBV–CS/ZnO–Ag exhibits antibacterial action, suggesting a potential substitute for the conventional synthetic plastics currently employed for food as well as poultry packaging. In comparison to the pure PHBV matrix, PHBV/SiO_2 nanocomposites exhibit a significant improvement in resistance to heat, thermal, mechanical and barrier properties.

Given the chain positioning that occurred during the rapid expansion in the barrier (water vapour and oxygen) and the rheological characteristics resulting from the superior dispersion of nano clay and its powerful bonding with the PHBV/PBAT matrix, the cast PHBV/PBAT/nano clay (1.2%) nanocomposite film demonstrated substantial enhancement when compared to compression moulded films. Additionally, because of its superior stress transferability, the cast extruded nanocomposite film with 1.2% nano clay showed promise for usage in flexible packaging, as seen by its significantly higher percent elongation at break (567.6 0.1) when compared to the PHBV/PBAT blend. Additionally, PHA nanocomposites, such as PHB4HB, and PHB have been created for food and pharmaceutical packaging purposes by reinforcing nanofillers such MWCN, montmorillonite and chitosan matrix. In comparison to PHBV, the clay polymer nano-composite that was created using vinyl triethoxysilane grafted sepiolite (VTES-g-SEP) and PHBV by the solution casting method exhibited superior biodegradability as well as mechanical and thermal properties, indicating that it may find use in packaging applications (Ajmal et al., 2018).

PHA has mostly been utilized in agriculture to produce mulching material, PHA-producing bacteria as plant growth accelerators and soil since it decomposes naturally. These polymers are employed in the chemical industry as releasing agents for insecticides and herbicides, biodegradable solvents, and dye manufacturing (Guimaraes et al., 2022). Given that the PHA synthesis route has recently been shown to be capable of producing bio-fuels, PHA has the potential to be a sustainable source of biofuel because it can be made from a variety of waste sources by using them as PHA precursors.

6 Challenges and Future Prospects

Reduced greenhouse gas emissions, a decreased reliance on fossil fuels and the possibility of biodegradability-which might aid in addressing the issues of pollution and plastic waste-are just a few benefits of the microbial creation of bio plastics. The creation of appropriate manufacturing hosts is one of the major obstacles in the manufacture of bio plastic. It is difficult to optimize the growing conditions in hosts in order to maximize production. Since existing bio plastic substitutes to fossil-based polymers tend to be more expensive and poorly handled, the greatest obstacle is establishing suitable waste disposal systems based on the circular economy concept. Most research efforts have concentrated on sustainable but profitable production systems. In addition, improper disposal methods brought on by ignorance of the end users regarding bio plastics may cause environmental leakage. International bio plastic label standardization is thus necessary to give consumers accurate details about the chemical composition and available recycling solutions. The main obstacle at the moment is the expense of producing PHA from polymers derived from fossil fuels. Pure or recombinant bacterial strains and sugar-rich substrates are needed for the large-scale manufacture of PHAs, which raises the cost of production. Therefore, the researchers' current top objective is to find less expensive substitutes for use as feedstock for bacterial strains that will create high-quality PHAs in large quantities. Effective disposal and management of waste can be achieved by using household garbage as a feedstock for PHA manufacturing. The process is good for the environment and helps reduce pollutants in the environment.

Simple labeling combined with deposit-refund schemes or additional incentives may motivate users to properly dispose of their bioplastic waste. In the coming years, in order to avoid bioplastic accumulation in the environment, effective recycling solutions must be devised in conjunction with their rising production. Nonetheless, there are still issues with scalability, cost-effectiveness and competing in the market with well-known petroleum-based polymers. In order to make microbial bioplastic manufacturing a more economically feasible and environmentally friendly alternative to traditional plastics, industry and researchers need to improve its efficiency (Bhat, 2023). Future research on biological and herbal PHA manufacturing can be done in great detail in order to create a multipurpose, high-performing green microbial plastic. Moreover, a few outstanding eukaryotic extremophiles ought to be investigated in order to maximize the economical PHA production characteristics. However, more investigation aimed at refining the suggested fermentation process may result in an increase in the production of bio-plastics, making the transformation simpler and affordable.

7 Conclusion

Although there is always a market for plastics, petrochemical-based polymers have multiple disadvantages, including the inability to biodegrade and significant harm to the environment. Compared to plastic made from petroleum, bioplastics are renewable, more friendly to the environment, biodegradable and everlasting and thus attracted a lot of attention from researchers as a prospective agent for a variety of uses. By using biodegradable polymers, greenhouse gas emissions are reduced and the damage caused by trash made of plastic is addressed. Recent advancements in PHB along with other PHA co-polymer nano composites serve as a dependable substitute for petro-chemical plastics made from petroleum and hold great promise for use in surgical implants, tissue engineering, drug delivery for the repair of blood vessels and wounds and bone scaffoldings. An outstanding clever and active packaging biomaterial has been discovered by blending PHA and its copolymers with other biopolymers (cellulose, chitosan), then impregnating the mixture with antimicrobial nanoparticles. As the manufacturing of bioplastics increases, effective recycling systems must be created to stop the plastics from building up in the natural world in the future.

References

Ajmal, A. W., Masood, F., & Yasin, T. (2018). Influence of sepiolite on thermal, mechanical and biodegradation properties of poly-3-hydroxybutyrate-co-3-hydroxyvalerate nano composites. *Applied Clay Science, 156*, 11–19.

Alsafadi, D., Ibrahim, M. I., Alamry, K. A., Hussein, M. A., & Mansour, A. (2020). Utilizing the crop waste of date palm fruit to biosynthesize polyhydroxyalkanoatebio plastic s with favorable properties. *Science of the Total Environment, 737*, 139716.

Alves, A. A., Siqueira, E. C., Barros, M. P. S., Silva, P. E. C., & Houllou, L. M. (2023). Polyhydroxyalkanoates: A review of microbial production and technology application. *International Journal of Environmental Science and Technology, 20*(3), 3409–3420.

Awasthi, K., Akhtar, S., & Khan, M. K. A. (2021). Bioplastic: An accost towards sustainable development. *NeuroPharmac Journal, 4*, 162–168.

Bezirhan, A., Bouchareb, E., Bouchareb, E. M., Yagcı, R., Dizge, N. (2021). Innovative technologies adopted for the production of bio plastic s at industrial level. *Bio plastics for Sustainable Development*, 83–102.

Bhatia, S. K., Gurav, R., Choi, T. R., Jung, H. R., Yang, S. Y., Moon, Y. M., Song, H. S., Jeon, J. M., Choi, K. Y., & Yang, Y. H. (2019). Bioconversion of plant biomass hydrolysate into bio plastic (polyhydroxyalkanoates) using *Ralstoniaeutropha* 5119. *Bioresource Technology, 271*, 306–315.

Bhatia, S. K., Otari, S. V., Jeon, J. M., Gurav, R., Choi, Y. K., Bhatia, R. K., Pugazhendhi, A., Kumar, V., Banu, J. R., Yoon, J. J., & Choi, K. Y. (2021). Biowaste-to-bio plastic (polyhydroxyalkanoates): Conversion technologies, strategies, challenges, and perspective. *Bioresource Technology, 326*, 124733.

Bhatt, M. S. (2023). Microbial production of bio plastics. *Journal of Agriculture Biotechnology and Applied Sciences, 1*(1), 1–4.

Bomfim, A. S. C. D., Oliveira, D. M. D., Voorwald, H. J. C., Benini, K. C. C. D. C., Dumont, M. J., & Rodrigue, D. (2022). Valorization of spent coffee grounds as precursors for biopolymers and composite production. *Polymers, 14*(3), 437.

Chen, G. Q., & Patel, M. K. (2012). Plastics derived from biological sources: Present and future: A technical and environmental review. *Chemical Reviews, 112*(4), 2082–2099.

Chia, W. Y., Tang, D. Y. Y., Khoo, K. S., Lup, A. N. K., & Chew, K. W. (2020). Nature's fight against plastic pollution: Algae for plastic biodegradation and bio plastic s production. *Environmental Science and Ecotechnology, 4*, 100065.

Choi, S. Y., Cho, I. J., Lee, Y., Kim, Y. J., Kim, K. J., & Lee, S. Y. (2020). Microbial polyhydroxyalkanoates and nonnatural polyesters. *Advanced Materials, 32*(35), 1907138.

Francis, D., Parayil, D. J. (2023). Microbial technology in bio plastic production and engineering. *Handbook of Bio Plastics and Biocomposites Engineering Applications*, 121–148.

Guimaraes, T. C., Araujo, E. S., Hernandez-Macedo, M. L., & Lopez, J. A. (2022). Polyhydroxyalkanoates: Biosynthesis from alternative carbon sources and analytic methods: A short review. *Journal of Polymers and the Environment, 30*(7), 2669–2684.

Karbowniczek, J. E., Kaniuk, Ł. Berniak, K., Gruszczyński, A., & Stachewicz, U. (2021). Enhanced cells anchoring to electrospun hybrid scaffolds with PHBV and HA particles for bone tissue regeneration. *Frontiers in Bioengineering and Biotechnology, 9*, 632029.

Khatami, K., Perez-Zabaleta, M., Owusu-Agyeman, I., Cetecioglu, Z. (2021). Waste to bio plastic s: How close are we to sustainable polyhydroxyalkanoates production? *Waste Management, 119*, 374–388.

Kilicay, E., Erdal, E., Hazer, B., Türk, M., & Denkbas, E. B. (2016). Antisense oligonucleotide delivery to cancer cell lines for the treatment of different cancer types. *Artificial Cells, Nanomedicine, and Biotechnology, 44*(8), 1938–1948.

Koller, M., Mukherjee, A. (2022). A new wave of industrialization of PHA biopolyesters. *Bioengineering, 9*(2), 74.

Kovalcik, A., Kucera, D., Matouskova, P., Pernicova, I., Obruca, S., Kalina, M., Enev, V., & Marova, I. (2018). Influence of removal of microbial inhibitors on PHA production from spent coffee grounds employing Halomonashalophila. *Journal of Environmental Chemical Engineering, 6*(2), 3495–3501.

Kumar, P., Singh, M., Mehariya, S., Patel, S. K., Lee, J. K., & Kalia, V. C. (2014). Ecobiotechnological approach for exploiting the abilities of Bacillus to produce co-polymer of polyhydroxyalkanoate. *Indian Journal of Microbiology, 54*, 151–157.

Li, M., & Wilkins, M. R. (2020). Recent advances in polyhydroxyalkanoate production: Feedstocks, strains and process developments. *International Journal of Biological Macromolecules, 156*, 691–703.

Li, Y., Wang, L. E., Liu, G., & Cheng, S. (2021). Rural household food waste characteristics and driving factors in China, Resources. *Conservation and Recycling, 164*, 105209.

Mohan, A., Girdhar, M., Kumar, R., Chaturvedi, H. S., Vadhel, A., Solanki, P. R., Kumar, A., Kumar, D., & Mamidi, N. (2021). Polyhydroxybutyrate-based nano composites for bone tissue engineering. *Pharmaceuticals, 14*(11), 1163.

Montemurro, M., Salvatori, G., Alfano, S., Martinelli, A., Verni, M., Pontonio, E., Villano, M., & Rizzello, C. G. (2022). Exploitation of wasted bread as substrate for polyhydroxyalkanoates production through the use of *Haloferaxmediterranei* and seawater. *Frontiers in Microbiology, 13*, 1000962.

Moodley, P., & Trois, C. (2022). Circular closed-loop waste biorefineries: Organic waste as an innovative feedstock for the production of bio plastic in South Africa. *South African Journal of Science, 118*, 1–6.

Muneer, F., Rasul, I., Azeem, F., Siddique, M. H., Zubair, M., & Nadeem, H. (2020). Microbial polyhydroxyalkanoates (PHAs): Efficient replacement of synthetic polymers. *Journal of Polymers and the Environment, 28*, 2301–2323.

Murugan, S., Duraisamy, S., Balakrishnan, S., Kumarasamy, A., Subramani, P., & Raju, A. (2021). Production of eco-friendly PHB-based bio plastic s by Pseudomonas aeruginosa CWS2020 isolate using poultry (chicken feather) waste. *BiologiaFutura, 72*, 497–508.

Narancic, T., Cerrone, F., Beagan, N., & O'Connor, K. E. (2020). Recent advances in bio plastic s: Application and biodegradation. *Polymers, 12*(4), 920.

Ojha, N., Das, N. (2021). Microbial production of bio plastics: Current trends and future perspectives. In *Bio plastics for sustainable development* (pp. 1–60).

Ojha, N., & Das, N. (2018). A statistical approach to optimize the production of polyhydroxyalkanoates from Wickerhamomycesanomalus VIT-NN01 using response surface methodology. *International Journal of Biological Macromolecules, 107*, 2157–2170.

Ojha, N., & Das, N. (2020). Process optimization and characterization of polyhydroxyalkanoate copolymers produced by marine *Pichiakudriavzevii* VIT-NN02 using banana peels and chicken feather hydrolysate. *Biocatalysis and Agricultural Biotechnology, 27*, 101616.

Pagliano, G., Ventorino, V., Panico, A., & Pepe, O. (2017). Integrated systems for biopolymers and bioenergy production from organic waste and by-products: A review of microbial processes. *Biotechnology for Biofuels, 10*, 1–24.

Pan, L., Li, J., Wang, R., Wang, Y., Lin, Q., Li, C., & Wang, Y. (2021). Biosynthesis of polyhydroxyalkanoate from food waste oil by Pseudomonas alcaligenes with simultaneous energy recovery from fermentation wastewater. *Waste Management, 131*, 268–276.

Pratap, J. K., Krishnan, K. (2023). Microbial production of polyhydroxyalkonates (Bioplastic) using cheap household waste resources and their biomedical applications: A systematic review. *Letters in Applied NanoBioScience, 12*(4), 174.

Rayasam, V., Chavan, P., & Kumar, T. (2020). Polyhydroxyalkanoate synthesis by bacteria isolated from landfill and ETP with pomegranate peels as carbon source. *Archives of Microbiology, 202*(10), 2799–2808.

Sharma, N. (2019). Polyhydroxybutyrate (PHB) production by bacteria and its application as biodegradable plastic in various industries. *Academic Journal of Polymer Science, 2*(3), 001–003.

Sinaei, N., Zare, D., & Azin, M. (2021). Production and characterization of poly 3-hydroxybutyrate-co-3-hydroxyvalerate in wheat starch wastewater and its potential for nanoparticle synthesis. *Brazilian Journal of Microbiology, 52*, 561–573.

Usurelu, C. D., Badila, S., Frone, A. N., & Panaitescu, D. M. (2022). Poly (3-hydroxybutyrate) nano composites with cellulose nanocrystals. *Polymers, 14*(10), 1974.

Vu, D. H., Åkesson, D., Taherzadeh, M. J., & Ferreira, J. A. (2020). Recycling strategies for polyhydroxyalkanoate-based waste materials: An overview. *Bioresource Technology, 298*, 122393.

Yanez, L., Conejeros, R., Vergara-Fernandez, A., & Scott, F. (2020). Beyond intracellular accumulation of polyhydroxyalkanoates: Chiral hydroxyalkanoic acids and polymer secretion. *Frontiers in Bioengineering and Biotechnology, 8*, 248.

Yu, J. (2007). Microbial production of bio plastic s from renewable resources. In *Bioprocessing for value-added products from renewable resources* (pp. 585–610). Elsevier.

Zhang, L., Jiang, Z., Tsui, T. H., Loh, K. C., Dai, Y., Tong, Y. W. (2022). A review on enhancing *Cupriavidusnecator* fermentation for poly (3-hydroxybutyrate) (PHB) production from low-cost carbon sources. *Frontiers in Bioengineering and Biotechnology, 10*, 946085.

Exploring Endophytes as a Treasure House of Bioprospecting Compounds, Emphasizing Their Significance in the Agricultural and Pharmaceutical Fields

Ankika Bhakat, Priyanka Das, Mitrabrata Goswami, Sujata Sen, and Keka Sarkar

Abstract

Endophytes are endosymbiont microorganisms that colonize internal plant tissues without eliciting any adverse effect on the host. This chapter aims at navigating the potential of endophytes as a treasure trove for agricultural and pharmaceutical purposes. Endophytes possess the ability to stimulate plant growth and optimize nutrient absorption, resulting in superior agricultural productivity. They can condition plants to develop tolerance to abiotic stresses such as salinity, drought, and arid conditions. Endophyte-derived bioactive compounds, namely acyl-homoserine lactones, diacetylphloroglucinol, and pyoluteorincan control phytopathogens and inhibit weed growth. Endophytes produce non-ribosomal peptides and polyketides that immunize the host plants to various pathogen-borne diseases. Moreover, endophytes serve as an environmentally friendly, cost-effective, and enduring substitute for conventional agricultural approaches by eliminating the reliance on artificial fertilizers and pesticides. Endophytes synthesize a diverse array of secondary metabolites, including flavonoids, carotenoids, melatonins, terpenoids, phenolics, alkaloids, and peptides that possess antimicrobial, anticancerous, and pesticidal attributes. Approximately 42% of clinically authorized medications have been derived from natural sources or their derivatives, with a sizeable number being acquired from endophytes. Endophytic fungi have emerged as a viable alternative to plant-extracted secondary metabolites owing to the risks associated with unregulated plant collection, which endangers numerous plant species. Despite their illimitable potential, additional research is necessary to comprehensively harness their capabilities. Further assessment of the pharmaceutical prospects of endophytes requires advanced biotechnological techniques like multi-locus sequence typing, metabolomics, metagenomics, and next-generation sequencing. Continued endeavors to recognize, describe, and comprehend endophyte-plant interactions will conceivably enable practical implementations in both domains.

Keywords

Endophytes · Bioactive compounds · Anticancer activity · Agricultural applications · Pharmaceutical applications

1 Introduction

Extensive research investigating interactions between pathogens and the plants they colonize has conjectured that merely a minute proportion of the microorganisms interacting with plants exhibit pathogenicity (Andreote et al., 2014). A majority of the plant-inhabiting microorganisms possess a mutualistic relationship with their host plants (Mandal et al., 2021). A single plant species may harbor innumerable microbes, which are categorized as either epiphyte (microbial inhabitants of the rhizosphere and phyllosphere, residing in close proximity of plant tissues) or endophytes (microbes inhabiting plant tissues in the stems, leaves, or roots) contingent upon their region of colonization in the host plants (Dastogeer et al., 2020). The prevalence of non-pathogenic microbes in plants was first documented by De Bary (1866), who coined the term endophyte and defined it as "any organism that grows within plant tissues." Subsequently, the definition has altered over time per the findings of various researchers (Gouda et al.,

A. Bhakat · P. Das · M. Goswami · S. Sen · K. Sarkar (✉)
Department of Microbiology, University of Kalyani, West Bengal, Kalyani, India
e-mail: keka@klyuniv.ac.in

K. Sharma (ed.), *Bio-prospecting of Novel Microbial Bioactive Compounds for Sustainable Development*, Sustainable Landscape Planning and Natural Resources Management, https://doi.org/10.1007/978-3-031-82178-3_3

2016). Nonetheless, endophytes are most appropriately defined as microorganisms that colonize internal plant tissues at a specific point in their life cycle without adversely affecting the host plant. Furthermore, a comprehensive assessment of endophytic microorganisms has assisted in their segregation into different subgroups, namely 'obligate' or 'facultative'. Endophytes depending on plant metabolism for sustenance and disseminated among plants by vectors or vertical transmission are deemed obligate endophytes (Hardoim et al., 2008). Contrarily, facultative endophytes reside on the external surfaces of the host body and are predominantly associated with neighboring plants (Gouda et al., 2016). Typically, specific endophyte-plant interactions and their strategy of existence and transmission are evidenced by their genome organization (Andreote et al., 2014). Research aimed at discovering the origin of endophytic organisms in different species correlated genome size with endophytic lifestyles (Andreote et al., 2014) and initially determined rhizosphere- or seed-borne microbial communities as the primary endophytic sources (Gouda et al., 2016). The definition of endophytes presumably corroborates the hypothesis that they reside inside plant species in environments that are more stable than the plant-growing soils. However, certain endophytes only appear in the plant during part of their lifecycle. Therefore, the endophytic community comprises organisms of distinct origins; moreover, the microbes possessing larger and smaller genomes preferably thrive in variable and stable environments, respectively (Mitter et al., 2013).

This chapter endeavors to explore the potential of plant-derived endophytes as a valuable repository of bioprospecting compounds and emphasize their significance in agricultural and pharmaceutical fields (Fig. 1). Following their establishment within plants, endophytes can positively influence plant growth via distinct mechanisms, including phytostimulation, biocontrol of phytopathogens, and biofertilization, both directly and indirectly, through the production of several bioactive metabolites. Endophyte-synthesized biofertilizers can be formulated as a substitute for synthetic agrochemicals to ensure crop productivity in an agro-ecological manner. Additionally, endophyte-secreted bioactive compounds exhibit diversified biological activities, including antibacterial, antifungal, anti-oxidant, anti-inflammatory, and anticancer effects (Wijesekara & Xu, 2023). The therapeutic potentiality of endophytes depicts promising prospects for both traditional medicine as well as modern pharmaceutical applications and deems them valuable for drug discovery and development. Thorough exploitation of endophytes as a source of pharmaceuticals warrants cutting-edge biotechnological and bioinformatics-based approaches to establish and improve their in vitro biosynthetic proficiency. The burgeoning

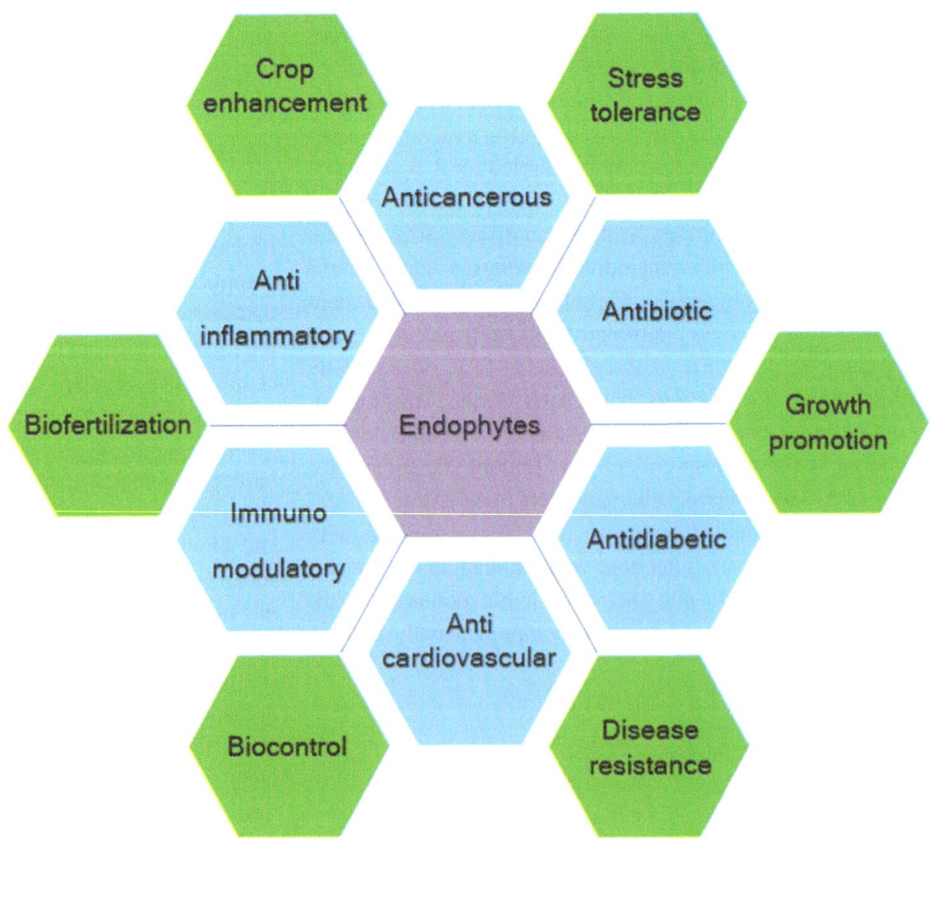

Fig. 1 Prospective applications of endophytes in the agricultural (green) and pharmaceutical (blue) fields

intrigue surrounding traditional herbal medicines necessitates the elucidation of the endophytic origin of therapeutic compounds for sustainable utilization.

2 Types of Endophytes

Endophytes that colonize plants include either bacteria (actinomycetes or mycoplasma) or fungi. Over 200 genera belonging to 16 phyla of bacterial species, particularly Actinobacteria, Proteobacteria, and Firmicutes, have demonstrated endophytic associations (Golinska et al., 2015), and range in diversity from gram-positive (*Bacillus, Brevibacterium, Mycobacterium*) to gram-negative bacteria (*Achromobacter, Acinetobacter, Agrobacterium*). Bacterial endophytes produce different metabolites that serve as a variety of bioactive compounds. Actinomycetes (phylum Actinobacteria) are prokaryotic microorganisms that possess fungi-like mycelium and form spores (Gouda et al., 2016) and represent a transitional form connecting fungi and bacteria (Barka et al., 2016). Endophytic actinomycetes evidently secrete various unique-structured chemical entities of considerable medicinal importance (Gouda et al., 2016). Streptomyces, commonly isolated as endophytic actinomycetes (Golinska et al., 2015), secrete compounds, including clethramycin, coronamycin, naphthomycin (A and K), cedarmycin (A and B), saadamycin, kakadumycins, munumbicins (A and B), and others such as *Kitasatospora* spp.-derived paclitaxel and *Emblica officinalis*-derived tyrosol, which reportedly inhibits foodborne microbes (Golinska et al., 2015). Endophytic mycoplasma species display a symbiotic relationship with certain red algae, namely *Bryopsis hypnoides*, *B. pennata*, and *Arcobacter* spp. (Hollants et al., 2011). Nevertheless, their source of extraction or application as antimicrobial agents is yet to be validated. Endophytic fungi have been classified into two broad categories depending on their life history and phylogenetic traits: clavicipitaceous (endophytes infecting certain grasses confined to frigid regions) and non-clavicipitaceous (endophytes colonizing the asymptomatic tissues of non-vascular plants, ferns, allies, conifers, and angiosperms and strictly belong to Ascomycota or Basidiomycota) (Bhardwaj & Agrawal, 2014). Endophytic fungi produce various extremely broad-spectrum antibiotics and anticancer drugs. Penicillenols, extracted from *Penicillium* spp., exhibit cytotoxicity against numerous cell lines. Taxol, isolated from *Taxomyces andreanae*, is evidently the most effective fungal endophyte-derived anticancer drug discovered to date. Furthermore, several other compounds acquired from endophytic fungi possess potent antifungal, antibacterial, and anti-oxidant properties (Gouda et al., 2016).

3 Prospective Application of Endophytes in Agriculture

The conventional strategies adopted for increasing agricultural productivity are neither sustainable nor ecologically balanced and pose severe threats to the environment. The primary factor among the aforementioned approaches is the extensive application of chemical fertilizers and intensive agricultural practices that are not only hazardous for the ecosystem but also adversely affect crop quality, soil salinity, consumer health, and nutrient disparity. This deductively modifies the native soil microbiota, which renders the soil unproductive and unfertile and exerts a strain on the global availability of sustenance (Hazra et al., 2024). Consequently, it is imperative to increase agricultural productivity sustainably so as to replenish soil fertility, increase food provisions, and amend the ecosystem. Endophytes inhabiting plant tissues benefit the host by stimulating growth, inhibiting diseases, strengthening immunity, and heightening stress tolerance. The prospective application of endophytes as a substitute for synthetic agrochemicals may guarantee crop productivity in an agroecological manner.

3.1 Role of Endophytes in Facilitating Adaptation of Agricultural Crops to Biotic and Abiotic Environmental Stress

Endophytes boost plant immunity to suppress different biotic and abiotic stress (Fig. 2), which aids in the acquisition of plant resistance and adaptation to unfamiliar environments (Lu et al., 2021). Endophytes have been utilized for seed coating and are beneficial during seed germination. Endophytes fortify plant defenses against pathogenic infections by secreting secondary metabolites, such as hydrolytic enzymes, siderophores, and antibiotics, and drastically augmenting oligomeric proanthocyanidins, secondary plant metabolites like phenylpropanoids, and antimicrobial chemicals in crops (Hazra et al., 2024). Endophytes promote plant growth and improve plant biomass and crop yield by stimulating biological nitrogen fixation, phosphate solubilization, and phytohormone synthesis, hindering ethylene biosynthesis (Wang et al., 2020), interfering with photosynthesis and carbon fixation, and inducing the production of phytohormones, namely auxin, adenine ribosides, gibberellin, polyamines, acetoin, cytokinin, adenine, 2,3-butanediol, and indole-3-butyric acid (Table 1) (Shahnawaz, 2021).

In addition to biotic stresses, endophytes assist plants in developing tolerance to abiotic stresses, including heat, salt, and drought. *Neotyphodium* spp. enhanced drought tolerance in grass plants and immunized them under

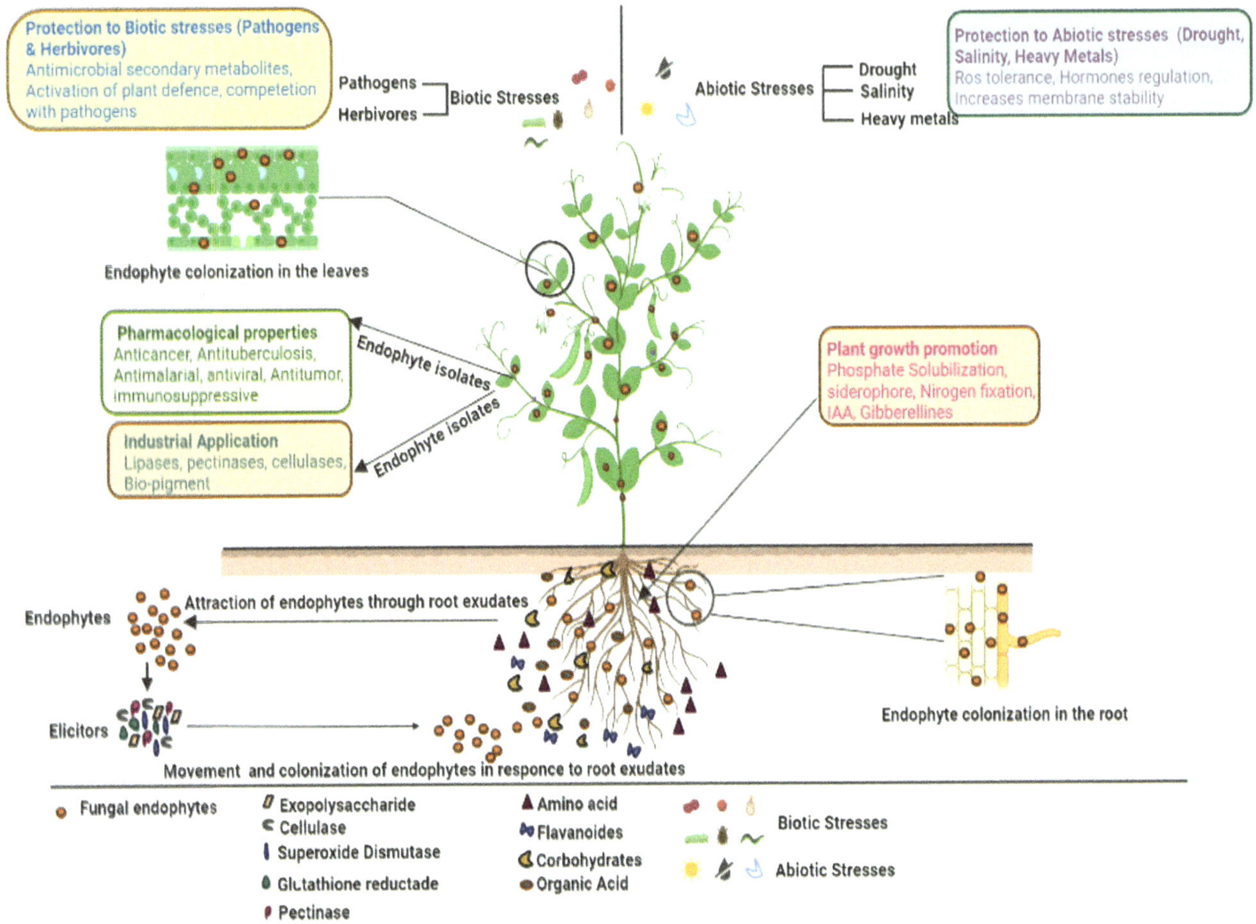

Fig. 2 Reprinted from "Secondary metabolites from endophytic fungi: Production, methods of analysis, and diverse pharmaceutical potential," by Singh, V. K., Kumar, A., 2023, *Symbiosis* 90, 111–125. https://doi.org/10.1007/s13199-023-00925-9

Table 1 Plant growth-promoting hormones secreted by endophytes

Plant hormone	Endophytes	Mechanism	References
Indole acetic acid	*Chaetomium globosum, Fusarium oxysporum, Phomopsis* spp., *Phoma glomerata, Penicillium verruculosum, P. funiculosum, Paecilomyces formosus*	Promotes plant growth and alters fundamental plant defense mechanisms	Ikram et al., (2018)
Gibberellin	*Aspergillus fumigatus, A. flavus, A. niger, Arthrinium phaeospermum, Chrysosporium pseudomerdarium, Cladosporium sphaerospermum, Exophiala* spp., *Fusarium oxysporum, Chaetomium globosum, Gliomastix murorum, Penicillium corylophilum*	Triggers seed germination, stem elongation, leaf development, flowering, and fruit ripening and aids in overcoming dormancy	Vardharajula et al., (2017)
Abscisic acid	*Azospirillum* spp., *Bacillus amyloliquefaciens*	Promotes plant growth and development, modulates plant water balance, and enhances tolerance to osmotic and salinity stress	Vardharajula et al., (2017)
Ethylene	*Methylobacterium oryzae*	Regulates plant stress response signaling	Miliute et al., (2015)

circumstances of nitrogen deficiency and water stress via osmotic and stomatal regulation (Lee et al., 2021). Correspondingly, Chinese cabbage, barley, and cacao plants exhibited drought and salt tolerance when inoculated with *Trichoderma* and *Serendipita indica* (Hasanuzzaman et al., 2019).

Under stress, *Thermomyces* spp. preserves the water content of leaves and improves water usage efficiency.

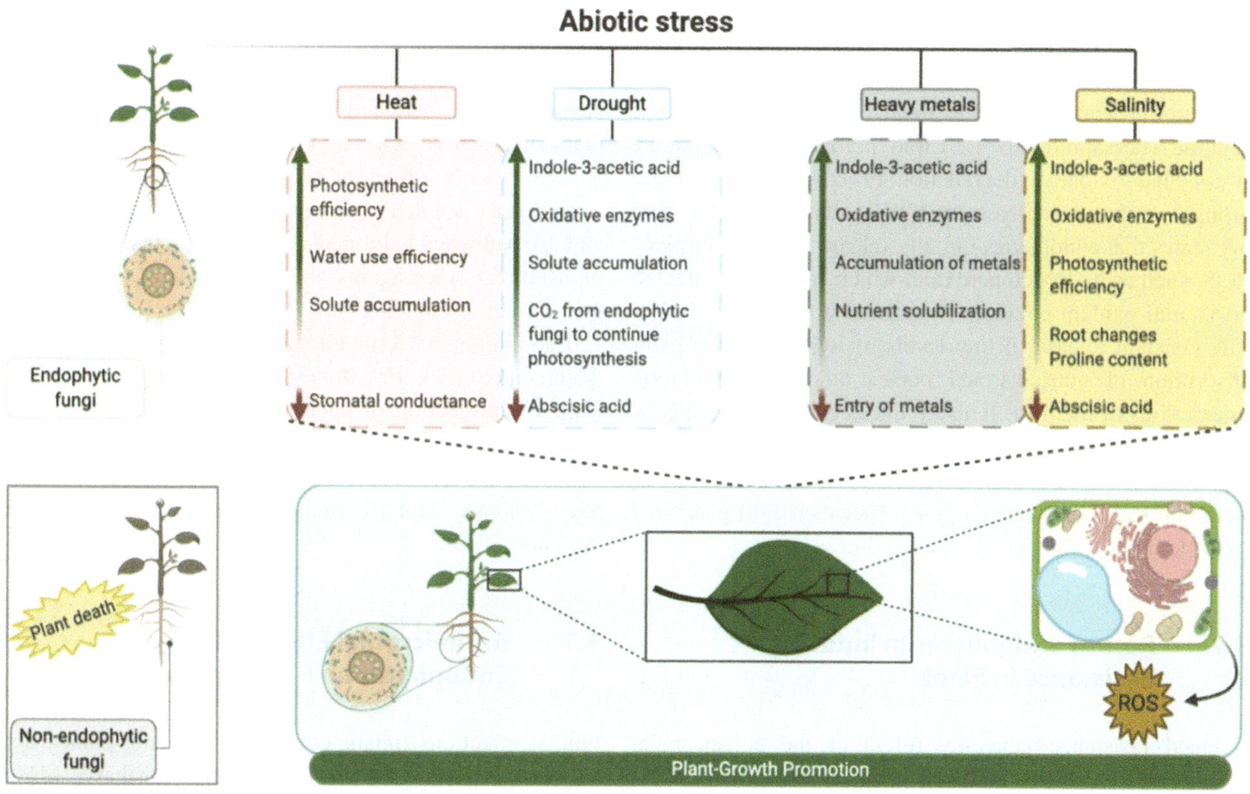

Fig. 3 Potential responses to abiotic stresses as portrayed by fungal endophyte-inhabiting plants. The green and red arrows denote an increase and decrease, respectively, in the listed characteristics. Reprinted from "Endophytic Fungi: Biological Control and Induced Resistance to Phytopathogens and Abiotic Stresses," by Fontana, D. C., de Paula, S., Torres, A. G., de Souza, V. H. M., Pascholati, S. F., Schmidt, D., & Dourado Neto, D. (2021). *Pathogens (Basel, Switzerland), 10*(5), 570. https://doi.org/10.3390/pathogens10050570

Primary and secondary metabolite buildup is encouraged by various endophytic fungi (Fontana et al., 2021). In certain plant-microbe interactions, the elevated concentration of sugars and flavonoids in plant tissues serves as a scavenger of reactive oxygen species (ROS) and a signaling molecule, which enables plant development and the resistance acquisition to abiotic stress. The alleviation of water stress effectuated by endophytes may be attributable to the overexpression of anti-oxidant enzymes, bioactive substances, chlorophyll, carotenoid, and fluorescence. *Penicillium citrinum*, *Aureobasidium pullulans*, and *Dothideomycetes* spp. promoted plant growth in *C. reticulata* plants in addition to altering all these factors (Sadeghi et al., 2020). Furthermore, endophyte-mediated mechanisms facilitate the adaptation to drought tolerance by producing volatile compounds, phytohormones, ROS, exopolysaccharides, 1-aminocyclopropane-1-carboxylate deaminase, and phytohormones and modify root morphology, synthesize anti-stress metabolites, and positively regulate stress-responsive genes in host plants (Fite et al., 2023).

In plants such as *Glycine max*, *Lycopersicon esculentum*, and *Triticum aestivum*, endophytes assume a crucial role in conferring tolerance to heavy metal stressors. For instance, *P. roqueforti* impedes heavy metal from being transferred from the soil to the plants, thereby inducing resistance in *T. aestivum* plants cultivated in heavy metal-contaminated soil (Ikram et al., 2018). Furthermore, these endophytic fungus-inoculated wheat plants that were irrigated with leftover water demonstrated reduced concentrations of heavy metals in the roots and shoots, increased growth, and improved nutrient absorption. Conversely, under significant metal stress, uninoculated wheat plants displayed stunted growth along with indications of chlorosis. Endophytes help their host plants by improving their capacity to take up vital nutrients from contaminated soil (Deng & Cao, 2017). These endophytes can also break down contaminants found in contaminated soil and change them into non-toxic forms. Positive physiological changes that help the host plant resist stress can be brought about by endophytes supplying phytohormones exogenously. Apart from phytohormones, the biofertilization ability of endophytes might augment the host plant's nutrient availability in heavy metal-contaminated soil through solubilization process (Rajkumar et al., 2009). Figure 3 illustrates possible mechanisms underlying endophyte-host plant interaction-mediated response modification under abiotic stressors.

Abiotic stressors such as heat stress, salinity, drought, flooding, and oxidative stress are connected and lead to the production of ROS, which harm cells and eventually cause them to die when exposed to them for an extended period of time (Pessarakli, 2014). Lipid peroxidation, oxidative stress, and oxidative degradation of DNA and RNA are brought on by an increase in ROS in plant cells. The theory states that endophytes initially release a tiny quantity of ROS, such as hydrogen peroxide, which causes the infected host's anti-oxidant enzymes to activate (Ogbe et al., 2020). The constant release of tiny levels of ROS enhances plant absorption of nutrients and boosts other endosymbiotic interactions of the host. It also reduces cell hypersensitivity to ROS. Plant tissues store proline, methionine, flavonoids, and other phenolic compounds to increase their resistance to the presence of reactive oxygen species (ROS) generated by endophytes (Chen et al., 2022).

3.2 Role of Endophytes in Inducing Resistance in Plants

Induced resistance generally refers to the enhancement of defense mechanisms specifically in the plant tissues exposed to the inducing agent; however, the term can also describe a systemic resistance that develops in parts of the plant distant from the initial treatment site. Plant defense responses can be triggered by compounds found in plant extracts (Fontana et al., 2021), yeast preparations (El-Beltagi et al., 2022), growth-promoting rhizobacteria, growth-promoting fungi, avirulent pathogens, endophytic fungi (Fontana et al., 2021), and others. Induced resistance involves the activation of the defense mechanisms of plants by external agents without altering their genome; it triggers a broad response involving the expression of defense-related genes, such as those responsible for oxidative bursts, hypersensitive responses, and the production of PR-proteins, phenolic compounds, and enzymes linked to the phenylpropanoid pathway, lipid peroxidation, and phytoalexin synthesis (Stadnik et al., 2000). The signaling pathways that stimulate defense expression in plants show that resistance can be triggered by non-pathogenic microorganisms or biotrophic pathogens, which release salicylic acid (SA) as the main signaling molecule, leading to the expression of PR-proteins; this resistance is known as systemic acquired resistance (SAR). In contrast, resistance activated by rhizosphere-associated growth-promoting microorganisms or necrotrophic pathogens is called induced systemic resistance (ISR) and is mainly regulated by jasmonic acid (JA) and ethylene (ET), functioning independently of SA (Jain et al., 2016). This segregation appears to be generalized ascertain pathogens, like *Phakopsora pachyrhizi* (the etiological agent of Asian soybean rust), supposedly modulate target gene expression upon penetrating the host tissue, which activates the JA pathway and suppresses SA-mediated defense (Fontana et al., 2021). A plausible positive crosstalk exists between SA, JA, ET, and gene expression effectors, which has been portrayed in an Arabidopsis-isolated rhizobacteria model wherein the SA and JA pathways exhibited combined effects in enhancing induced resistance against the *P. syringaepv.* tomato pathogen (Fontana et al., 2021).

Endophytic bacteria eliciting ISR in plants were first documented in the *Pseudomonas fluorescens* strain 89B-61-triggered protection of cucumber plants against cucumber anthracnose. Subsequent studies established that the ISR was induced by endophytic bacteria of genera *Bacillus*, *Pseudomonas*, and *Serratia* in different plant-pathogen systems.

3.3 Antimicrobial Effect of Fungal Endophytes on Phytopathogens

Fungal infections impair crops by destroying plants, deteriorating quality and yield, and inflicting postharvest losses (Zain, 2011). Synthetic chemical fungicides employed to manage fungal diseases have detrimental and unintended effects on the environment. Fungal endophytes, like *Ampelomyces*, serve as a more environment-friendly alternative to chemical fungicides by reducing pathogen prevalence and preserving mutualistic fungi. Therefore, incorporating these biological control agents (BCAs) into integrated pest management strategies can amend soil quality and promote agricultural sustainability.

A myriad of bioactive compounds, such as glucosides, azaphilones, cytochalasans, non-ribosomal peptides, terpenoids, macrolide polyketides, benzenoids, and lactones have been extracted and isolated from certain taxonomic orders of fungi using top-down techniques (Walker et al., 2014).

The fungal endophyte *Daldinia eschscholtzii*, isolated from the roots of *Zingiber officinale* (ginger) and *Stemona tuberosa*, produces 60 distinct compounds, including elemicin (24%), benzaldehyde dimethyl acetal (8%), ethyl sorbate (7%), methyl geranate (6%), trans-sabinene hydrate (5%), and 3,5-dimethyl-4-heptanone (5%), which inhibit the growth of the plant pathogens *Colletotrichum acutatum* and *Sclerotium rolfsii* (Suebrasri et al., 2020). *Colletotrichum gloeosporoides*, *C. nymphaeae*, and *C. musae* have all been found to respond favorably to the antifungal drug eleven (Khruengsai et al., 2021). Mixtures of 27 *Daldinia cf. concentrica*-generated volatile organic compounds (VOCs), including 3-methyl-1-butanol, (±)-2-methyl-1-butanol, 4-heptanone, isoamyl acetate, and trans-2-octenal, demonstrated a wide range of antifungal activity in solutions (Liarzi et al., 2016).

Secondary metabolites secreted by fungal endophytes belonging to the genera *Aspergillus*, *Phyllosticta*, *Diaporthe*, *Colletotrichum*, *Gliocladium*, *Lecanicillium*, *Phyllosticta*, and *Trichoderma* are being investigated for their antifungal qualities. *Trichoderma* spp. isolated from soybean (*Glycine max*) secretes the hydrolytic enzymes pectinase and chitinase, which are capable of degrading cell wall constituents; administration of solution formulations of *T. asperellum*, *T. atroviride*, and *T. longibrachiatum* empirically diminished *Rhizoctonia solani*-mediated seed infection by 64, 60, and 55%, respectively (Sallam et al., 2021). Furthermore, *Trichoderma* spp. produces IAA, which potently influences plant growth and siderophore activity that constrains the iron availability to pathogenic fungi. *T. erinaceum* extracted from ginger and Stemona root can suppress *S. rolfsii*-infected southern stem rot disease by 58% in pot experiments and 64% in dual culture (Suebrasri et al., 2020). Extract analysis demonstrated that *T. erinaceum* produces chitinase, β-1,3 glucanase, and 6-n-pentyl-2H-pyran-2-one, and polyketides that interrupt the growth of *S. rolfsii*.

3.4 Antimicrobial Effect of Bacterial Endophytes on Phytopathogens

A multitude of endophytic bacteria have displayed the ability to suppress microbe-borne diseases in various plant species. The majority of hostile endophytic bacteria are Gram-negative, fluorescent pseudomonads, which serve as potent BCAs owing to their prevalence in the endorhiza. Reportedly, *P. putida*, *P. graminis*, *P. chlororaphis*, *P. fluorescens*, *P. tolaasii*, and *P. veronii* are employed as antagonistic isolates and, under greenhouse circumstances, exhibit approximately 80% control potential as quantified by decreased disease severity (Mengoni et al., 2003). Various bacterial endophytes of Gram-positive genera, namely *Bacillus*, *Paenibacillus*, *Streptomyces*, *Microbispora*, and *Nocardioides*, markedly impede fungal infections. *Actinoplanesmis souriensis*, a streptomycete, has effectively suppressed infectivity of the causal agent of lupin root rot, *Plectosporium tabacinum*, both in greenhouse settings as well as field experiments (Tjamos et al., 2004). *Bacillus* spp. isolated from the stem of *Jasminum officinale* has been empirically evidenced to prevent chili anthracnose disease caused by the fungal phytopathogen *Colletotrichum siamense* through the production of lipopeptides such as surfactin, iturin, and fengycin (Bhakat et al., 2023). Further verification of their biocontrol efficacy in distinct climatic zones and cultural circumstances warrants additional fieldwork and the designing of formulations and applications that adhere to standard farming practices.

3.5 Future Prospects of Endophytes in Sustainable Agriculture

Comprehensive research has demonstrated that bacterial and fungal endophytes command striking potential as environment-friendly natural resources that may be implemented in agri-food production to intensify crop cultivation, augment agricultural productivity, and achieve environmental sustainability (Omomowo & Babalola, 2019). Owing to their prospective uses as bioeffectors, biofertilizers, anti-biopesticidal BCAs, and biostimulants, endophytes exhibit several advantages in agricultural practices. In order to mitigate the drawbacks of conventional agrochemicals and promote ecological sustainability, endophytic organisms, including synergistic consortia of both bacterial and fungal endophytes, should be developed and used as bioformulated inoculants in agriculture (Adeleke and Babalola, 2021). Effective implementation of these organisms in field applications may be achieved through the advanced biotechnological screening- and investigative survey-based elucidation of the genetic, metabolic, and physiological processes in endophytic microorganisms and their symbiotic relationships with plants (Sharma et al., 2023). However, certain ambiguities require further clarification:

Firstly, it is necessary to assess the endophytic microbial community and functions. Further exploratory research on the complete endomicrobiome of plant tissue may uncover novel endophytic bacteria with significant properties that can be effectively utilized to enhance crops in a sustainable manner. In order to achieve optimal outcomes, it is advisable to employ advanced biotechnological, omic-dependent, and next-generation molecular techniques.

Secondly, the identification and isolation of novel endophytes with specific traits that may benefit crop cultivation warrants a thorough bioprospecting investigation of endophytic microorganisms from diverse ecological niches, including marine habitats and harsh environments.

Lastly, to avoid recognized environmental obstacles, it is indispensable to efficiently screen for necessary metabolites that may be employed in agriculture.

4 Pharmaceutical Applications of Endophytes

4.1 Immunosuppressant Effects of Endophytes

Endophytic microorganisms associated with plants have assumed pivotal roles in the advancement of the pharmaceutical industry. In recent times, endophytes are being thoroughly surveyed as a clinical source of immunomodulatory compounds that may serve as prospective therapeutic interventions for autoimmune disorders, like insulin-dependent diabetes and rheumatoid arthritis, and precursors for averting allograft declination in transplant patients (Kim & Moudgil, 2017). Owing to the heterogeneous mechanisms of most synthetic immunosuppressive drugs, prolonged duration of their administration may elicit complicated effects, such as hyperlipidemia, nephrotoxicity, hypertension, neurotoxicity, and a heightened susceptibility to infection. Consequently, the endeavor to discover effective and safe immunosuppressive drugs is currently underway, and recent research has identified potent immunosuppressive therapeutics from plant-based endophytes, including sydoxanthone A and B, colutellin A, 13-O-acetylsydowinin B, dibenzofurane, methyl peniphenone, xanthone derivatives, subglutinol A and B, lipopeptide, peniphenone, benzophenone derivatives, (–) mycousnine, and polyketide benzannulated spiroketal (Leroy et al., 2015). The commercial immunosuppressant drug Cyclosporin A extracted from *Tolypocladium inflatum* and *Trichoderma polysporum* has evidenced remarkable inhibitory efficacy in mixed lymphocyte reaction and thymocyte proliferation assays (Hošková et al., 2017). Diterpene pyrenes, a non-cytotoxic compound, and Subglutinol A and B, produced by the *Tripterygium wilfordii*-colonizing endophyte *Fusarium subglutinan*, have been documented to possess considerable immunosuppressive efficiency; equimolar administration of Subglutinol A and B (0.1 μM) has yielded an equal percentage (50%) inhibitory concentration in both mixed lymphocyte reaction and thymocyte proliferation assays (Nalini & Prakash, 2017). Subglutinol A effectively hinders the proliferation and survival of T lymphocytes in vivo and in vitro. Mycophenolic acid synthesized by endophytic fungi of the genera *Busssochlamys*, *Penicillium*, *Septoria*, and *Aspergillus* functions as a potent clinical immunosuppressant that is used for treating autoimmune diseases and impeding transplantation rejection (Song et al., 2020).

4.2 Efficacy of Endophyte-Derived Compounds as Antibiotic Agents

Metabolic pathway-mediated antibiotic synthesis is considered an effective strategy for protecting plants against infection. A myriad of endophyte-derived bioactive substances, such as alkaloids, flavonoids, peptides, phenols, polyketides, quinones, steroids, and terpenoids, have demonstrated antimicrobial activity against pathogens (Singh & Kumar, 2023) (Table 2). Various endophytes secrete bioprospective compounds that serve as antiviral agents against a wide variety of viruses. *Cytonaema* spp.-secreted cytonic acid A and cytonic acid B serve as ptrideside isomers, which are novel inhibitors of the protease activity of human cytomegalovirus (Guo et al., 2000). Hinnuliquinone and altertoxins synthesized by *Quercus coccifera*-inhabiting endophytes and *Alternaria tenuissima* QUE1Se, respectively, demonstrate potent antiviral activity against HIV-1 virus (Singh et al., 2004). *Emericella* spp. (HKZJ)-generated compounds, including emerimidine (A and B), dehydroaustin, austinol, aspernidine (A and B), austin, emeriphenolicins (A and D), and acetoxydehydroaustin, exert therapeutic effect against the influenza A (H1N1) virus (Ambele et al., 2020).

4.3 Anti-Cardiovascular Effects of Endophytes

Cardiovascular diseases (CVDs) have garnered attention as a burgeoning cause of mortality in developing and industrialized nations. Hypercholesterolemia is the predominant contributing factor effectuating the progression of atherosclerosis and CVDs (Nelson, 2013). The fungal secondary metabolites statins are globally prescribed as competitive inhibitors of β-hydroxy β-methylglutaryl-CoA reductase, which is a vital enzyme modulating the cholesterol biosynthesis pathway (Maron et al., 2000). Lovastatin, produced by *Aspergillus terreus* and *Taxus baccata*-colonizing *A. niger*, serve as a potent HMG-CoA reductase inhibitor (Mouafi et al., 2016). Chartarlactams A-P synthesized by *Stachy botryschartarum* exhibited anti-hyperlipidemic activity in human hepatoma cells (Li et al., 2014). Piperine acquired from *Mycosphaerella* spp. PF13 can reduce mean arterial pressure in a dose-dependent manner (Chithra et al., 2014).

4.4 Anti-inflammatory and Anti-arthritic Efficacy of Endophytes

Rheumatoid arthritis (RA) represents a systemic, autoimmune, and inflammatory disease that manifests as swelling, discomfort, and bone and cartilage degradation and may eventually result in permanent disability. Intriguingly, the precise etiology underlying this disorder remains ambiguous, which limits its therapeutic regimens. Moreover, synthetic medications prescribed for the management of RA are expensive and may elicit adverse side effects. Consequently, extensive research has been dedicated to formulating alternative therapeutic strategies, including microorganism-derived compounds. Methanolic compounds extracted from *Talaromyces wortmannii*, an endophytic resident of aloe vera, possess anti-inflammatory properties (Pretsch et al., 2014). Additionally, the pathogenic fungal endophyte *Lepidosphaeria* spp. synthesizes

Table 2 Natural antibiotic compounds derived from plant-inhabiting endophytes

Endophytes	Host plants	Antibiotic property	Chemical class	References
Streptomyces spp.	*Glycinemax*	Antifungal	Alkaloids	Zhang et al., (2014)
Aeromicrobium poni	*Vochysia divergens*	Antibacterial		Gos et al., (2017)
Aspergillus spp.	*Bauhinia guianensis*	Antibacterial		Pinheiro et al., (2013)
Phomopsis spp.	*Ginko biloba*	Antifungal and antibacterial		Donald et al., (2005), Qin et al., (2009), Rukachaisirikul et al., (2008)
Acremonium zeas	*Zea mays*			
Chaetomium globosum	*Garcinia dulcis*			
Phoma spp.	*Cinnamomum*	Antibacterial and antifungal	Polyketides	Santiago et al., (2012)
Bacillus mojavensis, *B. atrophaeus*	*Glycyrrhiza uralensis*	Antifungal		Mohamad et al., (2018)
Actinosynema pretiosum	*Maytenus serrata*	Antibacterial		Siyu-Mao et al., (2013)
Geotrichum candidum	*Phyllanthus reticulatus*	Antifungal and antibacterial		Pai and Chandra (2018)
Nigrospora sphaerica	*Indigofera suffruticosa*	Antibacterial		Santos et al., (2015)
Streptomyces remosus	*Glycinemax*	Antifungal	Steroids	Nelson (2001)
Streptomyces noursei	*Grevillea pterdifolia*	Antifungal		Fjærvik and Zotchev (2005)
Phompsis spp.	*Aconitum carmichaelii*	Antifungal		Wu et al., (2013)
Streptomyces spp.	*Boesenber giarotunda*, *Grevillea pterdifolia*	Antibacterial	Peptides	Castillo et al., (2003)
Verrucosis poramaris	*Sonchus oleraceus*	Antibacterial		Fiedler et al., (2008)
Cryptosporiopsis spp.	*Pinus sylvestris*	Antifungal and antibacterial		Noble et al., (1991)
Pezicula spp.	*Fagussyl vatica*	Antifungal and antibacterial		
Bacillus subtilis	*Combretum molle*	Antibacterial		Singh et al., (2017)
Bacillus subtilis	*Allanmandas cathartica*	Antifungal	Terpenoids	Nithya and Muthumary (2011)
Streptomyces spp.	*Aucubaja ponica*	Antibacterial		Sasaki et al., (2001)
Nodulisporium spp.	*Juniperus cedre*	Antibacterial	Flavonoids	Dai et al., (2006)
Streptomyces spp.	*Boesenbergia rotunda*	Antibacterial		Taechowisan et al., (2014)

anti-inflammatory compounds which may be efficacious against RA (Younis et al., 2022).

4.5 Anti-diabetic Efficacy of Endophytes

Diabetes mellitus refers to a set of metabolic disorders characterized by elevated blood glucose levels due to either insufficient insulin production or impaired cellular response to insulin or both. Diabetes has emerged as a critical public health concern, and warrants management through the discovery of safe and efficacious anti-diabetic medications (American Diabetes Association, 2014). Multiple bioprospective compounds derived from endophytic fungi have demonstrated anti-diabetic and hypolipidemic properties (Nalin Rathnayake et al., 2019). The endophytic fungus *Alternaria* spp. produces N-acetylgalactosamine, which exhibits anti-diabetic activity in vivo and in vitro (Govindappa et al., 2015). The non-peptidal substance demethylasterriquinone B-1 secreted by *Pseudomassaria* spp. possesses insulin-like activity (Zhang et al., 1999). *Nigrospora oryzae* inhabiting the leaves of *Combretum dolichopetalum* produces the bioactive compounds (S)-(+)-2-cis-4-trans-abscisic acid, 7′-hydroxy-abscisic acid, and 4-des-hydroxyl altersolanol, which lowered blood sugar levels in diabetic mice (Uzor et al., 2017). Extracts from *Phoma* and *Aspergillus* spp. isolated from *Salvadora oleoides* visibly decrease blood sugar levels in aldohexose- and alloxan-administered Wistar anomaly rats (Ezekwesili & Ogbunugafor, 2015). Additionally, the endophytic fungi *Fusarium* spp. and *Alternaria* spp. secrete compounds that possess therapeutic potential as precursors of α-glucosidase inhibitors (Fadiji & Babalola, 2020).

4.6 Implementation of Endophytes in the Treatment of Cancer

Cancer represents a heterogeneous group of disorders characterized by the unrestricted proliferation and dissemination of atypical cells at distinct locations in the body. These aberrant cells, referred to as cancer cells, possess the ability to infiltrate and destroy surrounding tissues and metastasize to other parts of the body through the circulatory and lymphatic systems. Cancer can potentially manifest anywhere in the body and may develop in a variety of organs and tissues (Communications, 1985).

The human body comprises trillions of cells, which partake in the normal-state cell cycle encompassing growth, division, and programmed death or apoptosis, to maintain a healthy equilibrium (Pucci et al., 2000). Cancer disrupts this regular progression of events, leading to the accumulation of cells that fail to respond to typical growth- and cell death-governing signals (Na et al., 2018).

A myriad of different malignancies have been diagnosed to date and have been categorized into various histological subtypes, such as breast cancer, lung cancer, prostate cancer, pancreatic cancer, colorectal cancer, and leukemia, depending on the specific types of cells or tissue of proliferative and metastatic origin (Flaherty et al., 2020).

A wide range of etiological agents, such as genetic factors, carcinogen exposure, lifestyle choices, including substance abuse and diet, infections, and certain environmental determinants, are attributable to cancer (Shweta et al., 2013). Multidimensional research across the globe aims to elucidate the molecular mechanisms underlying cancer and develop more efficacious prevention strategies, diagnostic tools, and therapeutic interventions (Pulumati et al., 2023). Early detection and advanced cancer treatment have improved the survival rates of cancer patients. Nonetheless, the complexity and diversity associated with malignancies pose a substantial strain on global health (Pina-Sanchez et al., 2021).

Presently prescribed anticancer pharmaceuticals inflict nonspecific toxicity on normal cells surrounding the malignant cellular mass that exerts severe adverse effects and may even be ineffective against various cancer types (Pérez-Herrero, et al., 2015). Endophytes are currently being investigated as promising sources of candidate anticancer agents. Numerous bioactive compounds isolated from different endophytic organisms have evidently impeded cancer cell proliferation by inducing apoptosis and exerting anti-inflammatory effects, consequently validating them as viable alternatives to conventional chemotherapeutic agents (Table 3). The National Cancer Institute initiated a program emphasizing plants as primary sources of anticancer agents (Cragg & Pezzuto, 2016), with an estimated 60% of extant anticancer drugs being formulated from natural products and their derivatives.

The novel bioprospective compound, peptidendrocin, extracted from the *Dendrobium officinale*-associated endophyte *Streptomyces* spp., has demonstrated potent anticancer activity against lung and liver malignancies (Zhao et al., 2020). Secondary metabolites secreted by the endophytic bacteria *Pseudomonas putida* and *Bacillus safensis* isolated from the leaves of *Crinum macowanii* depicted anticancer activity against lung cancer and glioblastoma (Sebola et al., 2020).

The discovery of the fungal endophyte *Taxomyces andreanae* in the Yew plant (Strobel et al., 1993) prompted the quest to survey endophytic fungi with potential anticancer properties. Fungal endophytes reportedly secrete a myriad of secondary metabolites, namely paclitaxel, xanthones, ergoflavin, lignin, torreyanic acid, vincristine, camptothecin, and podophyllotoxin with different chemical nature like alkaloids, terpenoids, polyketides, and peptide (Tenguria & Firodiya, 2016); these bioprospective compounds exhibit potent anticancer, antiproliferative, and antioxidant activities (Hridoy et al., 2022) and hinder cancer cell proliferation and metastasis through different mechanisms, including apoptosis and suppression of associated signaling pathways.

Nevertheless, the identification, characterization, and safety verification of these natural endophyte-derived bioactive compounds is constrained by various challenges (Wijesekara & Xu, 2023) and warrants further research to translate the experimental findings into effective clinical treatments with due consideration of regulatory and safety concerns (Kristensen et al., 2015).

Table 3 Natural anticancer compounds derived from plant-inhabiting endophytes

Bioactive Compound	Chemical Class	Endophyte	Host Plant	Mode of Action	Implications	References
Camptothecin [$C_{20}H_{16}N_2O_4$]	Alkaloids	*Entrophospora infrequens*	*Nothapodytes foetida*	Inhibits DNA topoisomerase-I dissociation during replication	Liver, lung, ovarian cancer	Raj et al., (2023), Singh and Singh, (2023)
		Fusarium solani	*Camptotheca acuminata*			
Vincristine [$C_{46}H_{56}N_4O_{10}$] and Vinblastine [$C_{46}H_{58}N_4O_9$]		*Nigrosporaz immermanii*	*Catharanthus roseus*	Interrupts mitotic spindle formation and arrests tumor cell proliferation	Acute nephroblastoma and lymphoblastic leukemia	Almagro et al., (2015)
Swainsonine [$C_8H_{15}NO_3$]		*Alternaria* sect. *Undifilum oxytropis*	*Astragalus variabilis*	Affects carbohydrates, glycolipids, and glycoproteins synthesis that promotes apoptosis in tumor cells	Ehrlich ascites carcinoma, leukemia, colorectal and lymphatic cancer, human hepatoma	Bowlin et al., (1991)
Sclerotiorin [$C_{21}H_{23}ClO_5$]		*Cephalotheca faveolata*	*Eugenia jambolana*	Inhibits Grb2-Sch interactions, which hinders oncogenic Ras signaling	Leukemia, breast, lung, colon cancer	Tang et al., (2019)
Paclitaxel [$C_{47}H_{51}NO_{14}$]	Diterpenes	*Taxomyces anderane*	*Taxus brevifolia,*	Inhibits cancer cell proliferation by affecting tubulin dynamics and mitotic spindle formation	Lung, breast, ovarian cancer, and AIDS-related Kaposi's sarcoma	Tiwari and Bae (2022)
Cercosporene F		*Cercospora* spp.	*Fallopia japonica*	Inhibits ubiquitin–proteasome and NF-κB pathway and modulatesKeap1-Nrf2-ARE signaling	Lung, breast, cervical, prostate, colon, bladder cancer	Yang and Ping Dou (2010)
Chrysin (5,7-dihydroxyflavone) [$C_{15}H_{10}O_4$]	Flavonoids	*Alternaria alternata*	*Passiflora incarnata*	Induces apoptosis via X-box binding protein-1 splicing and GRP78overexpression	Liver, prostate, breast, gastric, colon cancer	Sood and et al., (2024)
Cajanol, [5-hydroxy-3-(4-hydroxy-2-methoxyphenyl)-7-methoxychroman-4-one] [$C_{17}H_{16}O_6$]		*Hypocrea lixii*	*Cajanus cajan*	Inhibits transcription and translation of P-gp through the PI3K/Akt/NF-κB pathway	Lung and ovarian cancer	Sui et al., (2021)
Brefeldin A $C_{16}H_{24}O_4$]	Lactones	*Penicillium brefeldianum*	*Houttuynia cordata*	Induces apoptosis, disrupts Golgi complex and protein transport, and inhibits small GTPases	Pancreatic adenocarcinoma, lung, breast, prostate, and colorectal cancer, leukemia	Bai et al., (2022)
Myrotheciumone A		*Myrothecium roridum*	*Ajuga decumbens*	Induces apoptosis and disrupts mitochondrial function	Hepatocellular carcinoma, lung and breast cancer	Lin et al., (2014)

(continued)

Table 3 (continued)

Bioactive Compound	Chemical Class	Endophyte	Host Plant	Mode of Action	Implications	References
Exserolide G [$C_{16}H_{20}O_6$]	Polyketides	*Exserohilum* spp.	*Acer truncatum*	Disrupts DNA replication and cytoskeletal organization and induces apoptosis	Lung, cervical, colorectal cancer	Alsufyani et al., (2023)
Rubrofusarin B [$C_{15}H_{12}O_5$]		*Aspergillus niger*	*Cynodon dactylon*	Inhibits topoisomerase II-α and induces apoptosis	Colorectal and liver cancer	Branco et al., (2008)
Torreyanic acid [$C_{38}H_{44}O_{12}$]	Quinones	*Pestaliopsis microspore*	*Torreya taxifolia*	Sensitizes PKC a induces apoptosis	Lung, breast, colorectal cancer	Hridoy et al., (2022)
Lapachol [$C_{15}H_{14}O_3$]		*Aspergillus niger*	*Tabebuia argentea*	Interferes with topoisomerase, which governs DNA replication and induces apoptosis	Prostate, breast, liver cancer	Kapoor et al., (2019)
Ergoflavin [$C_{30}H_{26}O_{14}$]	Xanthones	*Aspergillus* spp., *Phoma terrestris, Claviceps purpurea*	*Mimusops elengi,Secale cereale*	Inhibits IL-6 and TNF-α and induce apoptosis	Leukemia	Deshmukh et al., (2009)
Phomoxanthone A [$C_{38}H_{38}O_{16}$]		*Phomopsis* spp.	*Tectona grandis*	Induces apoptosis and cell cycle arrest, inhibits angiogenesis, and disrupts signaling pathways	Epithelial carcinoma, lymphoma	Bowlin et al., (1991)

5 Conclusion

Endophytes have emerged as a promising treasure trove for bioprospecting compounds with immense potential in the agricultural and pharmaceutical fields. They possess the remarkable ability to stimulate plant growth, enhance nutrient absorption, and condition plants to develop tolerance against various biotic and abiotic stresses. Endophyte-derived bioactive compounds have garnered significant attention in the pharmaceutical industry, as they offer antimicrobial properties and the potential for developing new antibiotics. They also have the potential for producing novel anticancer compounds, which could be used in treating various health conditions, including diabetes, inflammation, and immunomodulation.

In the future, the applicability of bioactive compounds may be enhanced with the incorporation of nanotechnology, which improves bioavailability, targeted delivery, therapeutic efficacy, and resistance to drug resistance in natural bioactive compounds, particularly in cancer therapy, protects compounds from degradation, improves therapeutic effects, and delays drug resistance. In the agricultural sector, nanoformulations optimize nutrient delivery, leading to increased crop yields; it promotes sustainable practices by reducing chemical use and allows for smart formulations to respond to environmental stimuli. Therefore, by harnessing the power of these endosymbiotic microorganisms, we can strive to create a more sustainable and healthier future for our society.

Acknowledgements The authors would like to acknowledge the University of Kalyani, for providing research facilities and academic support. The authors appreciate the constructive feedback provided by the reviewers, whose insightful comments aided in enhancing the quality of the manuscript.

Author Contribution All authors contributed equally to the construction of this manuscript.

References

Adeleke, B. S., & Babalola, O. O. (2021). The endosphere microbial communities, a great promise in agriculture. *International Microbiology, 24*(1), 1–17.

Almagro, L., Fernández-Pérez, F., & Pedreño, M. A. (2015). Indole alkaloids from Catharanthus roseus: bioproduction and their effect on human health. *Molecules, 20*(2), 2973–3000.

Alsufyani, T., Al-Otaibi, N., Alotaibi, N. J., M'sakni, N. H., & Alghamdi, E. M. (2023). GC analysis, anticancer, and antibacterial activities of secondary bioactive compounds from endosymbiotic bacteria of pomegranate aphid and its predator and protector. *Molecules, 28*(10), 4255.

Ambele, M. A., Dhanraj, P., Giles, R., & Pepper, M. S. (2020). Adipogenesis: a complex interplay of multiple molecular determinants and pathways. *International Journal of Molecular Sciences, 21*(12), 4283.

American Diabetes Association (2014). Diagnosis and classification of diabetes mellitus. *Diabetes Care, 37*(Supplement_1): S81-S90.

Andreote, F. D., Gumiere, T., & Durrer, A. (2014). Exploring interactions of plant microbiomes. *Scientia Agrícola, 71*, 528–539.

Bai, Y., Yi, P., Li, F., Hu, J., & Pan, H. (2022). Penipentenone A and brefeldin A derivatives potently inhibit KRAS mutant cancer cells from an endophytic fungus Penicillium brefeldianum F4a. *Phytochemistry, 200*, 113243.

Barka, E. A., Vatsa, P., Sanchez, L., Gaveau-Vaillant, N., Jacquard, C., Klenk, H. P., Clément, C., Ouhdouch, Y., & van Wezel, G. P. (2016). Taxonomy, physiology, and natural products of Actinobacteria. *Microbiology and Molecular Biology Reviews, 80*(1), 1–43.

Bhakat, A., Sen, S., Banerjee, S., & Sarkar, K. (2023). Plant growth promotion and lipopeptide-mediated biological control of chilli pathogen Colletotrichum siamense by endophytic Bacillus sp. *Physiological and Molecular Plant Pathology, 125*, 102026.

Bhardwaj, A., & Agrawal, P. (2014). A review fungal endophytes: as a store house of bioactive compound. *World Journal of Pharmaceutical Sciences, 3*(9).

Bowlin, T. L., Schroeder, K. K., & Fanger, B. O. (1991). Swainsonine, an inhibitor of mannosidase II during glycoprotein processing, enhances concanavalin A-induced T cell proliferation and interleukin 2 receptor expression exclusively via the T cell receptor complex. *Cellular immunology, 137*(1), 111–117.

Branco, A., Pinto, A. C., Braz-Filho, R., Silva, E. F., Grynberg, N. F., & Echevarria, A. (2008). Rubrofusarina, um policetídeo natural inibidor da topoisomerase II-α humana. *Revista Brasileira de Farmacognosia, 18*, 703–708.

Castillo, U., Harper, J. K., Strobel, G. A., Sears, J., Alesi, K., Ford, E., Lin, J., Hunter, M., Maranta, M., Ge, H., & Yaver, D. (2003). Kakadumycins, novel antibiotics from Streptomyces sp. NRRL 30566, an endophyte of Grevillea pteridifolia. *FEMS Microbiology Letters, 224*(2), 183–190.

Chen, D., Mubeen, B., Hasnain, A., Rizwan, M., Adrees, M., Naqvi, S. A. H., Iqbal, S., Kamran, M., El-Sabrout, A. M., Elansary, H. O., & Mahmoud, E. A. (2022). Role of promising secondary metabolites to confer resistance against environmental stresses in crop plants: Current scenario and future perspectives. *Frontiers in Plant Science, 13*, 881032.

Chithra, S., Jasim, B., Anisha, C., Mathew, J., & Radhakrishnan, E. K. (2014). LC-MS/MS based identification of piperine production by endophytic Mycosphaerella sp. PF13 from Piper nigrum. *Applied Biochemistry and Biotechnology, 173*, 30–35.

Communications, N. C. I. O. o. C. (1985). National Cancer Institute, National Cancer Institute, Office of Cancer Communications.

Cragg, G. M., & Pezzuto, J. M. (2016). Natural products as a vital source for the discovery of cancer chemotherapeutic and chemopreventive agents. *Medical Principles and Practice, 25*(Suppl. 2): 41–59.

Dai, J., Krohn, K., Flörke, U., Draeger, S., Schulz, B., Kiss-Szikszai, A., Antus, S., Kurtán, T., & Van Ree, T. (2006). *Metabolites from the endophytic fungus Nodulisporium sp. from Juniperus cedre.* Wiley Online Library.

Dastogeer, K., Tumpa, F., Sultana, A., Akter, M., & Chakraborty, A. (2020). Plant microbiome–an account of the factors that shape community composition and diversity. *Current Plant Biology, 23*, 100161.

Deng, Z., & Cao, L. (2017). Fungal endophytes and their interactions with plants in phytoremediation: A review. *Chemosphere, 168*, 1100–1106.

Deshmukh, S. K., Mishra, P. D., Kulkarni-Almeida, A., Verekar, S., Sahoo, M. R., Periyasamy, G., Goswami, H., Khanna, A., Balakrishnan, A., & Vishwakarma, R. (2009). Anti-inflammatory and anticancer activity of ergoflavin isolated from an endophytic fungus. *Chemistry and Biodiversity, 6*(5), 784–789.

Donald, T., Shoshannah, R. O. T. H., Deyrup, S. T., & Gloer, J. B. (2005). A protective endophyte of maize: Acremonium zeae antibiotics inhibitory to Aspergillus flavus and Fusarium verticillioides. *Mycological Research, 109*(5), 610–618.

El-Beltagi, H. S., Mohamed, H. I., Aldaej, M. I., Al-Khayri, J. M., Rezk, A. A., Al-Mssallem, M. Q., Sattar, M. N., & Ramadan, K. M. (2022). Production and antioxidant activity of secondary metabolites in Hassawi rice (Oryza sativa L.) cell suspension under salicylic acid, yeast extract, and pectin elicitation. *In Vitro Cellular and Developmental Biology-Plant, 58*(4), 615–629.

Ezekwesili, C. N., & Ogbunugafor, H. A. (2015). Blood glucose lowering activity of five Nigerian medicinal plants in alloxan-induced diabetic Wistar albino rats. *Animal Research International, 12*(2), 2150–2158.

Fadiji, A. E., & Babalola, O. O. (2020). Elucidating mechanisms of endophytes used in plant protection and other bioactivities with multifunctional prospects. *Frontiers in Bioengineering and Biotechnology, 8*, 467.

Fiedler, H. P., Bruntner, C., Riedlinger, J., Bull, A. T., Knutsen, G., Goodfellow, M., Jones, A., Maldonado, L., Pathom-Aree, W., Beil, W., & Schneider, K., (2008). Proximicin A, B and C, novel aminofuran antibiotic and anticancer compounds isolated from marine strains of the actinomycete Verrucosispora. *The Journal of Antibiotics, 61*(3), 158–163.

Fite, T., Kebede, E., Tefera, T., & Bekeko, Z. (2023). Endophytic fungi: versatile partners for pest biocontrol, growth promotion, and climate change resilience in plants. *Frontiers in Sustainable Food Systems, 7*, 1322861.

Fjærvik, E., & Zotchev, S. B. (2005). Biosynthesis of the polyene macrolide antibiotic nystatin in Streptomyces noursei. *Applied Microbiology and Biotechnology, 67*, 436–443.

Flaherty, K. T., Gray, R. J., Chen, A. P., Li, S., McShane, L. M., Patton, D., Hamilton, S. R., Williams, P. M., Iafrate, A. J., Sklar, J., & Mitchell, E. P. (2020). Molecular landscape and actionable alterations in a genomically guided cancer clinical trial: National Cancer Institute Molecular Analysis for Therapy Choice (NCI-MATCH). *Journal of Clinical Oncology, 38*(33), 3883–3894.

Fontana, D. C., de Paula, S., Torres, A. G., de Souza, V. H. M., Pascholati, S. F., Schmidt, D., & Dourado Neto, D. (2021). Endophytic fungi: Biological control and induced resistance to phytopathogens and abiotic stresses. *Pathogens, 10*(5), 570.

Golinska, P., Wypij, M., Agarkar, G., Rathod, D., Dahm, H., & Rai, M. (2015). Endophytic actinobacteria of medicinal plants: diversity and bioactivity, *Antonie Van Leeuwenhoek, 108*, 267–289.

Gos, F. M., Savi, D. C., Shaaban, K. A., Thorson, J. S., Aluizio, R., Possiede, Y. M., Rohr, J., & Glienke, C. (2017). Antibacterial activity of endophytic actinomycetes isolated from the medicinal plant Vochysia divergens (Pantanal, Brazil). *Frontiers in Microbiology, 8*, 1642.

Gouda, S., Das, G., Sen, S. K., Shin, H. S., & Patra, J. K. (2016). Endophytes: a treasure house of bioactive compounds of medicinal importance. *Frontiers in Microbiology, 7*, 1538.

Govindappa, M., Sadananda, T., Ramachandra, Y. L., Chandrappa, C. P., Rai, P. S., & Prasad, S. K. (2015). In vitro and in vivo antidiabetic activity of lectin (N-acetyl-galactosamine, 64 kDa) isolated from endophytic fungi, Alternaria species from Viscum album on alloxan induced diabetic rats. *Integr Diabetes Obesity, 1*(1), 11–19.

Guo, B., Dai, J. R., Ng, S., Huang, Y., Leong, C., Ong, W., & Carté, B. K. (2000). Cytonic acids A and B: novel tridepside inhibitors of hCMV protease from the endophytic fungus Cytonaema species. *Journal of Natural Products, 63*(5), 602–604.

Hardoim, P. R., van Overbeek, L. S., & van Elsas, J. D. (2008). Properties of bacterial endophytes and their proposed role in plant growth. *Trends in Microbiology, 16*(10), 463–471.

Hasanuzzaman, M., Fujita, M., Oku, H., & Islam, M. T. (2019). *Plant tolerance to environmental stress: Role of phytoprotectants*. CRC Press.

Hazra, S., Das, D., Moulick, D., & Hossain, A. (2024). *Endophytes: The treasure house of bioactive compounds with potential applications in sustainable agriculture and other sectors* (pp. 477–506). Elsevier.

Hollants, J., Leroux, O., Leliaert, F., Decleyre, H., De Clerck, O., & Willems, A. (2011). Who is in there? Exploration of endophytic bacteria within the siphonous green seaweed Bryopsis (Bryopsidales, Chlorophyta). *PLoS One, 6*(10), e26458.

Hošková, L., Málek, I., Kopkan, L., & Kautzner, J. (2017). Pathophysiological mechanisms of calcineurin inhibitor-induced nephrotoxicity and arterial hypertension. *Physiological Research, 66*(2), 167.

Hridoy, M., Gorapi, M. Z. H., Noor, S., Chowdhury, N. S., Rahman, M. M., Muscari, I., Masia, F., Adorisio, S., Delfino, D. V., Mazid, M. A. (2022). Putative anticancer compounds from plant-derived endophytic fungi: a review. *Molecules, 27*(1), 296.

Ikram, M., Ali, N., Jan, G., Jan, F. G., Rahman, I. U., Iqbal, A., & Hamayun, M. (2018). IAA producing fungal endophyte Penicillium roqueforti Thom., enhances stress tolerance and nutrients uptake in wheat plants grown on heavy metal contaminated soils. *PLoS One, 13*(11), e0208150.

Jain, S., Varma, A., Tuteja, N., & Choudhary, D. K. (2016). Bacteria-mediated elicitation of induced resistance in plants upon fungal phytopathogen. *Plant-Microbe Interaction: An Approach to Sustainable Agriculture, 249–269.*

Kapoor, N., Jamwal, V. L., & Gandhi, S. G. (2019). Endophytes as a source of high-value, bioactive metabolites. In *Endophytes and secondary metabolites. Reference series in phytochemistry*. Springer.

Khruengsai, S., Pripdeevech, P., Tanapichatsakul, C., Srisuwannapa, C., D'Souza, P. E., & Panuwet, P. (2021). Antifungal properties of volatile organic compounds produced by Daldinia eschscholtzii MFLUCC 19-0493 isolated from Barleria prionitis leaves against Colletotrichum acutatum and its post-harvest infections on strawberry fruits. *PeerJ, 9*, e11242.

Kim, E. Y., & Moudgil, K. D. (2017). Immunomodulation of autoimmune arthritis by pro-inflammatory cytokines. *Cytokine, 98*, 87–96.

Kristensen, N., Nymann, C., & Konradsen, H. (2015). Implementing research results in clinical practice-the experiences of healthcare professionals. *BMC Health Services Research, 16*, 1–10.

Lee, K., Missaoui, A., Mahmud, K., Presley, H., & Lonnee, M. (2021). Interaction between grasses and Epichloë endophytes and its significance to biotic and abiotic stress tolerance and the rhizosphere. *Microorganisms, 9*(11), 2186.

Leroy, C., Rigot, J. M., Leroy, M., Decanter, C., Le Mapihan, K., Parent, A. S., Le Guillou, A. C., Yakoub-Agha, I., Dharancy, S., Noel, C., & Vantyghem, M. C. (2015). Immunosuppressive drugs and fertility. *Orphanet Journal of Rare Diseases, 10*, 1–15.

Li, Y., Wu, C., Liu, D., Proksch, P., Guo, P., & Lin, W. (2014). Chartarlactams A–P, phenylspirodrimanes from the sponge-associated fungus Stachybotrys chartarum with antihyperlipidemic activities. *Journal of Natural Products, 77*(1), 138–147.

Liarzi, O., Bar, E., Lewinsohn, E., & Ezra, D. (2016). Use of the endophytic fungus Daldinia cf. concentrica and its volatiles as bio-control agents. *PLoS One, 11*(12), e0168242.

Lin, T., Wang, G., Shan, W., Zeng, D., Ding, R., Jiang, X., Zhu, D., Liu, X., Yang, S., & Chen, H. (2014). Myrotheciumones: Bicyclic cytotoxic lactones isolated from an endophytic fungus of Ajuga decumbens. *Bioorganic and Medicinal Chemistry Letters, 24*(11), 2504–2507.

Lu, H., Wei, T., Lou, H., Shu, X., & Chen, Q. (2021). A critical review on communication mechanism within plant-endophytic fungi interactions to cope with biotic and abiotic stresses. *Journal of Fungi, 7*(9), 719.

De Mandal, S., Sonali, Singh, S., Hussain, K., & Hussain, T. (2021). Plant–microbe association for mutual benefits for plant growth and soil health. *Current Trends in Microbial Biotechnology for Sustainable Agriculture*, 95–121.

Maron, D. J., Fazio, S., & Linton, M. F. (2000). Current perspectives on statins. *Circulation, 101*(2), 207–213.

Mengoni, A., Mocali, S., Surico, G., Tegli, S., & Fani, R. (2003). Fluctuation of endophytic bacteria and phytoplasmosis in elm trees. *Microbiological Research, 158*(4), 363–369.

Miliute, I., Buzaite, O., Baniulis, D., & Stanys, V. (2015). Bacterial endophytes in agricultural crops and their role in stress tolerance: a review.

Mitter, B., Petric, A., Shin, M. W., Chain, P. S., Hauberg-Lotte, L., Reinhold-Hurek, B., Nowak, J., & Sessitsch, A. (2013). Comparative genome analysis of Burkholderia phytofirmans PsJN reveals a wide spectrum of endophytic lifestyles based on interaction strategies with host plants. *Frontiers in Plant Science, 4*, 120.

Mohamad, O. A., Li, L., Ma, J. B., Hatab, S., Xu, L., Guo, J. W., Rasulov, B. A., Liu, Y. H., Hedlund, B. P., & Li, W. J. (2018). Evaluation of the antimicrobial activity of endophytic bacterial populations from Chinese traditional medicinal plant licorice and characterization of the bioactive secondary metabolites produced by Bacillus atrophaeus against Verticillium dahliae. *Frontiers in Microbiology, 9*, 924.

Mouafi, F. E., Ibrahim, G. S., & Elsoud, M. M. A. (2016). Optimization of lovastatin production from *Aspergillus fumigatus*. *Journal of Genetic Engineering and Biotechnology, 14*(2), 253–259.

Na, Z., Qiao, X., Hao, X., Fan, L., Xiao, Y., Shao, Y., Sun, M., Feng, Z., Guo, W., Li, J., & Li, J. (2018). The effects of beta-blocker use on cancer prognosis: a meta-analysis based on 319,006 patients. *OncoTargets and Therapy*, 4913–4944.

Nalin Rathnayake, G. R., Savitri Kumar, N., Jayasinghe, L., Araya, H., & Fujimoto, Y (2019). Secondary metabolites produced by an endophytic fungus Pestalotiopsis microspora. *Natural Products and Bioprospecting, 9*, 411–417.

Nalini, M. S., & Prakash, H. S. (2017). Diversity and bioprospecting of actinomycete endophytes from the medicinal plants. *Letters in Applied Microbiology, 64*(4), 261–270.

Nelson, M. L. (2001). The chemistry and cellular biology of the tetracyclines. In *Tetracyclines in biology, chemistry and medicine* (pp. 3–63). Springer

Nelson, R. H. (2013). Hyperlipidemia as a risk factor for cardiovascular disease. *Primary Care, 40*(1), 195–211.

Nithya, K., & Muthumary, J. (2011). Bioactive metabolite produced by *Phomopsis* sp., an endophytic fungus in *Allamanda cathartica* Linn. *Recent Research in Science and Technology, 3*(3), 44–48.

Noble, H. M., Langley, D., Sidebottom, P., Lane, S., & Fisher, P. (1991). An echinocandin from an endophytic Cryptosporiopsis sp. and Pezicula sp. in Pinus sylvestris and Fagus sylvatica.

Ogbe, A. A., Finnie, J. F., & Van Staden, J. (2020). The role of endophytes in secondary metabolites accumulation in medicinal plants under abiotic stress. *South African Journal of Botany, 134*, 126–134.

Omomowo, O. I., & Babalola, O. O (2019). Bacterial and fungal endophytes: tiny giants with immense beneficial potential for plant growth and sustainable agricultural productivity. *Microorganisms, 7*(11), 481.

Pai, G., & Chandra, M. (2018). Antimicrobial activity of endophytic fungi isolated from ethnomedicinal plant Phyllanthus reticulatus poir. *The International Journal of Science and Engineering Invention, 7*, 40–46.

Pérez-Herrero, E., & Fernández-Medarde, A. (2015). Advanced targeted therapies in cancer: Drug nanocarriers, the future of chemotherapy. *European Journal of Pharmaceutics and Biopharmaceutics, 93*, 52–79.

Pessarakli, M. (2014). *Handbook of plant and crop physiology*. CRC Press.

Pina-Sanchez, P., Chavez-Gonzalez, A., Ruiz-Tachiquin, M., Vadillo, E., Monroy-Garcia, A., Montesinos, J. J., Grajales, R., Gutierrez de la Barrera, M., & Mayani, H. (2021). Cancer biology, epidemiology, and treatment in the 21st century: current status and future challenges from a biomedical perspective. *Cancer Control, 28*, 10732748211038735.

Pinheiro, E. A. A., Carvalho, J. M., dos Santos, D. C. P., Feitosa, A. D. O., Marinho, P. S. B., Guilhon, G. M. S. P., de Souza, A. D. L., da Silva, F. M. A., & Marinho, A. M. D. R. (2013). Antibacterial activity of alkaloids produced by endophytic fungus Aspergillus sp. EJC08 isolated from medical plant Bauhinia guianensis. *Natural Product Research, 27*(18), 1633–1638.

Pretsch, A., Nagl, M., Schwendinger, K., Kreiseder, B., Wiederstein, M., Pretsch, D., Genov, M., Hollaus, R., Zinssmeister, D., Debbab, A., & Hundsberger, H. (2014). Antimicrobial and anti-inflammatory activities of endophytic fungi Talaromyces wortmannii extracts against acne-inducing bacteria. *PloS One, 9*(6), e97929.

Pucci, B., Kasten, M., & Giordano, A. J. N. (2000). Cell cycle and apoptosis. *2*(4), 291–299.

Pulumati, A., Pulumati, A., Dwarakanath, B. S., Verma, A., & Papineni, R. V. (2023). Technological advancements in cancer diagnostics: Improvements and limitations. *Cancer Reports, 6*(2), e1764.

Qin, J. C., Zhang, Y. M., Gao, J. M., Bai, M. S., Yang, S. X., Laatsch, H., & Zhang, A. L. (2009). Bioactive metabolites produced by Chaetomium globosum, an endophytic fungus isolated from Ginkgo biloba. *Bioorganic and Medicinal Chemistry Letters, 19*(6), 1572–1574.

Raj, A., Thomas, R. K., Vidya, L., Aparna, V. M., Neelima, S., & Sudarsanakumar, C. (2023). Exploring the cytotoxicity on human lung cancer cells and DNA binding stratagem of camptothecin functionalised silver nanoparticles through multi-spectroscopic, and calorimetric approach. *Scientific Reports, 13*(1), 9045.

Rajkumar, M., Ae, N., Freitas, H. J. C. (2009). Endophytic bacteria and their potential to enhance heavy metal phytoextraction. 77(2), 153–160.

Rukachaisirikul, V., Sommart, U., Phongpaichit, S., Sakayaroj, J., & Kirtikara, K. J. P. (2008). Metabolites from the endophytic fungus Phomopsis sp. PSU-D15. *69*(3), 783–787.

Sadeghi, F., Samsampour, D., Seyahooei, M. A., Bagheri, A., & Soltani, J. (2020). "Fungal endophytes alleviate drought-induced oxidative stress in mandarin (Citrus reticulata L.): toward regulating the ascorbate–glutathione cycle. *Scientia Horticulturae, 261*, 108991.

Sallam, N., Ali, E. F., Seleim, M. A., & Khalil Bagy, H. M. (2021). Endophytic fungi associated with soybean plants and their antagonistic activity against Rhizoctonia solani. *Egyptian Journal of Biological Pest Control, 31*, 1–9.

Santiago, C., Fitchett, C., Munro, M. H., Jalil, J., & Santhanam, J. (2012). Cytotoxic and antifungal activities of 5-hydroxyramulosin, a compound produced by an endophytic fungus isolated from Cinnamomum mollisimum. *Evidence-Based Complementary and Alternative Medicine, 2012*(1), 689310.

Santos, I. P. D., Silva, L. C. N. D., Silva, M. V. D., Araújo, J. M. D., Cavalcanti, M. D. S., & Lima, V. L. D. M. (2015). Antibacterial activity of endophytic fungi from leaves of Indigofera suffruticosa Miller (Fabaceae). *Frontiers in Microbiology, 6*, 350.

Sasaki, T., Igarashi, Y., Saito, N., & Furumai, T. (2001). Cedarmycins A and B, new antimicrobial antibiotics from Streptomyces sp. TP-A0456. *The Journal of Antibiotics, 54*(7), 567–572.

Sebola, T. E., Uche-Okereafor, N. C., Mekuto, L., Makatini, M. M., Green, E., & Mavumengwana, V. (2020). Antibacterial and anticancer activity and untargeted secondary metabolite profiling of crude bacterial endophyte extracts from *Crinum macowanii* Baker leaves. *International Journal of Microbiology, 2020*(1), 8839490.

Shahnawaz, M. (2021). *Biotechnological approaches to enhance plant secondary metabolites: Recent trends and future prospects.*

Sharma, I., Raina, A., Choudhary, M., Kaul, S., & Dhar, M. K. J. H. (2023). Fungal endophyte bioinoculants as a green alternative towards sustainable agriculture.

Shweta, S., Gurumurthy, B. R., Ravikanth, G., Ramanan, U. S., & Shivanna, M. B. J. P. (2013). Endophytic fungi from Miquelia dentata Bedd., produce the anti-cancer alkaloid, camptothecine. *20*(3–4), 337–342.

Singh, V. K., & Kumar, A. J. S. (2023). Secondary metabolites from endophytic fungi: Production, methods of analysis, and diverse pharmaceutical potential. *90*(2), 111–125.

Singh, A., & Singh, B. (2023). Bioactive compounds in cancer care and prevention. In Role of nutrigenomics in modern-day healthcare and drug discovery (pp. 439–468). Elsevier.

Singh, S. B., Ondeyka, J. G., Tsipouras, N., Ruby, C., Sardana, V., Schulman, M., Sanchez, M., Pelaez, F., Stahlhut, M. W., Munshi, S., & Olsen, D. B. (2004). Hinnuliquinone, a C2-symmetric dimeric non-peptide fungal metabolite inhibitor of HIV-1 protease. *Biochemical and Biophysical Research Communications, 324*(1), 108–113.

Singh, M., Kumar, A., Singh, R., & Pandey, K. D. J. B. (2017). Endophytic bacteria: a new source of bioactive compounds. *7*, 1–14.

Siyu-Mao, Hong-Chen, Li-Chen, Chuanxi-Wang, Wei-Jia, Xiaoming-Chen, Huangjian-Yang, Wei-Huang, & Wei-Zheng (2013). Two novel ansamitocin analogs from *Actinosynnema pretiosum. Natural Product Research, 27*(17), 1532–1536.

Song, X., Tu, R., Mei, X., Wu, S., Lan, B., Zhang, L., Luo, X., Liu, J., & Luo, M. (2020). A mycophenolic acid derivative from the fungus Penicillium sp. SCSIO sof101. *Natural Product Research, 34*(9), 1206–1212.

Sood, A., Mehrotra, A., Sharma, U., Aggarwal, D., Singh, T., Shahwan, M., Jairoun, A. A., Rani, I., Ramniwas, S., & Tuli, H. S. (2024). Advancements and recent explorations of anti-cancer activity of chrysin: from molecular targets to therapeutic perspective. *Exploration of Targeted Anti-tumor Therapy, 5*(3), 477–494.

Stadnik, M. J, Buchenauer, H. (2000). Inhibition of phenylalanine ammonia-lyase suppresses the resistance induced by benzothiadiazole in wheat to *Blumeria graminis* f. sp. *Tritici. Physiological and Molecular Plant Pathology, 57*(1), 25–34.

Strobel, G., Stierle, A., Stierle, D., & Hess, W. (1993). Taxomyces andreanae, a proposed new taxon for a bulbilliferous hyphomycete associated with Pacific yew (Taxus brevifolia).

Suebrasri, T., Somteds, A., Harada, H., Kanokmedhakul, S., Jogloy, S., Ekprasert, J., Lumyong, S., Boonlue, S. J. R. (2020). Novel endophytic fungi with fungicidal metabolites suppress sclerotium disease. 16, 100250.

Sui, M., Yang, H., Guo, M., Li, W., Gong, Z., Jiang, J., & Li, P. (2021). Cajanol sensitizes a2780/taxol cells to paclitaxel by inhibiting the PI3K/akt/NF-κB signaling pathway. *Frontiers in Pharmacology, 12*, 783317.

Taechowisan, T., Chanaphat, S., Ruensamran, W., & Phutdhawong, W. S. (2014). Antibacterial activity of new flavonoids from Streptomyces sp. BT01; an endophyte in Boesenbergia rotunda (L.) Mansf. *Journal of Applied Pharmaceutical Science, 4*(4), 008–013.

Tang, J. L., Zhou, Z. Y., Yang, T., Yao, C., Wu, L. W., & Li, G. Y. (2019). Azaphilone alkaloids with anti-inflammatory activity from fungus Penicillium sclerotiorum cib-411. *Journal of Agricultural and Food Chemistry, 67*(8): 2175–2182.

Tenguria, R. K., & Firodiya, A. D. (2016). Cytotoxic activities of endophytic fungi isolated from central region of Madhya Pradesh. *Journal of Innovations in Pharmaceuticals and Biological Sciences, 3*(3), 29–38.

Tiwari, P., & Bae, H. J. M. (2022). Endophytic fungi: Key insights, emerging prospects, and challenges in natural product drug discovery. *10*(2), 360.

Tjamos, E. C., Tsitsigiannis, D. I., Tjamos, S. E., Antoniou, P. P., & Katinakis, P. (2004). Selection and screening of endorhizosphere bacteria from solarized soils as biocontrol agents against Verticillium dahliae of solanaceous hosts. *110*, 35–44.

Uzor, P. F., Osadebe, P. O., & Nwodo, N. J. (2017). Antidiabetic activity of extract and compounds from an endophytic fungus *Nigrospora oryzae*. *67*(05), 308–311.

Vardharajula, S., SkZ, A., Shiva Krishna Prasad Vurukonda, S., & Shrivastava, M. (2017). Plant growth promoting endophytes and their interaction with plants to alleviate abiotic stress. *6*(3), 252–263.

Walker, D. M., Lawrence, B. R., Esterline, D., Graham, S. P., Edelbrock, M. A., & Wooten, J. A. (2014). A metagenomics-based approach to the top-down effect on the detritivore food web: A salamanders influence on fungal communities within a deciduous forest. *Ecology and Evolution, 4*(21), 4106–4116.

Wang, J., Hou, W., Christensen, M. J., Li, X., Xia, C., Li, C., & Nan, Z. (2020). Role of Epichloë endophytes in improving host grass resistance ability and soil properties. *Journal of Agricultural and Food Chemistry, 68*(26), 6944–6955.

Wijesekara, T., & Xu, B. (2023). Health-promoting effects of bioactive compounds from plant endophytic fungi. *9*(10), 997.

Wu, S. H., Huang, R., Miao, C. P., & Chen, Y. W. (2013). Two new steroids from an endophytic fungus Phomopsis sp. *10*(7), 1276–1283.

Yang, H., & Ping Dou, Q. (2010). Targeting apoptosis pathway with natural terpenoids: implications for treatment of breast and prostate cancer. *11*(6), 733–744.

Younis, T., Rahman, L., Rahman, S., Shinwari, A. K., Iqrar, I., & Shinwari, Z. K. (2022). Endophytes: Potential source of bioactive compounds of pharmaceutical importance: Pharmaceutical importance of bioactive compounds from endophytes. *Proceedings of the Pakistan Academy of Sciences: B. Life and Environmental Sciences, 59*(4), 1–13.

Zain, M. E. (2011). Impact of mycotoxins on humans and animals. *15*(2), 129–144.

Zhang, B., Salituro, G., Szalkowski, D., Li, Z., Zhang, Y., Royo, I., Vilella, D., Díez, M. T., Pelaez, F., & Ruby, C. J. S. (1999). Discovery of a small molecule insulin mimetic with antidiabetic activity in mice. *284*(5416), 974–977.

Zhang, J., Wang, J.-D., Liu, C.-X., Yuan, J.-H., Wang, X.-J., & Xiang, W.-S. (2014). A new prenylated indole derivative from endophytic actinobacteria Streptomyces sp. neau-D50. *28*(7), 431–437.

Zhao, H., Chen, X., Chen, X., Zhu, Y., Kong, Y., Zhang, S., Deng, X., Ouyang, P., Zhang, W., & Hou, S. (2020). New peptidendrocins and anticancer chartreusin from an endophytic bacterium of Dendrobium officinale. *8*(7).

Microbial Pigments: Fermentative Production and Biological Activities

Chi-Ching Lee

Abstract

Microbial pigments are valuable sources of natural colorants because they have a variety of functions and outstanding attributes, including being non-toxic, biodegradable, and environmentally friendly. Microorganisms are valuable biofactories because they can produce pigments and other significant primary and secondary metabolites. Various microbial pigments are identified and determined, such as orange carotenoids, yellow flavins, brown melanin, purple porphyrins, red betalains, green chlorophyll, and blue phycobiliproteins. These naturally occurring microbial pigments are highly promising because of their many health benefits and industrial applications compared to the adverse effects of synthetic colorants. Pigment-producing microbes are rapid-growing species with a larger production that may be increased in the fermentative process and remain constant under different climates. The primary factors influencing productive microbial pigment include microbial selection, temperature, oxygen supply, fermentative approaches, moisture level, pH, and nutrients from carbon or nitrogen sources. The microbial pigments are extremely promising since synthetic colorants might cause allergies and toxicity. Microbial pigments also have distinct biological effects, such as antimicrobial, antioxidant, anti-aging, photoprotection, anticancer activity, anti-inflammatory, cardiovascular disease prevention, anti-obesity, and antidiabetic effects. In this chapter, the major types of microbial pigments and their functions will be discussed. Different process approaches, laboratory to large-scale bioprocesses, and the factors influencing fermentative production will be also mentioned. Additionally, the potential biological activities and applications of microbial pigments are emphasized.

Keywords

Microbial pigments · Carotenoids · Flavins · Melanins · Porphyrins · Antimicrobial · Antioxidant · Anticancer · Anti-inflammatory

1 Introduction

A variety of naturally occurring substances known as microbial pigments can be produced by bacteria, fungi, and microalgae. The prospective applications of these pigments, which come in various colors and hues, have drawn much interest from the pharmaceutical, food, cosmetic, and textile industries (Paillié-Jiménez et al., 2020). Reduced dependency on petroleum-based commodities and a minimal ecological impact are just two benefits of using the fermentative production of microbial pigments as an ecologically friendly and sustainable alternative for synthetic pigments. Microbial pigments are recognized for their bright colors, which range from blue, green, and black to yellow, orange, and red. A yearly increase in market demand is observed for pigment. An annual growth rate of 4.9% is predicted for the increase, which may reach USD 54.64 billion by 2026 (Mohd Said, 2023). These pigments are important for microbial physiological functioning since they participate in metabolic activities, provide a defense system against environmental stresses, and contribute to the aesthetic appeal of microorganisms. Microbial pigments are much more significant than just their coloration. Potent biological effects such as antibacterial, anticancer, anti-inflammatory, and antioxidant have been attributed to these

C.-C. Lee (✉)
Faculty of Engineering and Natural Sciences, Department of Food Engineering, Istanbul Sabahattin Zaim University, Istanbul, Türkiye
e-mail: chi.lee@izu.edu.tr

Faculty of Technologies, Department of Food Technology and Nutrition, Klaipeda State University of Applied Sciences, Klaipėda, Lithuania

K. Sharma (ed.), *Bio-prospecting of Novel Microbial Bioactive Compounds for Sustainable Development*, Sustainable Landscape Planning and Natural Resources Management, https://doi.org/10.1007/978-3-031-82178-3_4

substances. Consequently, they have generated significant interest from experts and businesses looking for organic substitutes for synthetic products. The main goal of this chapter is to present a thorough introduction to microbial pigments, emphasizing their fermentative synthesis and biological activity. First, the pigments produced by microbes are classified and described along with their chemical characteristics, sources, and biological applications. Then, a comprehensive investigation of the steps involved in the fermentation production of microbial pigments, such as the downstream processing methods, microbial strain selection, and optimal fermentation. Afterward, many biological functions of microbial pigments are evaluated, and the possible positive effects and mechanisms for health are highlighted. The applications of microbial pigments are discussed in a range of sectors, including food, medicine, cosmetics, and textiles, highlighting their benefits for the environment and the economy. Current challenges and possible improvements in the future are also mentioned, such as biotechnology advancements and sustainability concerns (Demirci et al., 2020; Paillié-Jiménez et al., 2020).

2 Major Chemical Groups and Functions of Microbial Pigments

2.1 Carotenoids

A family of naturally occurring pigments known as carotenoids is produced by various microorganisms, including certain types of yeasts, fungi, bacteria, and algae. The bright yellow, orange, and red hues in many microbes are caused by these pigments. Eight isoprene units are composed of carotenoids, which are tetraterpenoids classified into two primary groups: xanthophylls, which are oxygenated derivatives including lutein and astaxanthin, and carotenes, which are pure hydrocarbons including β-carotene and lycopene. Carotenoids are a group of isoprenoid-derived substances derived from the precursor phytoene (C_{40}). Nevertheless, several bacterial species may synthesize C_{50} and C_{30} carotenoids from various precursors. Major production of carotenoids has been reported from bacteria like *Rhodococcus*, *Pseudomonas*, and *Dietzia*. Besides, β-carotene is abundantly produced by several species of fungi, such as *Neurospora crassa* and *Blakeslea trispora*. The high carotenoid concentration of microalgae, such as astaxanthin and β-carotene, is widely recognized. Examples of these species are *Haematococcus pluvialis* and *Dunaliella salina* (Liu et al., 2021). Carotenoids are reported to have biological functions, such as their chemical composition, which enables them to interact with cell membranes and scavenge free radicals. Carotenoids can protect against oxidative damage by integrating into the lipid bilayer of cell membranes. They also regulate the expression of genes and signaling pathways associated with immunological response, antioxidant activity, and cell proliferation (Joshi et al., 2023). Figure 1 shows the different chemical structures of carotenoids.

The various applications and health benefits of carotenoids have gained a lot of interest. Their potential to scavenge free radicals and mitigate oxidative stress serves as one of the most outstanding antioxidants. This lowers the

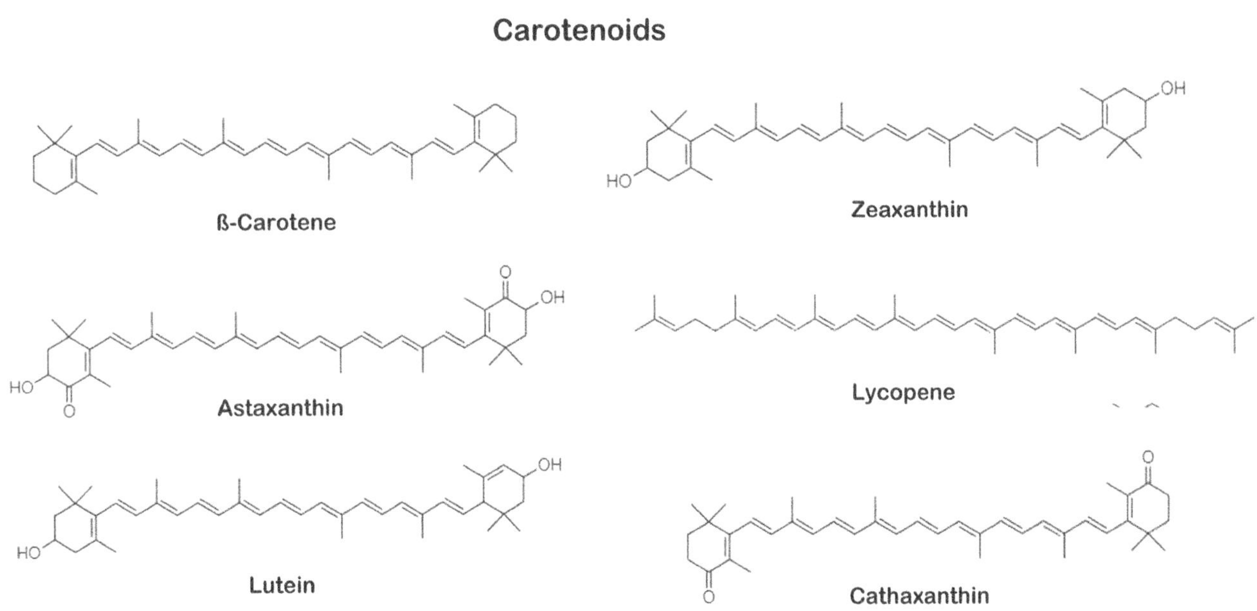

Fig. 1 Chemical structures of carotenoids

likelihood of developing chronic diseases, including cardiovascular disease, cancer, and age-related macular degeneration. In the food industry, carotenoids frequently serve as naturally occurring food colorants. Both their natural and vibrant colors draw in health-conscious customers and improve the visual attractiveness of food items (Mussagy et al., 2022). Carotenoids are used in dietary supplements because of their antioxidant qualities and possible health benefits. Studies have been conducted on the possible therapeutic benefits of carotenoids, such as β-carotene and astaxanthin, which include immunomodulatory, anticancer, and anti-inflammatory properties. The potential of carotenoids to increase the bioavailability of medicinal substances has been investigated (Lee, 2023b). Because carotenoids have skin-protective qualities, such as ultraviolet defense and anti-aging benefits, they are utilized in cosmetics. They support the health of the skin and minimize the visibility of age spots and wrinkles. The animal feed industry exploits carotenoids as feed additives to improve the color of products, including fish flesh, egg yolks, and chicken skin. Carotenoids also support the physiological health and productivity of livestock. Through microbial fermentation, higher production of carotenoids has been proven possible by developments in genetic engineering and biotechnology. Raising carotenoid yield while reducing production costs is being actively sought through methods such as metabolic pathway optimization, using high-yield microbial strains, and bioprocess optimization (Lee & Demirci, 2023; Lee, 2023a;).

2.2 Flavins

A class of yellow pigments called flavins is generated by a variety of microorganisms, such as yeasts, fungi, and bacteria. Flavins, such as flavin mononucleotide (FMN) and flavin adenine dinucleotide (FAD), are substances that are derived from riboflavin (vitamin B2). Especially in redox-related biochemical activities, these chemicals stand out for their fluorescence and vital functions as cofactors in a wide range of biological reactions. Several microbial species are well-known for producing riboflavin and its derivatives, including *Bacillus subtilis*, *Candida famata*, *Ashbya gossypii*, and *Escherichia coli*. The fermentation technique in which certain microbial strains are cultivated under ideal circumstances to promote riboflavin synthesis, is usually used to produce flavins (Nardo & Gilardi, 2020). Figure 2 presents the different chemical structures of flavins. To maximize the production, conditions during fermentation, including pH, temperature, and aeration, are carefully monitored. Regarding oxidoreductases, which catalyze critical redox events in metabolic pathways, including cellular respiration and photosynthesis, flavins are fundamental as enzyme cofactors. Flavins contribute to the electron transport chain during cellular respiration, which is essential for the synthesis of ATP (Ashaolu et al., 2023). Certain microbes also use flavin photoreceptors, which enable them to detect and react to light and support activities like DNA repair and phototaxis. The biological characteristics of flavins allow them to be applicable to a wide range of industries. Riboflavin is used in the food industry as a natural colorant to improve the appearance of food products and as a nutritional supplement to avoid vitamin B2 deficiency. Riboflavin's significance in phototherapy is utilized by the pharmaceutical industry to treat illnesses including neonatal jaundice and migraine prophylaxis. Moreover, flavins are investigated in therapeutic formulations to improve the effectiveness and stability of medicinal substances. Because flavins have antioxidant qualities that help prevent

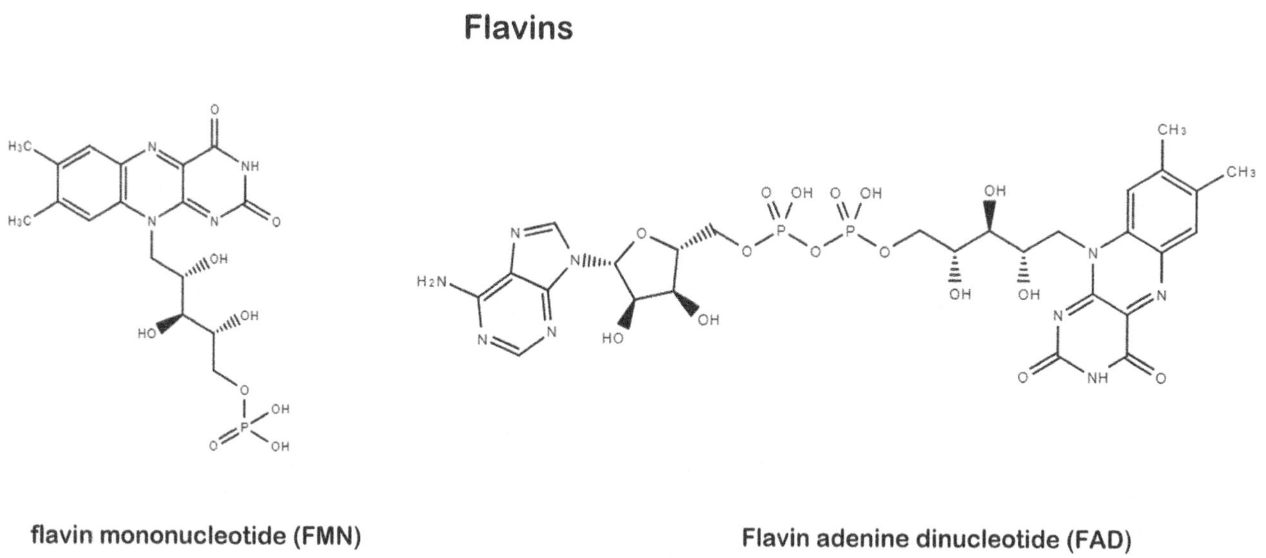

Fig. 2 Chemical structures of flavins

oxidative damage and enhance skin health, the cosmetics industry utilizes them in skin care products. Flavins are also used in anti-aging formulations to minimize wrinkles and improve a more youthful complexion. Utilizing their fluorescent features, flavins are used in biotechnology to produce biosensors that can detect a variety of biomolecules and environmental pollutants. Additionally, they have been investigated for their potential in the generation of bioenergy, namely in microbial fuel cells where they can help with electron transport. The primary way that flavins function is that they are redox-active cofactors that facilitate the transport of electrons in a variety of biological activities. Their capacity to alternate between reduced and oxidized states is crucial for metabolic functions. Their ability to absorb light and trigger photochemical reactions is another of their photochemical attributes, which increases their potential for use in industrial and medication (Dufossé, 2022).

2.3 Melanin

Melanin is usually black, brown, or dark brown in color and is produced by polymerizing phenolic or indolic chemicals. It displays specific physical and chemical characteristics due to its complex structure and large molecular weight. Among the most significant microorganisms that produce this pigment are *Streptomyces* spp., *Aspergillus niger*, and *Cryptococcus neoformans*. Melanin is essential to the survival and pathogenicity of these species by providing defense against environmental stresses, including UV radiation, oxidative stress, and severe temperatures (El-Naggar & Saber, 2022). Microorganisms produce this pigment as a defensive mechanism in response to various environmental stresses. Melanin is an essential protective agent because of its capacity to neutralize reactive oxygen species (ROS) and absorb a broad range of radiation. This pigment contributes to making pathogenic bacteria more virulent by protecting them from the host's immune system. For example, melanin synthesis is linked to enhanced resistance to phagocytosis and antimicrobial agents in the fungal pathogen *Cryptococcus neoformans*. Tyrosine and dihydroxyphenylalanine (DOPA) are examples of substrates that are oxidized and polymerized during the melanin production pathways in microorganisms. Tyrosinase and laccase are two of the enzymes that catalyze these reactions. Melanin's distinctive properties have prompted research into its potential uses in various industrial and biological fields. This pigment is presently studied in the pharmaceutical industry for its possible application in drug delivery systems. Its protective properties and biocompatibility can improve the stability and potency of medications. The potential of this pigment as a natural sunscreen ingredient in cosmetic compositions has also attracted attention due to its capacity to absorb ultraviolet light. It is also a promising ingredient in skincare products that prevent oxidative damage and premature aging because of its antioxidative functions. Its conductivity has been examined in biotechnology to develop bioelectronic systems and biosensors. Melanin is a desirable material for these applications based on its intrinsic electrical conductivity and stability in a range of environmental conditions. Additionally, environmental biotechnology applies this pigment in metal ion chelation for bioremediation, which allows heavy metals from polluted environments to be detoxified (Singh et al., 2021).

2.4 Porphyrins

Porphyrins are a family of colored macrocyclic compounds that originate from a variety of microorganisms, such as certain fungi, bacteria, and archaea. The massive, aromatic ring system is formed by the four pyrrole rings bonded through methine bridges that compose up these compounds' highly conjugated structure. Their capacity to absorb light gives them their characteristic red, purple, or green colors caused by their unique chemical structure and its metalions (Ansari et al., 2024). Coproporphyrin and uroporphyrin are microbial porphyrins that are crucial to microbial metabolism, especially for respiration and photosynthesis. Porphyrins are essential biomolecules that are precursors of heme and chlorophyll in microorganisms. Figure 3 exhibits the chemical structures of porphyrins. Heme is a necessary part of cytochromes, catalases, and peroxidases, all of which are vital for electron transportation and redox reactions in cellular respiration and metabolism. Heme includes an iron ion coordinated inside the porphyrin ring. Photosynthetic organisms depend on chlorophyll, which has a magnesium ion inside the porphyrin ring. Figure 4 demonstrates chemical structures of chlorophylls. This allows the organisms to absorb light energy, facilitating the synthesis of organic molecules from carbon dioxide and water. Microorganisms undergo a multi-step enzymatic process to biosynthesize porphyrins, beginning with basic precursors such succinyl-CoA and glycine. Porphyrins and their derivatives are associated with several chemical and photophysical characteristics. Porphyrins are utilized in photodynamic therapy (PDT) in medicine to treat many illnesses, including cancer. Patients receiving PDT are administered porphyrins, which concentrate primarily in tumor tissues. Certain porphyrins react with light at specific wavelengths to form reactive oxygen species, which target cancer cells only while allowing healthy tissues unharmed. This focused therapy technique has demonstrated potential in the treatment of various localized lesions as well as certain

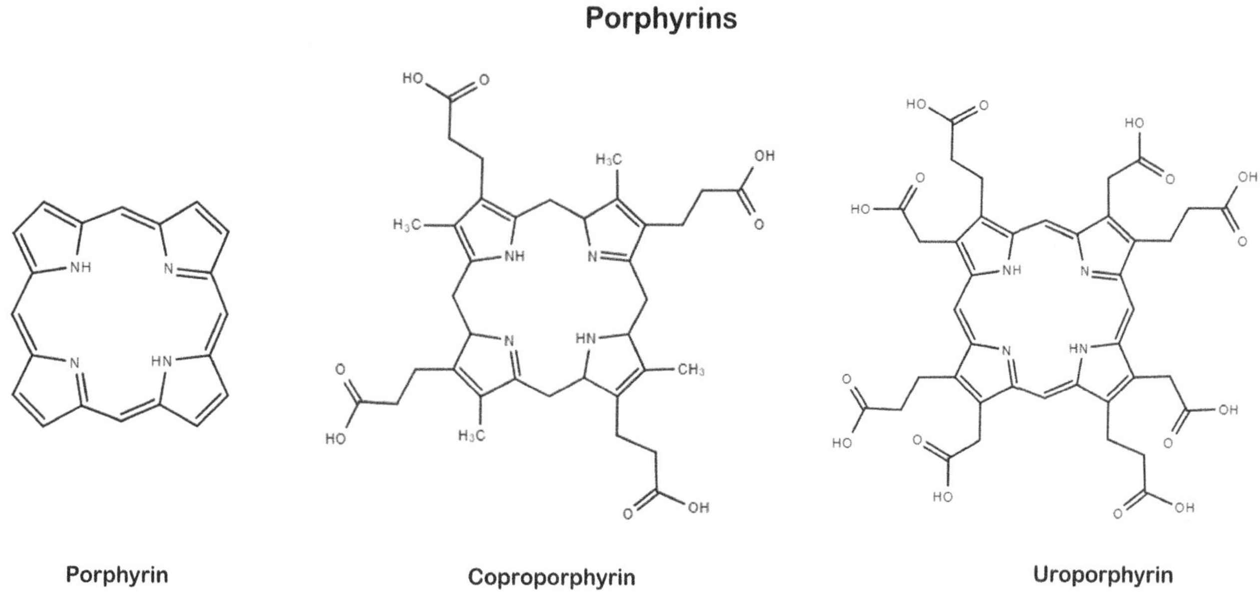

Fig. 3 Chemical structures of porphyrins

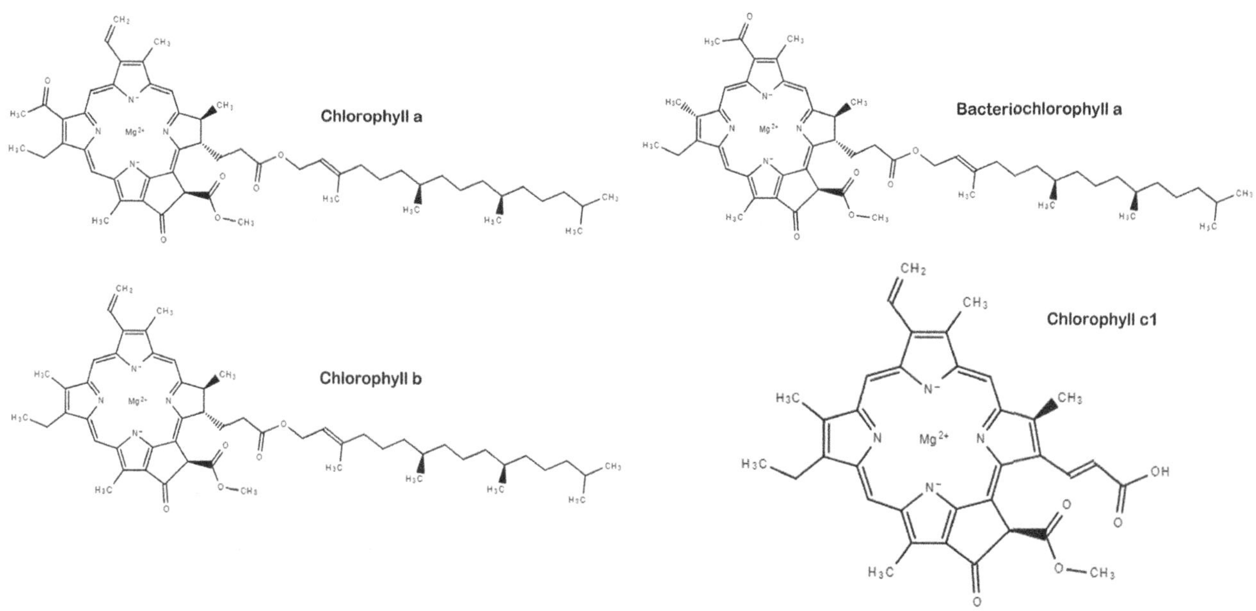

Fig. 4 Chemical structures of chlorophylls

malignancies (Guern et al., 2021). Porphyrins have been discovered for their potential in renewable energy technologies, including solar cells. They are appropriate for utilization in dye-sensitized solar cells (DSSCs), where they can improve the efficiency of light collection and conversion into electrical energy as a result of their potent light-absorbing properties and function in electron transfer processes. In addition, the examination of microbial porphyrins might offer valuable perspectives on the progression of metabolic pathways and the flexibility of microbes in many ecological niches. Microbial porphyrins are evaluated in bioremediation and biosensing applications. Because porphyrins may bind to different contaminants and catalyze their degradation, they could effectively decompose contamination. They can interact with organic contaminants and heavy metals to degrade or remove them. Moreover, biosensors for identifying environmental pollutants and tracking pollution levels are developed using porphyrins' intrinsic fluorescence and electrochemical abilities (Guterres et al., 2020).

2.5 Alkaloid Pigments

Alkaloids are a type of pigment found in some microorganisms. They comprise nitrogen-containing low molecular weight compounds where the nitrogen atom is obtained from an amino acid. The alkaloid class includes a number of pigment families, including betalains, tambjamines, and prodigiosines. Figures 5, 6 and 7 display different chemical structures of tambjamines, prodiginine and alkaloids. A class of reddish-pink pigments known as prodigiosines is mainly produced by certain actinomycetes, such as *Streptomyces* species, and some bacterial species, most notably *Serratia marcescens* (Hernández-Velasco et al., 2020). These pigments are distinguished by the unique tripyrrole structure that provides them with their vibrant coloration. Prodigiosines have several biological actitivties, such as their antibacterial, antifungal, antiprotozoal, immunosuppressive, and anticancer capabilities. The microbial cell's internal regulation of prodigiosin synthesis is tightly linked to external conditions, including pH, temperature, and

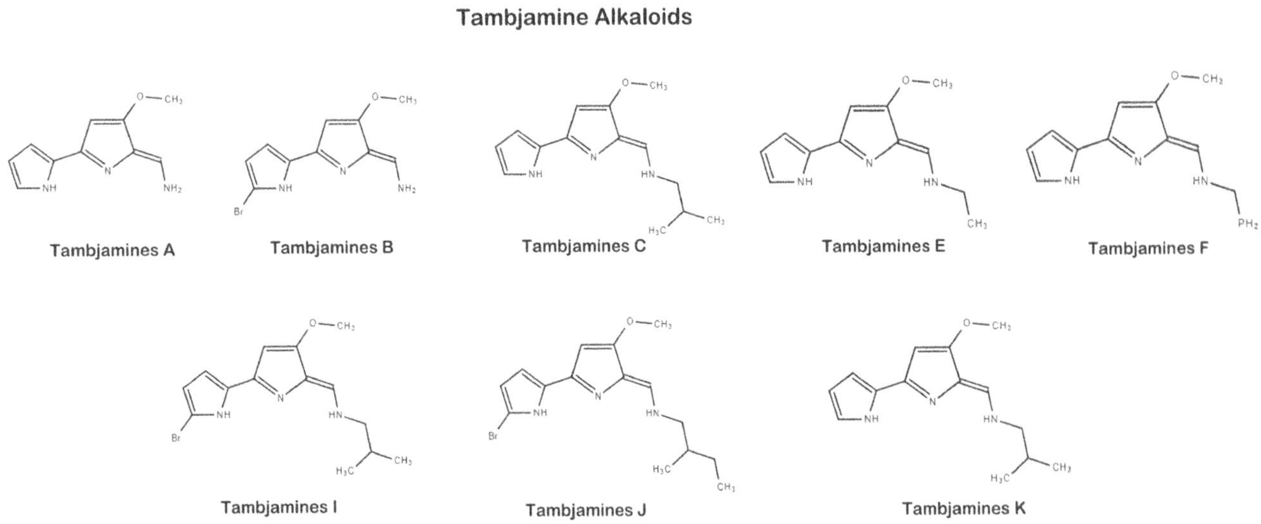

Fig. 5 Chemical structures of tambjamine alkaloids

Fig. 6 Chemical structures of prodiginine alkaloids

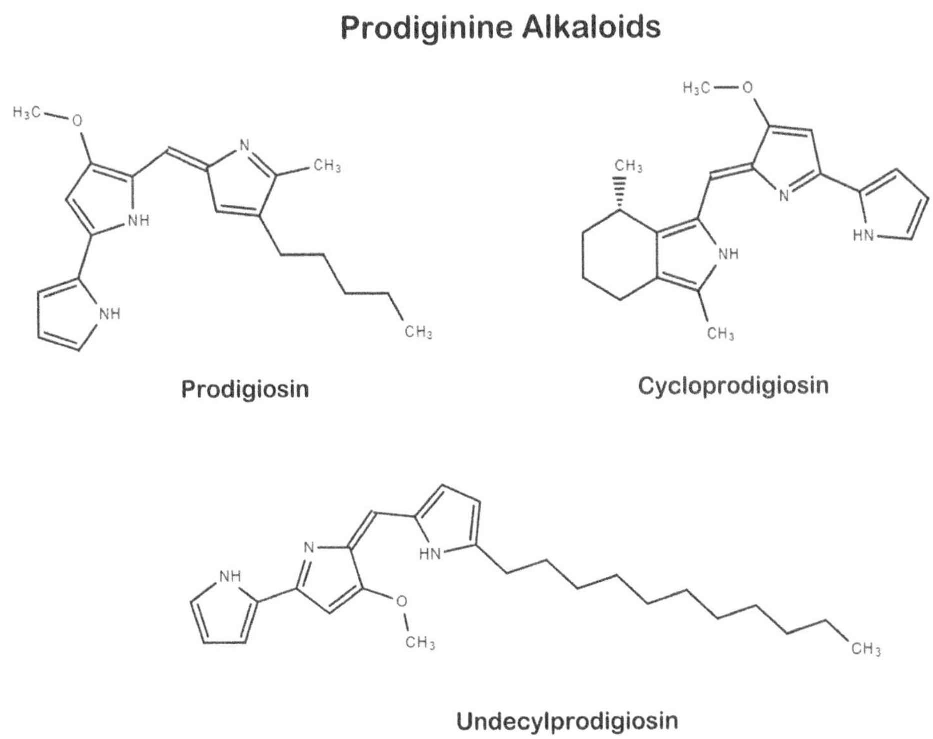

Betalains

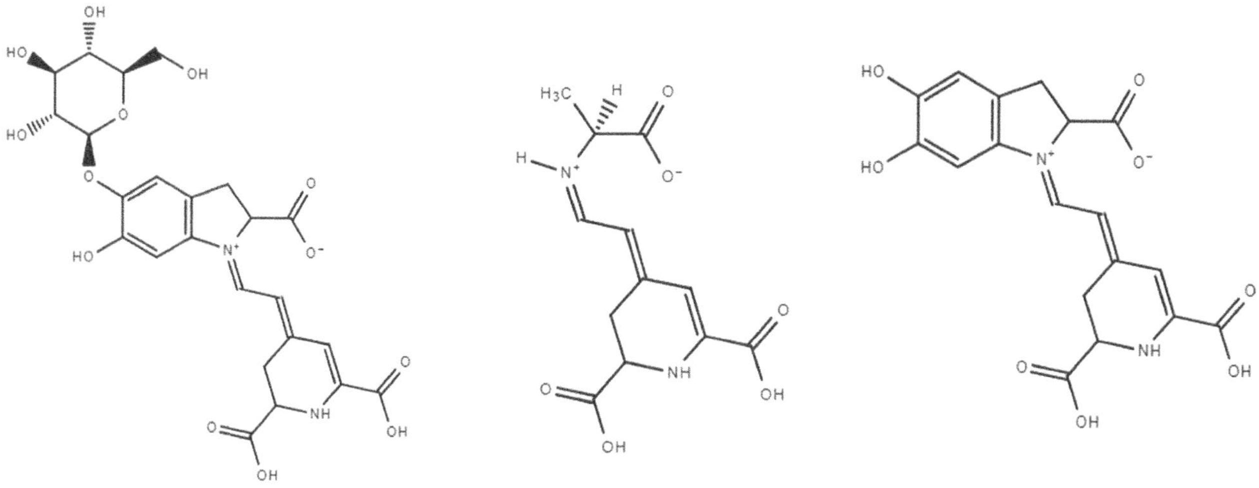

Betanin **Betaxanthins** **Betacyanins**

Fig. 7 Chemical structures of betalains alkaloids

nutrition availability. Prodigiosine biosynthesis is a complicated process involving the integration of several enzyme processes. This process involves several important enzymes, including those that create the pyrrole rings and begins with the condensation of primary metabolites generated from amino acids. The anticancer effects of Prodigiosines make them extremely promising for the pharmaceutical business. It has been demonstrated that they can cause cancer cells to undergo apoptosis by a number of different routes, including the production of ROS, damage to the mitochondria, and activation of caspase pathways. Prodigiosines also have immunosuppressive properties that may be useful in the management of autoimmune disorders and in organ transplantation. They are also potential candidates for the development of novel antibiotics due to their antibacterial abilities, particularly in considering the growing prevalence of antibiotic resistance (Pailliè-Jiménez et al., 2020).

Several marine bacteria and certain terrestrial actinomycetes produce a class of blue-green pigments called tambajamines. The structural characteristic of these pigments is their bispyrrolic core, which is often altered by different functional groups to produce a wide range of tambjamine compounds. Though they haven't been investigated as much as other microbial pigments, tampjamines have demonstrated biological properties, including cytotoxic and antibacterial behaviors. Although the exact ecological function of tambjamines in microbial communities is unknown, it is thought that they contribute to defensive mechanisms that shield the generating organisms from rival microbes and predators. A sequence of enzyme processes transforms basic ingredients into the intricate bispyrrolic structure during the production of tambjamines. This process, which involves important enzymes that catalyze the creation and alteration of the pyrrole rings, is assumed to be comparable to that of other pyrrole-based pigments. Apart from their antibacterial characteristics, tambjamines have demonstrated potential in studies related to cancer prevention. Tambjamines may inhibit the development of tumors and cause cancer cells to undergo apoptosis. Additionally, they are intriguing prospects for developing novel medications and bioactive substances due to their distinct chemical structure (Rao et al., 2021).

Betalains are water-soluble pigments produced by some bacteria and fungi in addition to being present in the *Caryophyllales* order of plants. The two primary groups of these pigments are called betacyanins (red to violet) and betaxanthins (yellow to orange). Betalains are derived from tyrosine, with betalamic acid serving as the biosynthetic precursor (Esquivel, 2024). The vast range of betalain colors is formed when this intermediate condenses with different amines or amino acids. In addition to their distinctive coloring, betalains are significant because of their potent antioxidant capabilities. It has been demonstrated that these pigments efficiently scavenge free radicals, offering advantages against oxidative stress. Betalacains are used in both medical and industrial applications because of their antioxidant activity as well as their anti-inflammatory, anticancer, and antimicrobial characteristics (Lee et al., 2016). They are especially desired as natural colorants in food production, where they are used to color a range of food products, such as beverages, desserts, and processed meals since they have water solubility and stability under different pH levels.

2.6 Polyketide

Polyketides are important not only as colors but also as medicines, antibiotics, and anticancer agents since they have many structures and biological functions. Quinones are a common class of these pigments made by fungi like *Penicillium*, *Aspergillus*, and *Helminthosporium*. Numerous tints and colors, such as yellow, orange, red, green, purple, brown, and blue, are produced by these pigments. A class of enzymes known as polyketide synthases (PKS) polymerizes the acetyl and propionyl subunits to produce these molecules. PKS is modular in nature, allowing them to integrate different extender components and modify the expanding polyketide chain by cyclization, reduction, and other chemical transformations, which results in the structural variety of polyketides. Because of their versatility, various compounds with various functional groups and ring configurations may be synthesized. Enzymatic processes play a complicated role in the production of these pigments, with PKS enzymes playing a facilitating role. Type I PKS (modular PKS), Type II PKS (iterative PKS), and Type III PKS (chalcone synthase-like PKS) are the three major enzymes. The broad chemical variety among these pigments results from the unique mechanisms each PKS uses to construct and modify the polyketide chain (Wang et al., 2020). Figure 8 shows different chemical structures of betalains alkaloids. These pigments are crucial to the physiology of the microbes that produce them because they protect them from oxidative stress, ultraviolet radiation, and predators. For example, certain pigments have the ability to absorb UV radiation, protecting the microbial cells' DNA from destruction. Furthermore, certain pigments' antioxidative capabilities facilitate the neutralization of ROS, improving the survival of microorganisms in adverse conditions. *Streptomyces coelicolor*, the bacterium that produces the blue-pigmented actinorhodin, is a well-known example of a polyketide pigment. The multifunctional characteristics of these pigments are demonstrated by actinorhodin, which not only adds a unique color but also has antimicrobial properties. Besides, fungal fermentation of *Monascus* spp. yields yellow–red pigments, which are also well-known microbial colorants. These pigments are highly valued for their antibacterial and antioxidant abilities and are frequently used in food coloring (Chaudhary et al., 2022). Azaphilones, which are fungal polyketide pigments with a highly oxygenated pyranoquinone bicyclic core, are determined for their potential uses in printing, dyeing, agro-alimentary, and pharmaceuticals. Six primary azaphilone pigment compounds are found in *Monascus* spp., including rubropunctatin, monascorubrin, monascin (monascoflavin), ankaflavin, rubropunctamine, and monascorubramine. These pigments have important functions as pigments, but they also have important pharmacological applications. Numerous polyketide antibiotics have been manufactured as necessary medications for the treatment of bacterial infections, including erythromycin, tetracycline, and rifamycin. Anticancer medications made from these pigments, such as doxorubicin and epothilone, are essential for cancer treatment because they may hinder cancer cells from proliferating and cause them to undergo apoptosis. Polyketides are promising agents in the discovery

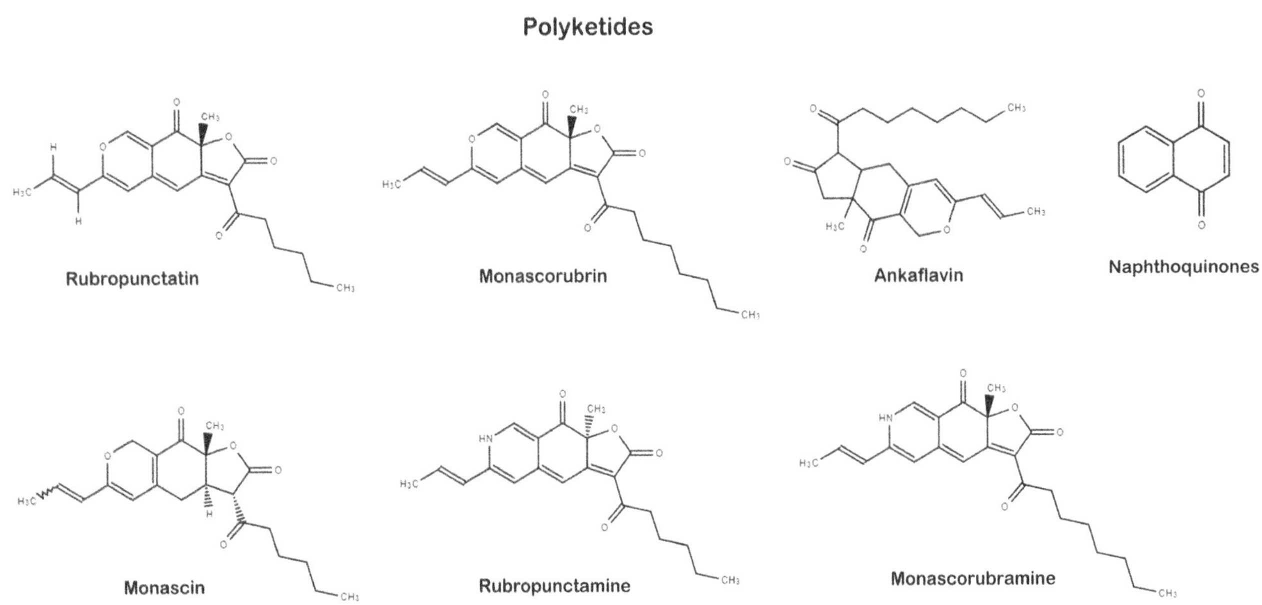

Fig. 8 Chemical structures of polyketides

and development of novel medicinal products on account of their efficacy and adaptability (Afroz Toma et al., 2023).

2.7 Phycobiliproteins

Phycobiliproteins are water-soluble fluorescent protein pigments found mainly by cyanobacteria, red algae, and cryptomonads. These protein pigments are crucial to phycobilisomes and light-harvesting complexes in photosynthetic organisms. Phycobilisomes collect light energy and transport it to chlorophyll to carry out photosynthesis. Because of their brilliant colors and fluorescent characteristics, phycobiliproteins are exceptionally valuable and have a variety of biomedical applications. Phycobilins, which are open-chain tetrapyrrole chromophores, are covalently bound to protein subunits to form phycobiliproteins. The ability of the proteins to absorb light is attributed to these phycobilins. The phycobiliproteins are arranged into phycobilisomes, higher-order complexes that are bonded to the thylakoid membranes of photosynthetic cells. Blue phycocyanin, red phycoerythrobilin, and light-blue allophycocyanin are three major pigments of phycobiliproteins (Kannaujiya et al., 2021) The phycoerythrin is commonly produced from *Porphyridium* spp. and phycocyanobili from *Spirulina* spp. in the pharmaceutical and biotechnological industries. Phycobiliproteins contribute to enhanced health, including immunological responses and antibacterial, anti-inflammatory, and antioxidant activities. The blue pigment phycocyanin absorbs orange and red light and is found in cyanobacteria and some red algae, including *Cyanidioschyzon merolae*, *Bangia atropurea*, *Acanthophora specifera*, *Gracilaria corticata*, *Laurencia papillosa* (Fernandes et al., 2023). Phycocyanin is associated with a number of desirable biological characteristics, including hepatic protection, anticancer activity, antioxidant and anti-inflammatory effects, and neuroprotection. Allophycocyanin, a blue colour water-soluble pigment, also absorbs red light and is typically found in cyanobacteria, acting as a bridge in the energy transfer from phycocyanin to chlorophyll. The common sources of allophycocyanin *are Gracilaria corticata*, *Laurencia papillosa*, and *Acanthophora specifera.* The red protein pigment phycoerythrin contains chromophore phycoerythrobilin to display various biological activities, such as antioxidant and anti-inflammatory. Because of its fluorescent feature, it can be used in immunodiagnostics for protein and cell labeling. Red algae including *Porphyra purpurea*, *Pyropia haitanensis*, *Gracilaria corticata*, *Bangia atropurpurea*, *Portieria hornemannii*, and *Aglaothamnion neglectum* are the main producers of this pigment (Osório et al., 2020).

3 Fermentative Production of Microbial Pigments

The fermentative production of microbial pigments involves a number of crucial steps, all with the purpose of maximizing the quantity and quality of the final pigment. In order to develop colors under regulated circumstances, this procedure utilizes the metabolic capacities of microorganisms, including bacteria, fungi, and algae. The fermentative production includes five main stages: microbial cultivation, fermentation, extraction, purification, and application. A number of crucial steps are involved in the fermentative process that produces microbial pigments, and their purpose is to maximize both pigment output and quality (Sen et al., 2019). Microbial culture is the first step in the process, in which particular pigment-producing microbes are cultivated in particular environments to start their metabolic processes. The core fermentation stage comes next, and depending on the required amount and efficiency, this can be carried out by using batch, fed-batch, or continuous fermentation techniques. Critical factors, including temperature, pH, aeration, and agitation, are continuously monitored to enhance the production of pigment. Following fermentation, a number of methods, including centrifugation, solvent solubilization, and cell lysis, are used to extract pigments from the microbial cells and culture medium. The extracted pigments are further subjected to purifying procedures such as solvent extraction, chromatography, and crystallization to achieve high purity and concentration. Ultimately, the refined pigments are mixed for a range of applications, including natural food coloring, medicinal compounds in pharmaceuticals, and coloring in cosmetics and other merchandise. Technological developments in metabolic engineering and biotechnology have allowed it to produce a wide range of microbial pigments with significant financial value in an environmentally friendly way by improving the capacity and efficiency of these processes (Ramesh et al., 2022).

3.1 Microbial Cultivation

A crucial stage in the fermentative synthesis of microbial pigments is the identification of relevant microbial strains. Numerous microorganisms, such as fungi, algae, and bacteria, are recognized for their capacity to produce pigments. In bacteria, *Streptomyces* species for actinorhodin and other polyketides, and *Serratia marcescens* for prodigiosin. In fungi, *Aspergillus* and *Penicillium* species are used for a range of secondary metabolites, and Monascus species are used for betalains. Besides, there is a variety of microalgae and cyanobacteria for phycobiliproteins and carotenoids. The desired pigment and its unique characteristics, such as tint, stability, and bioactivity, determine the appropriate microbial strain. The culture medium supplies the vital nutrients needed for microbial growth and pigment formation. The medium's composition is specifically designed to improve pigment production and satisfy the nutritional requirements of the specified microorganisms. Sucrose, glucose, or other sugars serve as carbon skeletons and sources of energy. Protein synthesis and other biological processes require nitrogen, which can be obtained via ammonium salts, peptones, or yeast extract. Minerals are required to sustain enzymatic processes, such as calcium, magnesium, and potassium. Also, the microbes could need particular vitamins and cofactors. By using pH adjusters, acids/bases, or buffers to maintain an optimal pH for the development of microorganisms and the synthesis of pigment. In order to enhance pigment production and quality, the medium's composition is frequently modified through experimental designs (Lagashetti et al., 2019).

Inoculation and incubation follow the preparation of the culture media. The selected microbial strain is inoculated into the culture medium during the inoculation process. The inoculum that has been pre-grown under monitored conditions to guarantee viability and activity can be used. A balance between rapid proliferation and effective pigment synthesis is usually achieved by optimizing the inoculum size, which is a crucial factor. There are several factors influencing microbial pigment production, such as moisture, carbon and nitrogen source, pH value, aeration rate, temperature, minerals and the type of fermentation. *Monascus ruber* produces a greater amount of color during solid-state fermentation at 70% moisture content. The types of carbon sources, such as glucose, fructose, maltose, lactose, galactose, etc., influence the mycelial development of pigment-producing microorganisms. The best nitrogen source for producing monascus pigment is ammonium chloride, which is followed by ammonium nitrate and glutamate. The worst nitrogen source is potassium nitrate, although glutamate serves best for producing color. The pH of the medium influences the type of pigment produced and the development of the microorganism. For *Monascus sp.* and *Rhodotorula*, the ideal pH ranges are 5.5–6.5 and 4.0–4.5, respectively. Lycopene production is preferred by a pH of neutral to slightly alkaline, while β-Carotene synthesis is enhanced by a low pH value (Darwesh et al., 2020) The bed of rice is continually aerated by mixing with humidified air under 95% relative humidity while the *Monascus* colors develop from the solid-state fermentation. The synthesis of pigments and biomass as a result of water loss from the bed is decreased by a forced aeration rate greater than 0.5 Liter per minute. From forced aeration rates of 0.05 to 0.2 Liter per minute, the highest amounts of pigments are produced. The incubation temperature is the primary variable dependent on the kind of microbe. *Pseudomonas* requires 35–36 °C for growth and pigment generation, while *Monascus* spp. need 25–28 °C for pigment formation (Agboyibor et al., 2018). Minerals are essential to the process of generating pigments. Zinc ions inhibited microbial growth in the liquid media, while substantial growth and coloration occurred in the solid medium. Compared to submerged fermentation, solid-state fermentation produces three times as much pigment. Although pigment synthesis in *Monascus purpureus* is also influenced by agitation, pH, and medium composition, solid cultures produce higher yields than submerged (He et al., 2022).

3.2 Fermentation Techniques

Solid-state fermentation (SSF) and submerged fermentation (SmF) are the two main fermentation processes that may be used for the fermentative production of microbial pigments. Each approach has its own advantages, and the selection of a fermentative technique is contingent upon the type of microbes, the properties of the pigment, and the particular demands of the application. Table 1 displays the fermentative production of microbial pigments involving SSF and SmF methods. SSF is an approach for manufacturing microbial pigments in which bacteria are grown on solid substrates without or almost without the presence of free water. SSF is an effective technique for obtaining colors from filamentous fungi and certain bacteria because it replicates their natural environment. Because it frequently uses low-cost agricultural waste and byproducts as substrates, the process is economical. In comparison to submerged fermentation, SSF also uses less energy and water, which lowers operating costs and has a less negative impact on the environment. The potential of SSF to promote the growth of filamentous fungi—which are frequently more efficient at producing pigment under these circumstances—is one of its primary advantages. Furthermore, the low water activity of SSF minimizes the possibility of contamination, enabling less complex fermentation procedures. Nevertheless, there are limitations to SSF, such as challenges in maintaining and monitoring crucial parameters like pH, temperature, and oxygen concentrations. Unequal distribution of nutrients and delayed microbial growth might lead to inconsistencies in production results and reduced biomass yields. Notwithstanding these difficulties, SSF is a useful material for a number of processes, most notably the synthesis of biodegradable and natural food coloring (Chen et al., 2021). Melanins from different fungi, carotenoids from *Blakeslea trispora*, and betalains from *Monascus* species are a few examples of pigments produced by SSF. The technique also has the advantage of producing very little waste, and

Table 1 The fermentative production of microbial pigments involves solid-state and submerged methods

Methodology	Advantages	Disadvantages	Applications	Examples
Solid-State Fermentation	• Low-cost with inexpensive substrates • Minimal waste generation; residues can be used as animal feed or compost • Lower water and energy requirements • Suitable for filamentous fungi and certain bacteria • Reduced risk of contamination due to low water activity • Emulates the native environment of many microorganisms	• Non-homogeneous nutrient distribution • Difficulties in substrate handling and processing • Limited scalability for industrial applications • Difficult to monitor and control parameters like pH and oxygen • Slower microbial growth and lower biomass yield	• Environmentally friendly textile dyes • Value-added products from agricultural waste • Organic pigments for pharmaceuticals and cosmetics • Food colorants • Pigments from filamentous fungi • Biodegradable dyes	• Betalains from *Monascus* species (Chaudhary et al., 2022; Darwesh et al., 2020) • Penicillin pigments from *Penicillium* species (Ziemons et al., 2017) • Actinorhodin from *Streptomyces coelicolor* (Finger et al., 2023; Lim et al., 2018) • Carotenoids from *Blakeslea trispora* (Cano-Lamadrid et al., 2023; Khan, 2023) • Melanins from various fungi (Tran-Ly et al., 2020)
Submerged Fermentation	• High reproducibility and consistency • Suitable for large-scale industrial applications • Homogeneous distribution of nutrients and oxygen • Applicable to a wide range of microorganisms • Easier control and monitoring of growth parameters • Rapid growth and high biomass yield	• Potential for higher risk of contamination • Disposal of large volumes of liquid waste • Some microorganisms do not grow in liquid culture • High-cost bioreactor and agitation equipment • High risk of contamination	• Textile dyes • Food colorants • Pharmaceutical and Cosmetic ingredients • Large-scale industrial production	• Astaxanthin from *Haematococcus pluvialis* (Mota et al., 2022) • Polyketides from *Streptomyces* species (Risdian et al., 2019; Wang et al., 2020b) • Carotenoids from *Phaffia rhodozyma* (Cassamo U. Mussagy et al., 2024) • Prodigiosin from *Serratia marcescens* (Ramesh et al., 2020; Tunca Koyun et al., 2022) • Flavins from various bacteria (Yurgel et al., 2014)

excess substrates may be turned into compost or animal feed. For small- to medium-sized companies in particular, SSF offers a practical and sustainable solution for the fermentative production of microbial pigments. A common approach that has been widely used in the industrial manufacturing of microbial pigments is SmF. Microorganisms in SmF grow in a liquid nutrient medium, facilitating rapid microbial growth and large biomass yields by allowing for the homogenous distribution of nutrients and oxygen. Large-scale industrial applications could gain greatly from this method since it provides accurate control and monitoring of numerous growth factors, including pH, temperature, and oxygen levels (Pranaw et al., 2022). The homogenous growth medium in SmF guarantees reliable pigment synthesis, which makes it a favored process for the manufacture of food colorants, cosmetic components, and medicinal pigments. Nonetheless, there are drawbacks to SmF. Higher operational expenses result from the technique's enormous energy and water consumption requirements. Furthermore, the high moisture content of the medium raises the possibility of contamination, requiring thorough sterilizing practices. There may be sustainability concerns associated with the disposal of huge amounts of liquid waste produced in SmF. Despite these limitations, SmF is a promising and effective strategy for manufacturing a wide variety of pigments, including prodigiosin from *Serratia marcescens*, carotenoids from *Phaffia rhodozyma*, and astaxanthin from *Haematococcus pluvialis* (Dev et al., 2022).

3.3 Extraction

The extraction of these natural substances is essential to the use of microbial pigments for many commercial uses, such as food colorants, medicines, and cosmetics. To obtain high yields and purity in the large-scale production of pigments from microbial sources, many techniques have been developed and optimized. Because of its ease of use and effectiveness, solvent extraction is the most popular technique. The introduction of natural pigments in the diet and a trend in the population toward healthier lifestyles have led to a rise in the extraction and recovery of pigments in recent years. Many pigment extraction techniques have been used to raise the extraction yield of these important bioactive compounds. Maceration, distillations, soxhlet extraction, and infusions are examples of traditional extraction techniques for pigment extraction. Because of its ease of use, low cost, and lower solvent concentration than other techniques like soaking, boiling, or macerating, soxhlet is the most often used approach to pigment extraction. Several solvents, including ethanol, methanol, benzene, chloroform, ethyl acetate, and others, were investigated in this extraction procedure in order to compare their effects on the extraction yield (Rather et al., 2023). The solvent is selected based on how soluble the particular pigment is. Although solvent extraction is a well-established technique, the use of hazardous solvents may result in risks to human health and the environment. Conventional extraction techniques like water infusion and maceration are still used. These techniques include the use of solvents such as methanol, ethanol, acetone, and water for the extraction of bioactive compounds. These conventional extraction techniques have higher solvent consumption, lower extraction yields, and longer extraction periods than alternative techniques. It has been discovered that applying heat extraction techniques may cause the extracted pigments' chemical structures to deteriorate, which would reduce their bioactivity. Tissue and cell barriers are broken down, solvents diffuse across membranes more quickly, and secondary metabolites are released when homogenization occurs. Similar to this, freezing and thawing had very little influence on cell wall breakage. The long freezing and thawing operations have relatively low pigment extraction rates. The search for alternative methods that would increase pigment extraction yields while lowering process costs was driven by the growing demand for natural microbial pigments. Furthermore, the techniques employed turned out to be more sustainable (Qin et al., 2023). Environmental-friendly extraction methods were developed for decades to use less energy and replace traditional harmful solvents with more eco-friendly options. The development of an ionic liquid has become increasingly popular in recent years since pigments have designable qualities, including low vapor pressure, high thermal stability, and dissolving capacity. The downstream engineering segment of fermentation methods can account for more than half of the overall cost. Much research has been done on substituting environmentally friendly solvents for dangerous ones. The twelve standards of sustainable solvent were developed, including price, renewability, flammability, toxicity, biodegradability, stability, synthesis, availability, storage recyclability, grade, and synthesis.

Since none of the solvents meet all of these characteristics, the sustainability of a solvent has to be assessed on an individual basis. Basic and acidic techniques are associated with low extraction efficiency, time-consuming, multi-step processes, corrosive solvents, and higher production costs. Pigment molecules are frequently degraded when acidic and basic solutions are used. Thus, sophisticated extraction techniques are required, including enzymatic digestion, microwave-assisted extraction, and ultrasonic-assisted extraction, pulsed electric field-assisted extraction, ionic liquid-assisted extraction, pressurized liquid extraction, and supercritical extraction. Ultrasound-assisted extraction with ultrasonic waves 20 kHz and 10 MHz can exhibit high extraction efficiency, rapid kinetics, uniform distribution of energy, and low cost. But it also causes pigment compound degradation (Zheng et al., 2021). Extraction with microwave under electromagnetic waves 300 MHz to 300 GHz can show rapid extraction within 12 to 18 min, a low quantity of solvent with 15–40 mL, and a substantial sample size of up to 100 g. However, costly equipment and cleanup is necessary (Mala et al., 2021). Enzymatic extraction requires specific enzymes under ideal conditions. It can increase the yield of extraction and improve the quality of the extract. Pressurized liquid extraction uses a liquid solvent and is carried out at high pressure and temperature. It presented decreased solvent usage, shorter extraction times, and increased extraction rates. Pulsed electric field-assisted extraction is the application of short pulses and high electric fields with 10 to 80 kV/cm to provide low temperatures and high yield. The disadvantages are expensive initial expenditure, production of free radicals, potential bioactive compound destructure, and enzymatic degradation of recovered pigments (Singh et al., 2021).

Supercritical fluids, most often carbon dioxide, are used as solvents in the sophisticated and eco-friendly process known as supercritical fluid extraction. These fluids are safe and reusable in their supercritical condition, and they have special qualities that improve pigment extraction efficiency. For the extraction of non-polar pigments such as astaxanthin and carotenoids, this method is especially effective (Lee, 2023c). By using high pressure and temperature during operation, the supercritical fluid is able to dissolve the pigments in microbial cells. High selectivity, low solvent residues, and the capacity to extract heat-sensitive substances without degradation are its primary advantages. Ionic liquid-assisted extraction uses ionic solvents containing cations and anions. Its advantages are high reproducibility, reduced time of extraction, minimum solvent, low vapor pressure, incombustible, wide liquid range, and high thermal stability.

3.4 Purification

The process involves removing the desired colors from the fermentation broth, purifying them, and then separating them to eliminate any contaminants. Typically, there are several phases involved in these purification procedures, including solvent extraction, solid–liquid extraction, solid–liquid separation, and chromatography. Before beginning the purification process, cell disruption breaks down microbial cells in order to extract intracellular pigments. Both chemical (detergent or enzymatic lysis) and mechanical (bead milling, ultrasonication, or high-pressure homogenization) can be used. By applying the proper solvents, the pigments are extracted after cell rupture. The pigments' chemical structure determines the solvent selection. Methanol, ethanol, acetone, and ethyl acetate are usually used as solvents. Through dissolution into the solvent phase, the pigments are intended to be separated from other insoluble components and cell debris in this stage. The mixture is filtered or centrifuged to remove cell debris and other particulate substances after extraction. This allows for pigment concentration within the solvent phase and extract purification (Saleem et al., 2021). Each crude pigment extract is subsequently concentrated (typically via solvent evaporation or a rotary evaporator) to preserve just a small amount of the extracting container packed with as little ethanol as possible and lots of pigment. It is critical to concentrate the pigments and achieve a high extraction rate. Several chromatographic procedures, like high-performance liquid chromatography (HPLC) or thin-layer chromatography, are utilized for further purification. Chromatographic procedures for separating pigments based on their affinity to the stationary phase and eluting rates. HPLC is beneficial for reaching high purity since it offers higher resolution and specificity. In some cases, pigments can be further purified by crystallization and recrystallization. This involves dissolving the crude extract in a suitable solvent and allowing the pigments to crystallize out. Adsorption using adsorbents like activated carbon, resins or silica gels can also be used to purify pigments. The pigments are adsorbed onto the adsorbent and then desorbed using the right solvents to get the purified product. Purification is important to ensure the final pigment product is free from microbial contaminants, residual solvents, and other impurities that can affect its quality and safety (Lee et al., 2024). The choice of purification method depends on the pigment properties, required purity level, and application. Advances in purification techniques and the development of more efficient and eco-friendly methods will continue to improve the yield and quality of microbial pigments.

4 Biological Activities of Microbial Pigments

Various biological activities are determined in microbial pigments, such as antimicrobial, antioxidant, anti-aging, anti-inflammatory, photoprotective effects, anti-obesity, anti-diabetic, cardiovascular disease prevention, and anticancer activity. Therefore, their application can be found in the food industry, pharmaceutical business, and medical treatment. Table 2 shows the overview of microbial pigments in terms of their color, microbial sources, and biological activities.

Table 2 Overview of microbial pigments in their color, microbial sources and biological activities

Pigments	Color	Microorganisms	Biological Activities	References
Actinorhodin	Blue	• *Streptomyces coelicolor*	• Antimicrobial	Verma et al., (2018), Xu et al., (2019)
Astaxanthin	Red or orange	• *Xanthophyllomyces dendrorhous* • *Haematococcus pluvialis* • Disease prevention • *Pontibacter korlensis* AG6	• Anti-Inflammatory • Antioxidant • Anti-aging • Cardiovascular disease prevention • Anti-obesity	Pachaiyappan et al., (2021), Pereira et al., (2020)
Carotenoid	Yellow or organge	• *Arthrobacter agilis* 50cyt • *Zobellia laminarie* 465 • *Arthrobacter psychrochitiniphilus* 366 • *Rubritalea squalenifaciens* • *Staphylococcus gallinarum* KX912244 • *Paracoccus marcusii*	• Antioxidant • Photoprotection • Antimicrobial • Anticancer • Anti-Inflammatory • Antidiabetic effect	Naik and Gupte, (2023), Regueiras et al., (2021), Silva et al., (2019)
Chlorophyll	Green	• *Chlorella* spp. • *Tetraselmis* spp. • *Arthospira* spp. • *Chlorococcum amblystomatis*	• Anticancer • Antioxidant • Anti-obesity	Martins et al., (2023)
Lutein, zeaxanthin	Yellow and orange	• *Flavobacterium* spp. • *Mesoflavibacter* spp. • *Muricauda* spp. • *Sphingomonas* spp. • *Algibacter*	• Anti-Inflammatory • Antioxidant • Cardiovascular disease prevention	Lee and Demirci (2023), Lee and Küçükata (2023), Zafar et al., (2021)
Melanin	Red-brown or black	• *Sporisorium reilianum* • *Streptomyces puniceus* RHPR9 • *Aureobasidium pullulans* • *Amorphotheca resine*	• Antioxidant • Antiproliferative activity • Antibacterial • Anticancer • Photoprotection	Fu et al., (2022), Ma et al., (2023)
Monascin and ankaflavin	Yellow	• *Monascus purpureus*	• Anti-inflammatory • Antioxidant • Cardiovascular disease prevention • Alcoholic liver disease prevention	Li et al., (2024), Lin et al., (2023)
Phycobiliproteins: Phycocyanin, Allophycocyanin, and Phycoerythrin	Light blue-greenish	• *Arthrospira platensis* • *Acanthophora specifera* • *Bangia atropurea* • *Cyanidioschyzon merolae* • *Gracilaria corticata* • *Laurencia papillosa* • *Porphyra purpurea* • *Pyropia haitanensis* • *Portieria hornemannii* • *Spirulina platensis*	• Antioxidants • Antibacterial • Anti-inflammatory • Anti-diabetic • Anticancer	Ji et al.,(2023), Puzorjov and McCormick (2020)
Polyketide	Orange or Yellow	• *Monascus sp.* KCCM 10,093	• Anti-inflammatory	Choe, Song and et al., (2020a, 2020b)
Prodigiosin	Red	• *Serratia marcescens* • *Serratia rubidaea* RAM_Alex	• Antitumor • Antioxidant • Antimalarial • Antimicrobial	Kancharla et al., (2021), Lapenda et al., (2020)
Pyocyanin	Blue	• *Pseudomonas aeruginosa* • *Pseudomonas fluorescens*	• Antimicrobial • Antioxidant • Anticancer • Anti-inflammatory	Marey et al., (2024), Mohamed and Hamad (2020)

(continued)

Table 2 (continued)

Pigments	Color	Microorganisms	Biological Activities	References
Staphyloxanthin	Yellow-Gold	• *Staphylococcus gallinarum*	• Antimicrobial • Antioxidant • Anticancer	Barretto and Vootla (2018), Elmesseri et al., (2022)
Violacein	Violet	• *Chromobacterium violaceum* • *Janthinobacterium lividum* • *Pseudoalteromonas tunicata* D2	• Antiparasitic • Antiprotozoan • Antibacterial • Antifungal • Antioxidant • Anticancer	Al Fahad et al., (2024), Gıda et al., (2020)

4.1 Antimicrobial Activity

Microbial pigments are valuable for their coloration and many biological activities, including their significance for medicine, agriculture, food production, and other sectors. Among these functions, the antimicrobial activity of microbial pigments is prominent. Microbial pigments have strong antimicrobial characteristics, making them useful against various pathogens (Lee & Taştemir, 2023; Liorančas & Lee, 2024). These features are caused by a variety of methods that interfere with the regular functioning of microbial cells (Lee et al., 2015). Many microbial pigments may integrate with pathogen cell membranes, producing structural changes that result in cellular leakage and death. For example, prodigiosin, a red pigment generated by *Serratia marcescens*, penetrates itself into cell membranes, enhancing their permeability and leading to cell lysis Prodigiosin indicated an excellent therapeutic index in antimalarial treatment by demonstrating strong asexual blood-stage antiplasmodial behavior against *Plasmodium falciparum* parasites at low nanomolar doses. Prodiginines also presented significant effects against erythrocytic *Plasmodium yoelii* in a mouse model when administered orally at a dose of 25 mg/kg for four days. The production of microbial cell wall components can be inhibited by some pigments. This process eventually results in cell rupture by degrading the cell wall and making it more vulnerable to osmotic pressure. Actinorhodin is an excellent pH-responsive antibiotic that triggers cells with at least three stress responses, such as those that control oxidative injury, protein degradation, and DNA damage. Actinorhodin, a blue pigment produced *by Streptomyces coelicolor*, has been demonstrated to prevent cell wall formation in Gram-positive bacteria (Gheni & Hasan, 2022). Certain pigments may bond to nucleic acids or inhibit enzymes responsible for DNA and RNA production. The disruption might restrict cell multiplication and transcription, causing apoptosis. The violet pigment violacein from *Chromobacterium violaceum* may integrate into DNA, interrupting replication and transcription. Microbial pigments may likewise suppress the protein synthesis of microorganisms. These pigments might prevent protein formation by attaching to ribosomes or suppressing necessary translation factors, both of which are required for cell growth and survival. *Clostridium beijerinckii* produce clostrubin A to inhibit protein synthesis in specific bacteria. Pigments such as pyocyanin cause oxidative stress by producing ROS, which damages cellular components and contributes to cell death. *Streptomyces tunisiensis* W4MT573222 pigment exhibited 78% suppression of hepatic virus in addition to its antibacterial and antifungal properties. Significant biological characteristics of the staphyloxanthin pigment from *Staphylococcus gallinarum* KX912244 were demonstrated in vitro antibacterial activity against the pathogens *Candida albicans*, *Escherichia coli*, and *Staphylococcus aureus* (Barretto & Vootla, 2018). Significant antibacterial activity of the red pigment from the fungus *Talaromyces assiutensis* was shown against the bacterial pathogens *Vibrio cholerae*, methicillin-resistant *Staphylococcus aureus*, and wild-type *Staphylococcus aureus* (Mishra et al., 2021). The pyocyanin extract from *Pseudomonas aeruginosa* revealed antibacterial activity against isolates of *Escherichia coli* and *Bacillus* spp., *Staphylococcus* spp., and *Streptococcus* spp. (Marey et al., 2024). Figure 9 exhibits different chemical structures of violacein, melanin, phycoerythrobilin, actinorhodin, pyocyanin, staphyloxanthin.

4.2 Antioxidant and Anti-Aging

High antioxidant ability can be found in many microbial pigments for foods, medicines, and cosmetics. These attributes mostly come from the capacity to eliminate free radicals to protect cells from oxidative stress. Microbial pigments may offer electrons to free radicals, neutralizing them and keeping them from causing cell

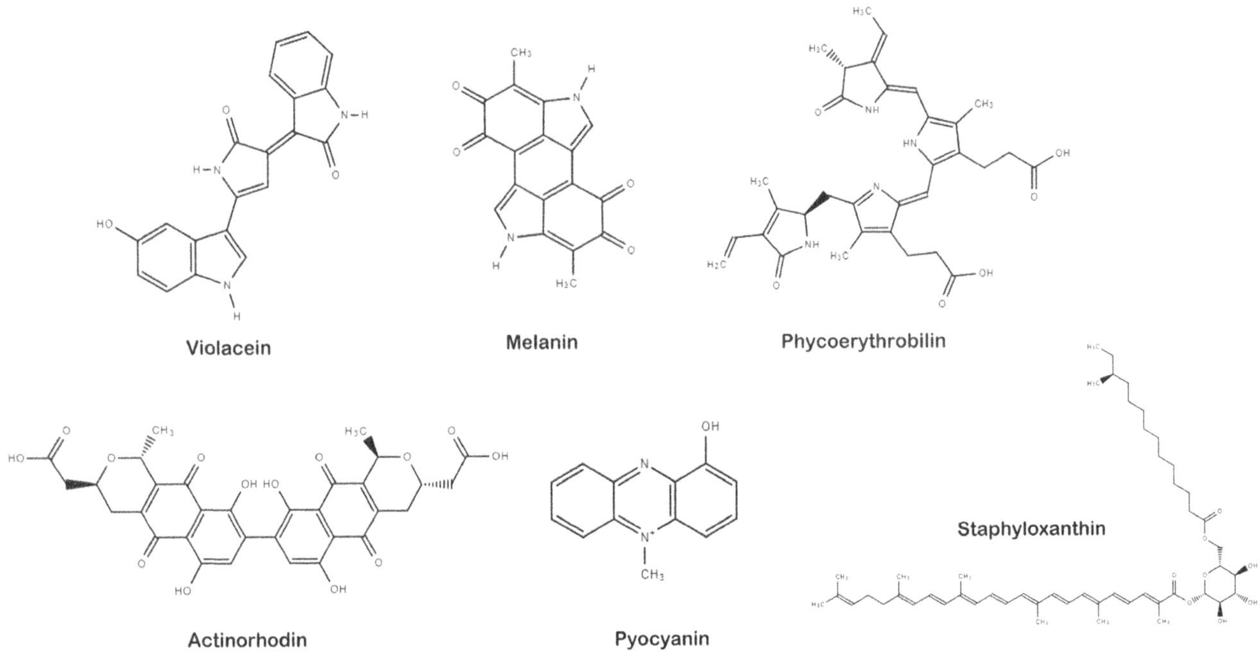

Fig. 9 Chemical structures of violacein, melanin, phycoerythrobilin, actinorhodin, pyocyanin, staphyloxanthin

destruction. Astaxanthin and β-carotene from microalgae and bacteria are recognized for their potent free radical scavenging effects. ROS levels within cells can be reduced by pigments. ROS are very reactive compounds that can negatively impact DNA and cell structure. Pigments contribute to the preservation of cellular integrity and function by lowering ROS concentrations. Some microbial pigments could bind to metal ions like iron and copper, which function as catalysts in the Fenton reactions that generate free radicals. Pigments prevent the production of free radicals by chelating these metals. Melanin is produced from fungi with potent metal-chelating ability. The antioxidant effects of carotenoids have been widely reported for decades. These pigments, which include substances like astaxanthin, lycopene, and β-carotene, are produced by various microorganisms. The microalga *Haematococcus pluvialis* produces astaxanthin, one of the most significant naturally occurring antioxidants that protect cells from oxidative damage and lipid peroxidation (Yao et al., 2023).

Melanin pigments exhibit antioxidant properties. Their ability to scavenge free radicals and chelate metal ions prevents cells from oxidative stress (Ansari et al., 2023). Melanin may be used for a number of purposes, including as skin protection and as an additive in food and cosmetics because of its antioxidant activity. Yellow-orange pigment carotenoids are also widely reported for their high antioxidant activity. The marine sponge *Halicaondria kadai* was the source of a unique carotenoid-producing bacteria later recognized as *Rubritalea squalenifaciens*. Squalene and novel acyclic C_{30} carotenoids, diapolycopenedioc acids, and xylosylesters A–C with potent antioxidant properties were produced by *Rubritalea squalenifaciens* to defend themselves against sunlight-produced activated oxygen. The antioxidant potential of the red pigment from the fungus *Talaromyces assiutensis* was also reported as an antioxidant assay against free radicals. The anti-inflammatory and antioxidant effects of astaxanthin in cardiovascular illnesses, including atherosclerosis were reported. The benefits and safety of astaxanthin might be implemented as supplementation to prevent the development of cardiovascular illnesses. Phycocyanin and allophycocyanin from *Arthrospira platensis* revealed heterologous fluorescence and antioxidant activity. The high antioxidant activity in vitro of staphyloxanthin pigment from *Staphylococcus gallinarum* KX912244 was measured by 2,2-diphenyl-1-picrylhydrazyl (DPPH) free radical scavenging activity with an IC_{50} value of 54.22 μg/mL (Barretto & Vootla, 2018).

4.3 Photoprotection

Because melanin has strong UV radiation protection, it may be used as a component in cosmetics like sunscreen. Many species of bacteria and fungi have been identified and investigated for their capacity to produce melanin. Melanin to shield microbial cells from UV radiation, may be used in cosmetics by extracting it and using it in a formulation that aims to be photoprotective while still being safe for human usage. The sunscreens were introduced by monascin and rubropunctamine, both of which are produced from the fungus *Monascus purpureus*, to increase their photoprotective activity by 13–36.5%, respectively (Bayram et al., 2023). Melanin produced from *Amorphotheca resine* provided a high antioxidant activity to present high sunscreen performance. Human keratinocytes would not be cytotoxically affected by the pigment. The yeast *Aureobasidium pullulans* can survive under stress due to its melanin content. The microbial melanin demonstrated great potential as a γ-ray shielding substance (El-Bialy et al., 2019). From obtained soil and water samples in the different locations, 19 pigment-producing bacteria were identified and purified. The 17 isolates' crude pigments demonstrated sunscreen potency ranging from 0.4 to 8.34. Only six of them, nevertheless, displayed evident antibacterial and antioxidant effects. Among the six possible pigments, the promising pigment is produced by one isolate. Using 16srRNA gene sequencing, the strain was identified as *Staphylococcus xylosus* MK770403 (Choksi et al., 2020). It is widely recognized that the melanins found in fungi provide protection against Ultraviolet and solar radiation and prevent the production of enzymes by other microbes that break down cell walls. Ultraviolet (UV)- B and UV-C radiation were not able to cause any damage to three Antarctic carotenoid-producing bacteria–*Arthrobacter agilis* 50cyt, *Zobellia laminarie* 465, and *Arthrobacter psychrochitiniphilus* 366. Zeaxanthin, β-cryptoxanthin, and β-carotene were found in *Zobellia laminarie* 465. Yellow pigmentation provides protection for *Thermus* strains found in naturally thermal locations that are exposed to direct sunlight (Silva et al., 2019).

4.4 Anticancer Activity

Chlorophylls produced from several microalgae, including *Chlorella* spp., *Tetraselmis* spp., and *Arthospira* spp., are frequently employed in the food sector as green colorants and also as anticancer therapy. Chlorophyllin is a semisynthetic form of chlorophyll that is produced by adding copper and sodium salt. Chlorophyllin-assisted photodynamic therapy indicated there is an anticancer effect on human cervical cancer by reducing apoptosis via oxidative stress. Green vegetable consumption may be associated with a lower incidence of cancer because of changes in the redox state of pancreatic cancer cells that are mediated by chlorophyll, which may have antiproliferative and anticancer effects. The outstanding antioxidant and anticancer properties of green tea and olive leaf were reported, and chlorophylls are crucial for eliminating tumors from plant-derived extracts. A yellowish pigment from the fungus *Monascus purpureus* H14 was investigated for use in a functional rice noodle because of its anti-inflammatory, anti-tumor, and anti-fatty activity; it gave the product high stability and quality (Gong et al., 2023). *Streptomyces tunisiensis* W4MT573222 pigment was safe for normal cells and demonstratednticancer effects against liver, lung, and pancreas cancer cell lines (Ibrahim et al., 2023). Staphyloxanthin exhibits potential anticancer therapeutic effects against 4 different cancer cell lines such as Dalton's lymphoma ascites, Ehrlich-Lettre ascites carcinoma, adenocarcinoma human alveolar basal epithelial cells A549 Lung carcinoma, and Mus mucus skin melanoma. A concentration-dependent anticancer effect of the red pigment from the fungus *Talaromyces assiutensis* was also observed on the HeLa cancer line (Mishra et al., 2021). The pyocyanin extract from Pseudomonas aeruginosa demonstrated significant anticancer activity against the cancer cell lines A549, MDA-MB-231, and Caco-2, respectively. The anticancer effect of phycocyanin from *Spirulina platensis* was found to be 68% at the concentration of 500 μg/ml and less potent at 100 μg/ml against Hep-G2 cell lines (Prabakaran et al., 2020).

4.5 Anti-inflammatory

Microbial pigments can provide potential anti-inflammatory to promote human health. These microorganisms and their metabolites have been shown to have a wide range of biological effects, such as the treatment of degenerative diseases and antihypertensive, antispasmodic, hepatoprotective, and anti-inflammatory. For instance, inflammation is one of the processes behind neurodegenerative disorders, and the central nervous system has received the potential positive effects of anti-inflammatory agents found in microbial pigments. One of the significant biological features displayed by many metabolites from *cyanobacteria* and microalgae, such as *Chlorella*, *Dunaliella*, and *Phaeodactylum*, is anti-inflammatory properties. These metabolites in their chemical compositions, are categorized as carotenoids. The anti-inflammatory and antioxidant functions of astaxanthin have been associated with a potential defense against stomach inflammation brought on by *Helicobacter Pylori*. The *Helicobacter Pylori*-induced stimulation of T helper cell type 1 response to T helper cell type 2 is modified in infected tissues by astaxanthin. Without lowering cytokine concentration, astaxanthin alleviates stomach inflammation and inhibits the development of *Helicobacter Pylor*. The activity of cytokines released by the *Helicobacter Pylori* T-lymphocyte was used to determine the impact of the antioxidant astaxanthin on mice infected with the bacteria. In mice infected with *Helicobacter pylori*, the antioxidant astaxanthin was shown to reduce the number of bacteria and gastrointestinal inflammation (Han et al., 2020).

The blue pigment pyocyanin from *Pseudomonas aeruginosa* suppresses the levels of nitric oxide, TNF-α, and IL-1β in mouse peritoneal macrophages through anti-inflammatory actions without affecting macrophage degradation. Intracellular ROS/reactive nitrogen species (RNS) caused by phenazine pyocyanin hampered the NLRP3 inflammasome in post-translational modification. This activation could be prevented exclusively by hydrogen peroxide or peroxynitrite. Because phenazine pyocyanin exhibited anti-inflammatory effects, it might support *Pseudomonas aeruginosa* in avoiding immunological identification (Virreira Winter & Zychlinsky, 2018). The 2-amino-4 picoline derivative of *Monascus* pigment showed 90.6% cell viability and 48.4% inhibition of NO generation. The derivative inhibited the expression of inducible NO synthase, a crucial enzyme in the NO production pathway. Consequently, this derivative displayed promise as an anti-inflammatory agent. Monascin and ankaflavin produced from *Monascus purpureu* are two yellow pigments to display anti-inflammatory impacts by lowering inflammatory factor concentrations like NF-κB (Lin et al., 2023) Maximum levels of the anti-inflammatory effects of phytocyanin from *Spirulina platensis* at 500 μg/ml have been observed, including inhibition of albumin denaturation, antiproteinase, hypotonicity-induced hemolysis, and anti-lipoxygenase effects (Prabakaran et al., 2020).

4.6 Cardiovascular Disease Prevention

People with high cholesterol levels associated with low-density lipoprotein (LDL), sometimes known as "bad cholesterol," are involved in developing arteriosclerosis, cardiovascular disease, peripheral artery disease, myocardial infarction, and cerebral thrombosis. Astaxanthin can enhance the levels of adiponectin and high-density lipoprotein, as well as decrease LDL cholesterol in the circulatory system of individuals. After administering astaxanthin supplements for three months, patients with heart failure and left ventricular systolic dysfunction showed reduced oxidative stress, cardiac contractility, and exercise tolerance. The protective effect of astaxanthin against cardiac damage and fibrosis in rats induced by a high-fat diet was determined by upregulating and activating sirtuin 1, a member of the histone deacetylase III family, as part of the therapeutic mechanism (Shatoor & Al Humayed, 2021). Supplementing with lutein, zeaxanthin, and meso-zeaxanthin can potentially reduce inflammatory signals and lipid oxidation in the cardiovascular system by decreasing serum IL-1β, TNF-α, and oxidative LDL. Two yellow pigments, monascin and ankaflavin, produced from *Monascus purpureus* fermented rice, are effective in preventing fatty liver, lipogenesis, and cardiovascular illnesses. Monascin and ankaflavin displayed the presentation of alcoholic liver disease by regulating the expression of sterol regulatory element-binding transcription factor-1 and peroxisome proliferator-activated receptor-γ in HepG2 cells when exposed to ethanol Peripheral artery disease, heart disease, strokes, and other severe illnesses are associated with hyperlipidemia which is atherosclerosis with the hardening of the arteries. The six pigments from monascus-fermented rice—monascin, ankaflavin, rubropunctatin, monascorubrin, rubropunctamine, and monascorubramine—can be developed as potential treatments for the control of hyperlipidemia (Sun et al., 2020).

4.7 Anti-obesity

The anti-obesity capacity and metabolism of monascin and ankaflavin were observed by using network pharmacology and computer modeling. The five major potential therapeutic obesity targets (MP9, BCL2, PPARG, ESR1, and STAT3) were identified using microarray data validation and protein–protein interaction analysis. Monascin and ankaflavin can be a promising anti-obesity agent for functional additives. The carotenoids and chlorophyll pigments from the microalgae *Chlorella vulgaris* and *Chlorococcum amblystomatis* showed anti-steatosis, anti-obesity, and anti-inflammatory effects to prove the promising in the nutraceuticals. Specifically, the ethanolic extract of the heterotrophic *C. vulgaris* facilitated by ultrasound significantly improved the anti-obesity activity (Regueiras et al., 2021). Three pigments, lycopene, amaranth, and sorghum red have the potential to mitigate the effects of a high-fat diet on obesity and regulate the gut microbiota in C57BL/6 J mice. Chlorophyll a and its derivatives from cyanobacterium *Arthrospira platensis* exhibited the enhancement of lipid reduction and appetite decrease in zebrafish larvae. It presented that chlorophyll could be potential in anti-obesity treatment. Astaxanthin produced by microalgae such as *Haematococcus pluvialisis* is a potential alternative to weight control supplements because it lowers inflammation, improves lipid metabolism, and suppresses adipogenesis (Lee, 2023a). Derivatives of fermented pigment from monascus have been shown to have high anti-obesity activity in C57BL/6 mice fed a high-fat diet. They are significantly effective in reducing body weight, epididymal adipose tissue weights, and fat accumulation (Choe et al., 2020a, 2020b).

4.8 Antidiabetic Effect

The postprandial blood glucose levels of Type II diabetes patients can be controlled by inhibiting pancreatic α-amylase and α-glucosidase in the human digestive system. This strategy can reduce the degradation and utilization of carbohydrates. Therefore, using acarbose as a positive control, anti-α glucosidase and anti-α amylase assays were used to assess antidiabetic effects. Bacterial carotenoids from the *Paracoccus marcusii* are identified and purified as η-carotene, β-zeacarotene, astaxanthin, and zeaxanthin. The IC_{50} of inhibition against α-amylase and α-glucosidase were 226 μg/ml and 0.7548 μg/ml, respectively. The results indicated that these four carotenoid compounds might be potential treatments for diabetes patients (Naik & Gupte, 2023). The potential of phycocyanin as an alternative medication with antidiabetic properties. According to an in-vitro inhibitory activity test, phycocyanin suppresses human salivary amylase by 51.13% on average. Tests on the durability of phycocyanin storage revealed that, when utilized as functional chemicals, bioactivity may be preserved by maintaining PC in a solid form, avoiding light, and maintaining a low temperature. The suppression of phycocyanin from Spirulina platensis to the enzymes α-amylase and β-glucosidase was observed to be dosage-dependent, with the highest inhibitory activity detected at 250 μg/ml. It showed that phycocyanin might be a promising antidiabetic agent (Prabakaran et al., 2020).

5 Challenges in Microbial Pigment Fermentative Production

Natural pigments can originate from a wide range of microorganisms, but it might be challenging to commercialize them in the food, cosmetics, textile, and pharmaceutical industries. Designing new substances for biological applications is a very challenging regulatory undertaking. In confectionery products, natural colors can cost as much as five to twenty times more than synthetic colorants. During fermentative production, the isolation of high-yield strains and genetic stability are of concern. Finding microbial strains that naturally generate large amounts of the desired color might take a lot of time and effort. It also might be difficult to guarantee that the selected or modified strains will continue to be able to produce pigment over the course of several generations. Besides, specific nutrients are needed by various microorganisms to survive and produce pigments optimally. It's critical to recognize and maximize these nutritional requirements. Establishing scalable, economically viable media compositions for high pigment yields without being unnecessarily costly. Besides, monitoring the fermentative condition required a lot of attention, such as pH, temperature, and humidity during cultivation. The large-scale bioreactors under aerobic fermentations need to offer an appropriate oxygen supply. While scaling up from laboratory to pilot and industrial scale, reproducibility refers to achieving consistent pigment output. The design of bioreactors capable of producing high-yield fermentative production ensures that they are designed with homogeneous mixing, sufficient oxygen transmission, and good heat removal (Lee & Taştemir, 2023). During extraction and purification, developing efficient and cost-effective methods from microbial biomass is also essential. Complicated regulatory approval procedures for novel microbial pigments are sometimes challenging to manage. Because current regulation is frequently based on local and traditional customs, there are regulatory concerns related to the use of microbial pigments in foods, nutraceuticals, and cosmetics (Lee & Neio Demirci, 2023). The primary obstacles to commercializing either native or non-native microbial pigments are regulatory obstacles and related consumer preferences, even if native pigment-producing microorganisms are more accepted by consumers and authorities. Furthermore, the distribution of such natural compounds with health advantages requires a sufficient toxicity examination for prospective bioactive microbial pigments. After toxicity assessment, the microbial pigments need to be safe and non-toxic in the human body. Moreover, waste management of fermentative production involves the ecologically sustainable handling of waste produced during downstream processes.

6 Conclusion and Future Prospects

Synthetic pigmentation has significantly harmed both human health and the environment. Microbial pigments derived from bacteria, fungi, yeasts, and microalgae are more valuable and in demand than artificial colorants due to their environmental-friendly attributes and health benefits. By finding alternative microbial sources, minimizing production costs through genetic engineering, strain development, and optimization of the fermentative process, the production of microbial pigment can be promising in large-scale industrial manufacture. A renewable and eco-friendly method of reducing pollution and decreasing costs in the development of assets with added value is utilizing agro-industrial waste as a feedstock for the synthesis of microbial pigments. Microbial pigments meet the increasing need for natural colorants while minimizing carbon burden. Moreover, most of these pigments produced from plant-based waste are vegan, eliminating the need for colors derived from animal products. Because numerous aquatic resources have been discovered to contain pigment-generating microbes and microalgae, pigmentation derived from marine sources has drawn much attention to discovering new and undiscovered natural pigments. It is strongly encouraged to expand the large-scale exploration into new habitats associated with terrestrial and particularly marine environments to uncover novel compounds from intriguing microorganisms. Microbial bioactive pigments have a distinctly interdisciplinary quality, bridging the fields of biotechnology, food science, and medicinal sciences to produce bioactive microbial pigments with possible beneficial health effects as well as colorant and nutraceutical properties. Numerous microbial pigments have been reported to have potential for various applications in bio-indicators, antioxidants, antimicrobials, and anticancers. It is crucial to obtain information on the physiological and biological properties of novel species and previously unexplored microorganisms in order to develop new medicinal compounds for biotechnological and medical applications.

Acknowledgements The author would like to appreciate Istanbul Sabahattin Zaim University, Department of Food Engineering.

References

Afroz Toma, M., Rahman, Md. H., Rahman, Md. S., Arif, M., Nazir, K. H. M. N. H., & Dufossé, L. (2023). Fungal pigments: carotenoids, riboflavin, and polyketides with diverse applications. *Journal of Fungi, 9*(4), 454. https://doi.org/10.3390/jof9040454

Agboyibor, C., Kong, W.-B., Chen, D., Zhang, A.-M., & Niu, S.-Q. (2018). Monascus pigments production, composition, bioactivity and its application: A review. *Biocatalysis and Agricultural Biotechnology, 16*, 433–447. https://doi.org/10.1016/j.bcab.2018.09.012

Al Fahad, M. A., Ahamed, R., Ahmed, T., Jahan, N., Mia, R., Toki, G. F. I., Mahmud, S. T., & Niloy, K. K. (2024). Pharmaceutical applications of natural dyes and pigments. *Renewable Dyes and Pigments*, 165–175. https://doi.org/10.1016/B978-0-443-15213-9.00011-9

Ansari, F., Lee, C.-C., Rashidimehr, A., Eskandari, S., Ashaolu, T. J., Mirzakhani, E., Pourjafar, H., & Jafari, S. M. (2024). The role of probiotics in improving food safety; detoxification of heavy metals and chemicals. *Toxin Reviews, 43*(1), 63–91. https://doi.org/10.1080/15569543.2023.2283768

Ansari, F., Lee, C.-C., Rashidimehr, A., Eskandari, S., Joshua Ashaolu, T., Mirzakhani, E., Pourjafar, H., & Jafari, S. M. (2023). The role of probiotics in ımproving food safety: Inactivation of pathogens and biological toxins. *Current Pharmaceutical Biotechnology, 24*. https://doi.org/10.2174/1389201024666230601141627

Ashaolu, T. J., Lee, C. C., Ashaolu, J. O., Tarhan, O., Pourjafar, H., & Jafari, S. M. (2023). Pepsin: An excellent proteolytic enzyme for the production of bioactive peptides. *Food Reviews International*, 1–38. https://doi.org/10.1080/87559129.2023.2238814

Barretto, D. A., & Vootla, S. K. (2018). In vitro anticancer activity of staphyloxanthin pigment extracted from staphylococcus gallinarum KX912244, a Gut Microbe of Bombyx mori. *Indian Journal of Microbiology, 58*(2), 146–158. https://doi.org/10.1007/s12088-018-0718-0

Bayram, S., Aygün, B., Karadayi, M., Alaylar, B., Güllüce, M., & Karabulut, A. (2023). Determination of toxicity and radioprotective properties of bacterial and fungal eumelanin pigments. *International Journal of Radiation Biology, 99*(11), 1785–1793. https://doi.org/10.1080/09553002.2023.2204957

Cano-Lamadrid, M., Martínez-Zamora, L., Mozafari, L., Bueso, M. C., Kessler, M., & Artés-Hernández, F. (2023). Response surface methodology to optimize the extraction of carotenoids from horticultural by-products—A systematic review. *Foods, 12*(24), 4456. https://doi.org/10.3390/foods12244456

Chaudhary, V., Katyal, P., Poonia, A. K., Kaur, J., Puniya, A. K., & Panwar, H. (2022). Natural pigment from Monascus: The production and therapeutic significance. *Journal of Applied Microbiology, 133*(1), 18–38. https://doi.org/10.1111/jam.15308

Chen, X., Yan, J., Chen, J., Gui, R., Wu, Y., & Li, N. (2021). Potato pomace: An efficient resource for Monascus pigments production through solid-state fermentation. *Journal of Bioscience and Bioengineering, 132*(2), 167–173. https://doi.org/10.1016/J.JBIOSC.2021.03.007

Choe, D., Jung, H. H., Kim, D., Shin, C. S., Johnston, T. V., & Ku, S. (2020a). In vivo evaluation of the anti-obesity effects of combinations of Monascus pigment derivatives. *RSC Advances, 10*(3), 1456–1462. https://doi.org/10.1039/C9RA08036H

Choe, D., Song, S. M., Shin, C. S., Johnston, T. V., Ahn, H. J., Kim, D., & Ku, S. (2020b). Production and characterization of anti-inflammatory monascus pigment derivatives. *Foods, 9*(7), 858. https://doi.org/10.3390/foods9070858

Choksi, J., Vora, J., & Shrivastava, N. (2020). Bioactive pigments from isolated bacteria and its antibacterial, antioxidant and sun protective application useful for cosmetic products. *Indian Journal of Microbiology, 60*(3), 379–382. https://doi.org/10.1007/S12088-020-00870-X/FIGURES/1

Darwesh, O. M., Matter, I. A., Almoallim, H. S., Alharbi, S. A., & Oh, Y.-K. (2020). Isolation and optimization of monascus ruber OMNRC45 for red pigment production and evaluation of the pigment as a food colorant. *Applied Sciences, 10*(24), 8867. https://doi.org/10.3390/app10248867

Demirci, M., Lee, C. C., Çavuş, M., & Çağlar, M. Y. (2020). Oleogels for food applications. *Biopolymer-Based Formulations: Biomedical and Food Applications*, 781–811. https://doi.org/10.1016/B978-0-12-816897-4.00031-X

Dev, M. J., Warke, R. G., Warke, G. M., Mahajan, G. B., Patil, T. A., & Singhal, R. S. (2022). Advances in fermentative production, purification, characterization and applications of gellan gum. *Bioresource Technology, 359*, 127498. https://doi.org/10.1016/J.BIORTECH.2022.127498

Di Nardo, G., & Gilardi, G. (2020). Natural compounds as pharmaceuticals: The key role of cytochromes P450 reactivity. *Trends in Biochemical Sciences, 45*(6), 511–525. https://doi.org/10.1016/J.TIBS.2020.03.004

Dufossé, L. (2022). *Back to nature, microbial production of pigments and colorants for food use* (pp. 93–122). https://doi.org/10.1016/bs.afnr.2022.04.008

El-Bialy, H. A., El-Gamal, M. S., Elsayed, M. A., Saudi, H. A., & Khalifa, M. A. (2019). Microbial melanin physiology under stress conditions and gamma radiation protection studies. *Radiation Physics and Chemistry, 162*, 178–186. https://doi.org/10.1016/j.radphyschem.2019.05.002

Elmesseri, R. A., Saleh, S. E., Elsherif, H. M., Yahia, I. S., & Aboshanab, K. M. (2022). Staphyloxanthin as a potential novel target for deciphering promising anti-staphylococcus aureus agents. *Antibiotics, 11*(3), 298. https://doi.org/10.3390/antibiotics11030298

El-Naggar, N. E. A., & Saber, W. I. A. (2022). Natural melanin: Current trends, and future approaches, with especial reference to microbial source. *Polymers, 14*(7), 1339. https://doi.org/10.3390/POLYM14071339

Esquivel, P. (2024). Betalains. In *Handbook on natural pigments in food and beverages* (pp. 147–167). Elsevier. https://doi.org/10.1016/B978-0-323-99608-2.00002-1

Fernandes, R., Campos, J., Serra, M., Fidalgo, J., Almeida, H., Casas, A., Toubarro, D., & Barros, A. I. R. N. A. (2023). Exploring the benefits of phycocyanin: from spirulina cultivation to its widespread applications. *Pharmaceuticals, 16*(4), 592. https://doi.org/10.3390/ph16040592

Finger, M., Sentek, F., Hartmann, L., Palacio-Barrera, A. M., Schlembach, I., Rosenbaum, M. A., & Büchs, J. (2023). Insights into *Streptomyces coelicolor* A3(2) growth and pigment formation with high-throughput online monitoring. *Engineering in Life Sciences, 23*(1). https://doi.org/10.1002/elsc.202100151

Fu, X., Xie, M., Lu, M., Shi, L., Shi, T., & Yu, M. (2022). Characterization of the physicochemical properties, antioxidant activity, and antiproliferative activity of natural melanin from S. reiliana. *Scientific Reports, 12*(1), 2110. https://doi.org/10.1038/s41598-022-05676-z

Gheni, A. I., & Hasan, A. H. (2022). Antibacterial screening and analysis of streptomyces coelicolor secondary metabolites. *Journal of Pharmaceutical Research International*, 26–39. https://doi.org/10.9734/jpri/2022/v34i7B35469

Gıda, A., Deniz Periz, Ç., Ulusoy, S., Tınaz, G., & Şekerler, T. (2020). Antibacterial and anticancer activities of violacein extracted through ultrasound-assisted extraction method. *Akademik Gıda, 18*(3), 241–246. https://doi.org/10.24323/akademik-gida.XXXXXX

Gong, Z., Jiao, P., Huang, F., Zhang, S., Zhou, B., Lin, Q., Liu, J., & Liang, Y. (2023). Separation and antioxidant activity of the water-soluble yellow monascus pigment and its application in the preparation of functional rice noodles. *LWT, 185*, 115172. https://doi.org/10.1016/j.lwt.2023.115172

Guern, F. Le, Ouk, T. S., Yerzhan, I., Nurlykyz, Y., Arnoux, P., Frochot, C., Leroy-Lhez, S., & Sol, V. (2021). Photophysical and bactericidal properties of pyridinium and ımidazolium porphyrins for photodynamic antimicrobial chemotherapy. *Molecules, 26*(4), 1122. https://doi.org/10.3390/MOLECULES26041122

Guterres, K. B., Rossi, G. G., de Campos, M. M. ko A., Moreira, K. S., Burgo, T. A. L., & Iglesias, B. A. (2020). Metal center ion effects on photoinactivating rapidly growing mycobacteria using water-soluble tetra-cationic porphyrins. *BioMetals, 33*(4–5), 269–282. https://doi.org/10.1007/S10534-020-00251-3/FIGURES/5

Han, H., Lim, J. W., & Kim, H. (2020). Astaxanthin inhibits helicobacter pylori-induced inflammatory and oncogenic responses in gastric mucosal tissues of mice. *Journal of Cancer Prevention, 25*(4), 244. https://doi.org/10.15430/JCP.2020.25.4.244

He, J., Jia, M., Li, W., Deng, J., Ren, J., Luo, F., Bai, J., & Liu, J. (2022). Toward improvements for enhancement the productivity and color value of *Monascus* pigments: A critical *review* with recent updates. *Critical Reviews in Food Science and Nutrition, 62*(26), 7139–7153. https://doi.org/10.1080/10408398.2021.1935443

Hernández-Velasco, P., Morales-Atilano, I., Rodríguez-Delgado, M., Rodríguez-Delgado, J. M., Luna-Moreno, D., Ávalos-Alanís, F. G., & Villarreal-Chiu, J. F. (2020). Photoelectric evaluation of dye-sensitized solar cells based on prodigiosin pigment derived from Serratia marcescens 11E. *Dyes and Pigments, 177*, 108278. https://doi.org/10.1016/J.DYEPIG.2020.108278

Ibrahim, W. M., Olama, Z. A., Abou-elela, G. M., Ramadan, H. S., Hegazy, G. E., & El Badan, D. E. S. (2023). Exploring the antimicrobial, antiviral, antioxidant, and antitumor potentials of marine Streptomyces tunisiensis W4MT573222 pigment isolated from Abu-Qir sediments, Egypt. *Microbial Cell Factories, 22*(1), 94. https://doi.org/10.1186/s12934-023-02106-1

Ji, L., Qiu, S., Wang, Z., Zhao, C., Tang, B., Gao, Z., & Fan, J. (2023). Phycobiliproteins from algae: Current updates in sustainable production and applications in food and health. *Food Research International, 167*, 112737. https://doi.org/10.1016/j.foodres.2023.112737

Joshi, K., Kumar, P., & Kataria, R. (2023). Microbial carotenoid production and their potential applications as antioxidants: A current update. *Process Biochemistry, 128*, 190–205. https://doi.org/10.1016/J.PROCBIO.2023.02.020

Kancharla, P., Li, Y., Yeluguri, M., Dodean, R. A., Reynolds, K. A., & Kelly, J. X. (2021). Total synthesis and antimalarial activity of 2-(p-Hydroxybenzyl)-prodigiosins, Isoheptylprodigiosin, and geometric isomers of tambjamine MYP1 isolated from marine bacteria. *Journal of Medicinal Chemistry, 64*(12), 8739–8754. https://doi.org/10.1021/acs.jmedchem.1c00748

Kannaujiya, V. K., Kumar, D., Singh, V., & Sinha, R. P. (2021). Advances in phycobiliproteins research: innovations and commercialization. In *Natural bioactive compounds* (pp. 57–81). Elsevier. https://doi.org/10.1016/B978-0-12-820655-3.00004-5

Khan, N. T. (2023). Carotenoids from *Blakeslea trispora*. *International Journal of Food Science and Agriculture, 7*(1), 29–32. https://doi.org/10.26855/ijfsa.2023.03.005

Lagashetti, A. C., Dufossé, L., Singh, S. K., & Singh, P. N. (2019). Fungal pigments and their prospects in different industries. *Microorganisms, 7*(12). https://doi.org/10.3390/MICROORGANISMS7120604

Lapenda, J. C. L., Alves, V. P., Adam, M. L., Rodrigues, M. D., & Nascimento, S. C. (2020). Cytotoxic effect of prodigiosin, natural red pigment, isolated from serratia marcescens UFPEDA 398. *Indian Journal of Microbiology, 60*(2), 182–195. https://doi.org/10.1007/S12088-020-00859-6/FIGURES/10

Lee, C. C., Chen, J., & Frank, J. F. (2015). Role of cellulose and colanic acid in attachment of shiga toxin-producing escherichia coli to lettuce and spinach in different water hardness environments. *Journal of Food Protection, 78*(8), 1461–1466. https://doi.org/10.4315/0362-028X.JFP-15-056

Lee, C. C., Chen, J., & Frank, J. F. (2016). Influence of extracellular cellulose and colanic acid production on the survival of shiga toxin-producing escherichia coli on spinach and lettuce after chlorine treatment. *Journal of Food Protection, 79*(4), 666–671. https://doi.org/10.4315/0362-028X.JFP-15-375

Lee, C. C., Suttikhana, I., & Ashaolu, T. J. (2024). Techno-functions and safety concerns of plant-based peptides in food matrices. *Journal of Agricultural and Food Chemistry*. https://doi.org/10.1021/ACS.JAFC.4C02464/ASSET/IMAGES/MEDIUM/JF4C02464_0002.GIF

Lee, C.-C., & Demirci, M. (2023). Zeaxanthin. In *Handbook of food bioactive ıngredients* (pp. 1–34). Springer International Publishing. https://doi.org/10.1007/978-3-030-81404-5_53-1

Lee, C.-C., & Küçükata, Y. Ş. (2023). Microbial production of zeaxanthin. In *Microbial production of food bioactive compounds* (pp. 1–38). Springer International Publishing. https://doi.org/10.1007/978-3-030-81403-8_36-1

Lee, C.-C., & Neio Demirci, M. (2023). Different types of fryers for the food industry. In *High-temperature processing of food products* (pp. 293–326). Elsevier. https://doi.org/10.1016/B978-0-12-818618-3.00003-3

Lee, C.-C., & Taştemir, İ. A. (2023). Direct heat exchangers in the food industry. In *Thermal processing of food products by steam and hot water* (pp. 181–208). Elsevier. https://doi.org/10.1016/B978-0-12-818616-9.00008-0

Lee, C.-C. (2023a). Astaxanthin. In *Handbook of food bioactive ıngredients* (pp. 687–727). Springer International Publishing. https://doi.org/10.1007/978-3-031-28109-9_54

Lee, C.-C. (2023b). Microbial production of astaxanthin. In *Microbial production of food bioactive compounds* (pp. 1–38). Springer International Publishing. https://doi.org/10.1007/978-3-030-81403-8_37-1
Lee, C.-C. (2023c). Microbial production of astaxanthin. In *Microbial production of food bioactive compounds* (pp. 1–38). Springer International Publishing. https://doi.org/10.1007/978-3-030-81403-8_37-1
Li, J., Zhang, H., Xu, L., Liu, H., Qi, C., Wang, C., & Chen, W. (2024). An integrated investigation combining network pharmacology and computer simulation dissects the potential mechanism of anti-obesity of Monascus pigments (MPs). *Food Bioscience, 60*, 104459. https://doi.org/10.1016/j.fbio.2024.104459
Lim, Y., Jung, E. S., Lee, J. H., Kim, E. J., Hong, S. J., Lee, Y. H., & Lee, C. H. (2018). Non-targeted metabolomics unravels a media-dependent prodiginines production pathway in Streptomyces coelicolor A3(2). *PLoS ONE, 13*(11), e0207541. https://doi.org/10.1371/journal.pone.0207541
Lin, C.-W., Lin, C.-H., Hsu, Y.-W., Pan, T.-M., & Lee, C.-L. (2023). Monascin and ankaflavin prevents metabolic disorder by blood glucose regulatory, hypolipidemic, and anti-inflammatory effects in high fructose and high fat diet-induced hyperglycemic rat. *Journal of Functional Foods, 104*, 105537. https://doi.org/10.1016/j.jff.2023.105537
Liorančas, V., & Lee, C.-C. (2024). Antimicrobial role and sensory influence of oregano essential oil on the shelf-life of atlantic cod loins. *Darnios Aplinkos Vystymas, 21*(1), 121–126. https://doi.org/10.52320/dav.v21i1.300
Liu, C., Hu, B., Cheng, Y., Guo, Y., Yao, W., & Qian, H. (2021). Carotenoids from fungi and microalgae: A review on their recent production, extraction, and developments. *Bioresource Technology, 337*. https://doi.org/10.1016/J.BIORTECH.2021.125398
Ma, Y., Zhang, P., Dai, X., Yao, X., Zhou, S., Ma, Q., Liu, J., Tian, S., Zhu, J., Zhang, J., Kong, X., & Bao, Y. (2023). Extraction, physicochemical properties, and antioxidant activity of natural melanin from Auricularia heimuer fermentation. *Frontiers in Nutrition, 10*. https://doi.org/10.3389/fnut.2023.1131542
Mala, T., Sadiq, M. B., & Anal, A. K. (2021). Comparative extraction of bromelain and bioactive peptides from pineapple byproducts by ultrasonic- and microwave-assisted extractions. *Journal of Food Process Engineering, 44*(6). https://doi.org/10.1111/jfpe.13709
Marey, M. A., Abozahra, R., El-Nikhely, N. A., Kamal, M. F., Abdelhamid, S. M., & El-Kholy, M. A. (2024). Transforming microbial pigment into therapeutic revelation: Extraction and characterization of pyocyanin from Pseudomonas aeruginosa and its therapeutic potential as an antibacterial and anticancer agent. *Microbial Cell Factories, 23*(1), 174. https://doi.org/10.1186/s12934-024-02438-6
Martins, T., Barros, A. N., Rosa, E., & Antunes, L. (2023). Enhancing health benefits through chlorophylls and chlorophyll-rich agro-food: a comprehensive review. *Molecules, 28*(14), 5344. https://doi.org/10.3390/molecules28145344
Mishra, R. C., Kalra, R., Dilawari, R., Deshmukh, S. K., Barrow, C. J., & Goel, M. (2021). Characterization of an endophytic strain talaromyces assiutensis, CPEF04 with evaluation of production medium for extracellular red pigments having antimicrobial and anticancer properties. *Frontiers in Microbiology, 12*. https://doi.org/10.3389/fmicb.2021.665702
Mohamed, N. F., & Hamad, D.A.M.S.M.R.E.-S. (2020). Toxicity evaluation and antimicrobial activity of purified pyocyanin from pseudomonas aeruginosa. *Biointerface Research in Applied Chemistry, 10*(6), 6974–6990. https://doi.org/10.33263/BRIAC106.69746990
Mohd Said, F. (2023). Production of microbial pigment in a bioreactor (pp. 1–19). https://doi.org/10.1007/978-3-031-30037-0_40-1
Mota, G. C. P., de Moraes, L. B. S., Oliveira, C. Y. B., Oliveira, D. W. S., de Abreu, J. L., Dantas, D. M. M., & Gálvez, A. O. (2022). Astaxanthin from *Haematococcus pluvialis* : Processes, applications, and market. *Preparative Biochemistry and Biotechnology, 52*(5), 598–609. https://doi.org/10.1080/10826068.2021.1966802
Mussagy, C. U., Khan, S., & Kot, A. M. (2022). Current developments on the application of microbial carotenoids as an alternative to synthetic pigments. *Critical Reviews in Food Science and Nutrition, 62*(25), 6932–6946. https://doi.org/10.1080/10408398.2021.1908222
Mussagy, C. U., Dias, A. C. R. V., Santos-Ebinuma, V. C., Shaaban Sadek, M., Ahmad, M., de Andrade, C. R., Haddad, F. F., dos Santos, J. L., Scarim, C. B., Pereira, J. F. B., Floriano, J. F., Herculano, R. D., & Mustafa, A. (2024). Is the carotenoid production from Phaffia rhodozyma yeast genuinely sustainable? a comprehensive analysis of biocompatibility, environmental assessment, and techno-economic constraints. *Bioresource Technology, 397*, 130456. https://doi.org/10.1016/j.biortech.2024.130456
Naik, R., & Gupte, S. (2023). Characterization of pigment produced by high carotenoid yielding bacteria Paracoccus marcusii RSPO1 and evaluation of its anti-diabetic, anti-microbial and antioxidant properties. *Natural Product Research*, 1–10. https://doi.org/10.1080/14786419.2023.2208358
Osório, C., Machado, S., Peixoto, J., Bessada, S., Pimentel, F. B., Alves, C. R., & Oliveira, M. B. P. P. (2020). Pigments content (Chlorophylls, Fucoxanthin and Phycobiliproteins) of different commercial dried algae. *Separations, 7*(2), 33. https://doi.org/10.3390/separations7020033
Pachaiyappan, A., Sadhasivam, G., Kumar, M., & Muthuvel, A. (2021). Biomedical potential of astaxanthin from novel endophytic pigment producing bacteria pontibacter korlensis AG6. *Waste and Biomass Valorization, 12*(4), 2119–2129. https://doi.org/10.1007/s12649-020-01169-0
Paillié-Jiménez, M. E., Stincone, P., & Brandelli, A. (2020). Natural Pigments of Microbial Origin. *Frontiers in Sustainable Food Systems, 4*, 590439. https://doi.org/10.3389/FSUFS.2020.590439/BIBTEX
Pereira, C., Souza, A., Vasconcelos, A., Prado, P., & Name, J. (2020). Antioxidant and anti-inflammatory mechanisms of action of astaxanthin in cardiovascular diseases (Review). *International Journal of Molecular Medicine, 47*(1), 37–48. https://doi.org/10.3892/ijmm.2020.4783
Prabakaran, G., Sampathkumar, P., Kavisri, M., & Moovendhan, M. (2020). Extraction and characterization of phycocyanin from Spirulina platensis and evaluation of its anticancer, antidiabetic and antiinflammatory effect. *International Journal of Biological Macromolecules, 153*, 256–263. https://doi.org/10.1016/j.ijbiomac.2020.03.009
Pranaw, K., Drewniak, L., Nain, L., & Singh, S. (2022). Editorial: Waste to wealth: A sustainable circular bioeconomy approach. *Frontiers in Bioengineering and Biotechnology, 10*. https://doi.org/10.3389/FBIOE.2022.1010811/FULL
Puzorjov, A., & McCormick, A. J. (2020). Phycobiliproteins from extreme environments and their potential applications. *Journal of Experimental Botany, 71*(13), 3827–3842. https://doi.org/10.1093/jxb/eraa139
Qin, Z., Wang, X., Gao, S., Li, D., & Zhou, J. (2023). Production of Natural Pigments Using Microorganisms. *Journal of Agricultural and Food Chemistry, 71*(24), 9243–9254. https://doi.org/10.1021/ACS.JAFC.3C02222/ASSET/IMAGES/MEDIUM/JF3C02222_0007.GIF
Ramesh, C., Prasastha, V. R., Venkatachalam, M., & Dufossé, L. (2022). Natural Substrates and Culture Conditions to Produce Pigments from Potential Microbes in Submerged

Fermentation. *Fermentation, 8*(9), 460. https://doi.org/10.3390/fermentation8090460

Ramesh, C., Vinithkumar, N. V., Kirubagaran, R., Venil, C. K., & Dufossé, L. (2020). Applications of Prodigiosin Extracted from Marine Red Pigmented Bacteria Zooshikella sp. and Actinomycete Streptomyces sp. *Microorganisms, 8*(4), 556. https://doi.org/10.3390/microorganisms8040556

Rao, A. S., Deka, S. P., More, S. S., Nair, A., More, V. S., & Ananthjaraju, K. S. (2021). A comprehensive review on different microbial-derived pigments and their multipurpose activities. *Microbial Polymers: Applications and Ecological Perspectives*, 479–519. https://doi.org/10.1007/978-981-16-0045-6_20

Rather, L. J., Mir, S. S., Ganie, S. A., & Shahid-ul-Islam, & Li, Q. (2023). Research progress, challenges, and perspectives in microbial pigment production for industrial applications - A review. *Dyes and Pigments, 210*, 110989. https://doi.org/10.1016/j.dyepig.2022.110989

Regueiras, A., Huguet, Á., Conde, T., Couto, D., Domingues, P., Domingues, M. R., Costa, A. M., da Silva, J. L., Vasconcelos, V., & Urbatzka, R. (2021). Potential Anti-Obesity, Anti-Steatosis, and Anti-Inflammatory Properties of Extracts from the Microalgae Chlorella vulgaris and Chlorococcum amblystomatis under Different Growth Conditions. *Marine Drugs, 20*(1), 9. https://doi.org/10.3390/md20010009

Risdian, C., Mozef, T., & Wink, J. (2019). Biosynthesis of polyketides in streptomyces. *Microorganisms, 7*(5), 124. https://doi.org/10.3390/microorganisms7050124

Saleem, H., Mazhar, S., Syed, Q., Javed, M. Q., & Adnan, A. (2021). Bio-characterization of food grade pyocyanin bio-pigment extracted from chromogenic Pseudomonas species found in Pakistani native flora. *Arabian Journal of Chemistry, 14*(3), 103005. https://doi.org/10.1016/j.arabjc.2021.103005

Sen, T., Barrow, C. J., & Deshmukh, S. K. (2019). Microbial pigments in the food ındustry—challenges and the way forward. *Frontiers in Nutrition, 6*. https://doi.org/10.3389/fnut.2019.00007

Shatoor, A. S., & Al Humayed, S. (2021). Astaxanthin Ameliorates high-fat diet-induced cardiac damage and fibrosis by upregulating and activating SIRT1. *Saudi Journal of Biological Sciences, 28*(12), 7012–7021. https://doi.org/10.1016/j.sjbs.2021.07.079

Silva, T. R., Tavares, R. S. N., Canela-Garayoa, R., Eras, J., Rodrigues, M. V. N., Neri-Numa, I. A., Pastore, G. M., Rosa, L. H., Schultz, J. A. A., Debonsi, H. M., Cordeiro, L. R. G., & Oliveira, V. M. (2019). Chemical Characterization and Biotechnological Applicability of Pigments Isolated from Antarctic Bacteria. *Marine Biotechnology, 21*(3), 416–429. https://doi.org/10.1007/s10126-019-09892-z

Singh, S., Nimse, S. B., Mathew, D. E., Dhimmar, A., Sahastrabudhe, H., Gajjar, A., Ghadge, V. A., Kumar, P., & Shinde, P. B. (2021). Microbial melanin: Recent advances in biosynthesis, extraction, characterization, and applications. *Biotechnology Advances, 53*. https://doi.org/10.1016/J.BIOTECHADV.2021.107773

Sun, N., Agyei, D., & Ji, D. (2020). Medicinal chemistry friendliness of pigments from monascus-fermented rice and the molecular docking analysis of their anti-hyperlipidemia properties. *Fermentation, 6*(4), 111. https://doi.org/10.3390/fermentation6040111

Tran-Ly, A. N., Reyes, C., Schwarze, F. W. M. R., & Ribera, J. (2020). Microbial production of melanin and its various applications. *World Journal of Microbiology and Biotechnology, 36*(11), 170. https://doi.org/10.1007/s11274-020-02941-z

Tunca Koyun, M., Sirin, S., Aslim, B., Taner, G., & Nigdelioglu Dolanbay, S. (2022). Characterization of prodigiosin pigment by Serratia marcescens and the evaluation of its bioactivities. *Toxicology in Vitro, 82*, 105368. https://doi.org/10.1016/j.tiv.2022.105368

Verma, E., Chakraborty, S., Tiwari, B., & Mishra, A. K. (2018). Antimicrobial Compounds From Actinobacteria. In New and Future Developments in Microbial Biotechnology and Bioengineering (pp. 277–295). Elsevier. https://doi.org/10.1016/B978-0-444-63994-3.00019-9

Virreira Winter, S., & Zychlinsky, A. (2018). The bacterial pigment pyocyanin inhibits the NLRP3 inflammasome through intracellular reactive oxygen and nitrogen species. *Journal of Biological Chemistry, 293*(13), 4893–4900. https://doi.org/10.1074/jbc.RA117.001105

Wang, J., Zhang, R., Chen, X., Sun, X., Yan, Y., Shen, X., & Yuan, Q. (2020). Biosynthesis of aromatic polyketides in microorganisms using type II polyketide synthases. *Microbial Cell Factories, 19*(1), 110. https://doi.org/10.1186/s12934-020-01367-4

Xu, Z., Li, Y., Wang, Y., Deng, Z., & Tao, M. (2019). Genome-Wide Mutagenesis Links Multiple Metabolic Pathways with Actinorhodin Production in Streptomyces coelicolor. *Applied and Environmental Microbiology, 85*(7). https://doi.org/10.1128/AEM.03005-18

Yao, Q., Ma, J., Chen, X., Zhao, G., & Zang, J. (2023). A natural strategy for astaxanthin stabilization and color regulation: Interaction with proteins. *Food Chemistry, 402*, 134343. https://doi.org/10.1016/j.foodchem.2022.134343

Yurgel, S., Lynch, J., Rice, J., Adhikari, N., & Roje, S. (2014). Quantification of Flavin Production by Bacteria. *BIO-PROTOCOL, 4*(15). https://doi.org/10.21769/BioProtoc.1197

Zafar, J., Aqeel, A., Shah, F. I., Ehsan, N., Gohar, U. F., Moga, M. A., Festila, D., Ciurea, C., Irimie, M., & Chicea, R. (2021). Biochemical and Immunological implications of Lutein and Zeaxanthin. *International Journal of Molecular Sciences, 22*(20), 10910. https://doi.org/10.3390/ijms222010910

Zheng, S., Zhang, G., Wang, H., Long, Z., Wei, T., & Li, Q. (2021). Progress in ultrasound-assisted extraction of the value-added products from microorganisms. *World Journal of Microbiology and Biotechnology, 37*(4), 71. https://doi.org/10.1007/s11274-021-03037-y

Ziemons, S., Koutsantas, K., Becker, K., Dahlmann, T., & Kück, U. (2017). Penicillin production in industrial strain Penicillium chrysogenum P2niaD18 is not dependent on the copy number of biosynthesis genes. *BMC Biotechnology, 17*(1), 16. https://doi.org/10.1186/s12896-017-0335-8

Microbial Production of Herbicides, Fungicides and Insecticides

Salvador A. Saldaña-Mendoza, Israel Bautista-Hernández, Gloria A. Martinez-Medina, Muyideen Olaitan Bamidele, Karen Nathiely Ramírez-Guzmán, Olga Berenice Alvarez-Pérez, Alejandro Yamaguchi Torres-Valdez, and Cristóbal N. Aguilar

Abstract

Since ancient times, pests have been a problem for human beings, especially in agriculture, where their presence currently generates enormous economic losses. Using synthetic products to protect crops has been the preferred choice of farmers for years. However, the excessive use of these substances has generated problems of environmental contamination, damage to health, and even resistance in organisms to fight. Today, at an international level, enormous efforts are being made to replace these strategies with safer alternatives for health and the environment. The research, development, and application of biotechnological products for crop protection are strongly driven by international policies seeking a change to the use of more eco-friendly technologies where the use of microorganisms as a fundamental basis for their development is gaining more and more strength. The present work provides a brief but precise explanation of the negative impact associated with the excessive use of synthetic pesticides. Likewise, it offers detailed information on biopesticides and their classification, highlighting the use of microorganisms as a source for these products through the development of biotechnological processes. In addition, some of the mechanisms of action of biopesticides are described, and some commercial products are mentioned. Finally, a vision of the international panorama and future perspectives in developing and applying these eco-friendly technologies is presented.

Keywords

Biopesticides · Biocontrol · Biological control · Microbial biopesticides · Synthetic pesticides

1 Introduction

Current agriculture encounters multiple challenges to obtain successful production; some problems have been highlighted since the beginning of agricultural production, such as soil fertility, pests, and fluctuating climatic conditions. In the specific case of pests, the introduction of chemical pesticides (1940s) provided an attractive tool for the reduction of yield losses by pests. However, the bad management of chemical pesticides has promoted a rising contamination level that directly impacts the environment

S. A. Saldaña-Mendoza (✉) · M. O. Bamidele · C. N. Aguilar
Bioprocesses and Bioproducts Research Group (BBG-UAdeC), School of Chemistry, Autonomous University of Coahuila, Saltillo, Mexico
e-mail: salvadorsaldana@uadec.edu.mx

M. O. Bamidele
e-mail: mbamidele@uadec.edu.mx

C. N. Aguilar
e-mail: cristobal.aguilar@uadec.edu.mx

I. Bautista-Hernández
Universidade Católica Portuguesa, CBQF- Centro de Biotecnologia E Química Fina - Laboratório Associado, Escola Superior de Biotecnologia, Porto, Portugal
e-mail: s-ibhernandez@ucp.pt

G. A. Martinez-Medina
Technological University of Parras de La Fuente, Parras de La Fuente, Mexico
e-mail: gloria.martinez@utparras.edu.mx

K. N. Ramírez-Guzmán
Agri-Food and Agro-Industrial Bioeconomy Research Group, Faculty of Biological Sciences, Universidad Autónoma de Coahuila, Torreón, Mexico
e-mail: nathiely.ramirez@uadec.edu.mx

O. B. Alvarez-Pérez
GreenCorp Biorganiks de Mexico S.A. de C.V, Residencial San Patricio, Saltillo, Mexico
e-mail: b.alvarez@greencorp.mx

A. Y. Torres-Valdez
School of Chemistry, Autonomous University of Coahuila, Saltillo, Mexico
e-mail: alejandrotorres@uadec.edu.mx

K. Sharma (ed.), *Bio-prospecting of Novel Microbial Bioactive Compounds for Sustainable Development*, Sustainable Landscape Planning and Natural Resources Management, https://doi.org/10.1007/978-3-031-82178-3_5

or the prevalence of chemicals in food. Thus, recently, a novel approach for pesticides englobes the production of alternative green pesticides, "Biopesticides," which can be defined as natural substances (or metabolic products) that are applied for the control of pest populations (Tomar et al., 2024). Nowadays, research has evidenced the negative impact of conventional pesticides (effects on beneficial insects, chemical residues, a health risk for human health, among others) on the environment and highlighted the current requirement for new sustainable alternatives (Damalas et al., 2024). It has been reported that each year, around 385 million cases of acute unintentional poisoning by pesticides occur worldwide, with nearly 11,000 deaths (Devrnja et al., 2022). As a response, novel green alternatives have been taken into account. For example, bioproducts from microbial metabolism are a current option for making the transition from conventional chemical pesticides to green ones.

2 Pesticides of Synthetic Origin

In recent years, agriculture has presented economic losses of around 30%, mainly due to the diseases and pests that attack crops during the pre- and post-harvest stages since they cause quantitative and/or qualitative damage (flavor, texture, color, and shape). Among the main methods of post-harvest disease control is the use of synthetic pesticides, although in recent years, it has generated various controversies due to the negative consequences on health, environmental contamination, residuality, and the development of resistant pests and phytopathogens (Ramírez-Guzmán et al., 2023).

2.1 Types of Pesticides

Pesticides control various pests and disease transmitters, caused by microscopic organisms such as fungi, bacteria, and viroids, but also larger organisms such as insects such as mosquitoes, ticks, or even animals such as rats and mice. Therefore, these synthetic substances are used in agriculture to control weeds, insect infestation, and microorganisms and prevent or counteract diseases (Ramírez-Guzmán et al., 2020).

Pesticides are so diverse and combat so many pathogenic agents that are harmful to plants and crops that there is a wide range, each one being specific for a specific threat, for example, Antimicrobials to control germs and microbes such as bacteria and viruses responsible for damages such as leaf blights, soft rots of fruits, roots and stored organs (Shields et al., 2019). Within these, there are also Fungicides to control fungal problems such as *Colletotrichum, Fusarium,* and *Alternaria,* very aggressive genera for fruits and vegetables; there are also Herbicides to kill or inhibit the growth of unwanted plants that can compete for nutrients and space with the crops of interest, also known as weeds (Ramírez-Guzmán et al., 2020). Algaecides kill algae and/or slow down their proliferation, although algae are not organisms that commonly cause disease in plants. However, in recent years, a series of diseases have been discovered that seem, with a reasonable degree of certainty, to be caused by organisms belonging to the *Cephaleuros* genera that can appear in moist, bare, or very light areas of lawns, preventing the emergence of replanted seeds and damaging the structure of the soil, whose surface becomes impermeable. And, of course, there are Insecticides, whose purpose is to control insects by altering their metabolic functions, growth, and reproduction, among others (Barathi et al., 2024).

2.2 The Problem of the Indiscriminate Use of Synthetic Pesticides

Synthetic pesticides, being so varied and specific, have played a very important role in the control of plant diseases and have also influenced agricultural development since insects, weeds, and other pests destroy a large part of the crops and compete with them due to limiting factors such as space, humidity, and nutrients (2019 de la Cruz Quiroz et al., 2019). However, between what is sought from these products and the price that can be paid in the long term, even after years and cycles of good results, it is still something unimaginable since, although problems and consequences derived from indiscriminate use are already observed, excessive and inadequate products have caused severe environmental damage in all places where they have been applied, which, in many cases, have been irreversible or hardly reversible; the future could be worse (Fig. 1) (Lerro et al., 2021).

The use of these chemicals has generated pollution problems in the soil, water, biota, and sediments through discharges of toxic substances to lagoon systems via drains, irrigation, and rain, which constitutes a risk factor of contamination for terrestrial ecosystems and marine. These damages have occurred mainly due to the lack of knowledge about the risks to the environment and associated living beings and, consequently, the use of these substances (Ramírez-Guzmán et al., 2018).

Undesirable environmental effects can be classified according to the time of the impact: short-term, long-term in the near environment, and long-term in the distant environment. Speaking of the short-term consequences, we can refer to the instantaneous effects when they are applied, which is the contamination of the biotic and abiotic environment, affecting including humans; however, it mainly affects insects that are natural controllers of plant pathogens

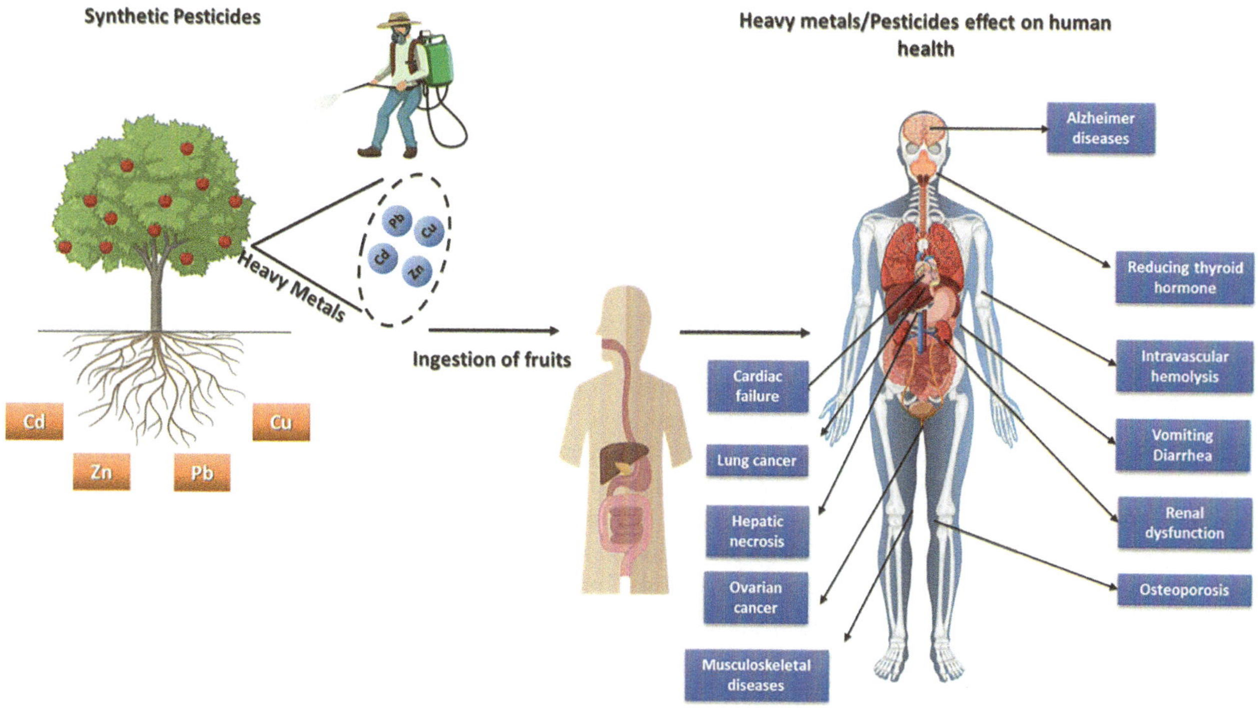

Fig. 1 Adverse effect of prolonged use of pesticides (Alengebawy et al., 2021)

such as ladybugs (*Coccinellidae*), which are highly appreciated for being natural predators of aphids (aphids), cocci or scale insects, mites, larvae of flies and other agricultural pests (Ramírez-Guzmán et al., 2023).

In addition to affecting a very important group of organisms called pollinators, which feed on the nectar or pollen of flowers, during their visits, they accidentally transport pollen from one flower to another, allowing the reproduction of plants and fruit production. The most famous example in recent years is mainly bees.

When these chemicals are used frequently, the problem increases since, in addition to the damage already mentioned, they accumulate in each application, which erodes soils, affecting sediments and groundwater tables, in addition to affecting the soil microbiota, affecting fungi and bacteria that act as rooting biostimulants such as the genera *Enterobacter, Pseudomonas sp., Bradyrizobium, Trichoderma* among others, in addition to creating resistant pests since they develop immunity to the substances, which means that more and more chemicals have to be poured (de la Cruz Quiroz et al., 2019).

In the long-term case in the distant environment, the most alarming problems are the presence of these in foods, not only in fruits and vegetables but also in products of animal origin such as meat, eggs, milk, and even fish and shellfish. When consumed by living beings, these are integrated into our organisms. The presence of these chemical pesticides has been found in breast milk, and various studies point to these as responsible for various types of cancers, mainly those related to the digestive system (Avila Vazquez et al., 2017).

3 Biopesticides and Classification

Biopesticides are recognized as a group of molecules naturally produced by a large group of organisms, including plants, animals, or microorganisms (fungi, bacteria, microalgae, among others), despite some authors also include multiple mineral substances, which could manage the dissemination or growth of pest or pathogens (Ayilara et al., 2023; EPA, 2024). The nature of biopesticides is extensive, including secondary metabolites, exudates, essential oils, crude extracts from leaves, roots, or other plant parts, or even be more sophisticated when integrated into special formulations such as nanoparticles (Ayilara et al., 2023). Until August 2020, nearly 390 biopesticide molecules have been registered by the United States Environmental Protection Agency (EPA, 2024).

Multiple biopesticide categories exist, normally separated by the origin, mechanisms, or class of employed molecules. Biopesticides have a wide working spectrum; nevertheless, each class has a preferred target. Botanical and biochemical pesticides are preferred against insects, while microbial pesticides work better against insects, fungi, oomycetes, and bacteria, and plant viruses are controlled better by employing essential oils (Fenibo et al., 2021). For this chapter, the categorization of biopesticides is cited below (Fig. 2):

3.1 Botanical Biopesticides

Botanical pesticides are derived from plants. Over time, multiple plant species have been reported for their repellent, inhibit growth, or kill pests actions, being one of the first and traditional mechanisms to pest control in agriculture (Gupta et al., 2023). The botanical pesticides include crude, partially or purified plant extracts and essential oils. The extracts could come from diverse plant parts such as barks, leaves, roots, seeds, flowers, fruits, rhizomes, steams, or cloves (Lengai et al., 2020). Their methodology production includes a solid/liquid extraction with different types of solvent and the grounded plant material (Gupta et al., 2023). Along the evolution, plants develop multiple defense mechanisms against insects and other stressors, derived from phytochemicals as alkaloids, terpenes, saponins, and flavonoids, among others, acting as antifeedants and interacting with chemoreceptors of pests (Stankovic et al., 2020). One of the most important issues during plant-based pesticides is related to the fact that they are not aggressive against beneficial insects as pollinators (Lengai et al., 2020).

3.2 Biochemical Biopesticides

Biochemical pesticides are a group of natural compounds, or their synthetic analogs, that possess activity against pest control with null or low toxicity against host plants, environment, or humans. Among these molecules, it is possible to mention plant-, animal-, bacterial-, or mineral-derived compounds, such as essential oils, secondary metabolites, semiochemicals, and plant- and insect-growth regulators (Fenibo et al., 2021; Saldaña-Mendoza et al., 2022).

Semiochemicals are compounds that interfere with insect natural living activities, such as mating, egg disposal, or feeding; in this category, the pheromones could be included; naturally, these pheromones help to attract, kill, or trap pests (Essiedu et al., 2020). Pheromones are produced to modulate behavior in organisms, playing a vital role in reproduction, survival from predators, feeding, and other interactions. The incorporation of pheromones into biopesticides possesses advantages, as they are species-specific, effective in low dosage, and do not affect non-targeted organisms (Villarreal et al., 2023).

The pheromones could be added to traps; nevertheless, the pest could survive in different stages (larvae, caterpillar, egg). Traps need certain abilities to be correctly employed and, therefore, special training. Also, pheromones could not be used in all weathers due to novel release dispositive needs to be developed (Villarreal et al., 2023).

Another biochemical biopesticide class is enzyme inhibitors; as enzymes are key drivers of plenty of biochemical reactions, these constitute one of the main targets. The digestive enzyme inhibitors cause a reduction in nutrient gain and, therefore, weakness, inhibition of growth, and death. Many enzyme inhibitors are present in plant components (Essiedu et al., 2020). On multiple occasions, botanical biopesticide components possess a biochemical action mode.

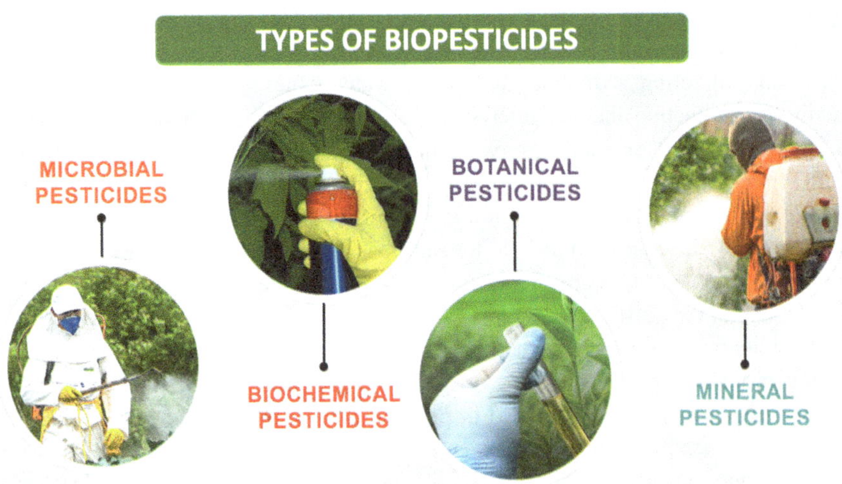

Fig. 2 Classification of biopesticides

3.3 Microbial Biopesticides

Microbial-based biopesticides are formulations that employ microorganisms as control agents for pests. Microbial biopesticides are constituted by fungi, bacteria, algae, viruses, or protozoan, natural or genetically modified (Archana et al., 2022). On many occasions, the microbial pesticide properties came from metabolic compounds; these compounds act as toxins for pests or also for microbes and could act as competitors avoiding their growth (Archana et al., 2022). The bacteria containing pesticides are the most common and cheaper products. This class of biopesticides includes in their formulation bacteria from genera *Bacillus* spp., *Serratia* spp., *Pseudomonas* spp., *Xanthomonas* spp., *Streptomyces* spp., *Yersinia* spp., among others, being *Bacillus thuringiensis,* the most used strain (Archana et al., 2022; Ortiz & Sansinenea, 2023). *B. thuringiensis* generates a protein that presents a toxic behavior when consumed and dissolved in gut insects; the disadvantages are related to their elevated instability under sunlight, demanding multiple applications (Ortiz & Sansinenea, 2023). Other microbial pesticides employ fungi spores or mycelia as key ingredients (Saldaña-Mendoza et al., 2023a, 2023b, 2023c, 2023d); the most used fungi are *Metarhizium anisopliae* and *Beauveria bassiana*; the fungal biopesticides also could act as a biocontrol agent (Archana et al., 2022). The key techniques for pest control derived from fungi are direct competition (food and space), mycoparasitism, and antimicrobial substance production (Saldaña-Mendoza et al., 2023a, 2023b, 2023c, 2023d). The search for novel strains with pest control potential is continuously undergone as Booth et al., (2022) demonstrated in 2022, where the strains *Corynebacterium flavescens*, *Sporosarcina aquimarina,* and *Sporosarcina saromensis* a group of rhizobacteria, inhibits bacterial, fungal, and oomycete infections in crops with significative results, also the combination of multiple pesticide strategies could enhance the pest control as demonstrate (Abdellaoui et al., 2020), who employs *M. anisopliae* and a pheromone (Phenylacetonitrile) to avoid *Schistocerca gregaria* develop.

4 Biofungicides of Microbial Origin

Fungicides of microbial origin, also known as biofungicides, represent an innovative and environmentally friendly approach for controlling plant diseases caused by fungal pathogens (Saldaña-Mendoza et al., 2023a, 2023b, 2023c, 2023d). Biofungicides are derived from microorganisms, such as bacteria, fungi, and viruses, or their byproducts, such as enzymes and metabolites, which possess antagonistic properties against harmful fungi (Singh et al., 2024). Fungicides of microbial origin represent a promising approach to combating fungal diseases in agriculture and other sectors. The history of microbial fungicides dates to the discovery of natural compounds produced by microorganisms that exhibit antifungal properties. Some fungicides inhibit fungal cell wall synthesis or disrupt cellular processes critical for fungal growth and reproduction, whereas others induce systemic acquired resistance in plants (Saldaña-Mendoza et al., 2023a, 2023b, 2023c, 2023d). Microbial fungicides are often biodegradable and pose minimal risks to non-target organisms and ecosystems, thus offering a more environmentally sustainable approach to disease management (Salazar et al., 2023). They also provide diverse compounds with potential fungicidal activities, contributing significantly to integrated pest management strategies and sustainable agriculture. They offer a sustainable, safe, and efficient alternative to synthetic fungicides in agricultural settings. Utilizing the inherent antagonistic characteristics of microorganisms to curb fungal pathogens, these fungicides offer several advantages, including environmental sustainability, a reduced likelihood of resistance development, and compatibility with organic farming practices (Nathiely Ramírez-Guzmán et al., 2018; Pacios-Michelena et al., 2024). However, challenges such as fluctuating effectiveness, regulatory obstacles, and production expenses persist. However, continuous research and technological progress offers potential solutions to these obstacles. The future of microbial fungicides lies in the development of innovative formulations, integration into integrated pest management strategies, and continued exploration of microbial diversity, all of which contribute to creating more resilient and sustainable agricultural systems.

4.1 Mode of Action of Biofungicides

Biofungicides are a vital component of sustainable agriculture as they provide effective disease control while minimizing negative environmental impacts. To optimize their use and develop new formulations, it is essential to understand the mechanisms by which they function. Biofungicides operate through various processes, such as competition for nutrients and space, production of antimicrobial compounds, including antibiotics and volatile organic compounds, stimulation of systemic resistance in plants, and disruption of fungal cell structures or functions (Saldaña-Mendoza et al., 2023a, 2023b, 2023c, 2023d).

4.1.1 Microbial Competition

Biofungicides function via microbial competition to control plant pathogens. They compete for essential nutrients, limiting resources for pathogen growth and reproduction. They can occur at infection sites or within the plant rhizosphere, where pathogens thrive. Microbial biocontrol agents

(MBCAs) occupy niches in the plant environment, preventing pathogens from establishing themselves and reducing their ability to colonize and cause infections. Biofungicides produce antimicrobial compounds, including enzymes, secondary metabolites, and volatile organic compounds, that inhibit pathogen growth (Saldaña-Mendoza et al., 2023a, 2023b, 2023c, 2023d). They can also parasitize pathogens and weaken their ability to cause diseases. Biofungicides can induce systemic resistance in plants, making them more resilient to pathogens. Biochemical interference with pathogen biochemical pathways disrupts vital processes, such as respiration or sterol biosynthesis, impairs pathogen growth and viability, and contributes to disease control (Chen et al., 2022). These mechanisms enhance plant health and reduce disease incidence by targeting pathogens through microbial competition, making biofungicides a valuable tool for sustainable agriculture.

4.1.2 Induced Systemic Resistance (ISR)

Biofungicides can trigger Induced Systemic Resistance (ISR), which enhances plant defense mechanisms against pathogens (Saldaña-Mendoza et al., 2023a, 2023b, 2023c, 2023d) It involves priming plant defense responses without directly interacting with the pathogen, thereby enhancing the ability of the plant to defend itself. Upon exposure to biofungicides, plants produce defense compounds such as phytoalexins, pathogenesis-related proteins, and enzymes, which are crucial for combating pathogen attacks (Santra & Banerjee, 2020). Biofungicides also activate signaling pathways in plants, such as salicylic acid and jasmonic acid pathways, which regulate defense responses against pathogens. ISR provides long-lasting protection against various pathogens and prepares plants for future attacks. It also enhances resistance to diseases caused by fungi, bacteria, and other pathogens, thus strengthening the immune system of plants. ISR-based biofungicides contribute to environmental sustainability by reducing reliance on synthetic chemicals and promoting eco-friendly pest management practices (Hossain, 2024).

4.1.3 Antifungal Compounds

Biofungicides are used to combat plant pathogens through the production of antifungal compounds. These compounds inhibit the growth and activity of fungal pathogens by disrupting vital fungal processes such as cell wall synthesis and enzyme function (Ahmed et al., 2023). Biofungicides secrete enzymes that break down fungal cell walls and destroy pathogens. They also compete with pathogens for essential nutrients, space, and infection sites, thereby limiting their growth and disease-causing potential. They can also induce systemic resistance in plants, rendering them more resistant to fungal infections. Also, biofungicides act as parasites of fungal pathogens by directly parasitizing them or releasing substances that disrupt fungal growth and development. This parasitic activity contributes to the antifungal efficacy of biofungicides (Pacios-Michelena et al., 2024; Saldaña-Mendoza et al., 2023a, 2023b, 2023c, 2023d).

4.1.4 Siderophore Production

Biofungicides utilize siderophores, small molecules secreted by them, to combat plant pathogens. Siderophores are characterized by their high affinity for iron and their ability to sequester iron ions in environments with limited iron availability (Alamoudi, 2023). This competition for iron between beneficial microbes and fungal pathogens limits their growth and development, ultimately inhibiting pathogens. Additionally, siderophores have antagonistic effects on pathogens by indirectly inhibiting their growth and contributing to their biocontrol activity (Deb & Tatung, 2024). Furthermore, siderophores enhance plant growth by creating a beneficial microbial environment around the plant roots, promoting nutrient uptake, and protecting plants from pathogens. Siderophorogenic preparations of certain bacteria exhibit potent fungicidal activity against phytopathogens, highlighting the importance of siderophore production in the production of biofungicides (Pradhan et al., 2023).

4.1.5 Biofilm Disruption

Biofungicides are essential for combating plant pathogens by disrupting biofilms. They interfere with quorum-sensing pathways used by bacteria to coordinate biofilm formation and virulence (Negi et al., 2023). They also target the adhesion mechanisms of biofilm-forming pathogens, preventing bacteria from adhering to surfaces (Han et al., 2023). Biofungicides can also disrupt established biofilms, reduce biofilm mass and metabolic activity, and lead to the breakdown of the biofilm structure. They use extracellular proteins to degrade biofilms and weaken their structural integrity (Yin et al., 2023). Additionally, biofilm disruption can enhance microbial growth rate by making nutrients more accessible, leading to increased growth and metabolic activity (Ray, 2024).

4.2 Advantages of Biofungicides

4.2.1 Environmentally Friendly

Biofungicides are environmentally friendly fungicides derived from natural sources, such as beneficial microorganisms, plant extracts, and biochemical compounds. They break down naturally into harmless components, reducing their environmental impacts (Khursheed et al., 2022). Biofungicides leave little to no toxic residues, promoting food safety and environmental sustainability. They are often specific to target pathogens, sparing beneficial organisms such as pollinators and soil microbes. This reduces the risk of harming non-target organisms and promotes an ecological balance in agricultural ecosystems (Hernández-Fernández et al., 2021). Biofungicides contribute to minimal environmental pollution, sustainable agricultural practices, and long-term environmental health. Biofungicides also contain beneficial microorganisms that can improve soil health and fertility, contributing to sustainable farming practices and enhanced crop productivity. They also align with sustainable agricultural practices and often meet the regulatory requirements for reduced chemical inputs and environmental impacts, fostering a more environmentally conscious approach to crop protection (Saldaña-Mendoza et al., 2023a, 2023b, 2023c, 2023d).

4.2.2 Safe for Humans and Animals

Biofungicides are natural substances with low toxicity levels compared with synthetic chemical fungicides, posing minimal risks to human health and safety during handling, application, and consumption (Saha et al., 2023). They degrade rapidly after application, leaving little to no residue on crops or the environment, reducing bioaccumulation in the food chain and minimizing long-term exposure risks. Biofungicides contribute to environment-friendly farming practices, reduce pollution, protect beneficial organisms, and support biodiversity in agricultural ecosystems (Tailor et al., 2023). They align well with Integrated Pest Management (IPM) strategies, reducing reliance on high-risk chemical pesticides and enhancing safety and the environment (Chavana & Joshi, 2024). Biofungicides often target specific pathogens, sparing beneficial insects, pollinators, and natural predators and minimizing negative impacts on beneficial organisms that are crucial for ecosystem balance and agricultural sustainability.

5 Bioinsecticides of Microbial Origin

Insects pose a major threat to crops at both the pre-harvest and post-harvest stages, as they can infest and contaminate agricultural resources throughout the entire food supply chain. Furthermore, in some cases, they are vectors of parasites and diseases of medical importance (Urrutia et al., 2023). Modern agricultural practices, such as international trade and transportation, have facilitated the dispersal of insects outside their native areas, causing ecological disruptions. Although farmers often use synthetic insecticides for pest control, they have lost popularity due to their negative impact on human health, domestic animals, pollinating insects, and the environment (Devrnja et al., 2022). Insects possess biochemical defense mechanisms, such as detoxifying enzymes, which also contribute to developing resistance to synthetic pesticides (Yang et al., 2021). This situation, together with the excessive application of pesticides, the rapid life cycle of insects, and the development of cross-resistance, drives the need to look for sustainable and safe alternatives for their control, such as products of biological origin.

5.1 Bacterial Sources

Recently, new microorganisms have been studied as a new source of insecticide compounds such as insect toxins, for example, bacteria strains like *Bacillus subtilis*, *Bacillus cereus*, and *Clostridium difficile*, among others (Tomar et al., 2024). The *B. subtilis* strain was reported as a source of insecticidal activity against Lepidoptera order, and this activity could be related to the presence of bioactive peptides, lipopeptides, iturin, and surfactin (Rocha et al., 2023). Additionally, Álvarez-Lagazzi et al. (2021) showed a reduction in the aphids population (*Sitobion avenue*) by the ingestion of *B. subtilis* spores; some possible mechanisms mentioned by the authors englobes (1) Septicemia or bacteremia and (2) Spores activate the defense mechanism of the plants. Also, it has been reported the potential inhibition of detoxifying enzymes like GST and CYP450 (*Aedes aegypti*, mosquito), which play a crucial role in metabolic regulation, and cell differentiation, among other processes (Ramasubramanian et al., 2022). Otherwise, some bacterial genera such as *Xenorhabdus* and *Photorhabdus* have been described as potential biological agents for the control of insect species, especially for the control of *A. aegypti*; the mechanism involved in their insecticide activity encompasses: (1) immunosuppression, toxin complexes with the capacity to destroy epithelial cell from the intestine and promote the apoptosis (Da Silva et al., 2020). Thus, it is important to highlight that bacterial pesticides' action mechanism mainly encompasses intestinal infection and damage. Otherwise, besides the remarkable insecticidal activity from bacterial metabolism, it is necessary to understand the potential application, possible interaction, and prevalence in food products.

5.2 Fungal Sources

Recently, the fungal microorganism with the greatest attention is *Trichoderma* spp. Numerous studies have been developed under the premise of evaluating and understanding the potential application of *Trichoderma* for insect control in the agricultural industry. Its effectiveness lies in the multiple plant-fungus-insect mechanisms that protect the crop, for example, the production of toxins, increase in the plant's defenses, or attraction of predators (Caccavo et al.,

2022). For example, strong negative perturbations have been reported in the transcriptional activity of the insect microbiota under treatment with *Trichoderma* spp., which would influence the metabolism of amino acids and sugars (Monte, 2023). On the other hand, the application of *T. harzianum* has shown changes in the levels of jasmonic acid (involved in defense processes against herbivores) and an increase in predator-attracting compounds "(z)-3-hexen-1-ol" in plants of maize (Contreras-Cornejo et al., 2021). Thus, different fungal species of the genus *Trichoderma* are attractive sources of molecules for developing effective biopesticides for the agricultural market.

However, other fungal species comprise a biotechnological tool of vital importance due to the capacity for growth, enzymatic production, and secondary metabolites (Destruxin A, Cordycepin, Beauvericin, Pseurotin A, among others) (Hong et al., 2024). Another fungal species of interest is *Lecanicillium lecanii*, considered a viable biological control species that can break the insect cuticle, enter, initiate nutrient uptake, generate toxins, and destroy cells (Keppanan et al., 2019). Similarly, this method of action has been reported for other fungal species, such as *Hirsutella citriformis*, a species of vital importance for pest control in the citrus industry (Pérez-González et al., 2022). Additionally, the relevance of fungi as a source of biopesticides has expanded their applications to the veterinary field. For example, Bayramoğlu et al., (2024) has successfully evaluated the application of entomopathogenic species (*Metharizium flavoviride* and *Beauveria bassiana*) against poultry red mite (*Dermanyssus gallinae*), a pest of great importance in egg production. However, environmental aspects must be considered to maintain adequate growth of the microorganism and not reduce the effectiveness of its application (Hong et al., 2024).

6 Bioherbicides

Weed infestations are critical to the agricultural economy and food security; crop losses due to undesirable weeds represent more than 30% of losses worldwide (Kostina-Bednarz et al., 2023). Weed populations impede crop development by depleting sunlight, water, and soil nutrients, and the investment to control them in the field represents around one-third of the total cost of production (Monteiro & Santos, 2022). Weed control in the field is commonly carried out by farmers using synthetic products due to their profitability at the expense of the collateral effects they produce, such as the presence of residues in crops, pollution, and soil erosion (Ammar et al., 2023). In addition, it has been shown through toxicological studies that synthetic herbicides can cause health damage, such as changes in DNA, teratogenesis, autism, and cancer (Maldaner et al., 2023).

Research into developing safer and more eco-friendly resources for crop protection has promoted bioherbicides, biological products for weed control composed of phytopathogenic microorganisms, microbial phytotoxins, or plant extracts. These bioproducts are characterized by commonly presenting more than one route of action, reducing the possibility of the development of resistance by weeds, not having prolonged persistence in the field, being less likely to damage water and soil, and being safe for non-target organisms (Camargo et al., 2019).

6.1 Bioherbicides of Microbial Origin

The research and development of biological sources for weed control from microorganisms has allowed the development of bioproducts with bioherbicidal activity, establishing environmentally friendly alternatives in current weed management (Camargo et al., 2019). There are two biological control approaches, classical or inoculative methods, where the importation and release of a highly targeted host-specific pathogen suppresses weed population and weed reproduction. Despite the inoculative approach, inundative methods, where pathogens isolated from weeds are cultured to produce infective biomass used as bioherbicides, are preferred due to versatility in agricultural systems (Cai & Gu, 2016).

Microorganisms as bioherbicides present high toxicity against weeds due to phytotoxic activity; microorganism metabolites can produce plant diseases such as necrosis, chlorosis, deformation, and stunting (Table 1). These compounds can help formulate bioherbicides for the biocontrol of weeds (Bo et al., 2019). Inoculating microorganisms can affect seed germination, growth inhibition, root elongation, and root deformation (Sindhu & Sehrawat, 2017). Formulations of bioherbicides with microbial enzymes are well-known (such as lipases and cellulases), allowing the entry of microorganisms or phytochemicals into target plants (Bordin et al., 2018). A biochemical approach occurs when enzymes can block the active sites of the process of cellulose biosynthesis, serine/threonine protein phosphatases, and deoxyxylulose-5-phosphate synthase (Ammar et al., 2023).

Weed infected with *Phoma Macrostoma* colonizes the root, and the proliferation of mycelium prevents the supply of nutrients by infecting the vascular trachea with the phytotoxins α, β-dehydrocurvularin (Radhakrishnan et al., 2018). *Colletotrichum* genus has been studied as a mycoherbicide *Colletotrichum echinochloa* shows bioherbicide activity against barnyard grass, with an effectivity of 80% in fresh weight (Gu et al., 2023).

Studies for Morning glories (*Ipomoea spp.*) control in sugar cane farms inoculated with a suspension of

Table 1 Fungi and bacteria involved in bioherbicides formulation

Microorganism	Target	References
Streptomyces	Digitaria sanguinalis, Sorghum bicolor, Echinochloa crus-galli, Agropyron smithii	Bo et al. (2019)
Thrichoderma Koningiopsis	*Heuphorbia heterophylla*	Bordin et al. (2018)
Diaporthe sp.	Lolium multiflorum, Conyza sp. Echinochloa sp. Glycine max, Triticum aestivum Oryza sativa	Pes et al. (2016)
Fusarium	*Striga hermonthica*	Anteyi et al. (2022)
Alternaria alternata	*Echinochola crus-galli*	Motlagh (2012)
Alternaria cassiae	*Senna obtusifolia, Crotolaria spectabilis y Senna occidentalis*	Hoagland and Boyette (2021)
Myrothecium	Carduus acanthoides L Euphorbia esula L Puereria lobata Sesbania exaltata	Daba (2019)

Cercospora leaf spot (*Cercospora* aff. *canescens* Ellis & G. Martin) of 2×10^7 conidia mL^{-1} show significant differences induces defloration in the first 24 h and leaf spots disease was observed in its phenological stages (Nechet & Halfeld-Vieira, 2019). Metabolites from solid-state fermentation of *Diaporthe sp.* against (*Cucumis Sativus*) with a mix of adjuvants (improving effects agents) show post-emergence herbicidal activity where chlorosis was visible at the site of herbicidal application, and plant height and dry mass weight was suppressed close to 40 and 30% respectively (Bastos et al., 2017).

In 2021, Ulrich et al. carried out studies on the production and characterization of *Trichoderma koningiopsis* metabolites produced by submerged fermentation in the model plant *Cucumis sativus*. They found that the extracts filtered with gauze achieved control of 80% of the plants after 15 days of application. The application of the extracts generated a change in color and necrosis of the evaluated model. In turn, these did not present ecotoxicity when assessed in the earthworm model (*Eisenia andrei*), highlighting the advantages of applying products of biological origin.

Applying bioherbicides in conjunction with synthetic products for weed control is a strategy that allows for efficient control by reducing the doses of synthetic pesticides used. Ulrich et al., reported in 2023 that the application of a bioherbicide based on *T. koningiopsis* in association with the pesticides Roundup Original®, Zapp Qi®, and Roundup WG® presented a synergistic effect, controlling the weed *E. heterophylla* by up to 90% using only 10% of the concentration recommended by the manufacturer and in the case of *B. plantaginea,* 100% control was achieved using half the recommended concentration dose. Furthermore, these applications did not harm soybean crop plants, highlighting the benefits of this combination as it does not affect crops of commercial interest. De Souza et al., (2015) produced a bioherbicide using *Diaporthe sp.* isolated from the Brazilian Pampa biome by liquid fermentation using sucrose corn steep liquor as substrate. They evaluated the effect of the bioproduct on seeds of the *Cucumis sativus* L. var. Wisconsin and *Sorghum sp.* In both cases, the control efficiency was 100%, its effect being attributed to the compounds Phomentrioloxin B and gulypyrone A, which are responsible for developing necrosis in plants. This work also represents an example of yet another benefit of the production of bioherbicides of microbial origin, which is the ability to use plant residues as substrates for the generation of these bioproducts through fermentation processes that could not be considered for human consumption because they do not meet the ideal commercial characteristics, among other various circumstances, which contributes to the zero-waste stream.

7 International Overview of the Production of Biopesticides of Microbial Origin

Currently, the global market for biopesticides is driven by the concept of carrying out all practices towards sustainability and care for the environment, and as a consequence of these practices, having efficient agriculture. All of this arises from the knowledge that pesticides of synthetic origin cause damage to human health, not only being able to reach the final consumer but also throughout the production chain: application and collection. This is why most organizations that regulate this issue at the international and national level are applying very rigorous protocols, which is why there is a demand for environmentally friendly solutions. In addition to health problems, the issue of generating pest resistance to these types of products also comes into play, so finding natural solutions that do not generate this problem gives them added value, which would be reflected in international marketing. By 2022, the global biopesticides market will reach 6.2 billion dollars, estimating that 2028 demand will reach 13.6 billion dollars, with a growth rate of 14.1% from 2023 to 2028 (Market Research Report, 2024). Within this commercial value of biopesticides, we find different categories contributing to global production, where 47% is occupied by bioinsecticides, 28% by biofungicides, 20% by bioherbicides, and in a smaller proportion is the production of bionematicides (4%) and biobactericides (1%). It is important to mention that we are also in a strong transition of changing practices, tipping the balance towards the organic side since the world population demands it, impacting the market even more. However, the increase in the value of these products began in 1990, when their demand began to compete with agrochemicals, being more notable from the years 2010–2015.

7.1 Biopesticides in the Global Market by Region

Biopesticides generate a high commercial value in different regions (Table 2). First of all, there is North America, with a production in 2023 of 2.51 billion dollars, estimated for 2028 a value of $5.22 billion, Latin America $1.60 billion, estimated to reach $3.64 in 2028; Europe (2023 $2.11–2028 $4.5), the Asia–Pacific region (2023 $1.07–2028 $2.3) and the rest of the world with a production in 2023 of $0.17 estimated to reach 0.36 billion dollars in 2028.

7.1.1 North America

In 2023, its market value was 2.5 trillion dollars, contributing 33% of the global market. Between the years 2020 to 2025, a compound annual growth rate (CAGR) of 15.8% is expected. The United States is the country with the fastest growing level in the market, dominating the use of biological control agents, making microorganisms the most popular in the segment, mainly fungi and bacteria, with a market share of 62.8% in the year 2023. The main competitors in this region are AMVAC, Andermatt Group, BASF, Bayer AG, Syngenta, BioWorks Inc., Corteva Agriscience, FMC Corporation, Gowan, Lallemand Inc., Marrone Bio Innovations, Novozymes and Valent Bioscience LLC.

The bioinsecticides segment has the largest market share due to the growing demand for natural and environmentally friendly pesticides with a broad spectrum of pathogenic action.

Table 2 Main competitors in the production of biopesticides

United States and Canada	Europe	Asia–Pacific	Brazil	Latin America	Middle East
AMVAC Andermatt Group BASF Bayer AG BioWorks Inc Corteva Agriscience FMC Corporation Gowan Koppert Biological Systems Lallemand Inc Marrone Bio Innovations Novozymes Valent Bioscience LLC	Certis Corteva Agriscience Koppert Biological Systems Seipasa Valent Bioscience LLC	Andermatt Group BASF Bayer AG Biobest Group NV-Agritop Bioherbicides Australia Biolchim Spa Coromandel Int. Limited Gowan Gujarat State Fertilizers/Chemicals Ltd Henan Jiyuan Baiyun Industry Co., Ltd IPL Biologicals Koppert Biological Systems Marrone Bio Innovations Syngenta Ag T. Stanes and Company Limited Valent Bioscience LLC	Andermatt Group Atlantica AgricolaBASF Bayer AG Bioworks Inc Certis Coromandel Int. Limited Corteva Agriscience FMC Corporation Isagro Agrochemicals Koppert Biological Systems Lallemand Inc Marrone Bio Innovations Rovensa Group Syngenta Ag UPL Valent Bioscience LLC Vittia Group	Advan LLC AMVAC Ballagro Agro Tecnologia BASF Bayer AG Certis Gowan Jiangsu Dongbao Agrochemical Co., Ltd Sumitomo Chemical Syngenta Ag Valent Bioscience LLC	Andermatt Group BASF Bayer AG Biobest Group NV-Agritop Bioherbicides Australia Biolchim Spa Coromandel Int. Limited Gowan Gujarat State Fertilizers/Chemicals Ltd Henan Jiyuan Baiyun Industry Co, Ltd IPL Biologicals Koppert Biological Systems Marrone Bio Innovations Syngenta Ag T. Stanes and Company Limited Valent Bioscience LLC

7.1.2 Europe

France has the largest market for these products in the European region, with biofungicides being the most marketed, dominating the European market. This segment had a market share of 59.3% in 2022, with a value of 971.3 million dollars, with a CAGR rate of 15.9% (2020–2025). Cereal crops abound in this region, such as oats, corn, and soybeans, among others. In Europe, there is the "Green Deal" with the ambitious objective of reducing the use and risk caused by chemical pesticides by 50% by 2030.

7.1.3 Pacific Asia

Within this region, China occupies first place in the market, being the most populous country in the world; it has been at the forefront of the production of biopesticides since 1960. The use of biopesticides in crops such as rice, corn, fruits, and vegetables has increased considerably in the region, in addition to the growing demand for organic and healthy products, which has further driven the use of organic biopesticides. Its CAGR rate is 16.6% (2020–2025). The main regions that use them are China, India, Japan, and Australia, with Japan standing out as a pioneer in using this product type. Its main competitors are: BASF, Valent, Bayer CropScience, Emery Oleochemicals, and Ishihara Sangyo Kaisha.

7.1.4 Latin America

In 2023, 1.6 trillion dollars were generated, and it is expected to reach a total of 3.64 trillion dollars by 2028, with a CAGR rate of 17.9% (2020–2025). Brazil has the largest market for biopesticides, mainly of the biocontrol type, contributing 35% of the total production, with 107 registrations only in 20–23. By 2022, Brazil registered 433 biopesticides, which increased marketing by 404% from 2013 to 2022. Brazil, Colombia, and Peru represent 60% of this market in Latin America, and this continues to increase. The main competitors in this region are Bayer CropScience AG, Isagro SpA, Rovensa Group, Valent BioSciences Corporation, and IPL Biologicals Limited.

7.1.5 Mexico

In Mexico, more than 100 companies are involved in the registration of biopesticides in the country, with a total of products registered in different categories: biofungicides (145), bioinsecticides (199), and bionematicides (39). According to a market study by Kline, the Value of the Mexican Biopesticide Market was 50 million dollars in 2019. Dunham Trimmer estimates a value of 124.6 million dollars in 2023. Its CAGR rate is 13.3% (2020–2025); however, some biopesticides do not have registration and are currently applied to crops. If we considered them, the market value by 2025 would be 405 million dollars. It is the third country to produce and export many agronomic products, mainly tomatoes and avocados.

7.1.6 India

By 2022, 69.62 million dollars will be generated, estimated that by 2029 it will be 130.37 million dollars. They have registered 970 products, of which 90% are foliar biopesticides. The bioinsecticides segment is growing faster than biofungicides and bioherbicides due to the country's tropical climate, and 70% of the available products are covered by a formulation based on *Trichoderma harzianum*. The use of nematicides is also increasing in greenhouses and the floriculture industry.

8 Conclusions

The global needs in the agricultural sector present a trend towards substituting synthetic products that negatively impact the environment and the population's health, using biotechnological alternatives to protect crops. Bioproducts of microbial origin, in addition to controlling undesirable organisms, present interesting advantages, among which are the absence of toxicity in non-target organisms, their production through fermentation processes provides an alternative for the revalorization of agro-industrial waste as a substrate, improvement of production yields through the stimulation of crops and even in some cases, their association with the fortification of the products generated, offering a sustainable and efficient solution for modern agriculture.

In summary, producing biopesticides of microbial origin not only represents a technological advance in pest and disease management but also contributes to a more sustainable and environmentally responsible approach. Implementing these bioproducts in agriculture can transform agricultural practices, promoting ecosystem health and ensuring long-term food security.

References

Abdellaoui, K., Miladi, M., Mkhinini, M., Boughattas, I., Ben Hamouda, A., Hajji-Hedfi, L., Tlili, H., & Acheuk, F. (2020). The aggregation pheromone phenylacetonitrile: Joint action with the entomopathogenic fungus Metarhizium anisopliae var. acridum and physiological and transcriptomic effects on Schistocerca gregaria nymphs. *Pesticide Biochemistry and Physiology*, *167*(2019). https://doi.org/10.1016/j.pestbp.2020.104594

Ahmed, H. F. A., Seleiman, M. F., Mohamed, I. A. A., Taha, R. S., Wasonga, D. O., & Battaglia, M. L. (2023). Activity of essential oils and plant extracts as biofungicides for suppression of soil-borne fungi associated with root rot and wilt of marigold (Calendula officinalis L.). *Horticulturae*, *9*(2), 222. https://doi.org/10.3390/horticulturae9020222

Alamoudi, S. A. (2023). Using some microorganisms as biocontrol agents to manage phytopathogenic fungi: A comprehensive review. *Journal of Plant Pathology, 106*(1), 3–21. https://doi.org/10.1007/s42161-023-01542-7

Alengebawy, A., Abdelkhalek, S. T., Qureshi, S. R., & Wang, M.-Q. (2021). Heavy metals and pesticides toxicity in agricultural soil and plants: Ecological risks and human health implications. *Toxics, 9*(3), 42. https://doi.org/10.3390/toxics9030042

Álvarez-Lagazzi, A. P., Cabrera, N., Francis, F., & Ramírez, C. C. (2021). Bacillus subtilis (Bacillales, Bacillaceae) spores affect survival and population growth in the grain Aphid Sitobion avenae (Hemiptera, Aphididae) in Relation to the Presence of the Facultative Bacterial Endosymbiont Regiella insecticola (Enterobacteriale. *Journal of Economic Entomology, 114*(5), 2043–2050. https://doi.org/10.1093/jee/toab164

Ammar, E. E., Elmasry, S. A., Ghosh, S., Al-Farga, A., Ghallab, Y. K., Fkr Eldeen, A. M., El-Shershaby, N. A., & Aioub, A. A. A. (2023). Insightful review of bioherbicides derived from plants (PHYTO-HERBICIDES). *Journal of the Chilean Chemical Society, 68*.

Anteyi, W. O., Klaiber, I., & Rasche, F. (2022). Diacetoxyscirpenol, a Fusarium exometabolite, prevents efficiently the incidence of the parasitic weed Striga hermonthica. *BMC Plant Biology, 22*(1). https://doi.org/10.1186/s12870-022-03471-6

Archana, H. R., Darshan, K., Lakshmi, M. A., Ghoshal, T., Bashayal, B. M., & Aggarwal, R. (2022). Biopesticides: A key player in agro-environmental sustainability. In *Trends of applied microbiology for sustainable economy* (pp. 613–653). Elsevier.

Avila Vazquez, M., Maturano, E., Etchegoyen, A., Difilippo, F. S., & Maclean, B. (2017). Asociación entre cáncer y exposición ambiental a glifosato. *International Journal of Clinical Medicine, 8*(2), 73–85.

Ayilara, M. S., Adeleke, B. S., Akinola, S. A., Fayose, C. A., Adeyemi, U. T., Gbadegesin, L. A., Omole, R. K., Johnson, R. M., Uthman, Q. O., & Babalola, O. O. (2023). Biopesticides as a promising alternative to synthetic pesticides: A case for microbial pesticides, phytopesticides, and nanobiopesticides. *Frontiers in Microbiology, 14*(February), 1–16. https://doi.org/10.3389/fmicb.2023.1040901

Barathi, S., Sabapathi, N., Kandasamy, S., & Lee, J. (2024). Present status of insecticide impacts and eco-friendly approaches for remediation-a review. In *Environmental research* (Vol. 240). Academic. https://doi.org/10.1016/j.envres.2023.117432

Bastos, B. de O., Deobald, G. A., Brun, T., Dal Prá, V., Junges, E., Kuhn, R. C., Pinto, A. K., & Mazutti, M. A. (2017). Solid-state fermentation for production of a bioherbicide from Diaporthe sp. and its formulation to enhance the efficacy. *3 Biotech, 7*(2). https://doi.org/10.1007/s13205-017-0751-4

Bayramoğlu, M., Bayramoğlu, Z., Aydın, L., Zengin, S. A., Çırak, V. Y., Demirbağ, Z., & Demir, İ. (2024). Entomopathogenic fungi with biological control potential against poultry red mite (Dermanyssus gallinae, Arachnida: Dermanyssidae). *Veterinary Parasitology, 328*(March). https://doi.org/10.1016/j.vetpar.2024.110155

Bo, A. B., Kim, J. D., Kim, Y. S., Sin, H. T., Kim, H. J., Khaitov, B., Ko, Y. K., Park, K. W., & Choi, J. S. (2019). Isolation, identification and characterization of Streptomyces metabolites as a potential bioherbicide. *PLoS ONE, 14*(9). https://doi.org/10.1371/journal.pone.0222933

Booth, J., Schenk, P. M., & Mirzaee, H. (2022). Microbial Biopesticides against bacterial, fungal and oomycete pathogens of tomato, cabbage and chickpea. *Applied Microbiology, 2*(1), 288–301. https://doi.org/10.3390/applmicrobiol2010021

Bordin, E. R., Frumi Camargo, A., Rossetto, V., Scapini, T., Modkovski, T. A., Weirich, S., Carezia, C., Barretta Franceschetti, M., Balem, A., Golunski, S. M., Galon, L., Funghetto Fuzinatto, C., Reichert Júnior, F. W., Fongaro, G., Mossi, A. J., & Treichel, H. (2018). Non-toxic bioherbicides obtained from trichoderma koningiopsis can be applied to the control of weeds in agriculture crops. *Industrial Biotechnology, 14*(3), 157–163. https://doi.org/10.1089/ind.2018.0007

Caccavo, V., Forlano, P., Mang, S. M., Fanti, P., Nuzzaci, M., Battaglia, D., & Trotta, V. (2022). Effects of Trichoderma harzianum Strain T22 on the arthropod community associated with tomato plants and on the crop performance in an experimental field. *Insects, 13*(5), 1–20. https://doi.org/10.3390/insects13050418

Cai, X., & Gu, M. (2016). Bioherbicides in organic horticulture. *Horticulturae, 2*(3). https://doi.org/10.3390/horticulturae2020003

Camargo, A. F., Stefanski, F. S., Scapini, T., Weirich, S. N., Ulkovski, C., Carezia, C., Bordin, E. R., Rossetto, V., Júnior, F. R., Galon, L., Fongaro, G., Mossi, A. J., & Treichel, H. (2019). Resistant weeds were controlled by the combined use of herbicides and bioherbicides. *Environmental Quality Management, 29*(1), 1–6. https://doi.org/10.1002/tqem.21643

Chavana, J., & Joshi, N. K. (2024). Toxicity and risk of biopesticides to insect pollinators in urban and agricultural landscapes. *Agrochemicals, 3*(1), 70–93. https://doi.org/10.3390/agrochemicals3010007

Chen, D., Mubeen, B., Hasnain, A., Rizwan, M., Adrees, M., Naqvi, S. A. H., Iqbal, S., Kamran, M., El-Sabrout, A. M., Elansary, H. O., Mahmoud, E. A., Alaklabi, A., Sathish, M., & Din, G. M. U. (2022). Role of promising secondary metabolites to confer resistance against environmental stresses in crop plants: Current scenario and future perspectives. *Frontiers in Plant Science, 13*. https://doi.org/10.3389/fpls.2022.881032

Contreras-Cornejo, H. A., Viveros-Bremauntz, F., del-Val, E., Macías-Rodríguez, L., López-Carmona, D. A., Alarcón, A., González-Esquivel, C. E., & Larsen, J. (2021). Alterations of foliar arthropod communities in a maize agroecosystem induced by the root-associated fungus Trichoderma harzianum. *Journal of Pest Science, 94*(2), 363–374. https://doi.org/10.1007/s10340-020-01261-3

de la Cruz Quiroz, R., Cruz Maldonado, J. J., Rostro Alanis, M. de J., Torres, J. A., & Parra Saldívar, R. (2019). Fungi-based biopesticides: shelf-life preservation technologies used in commercial products. *Journal of Pest Science, 92*(3), 1003–1015. https://doi.org/10.1007/s10340-019-01117-5

Da Silva, W. J., Pilz-Júnior, H. L., Heermann, R., & Da Silva, O. S. (2020). The great potential of entomopathogenic bacteria Xenorhabdus and Photorhabdus for mosquito control: A review. *Parasites and Vectors, 13*(1), 1–14. https://doi.org/10.1186/s13071-020-04236-6

Daba, G. M. (2019). Myrothecium as promising model for biotechnological applications, potentials and challenges. *Biomedical Journal of Scientific and Technical Research, 16*(3). https://doi.org/10.26717/bjstr.2019.16.002869

Damalas, C. A., Koutroubas, S. D., & Abdollahzadeh, G. (2024). Farmers' willingness to use lower risk pesticides for pest control: Barriers and facilitators in northern Greece. *Environmental Challenges, 14*(2023), 100871. https://doi.org/10.1016/j.envc.2024.100871

Deb, C. R., & Tatung, M. (2024). Siderophore producing bacteria as biocontrol agent against phytopathogens for a better environment: A review. *South African Journal of Botany, 165*, 153–162. https://doi.org/10.1016/j.sajb.2023.12.031

Devrnja, N., Milutinović, M., & Savić, J. (2022). When scent becomes a weapon—plant essential oils as potent bioinsecticides. *Sustainability (Switzerland), 14*(11). MDPI. https://doi.org/10.3390/su14116847

EPA. (2024). *What are biopesticides?* https://www.epa.gov/ingredients-used-pesticide-products/what-are-biopesticides

Essiedu, J. A., Adepoju, F. O., & Ivantsova, M. N. (2020). Benefits and limitations in using biopesticides: A review. *AIP conference proceedings, 2313*(December). https://doi.org/10.1063/5.0032223

Fenibo, E. O., Ijoma, G. N., & Matambo, T. (2021). Biopesticides in sustainable agriculture: A critical sustainable development driver governed by green chemistry principles. *Frontiers in Sustainable Food Systems, 5*, 1–6. https://doi.org/10.3389/fsufs.2021.619058

Gu, Q., Chu, S., Huang, Q., Chen, A., Li, L., & Li, R. (2023). Colletotrichum echinochloae: A potential bioherbicide agent for control of barnyardgrass (Echinochloa crus-galli (L.) Beauv.). *Plants, 12*(3). https://doi.org/10.3390/plants12030421

Gupta, I., Singh, R., Muthusamy, S., Sharma, M., Grewal, K., Singh, H. P., & Batish, D. R. (2023). Plant essential oils as biopesticides: Applications, mechanisms, innovations, and constraints. *Plants, 12*(16), 1–29. https://doi.org/10.3390/plants12162916

Han, V.-C., Michael, P. J., Swift, B., & Bennett, S. J. (2023). Biological control of Sclerotinia sclerotiorum: Modes of action of biocontrol agents, soil organic amendments, and soil microbiome manipulation. *Biological Control, 186*, 105346. https://doi.org/10.1016/j.biocontrol.2023.105346

Hernández-Fernández, M., Cordero-Bueso, G., Ruiz-Muñoz, M., & Cantoral, J. M. (2021). Culturable yeasts as biofertilizers and biopesticides for a sustainable agriculture: A comprehensive review. *Plants, 10*(5), 822. https://doi.org/10.3390/plants10050822

Hoagland, R. E., & Boyette, C. D. (2021). Effects of the fungal bioherbicide, alternaria cassia on peroxidase, pectinolytic and proteolytic activities in sicklepod seedlings. *Journal of Fungi, 7*(12). https://doi.org/10.3390/jof7121032

Hong, S., Shang, J., Sun, Y., Tang, G., & Wang, C. (2024). Fungal infection of insects: Molecular insights and prospects. *Trends in Microbiology, 32*(3), 302–316. https://doi.org/10.1016/j.tim.2023.09.005

Hossain, Md. M. (2024). Upscaling plant defense system through the application of plant growth-promoting fungi (PGPF). In *Microbial technology for agro-ecosystems* (pp. 61–95). Elsevier. https://doi.org/10.1016/B978-0-443-18446-8.00018-8

Keppanan, R., Sivaperumal, S., Hussain, M., Bamisile, B. S., Aguila, L. C. R., Qasim, M., Mekchay, S., Wang, L., & Krutmuang, P. (2019). Molecular characterization of pathogenesis involving the GAS 1 gene from Entomopathogenic fungus Lecanicillium lecanii and its virulence against the insect host Diaphorina citri. *Pesticide Biochemistry and Physiology, 157*, 99–107. https://doi.org/10.1016/j.pestbp.2019.03.012

Khursheed, A., Rather, M. A., Jain, V., Wani, A. R., Rasool, S., Nazir, R., Malik, N. A., & Majid, S. A. (2022). Plant based natural products as potential ecofriendly and safer biopesticides: A comprehensive overview of their advantages over conventional pesticides, limitations and regulatory aspects. *Microbial Pathogenesis, 173*, 105854. https://doi.org/10.1016/j.micpath.2022.105854

Kostina-Bednarz, M., Płonka, J., & Barchanska, H. (2023). Allelopathy as a source of bioherbicides: Challenges and prospects for sustainable agriculture. *Reviews in Environmental Science and Biotechnology, 22*(2), 471–504. https://doi.org/10.1007/s11157-023-09656-1

Lengai, G. M. W., Muthomi, J. W., & Mbega, E. R. (2020). Phytochemical activity and role of botanical pesticides in pest management for sustainable agricultural crop production. *Scientific African, 7*. https://doi.org/10.1016/j.sciaf.2019.e00239

Lerro, C. C., Beane, L. E., Dellavalle, C. T., Andreotti, G., Hofmann, J. N., Koutros, S., Parks, C. G., Shrestha, S., Michael, C., Alavanja, R., Blair, A., Lubin, J. H., Sandler, D. P., & Ward, M. H. (2021). Pesticide exposure and incident thyroid cancer among male pesticide applicators in agricultural health study. *Environment International, 146*(2020), 106187. https://doi.org/10.1016/j.envint.2020.106187

Maldaner, J., Silva, F. S., Santos, D. de A., Silva, S. Y. S., da Cruz Silva, S., Lima, T. da C., Goulart, S. L., Tomchinsky, B., & Oliveira, M. N. (2023). Bioherbicide potential of essential oils from plant species used in folk medicine and cuisine in the Amazon region. *South African Journal of Botany, 162*, 761–766. https://doi.org/10.1016/j.sajb.2023.10.004

Monte, E. (2023). The sophisticated evolution of Trichoderma to control insect pests. *Proceedings of the National Academy of Sciences of the United States of America, 120*(12), 10–12. https://doi.org/10.1073/pnas.2301971120

Monteiro, A., & Santos, S. (2022). Sustainable approach to weed management: The role of precision weed management. *Agronomy, 12*(118). https://doi.org/10.3390/agronomy12010118

Motlagh, M. R. S. (2012). Evaluation of Alternaria alternata causing leaf spot of barnyardgrass grown in rice fields. *African Journal of Microbiology Research, 6*(21). https://doi.org/10.5897/ajmr11.717

Nathiely Ramírez-Guzmán, K., Torres-León, C., Saldaña-Mendoza, S., Martinez-Medina, G., Tranier, M., Roussos, S., De la Cruz-Quiroz, R., Parra-Saldívar, R., Flores-Gallegos, A. C., Rodríguez-Herrera, R., & Aguilar, C. N. (2018). Biocontrol agents as strategy of agro-ecosystem management to restitution of productive soils for food production. In V. Kumar, M. Kumar, & R. Prasad (Eds.), *Phytobiont and ecosystem restitution* (pp. 185–213). Springer. https://doi.org/10.1007/978-981-13-1187-1_10

Nechet, K. L., & Halfeld-Vieira, B. A. (2019). Development of Cercospora leaf spot on Ipomoea weed species for biological control. *BioControl, 64*(2), 185–195. https://doi.org/10.1007/s10526-018-09918-w

Negi, R., Sharma, B., Kaur, S., Kaur, T., Khan, S. S., Kumar, S., Ramniwas, S., Rustagi, S., Singh, S., Rai, A. K., Kour, D., Thakur, N., & Yadav, A. N. (2023). Microbial antagonists: Diversity, formulation and applications for management of pest–pathogens. *Egyptian Journal of Biological Pest Control, 33*(1), 105. https://doi.org/10.1186/s41938-023-00748-2

Ortiz, A., & Sansinenea, E. (2023). Microbial-based biopesticides: Commercialization and regulatory perspectives. In *Development and commercialization of biopesticides* (pp. 103–118). Elsevier.

Pacios-Michelena, S., Valdés, R. A., Saldaña-Mendoza, S. A., Tucuch-Pérez, M. A., González, C. N. A., Alvarez-Perez, O. B., & Ilina, A. (2024). Encapsulation and application of cell-free extracts produced by Penicillium chrysogenum R1 in tomato plants for the control of Fusarium oxysporum. *Environmental Quality Management*. https://doi.org/10.1002/tqem.22182

Pérez-González, O., Gomez-Flores, R., & Tamez-Guerra, P. (2022). Insight into biological control potential of Hirsutella citriformis against Asian Citrus psyllid as a vector of citrus huanglongbing disease in America. *Journal of Fungi, 8*(6). https://doi.org/10.3390/jof8060573

Pradhan, S., Choudhury, A., Dey, S., Hossain, M. F., Saha, A., & Saha, D. (2023). Siderophore-producing *Bacillus amyloliquefaciens* BM3 mitigate arsenic contamination and suppress *Fusarium* wilt in brinjal plants. *Journal of Applied Microbiology, 134*(10). https://doi.org/10.1093/jambio/lxad217

Radhakrishnan, R., Alqarawi, A. A., & Abd Allah, E. F. (2018). Bioherbicides: Current knowledge on weed control mechanism. In *Ecotoxicology and environmental safety* (Vol. 158, pp. 131–138). Academic. https://doi.org/10.1016/j.ecoenv.2018.04.018

Ramasubramanian, R., Karthi, S., Senthil-Nathan, S Sivanesh, H., Shyam Sundar, N., & Stanley-Raja, V., Krutmuang, P. (2022). Effect of bacterial toxin identified from the Bacillus subtilis against the Cnaphalocrocis medinalis Guenée (Lepidoptera: Crambidae). *Toxin Reviews, 42*(1), 264–274. https://doi.org/10.1080/15569543.2022.2111444

Ramírez-Guzmán, N., Torres-León, C., Martínez-Terrazas, E., De la Cruz-Quiroz, R., Flores-Gallegos, A. C., Rodríguez-Herrera, R., & Aguilar, C. N. (2018). Biocontrol as an efficient tool for food control and biosecurity. In *Food Safety and Preservation* (pp. 167–193). https://doi.org/10.1016/b978-0-12-814956-0.00007-x

Ramírez-Guzmán, N., Chávez-González, M., Sepúlveda-Torre, L., Torres-León, C., Cintra, A., Angulo-López, J., Martínez-Hernández, J. L., & Aguilar, C. N. (2020). In J. S. Singh & S. R. B. T.-M. S. in R. E. Vimal (Eds.) *Chapter 1—Significant advances in biopesticide production: strategies for high-density bio-inoculant cultivation* (pp. 1–11). Elsevier. https://doi.org/10.1016/B978-0-12-819978-7.00001-4

Ramírez-Guzmán, N., Roussos, S., Martinez-Medina, G. A., Rodríguez, R., & Aguilar, C. N. (2023). Selection of a biocontrol agent based on a comparative spore production evaluation. *Systems Microbiology and Biomanufacturing, 0123456789*. https://doi.org/10.1007/s43393-023-00198-9

Ray, R. V. (2024). Effects of pathogens and disease on plant physiology. In *Agrios' plant pathology* (pp. 63–92). Elsevier. https://doi.org/10.1016/B978-0-12-822429-8.00002-9

Rocha, G. T., Queiroz, P. R. M., Grynberg, P., Togawa, R. C., de Lima Ferreira, A. D. C., Do Nascimento, I. N., Gomes, A. C. M. M., & Monnerat R. (2023). Biocontrol potential of bacteria belonging to the Bacillus subtilis group against pests and diseases of agricultural interest through genome exploration. *Antonie van Leeuwenhoek, 116*, 599–614. https://doi.org/10.1007/s10482-023-01822-3

Saha, S., Thosar, R., Kabade, S., Pawar, K., & Banerjee, K. (2023). Biopesticides. In *Trajectory of 75 years of Indian Agriculture after independence* (pp. 675–703). Springer Nature. https://doi.org/10.1007/978-981-19-7997-2_25

Salazar, B., Ortiz, A., Keswani, C., Minkina, T., Mandzhieva, S., Pratap Singh, S., Rekadwad, B., Borriss, R., Jain, A., Singh, H. B., & Sansinenea, E. (2023). Bacillus spp. as bio-factories for antifungal secondary metabolites: innovation beyond whole organism formulations. *Microbial Ecology, 86*(1), 1–24. https://doi.org/10.1007/s00248-022-02044-2

Saldaña-Mendoza, S. A., Pacios-Michelena, S., Palacios-Ponce, A. S., Chávez-González, M. L., & Aguilar, C. N. (2023d). Trichoderma as a biological control agent: Mechanisms of action, benefits for crops and development of formulations. *World Journal of Microbiology and Biotechnology, 39*(10), 269. https://doi.org/10.1007/s11274-023-03695-0

Saldaña-Mendoza, S. A., Ramírez-Guzman, K. N., Chávez-González, M. L., & Aguilar-González, C. N. (2022). Enzyme-assisted extraction of mint essential oil and evaluation of its antimicrobial activity against food-borne pathogens and fusarium sp. Extracción por asistencia enzimática de aceite esencial de menta y evaluación de su actividad antimicrobiana contr. *Journal of BioProcess and Chemical Technology, 14*(28), 31–38.

Saldaña-Mendoza, S. A., Chavez-González, M. L., & Aguilar Gonzalez, C. N. (2023). Protocol for the production of trichoderma spores for use as a biological control agent through the revalorization of agro-industrial waste. In *Food waste conversion* (pp. 169–176). Springer Link. https://doi.org/10.1007/978-1-0716-3303-8_12

Saldaña-Mendoza, S. A., Chávez-González, M. L., Ramírez-Guzmán, K. N., & Aguilar, C. N. (2023). Evaluation of the biocontrol effect of Trichoderma sp. Trs GT2 extracts produced by fermentation processes against Botrytis cinerea Btcr GM3. *Environmental Quality Management*, 1–12. https://doi.org/10.1002/tqem.22071

Saldaña-Mendoza, S. A., Pacios-Michelena, S., Palacios-Ponce, A. S., Chávez-González, M. L., & Aguilar, C. N. (2023a). Trichoderma as a biological control agent: mechanisms of action, benefits for crops and development of formulations. *World Journal of Microbiology and Biotechnology, 39*(10). Springer Science and Business Media B.V. https://doi.org/10.1007/s11274-023-03695-0

Santra, H. K., & Banerjee, D. (2020). Natural products as fungicide and their role in crop protection. In *Natural bioactive products in sustainable agriculture* (pp. 131–219). Springer. https://doi.org/10.1007/978-981-15-3024-1_9

Shields, M. W., Johnson, A. C., Pandey, S., Cullen, R., González-Chang, M., Wratten, S. D., & Gurr, G. M. (2019). History, current situation and challenges for conservation biological control. *Biological Control, 131*(2018), 25–35. https://doi.org/10.1016/j.biocontrol.2018.12.010

Sindhu, S. S., & Sehrawat, A. (2017). Rhizosphere microorganisms: Application of plant beneficial microbes in biological control of weeds. In *Microorganisms for sustainability* (Vol. 6, pp. 391–430). Springer. https://doi.org/10.1007/978-981-10-6241-4_19

Singh, J., Mishra, S., & Singh, V. (2024). Fungal metabolites as novel plant pathogen antagonists. In *Nanohybrid Fungicides* (pp. 209–237). Elsevier. https://doi.org/10.1016/B978-0-443-23950-2.00012-6

de Souza, A. R. C., Baldoni, D. B., Lima, J., Porto, V., Marcuz, C., Ferraz, R. C., Kuhn, R. C., Jacques, R. J. S., Guedes, J. V. C., & Mazutti, M. A. (2015). Bioherbicide production by Diaporthe sp. isolated from the Brazilian Pampa biome. *Biocatalysis and Agricultural Biotechnology, 4*(4), 575–578. https://doi.org/10.1016/j.bcab.2015.09.005

Stankovic, S., Kostic, M., Kostic, I., & Krnjajic, S. (2020). Practical approaches to pest control: The use of natural compounds. In *Pests, weeds and diseases in agricultural crop and animal husbandry production* (pp. 1–18). https://doi.org/10.5772/intechopen.91792

Tailor, S., Jain, K., Marwal, A., Meena, M., & Mishra, A. (2023). Agricultural management by improving beneficial microflora. In *Plant-microbe interaction—recent advances in molecular and biochemical approaches* (pp. 95–116). Elsevier. https://doi.org/10.1016/B978-0-323-91876-3.00004-X

Tomar, P., Thakur, N., Jhamta, S., Chowdhury, S., Kapoor, M., Singh, S., Shreaz, S., Rustagi, S., Rai, P. K., Rai, A. K., & Yadav, A. N. (2024). Bacterial biopesticides: Diversity, role in pest management and beneficial impact for agricultural and environmental sustainability. *Heliyon, 10*(11), e31550. https://doi.org/10.1016/j.heliyon.2024.e31550

Ulrich, A., Lerin, L. A., Camargo, A. F., Scapini, T., Diering, N. L., Bonafin, F., Gasparetto, I. G., Confortin, T. C., Sansonovicz, P. F., Fabian, R. L., Reichert Júnior, F. W., Treichel, H., Müller, C., & Mossi, A. J. (2021). Alternative bioherbicide based on Trichoderma koningiopsis: Enzymatic characterization and its effect on cucumber plants and soil organism. *Biocatalysis and Agricultural Biotechnology, 36*. https://doi.org/10.1016/j.bcab.2021.102127

Ulrich, A., Müller, C., Gasparetto, I. G., Bonafin, F., Diering, N. L., Camargo, A. F., Reichert Júnior, F. W., Paudel, S. R., Treichel, H., & Mossi, A. J. (2023). Bioherbicide effects of Trichoderma koningiopsis associated with commercial formulations of glyphosate in weeds and soybean plants. *Crop Protection, 172*. https://doi.org/10.1016/j.cropro.2023.106346

Urrutia, R. I., Jesser, E. N., Gutierrez, V. S., Rodriguez, S., Gumilar, F., Murray, A. P., Volpe, M. A., & Werdin-González, J. O. (2023). From waste to food and bioinsecticides: An innovative system integrating Tenebrio molitor bioconversion and pyrolysis bio-oil production. *Chemosphere, 340*. https://doi.org/10.1016/j.chemosphere.2023.139847

Villarreal, G. U., Campos, E. V. R., de Oliveira, J. L., & Fraceto, L. F. (2023). Development and commercialization of pheromone-based biopesticides: a global perspective. In *Development and commercialization of biopesticides* (pp. 37–56). Elsevier.

Yang, X., Wei, X., Yang, J., Du, T., Yin, C., Fu, B., Huang, M., Liang, J., Gong, P., Liu, S., Xie, W., Guo, Z., Wang, S., Wu, Q., Nauen, R., Zhou, X., Bass, C., & Zhang, Y. (2021). Epitranscriptomic regulation of insecticide resistance. *Science Advances, 7*(5). https://www.science.org

Yin, Y., Miao, J., Shao, W., Liu, X., Zhao, Y., & Ma, Z. (2023). Fungicide resistance: Progress in understanding mechanism, monitoring, and management. *Phytopathology®, 113*(4), 707–718. https://doi.org/10.1094/PHYTO-10-22-0370-KD

Microbial Pigments Production from Agro-Industrial Waste

Alessia Tropea, Daniele Giuffrida, Cassamo U. Mussagy, Laurent Dufossé, and Luigi Mondello

Abstract

The agro-industrial processing and post-harvesting downstream products generate massive volumes of waste which, if disposed and released unprocessed, can cause environmental pollution. Due to their rich chemical composition, these residues could be implemented as fermentative substrates for obtaining value-added products. Among these, natural microbial pigments are gaining more attraction as alternatives to synthetic ones, with their production considered eco-friendly and biocompatible. Agro-industrial waste-based bio-refineries to produce microbial pigments can represent a suitable approach for both the reduction of environmental pollution, as well as for providing a sustainable framework for addressing bioeconomy. This chapter aims to give an overview of the main microorganisms used for agro-industrial waste valorization for pigment production through a biotechnological process from a circular economy point of view. Moreover, the main drawbacks of the implementation of these residues are discussed.

Keywords

Microorganisms · Pigments · Agro-industrial waste · Fermentation · Circular bioeconomy

A. Tropea (✉) · L. Mondello
Messina Institute of Technology C/o Department of Chemical, Biological, Pharmaceutical and Environmental Sciences, Former Veterinary School, University of Messina, Messina, Italy
e-mail: atropea@unime.it

D. Giuffrida
Department of Biomedical, Dental, Morphological and Functional Imaging Sciences, University of Messina, Messina, Italy

C. U. Mussagy
Escuela de Agronomía, Facultad de Ciencias Agronomicas y de los Alimentos, Pontificia Universidad Catòlica de Valparaíso, Quillota, Chile

L. Dufossé
Chemistry and Biotechnology of Natural Products, CHEMBIOPRO, ESIROI Agroalimentaire, Université de La Réunion, Saint-Denis, France

L. Mondello
Chromaleont S.R.L., c/o Department of Chemical, Biological, Pharmaceutical and Environmental Sciences, Former Veterinary School, University of Messina, Messina, Italy

1 Introduction

Agro-industrial processing generates massive amounts of waste, which are recognized as the most copious renewable resources on earth. It has been reported that by 2050 the total amount of this waste will reach 3.40 billion metric tons.

These wastes, if disposed of untreated can cause environmental pollution, causing several health issues for living organisms. The chemical composition of the agro-industrial waste is affected by the processed starting material. Overall, the main constituents are fibers, sugars, proteins, and minerals. Moreover, these wastes are characterized by a high biological oxygen demand (BOD), and chemical oxygen demand (COD), which can determine environmental pollution if not properly disposed of (Wu et al., 2018).

The most applied methods for agricultural waste management are represented by landfilling, open dumping, burning, or incineration, increasing noteworthy environmental issues such as air and groundwater contaminations (Grewal et al., 2022; Maglione et al., 2024). To overcome these concerns a great interest has been focused on their reuse allowing them to get added-value products from their management, in accordance with the Sustainable Development Goals (SDGs) of the United Nations, in particular, SDGs 12 "*Ensure sustainable consumption and production patterns*" and 13 "*Take urgent action to combat*

K. Sharma (ed.), *Bio-prospecting of Novel Microbial Bioactive Compounds for Sustainable Development*, Sustainable Landscape Planning and Natural Resources Management, https://doi.org/10.1007/978-3-031-82178-3_6

climate change and its impacts". In recent years, agricultural waste has been largely applied through biotechnological techniques as promising substrates for the production of biofertilizers, enzymes, biomolecules, feed, biofuel, bioenergy, etc. (Tropea, 2022a), providing at the same time a sustainable and cost-effective process oriented to functionalize the circular bioeconomy model (Balbino et al., 2023; Tropea, 2022b).

Among the several added-value products that can be obtained by agro-industrial waste valorization, their implementation for microbial pigment production is gaining more interest from both scientific and industrial points of view. In fact, the dyes and pigments market, mainly represented by the textile, feed and food, nutraceutical, and pharmaceutical industrial sectors, is expected to reach 49.1 billion dollars by 2027 (Grewal et al., 2022). To date, most of them are still chemically synthesized, producing molecules potentially toxic, carcinogenic, and characterized by low biodegradability, causing further concerns for the environment, and the health of both humans and animals. Therefore, there is now growing interest in natural pigments because of their biodegradability, the absence of toxicity, carcinogenicity, and allergenicity properties, resulting in a growing acceptance by consumers, and a decrease in health hazards and environmental pollution (Grewal et al., 2022).

Among natural sources, microbial pigments show several major advantages over pigments obtained from plants, such as faster growth, no seasonal restrictions, independence from weather conditions, and large-scale production in bioreactors through fermentation processes, together with their high stability and solubility (Grewal et al., 2022).

Moreover, further important advantages are represented by the possibility of obtaining different pigments by changing the medium composition and the fermentation parameters as well as via microbial gene manipulations. In order to be employed for pigment production, microorganisms have to be non-pathogenic, nontoxic, classified as GRAS (generally recognized as safe), capable of metabolizing different carbon and nitrogen sources, as well as to produce high pigment yield to be extracted easily.

However, the main key concerns regarding microbial pigment production at a commercial scale are still represented by economic matters (Aman et al., 2021). In fact, the synthetic growth media are recognized as the main factors affecting the biotechnological synthesis costs. Therefore, the employment of inexpensive substrates can represent a possible solution for reducing the cost of natural microbial pigment production. In this scenario, the agro-industrial residues can represent a potential alternative to be considered as an ideal substrate for this purpose. These waste valorizations by their implementation as nutritional substrates for pigment production help in reducing environmental and health hazards as well as in supporting the development of economical bioprocess. This chapter is focused on providing a global overview regarding the main microorganisms involved in agro-industrial waste valorization for pigment production through the most applied biotechnological strategies. The main microbial pigment produced and their industrial application have been reported as well as the positive impact in terms of the circular economy offered by this biorefinery strategy.

2 Microorganisms Involved

Microorganisms like bacteria, fungi, microalgae, and yeasts are able to produce different types of pigments of various chemical classes, such as carotenoids, anthocyanins, and prodigiosin through fermentation processes of different substrates (Barreto et al., 2023). In this regard, the substrate composition plays a crucial role. Although the pigment production is genetically determined and it is strictly correlated to the microbial strain employed, it can be also influenced by nutritional factors such as carbon and nitrogen sources, which when represented by synthetic culture media can account for up to 70% of the whole bioprocess costs. Thus, the implementation of cost-effective microbial nutrients could reduce the process costs as well as improve the final fermentation efficiency in terms of the production of pigment.

Agro-industrial wastes, due to their chemical composition, have been investigated as suitable substrates for microbial pigment production by different microbial wild-type and modified strains (Table 1). Usually, the wild-type strains are not suitable for pigment production from a marketable point of view due to the low product yields that are often not economically viable. To improve the microbial strains, enabling them to increase the product yield, use different substrates for growing, as well as to select the gene expression of the desired products, molecular biology techniques and metabolic engineering strategies have been applied for the development of engineering microorganisms (Barreto et al., 2023; Grewal et al., 2022).

Bacteria

The bacteria employed for pigment production shows several advantages over fungi, due to their short life cycle, because of their speedy growth patterns that can be genetically controlled for promoting higher pigment yield, and easy maintenance. Moreover, the nontoxic nature of the pigments produced, together with their easy separation from cell biomass, make bacteria interesting sources to be further investigated for biotechnological pigments

Table 1 Production of microbial pigments using agro-industrial wastes by wild and modified strain of microbes

	Microorganisms	Strain	Agro-industrial Waste	Pigments	References
Wild type	Bacteria	*Bacillus safensis*	Fruit waste	Melanin	Tarangini and Mishra (2014)
		Chryseobacterium artocarpi	Liquid pineapple waste	Yellowish orange flexirubin type structure	Aruldass et al. (2015)
		Chromobacterium vaccinii	Rapeseed cake	Violacein	Cassarini et al. (2022)
		Chromobacterium violaceum	Liquid pineapple waste	Violacein and deoxyviolacein	Aruldass et al. (2015)
		Dietzia maris	Sugarcane bagasse hydrolysate	Trans-canthaxanthin	Goswami et al. (2015)
		Pseudomonas aeruginosa	Grape seed	Pyocyanin	El-Fouly et al. (2015)
		Serratia marcescens	Cassava waste, Kitchen waste; Groundnut cake	Prodigiosin	Casullo de Araújo et al. (2010), Nguyen et al. (2022)
Wild type	Fungi	*Aspergillus carbonarius*	Pomegranate pulp	Yellow bio-based pigmets	Arikan et al. (2020)
		Aspergillus niger	Tomato waste	Carotenoid	Gonzalez et al. (2023)
		Aureobasidium pullulans	Carrot Peel	Melanin	Müjdeci (2022)
		Blakeslea trispora	Fruits and vegetables waste	Carotenoids	Kaur et al. (2019)
		Monascus purpureus	Corn bran byproduct; whey; bakery waste; Orange processing waste; rice straw; Brewer's spent grain	Red extracellular pigments: monascorubramine and rubropuntamine	Mehri et al. (2021), Kantifedaki et al. (2018)
		Monascus sanguineus	Broken rice	Red extracellular pigments: monascorubramine and rubropuntamine	Kumar et al. (2021)
		Monascus ruber	Broken rice	Red and yellow pigments	Vidyalakshmi et al. (2009)
		Penicillium purpurogenum	Orange peels	Azaphilones	Kantifedaki et al. (2018)
		Talaromyces atroroseus	Corncob	Azaphilones	Morales-Oyervides et al. (2020)
Wild type	Microalgae	*Arthrospira platensis*	Industrial wastewater	Phycobiliproteins	Arashiro et al. (2020)
		Chlorella protothecoides	Ricotta production waste, scotta	Astaxanthin	Ribeiro et al. (2017)
		Dunaliella salina	Food waste hydrolysate; Poultry litter	Carotenoids	Wang et al. (2021), Han et al. (2019)
		Mychonasterotundus sp.	Palm oil mill	β-carotene	Sekatresna et al. (2015)
		Nostoc sp.	Industrial wastewater	Phycobiliproteins	Arashiro et al. (2020)
		Haematococcus pluvialis	Primary-treated piggery wastewater	Astaxanthin	Coutinho Rodrigues et al. (2021)
		Haematococcus sp.	Sugarcane vinasse	Astaxanthin	Rishu et al. (2021)
		Phormidium autumnale	Agro-industrial wastewater	Carotenoid	Rodrigues et al. (2014)
		Porphyridium purpureum	Industrial wastewater	Phycobiliproteins	Arashiro et al. (2020)

(continued)

Table 1 (continued)

	Microorganisms	Strain	Agro-industrial Waste	Pigments	References
Wild type	Yeasts	*Phaffia rhodozyma/ Xanthophyllomyces dendrorhous*	Soy molasses, molasses, sugarcane juice, Mustard waste; wheat wastes	Astaxanthin	de Carvalho et al. (2022), Dursun and Dalgic (2016)
		Pichia fermentans	Sugar cane molasses	Lutein and β-carotene	Cipolatti et al. (2019)
		Rhodotorula glutinis	Corncob; olive mill wastewater; mung bean waste; goat milk whey	Torulene; β-carotene; total carotenoids	Rane et al. (2023), Keskin et al. (2023)
		Rhodotorula marina	Spent brewer's yeast	Carotenoids	Rodrigues et al. (2019)
		Rhodotorula minuta	Spent brewer's yeast	Carotenoids	Rodrigues et al. (2019)
		Rhodotorula mucilaginosa	spent brewer's yeast; coffee waste, rice straw; waste ketchup and sugar-cane molasses; onion peels and mung bean husk	β-carotene, torulene, torularhodin	Rodrigues et al. (2019), Bardhan et al. (2022)
		Rhodotorula pacifica	Sugarcane molasses, bagasse	β-carotene	Deeba et al. (2022)
		Rhodotorula pallida	Spent brewer's yeast	Carotenoids	Rodrigues et al. (2019)
		Sporidiobolus pararoseus	Sugar cane molasses and corn steep liquor; sugar cane molasses	Carotenoids; β-carotene	Borba et al. (2018), Cipolatti et al. (2019)
		Sporidiobolus salmonicolor	Wheat wastes	Astaxanthin	Dursun and Dalgıç (2016)
		Sporobolomyces roseus	Pasta processing waste	Carotenoids	Villegas-Mendez et al., (2023)
		Yamadazyma guilliermondii	Wheat wastes	Astaxanthin	Dursun and Dalgıç (2016)
		Yarrowia lipolytica	Wheat wastes	Astaxanthin	Dursun and Dalgıç (2016)
Modified	Bacteria	*Paracoccus aminophilus and Paracoccus kondratievae*	Wastewater supplemented with molasses, wheat straw	Carotenoids	Maj et al. (2020), Pyter et al. (2022)
		Rhodobacter sphaeroides	Corn steep liquor	Lycopene	Su et al. (2018)
Modified	Fungus	*Monascus purpureus*	Rice husk	Yellow and orange *Monascus* pigments	Zhang et al. (2022)
Modified	Microalgae	*Chlorella zofingiensis*	Pretreated cane molasses	Astaxanthin	Liu et al. (2013)
Modified	Yeast	*Rhodosporidium toruloides*	Sugarcane molasses	Torulene, torularhodin, and β-carotene, astaxanthin	Zheng et al. (2021), Tran et al. (2023)

production. Carotenoids, violacein, melanin, pyocyanin, prodigiosin, actinorhodin, and zeaxanthin are examples of the wide range of pigments that can be produced by these microorganisms.

Despite bacterial pigments have been recognized among the key bio-products to be largely implemented in several industrial sectors, successful commercialization is still challenging due to the high cost of synthetic growth media (Venil et al., 2014). In order to overcome this issue, several research have been carried out in order to evaluate the chance of implementing cheaper growth media such as agro-industrial waste. Among the pigment-producing bacteria, *Serratia marcescens* and *Chromobacterium sp* are the most employed for prodigiosin and violacein production respectively by using different substrates such as liquid pineapple waste, cassava waste, kitchen waste or groundnut cake (Table 1).

Yeasts

Yeasts have been used for agro-industrial waste valorization into biotechnologically important pigments, mainly represented by carotenoids such as lutein, β-carotene, astaxanthin, and torulen (Usmani, 2020). They are versatile organisms able to metabolize a large spectrum of carbon sources. Due to their easy growth and the possibility to easily optimize the fermentation process for modulating the type and the stability of pigments produced, yeasts are reported to be a cost-effective and suitable option for large-scale pigment production (Balbino et al., 2023). Nevertheless, yeast-derived pigments are usually synthesized and intracellularly stored, representing a physical and tough limitation on the whole process (Balbino et al., 2023).

Among yeasts, *Rhodotorula sp.* has been the most employed due to its enhanced carotenogenic ability to use agro-industrial waste. Several studies have been reported on the implementation of spent brewer's yeast, coffee waste, rice straw, waste ketchup, sugarcane molasses, etc. for carotenoid production (Table 1).

Fungi

Filamentous fungi are microorganisms able to produce a remarkable variety of pigments represented by monascins, carotenoids, melanins, quinones, flavins, violacein, and indigo (Dufossé et al., 2014). Most of these compounds are produced as extracellular pigments, representing a significant requisite from an industrial point of view since the pigment extraction procedures cause an increase in the final process costs. The main drawback of their commercial applications is represented by their ability to secrete secondary toxic metabolites or mycotoxins, causing health and environmental issues (Balbino et al., 2023; Grewal et al., 2022). The most representative example is *Monascus* sp, largely used in Asia but forbidden in the United States and Europe because of the simultaneous production of pigments and mycotoxins. In order to overcome this issue, methods have been developed to avoid toxin production such as the identification of non-pathogenic strains, the control of biosynthesis and media selection. This fungal culture is currently the most usually employed for bio-colors production starting from agro-industrial waste. It is responsible for the production of three kinds of pigments, classified according to the colors in yellow, represented by ankaflavins and monascin, orange, due to rubropunctatin and monascorubin, and red color given by rubropunctamine and monascorubramine (Usmani, 2020) (Table 1).

Microalgae

Microalgae are recognized as an important biotechnological platform for pigment production because of their biodiversity, fast growth, capacity to grow in wastewater or wasteland, and flexible metabolisms.

In particular, they show the innate capacity to utilize a wide range of carbon and nitrogen sources in organic forms, such as reducing and non-reducing sugars, proteins, nucleic acids, and urea, that can be found in wastewater and use these molecules for value-added products, such as pigments, mainly represented by carotenoids and phycobiliproteins (Rishu et al., 2021; Arashiro 2020). Among microalgae, *Haematococcus pluvialis, Arthrospira platensis, Chlorella protothecoides* and *Dunaliella salina* are the most used for carotenoids and phycobiliproteins production by using agro-industrial waste (Table 1).

3 Agro-Industrial Waste Fermentation Techniques for Microbes' Pigment Production

Agro-industrial waste and byproducts bioconversion into value-added products are mainly based on biotechnological approaches, which the most applied are represented by fermentation processes, occurring by the implementation of microorganisms able to break down organic compounds for obtaining energy for their metabolisms, and at the same time for producing functional products. Fermentation strategies are the biotechnological techniques applied in order to meet the market demand for microbial pigment production, having a large array of industrial applications. Agro-industrial waste fermentation is becoming a profitable and promising technology for microbial pigment production, since microorganisms are cultivated using a nutritive substrate that allows fast microbial growth, enhancing at the same time their pigment production, reducing the media costs, representing at the same time an innovative waste management tool.

Depending on the kind of agro-industrial waste to be employed as fermentation substrate, it is possible to identify solid-state (SSF) and submerged fermentations (SmF) as the most applied in novel research and industry for obtaining bioactive compounds, such as antioxidants, protein hydrolysates, enzymes, single cell protein, single cell oils, oligosaccharides, growth-promoting substrates and pigments. The nutrient utilization rates are depending on the different substrates, as well as the final product yields. In solid-state fermentation techniques, carried out mainly by fungal strains for less moisture content requirements, efficacy and enzymatic machinery, the substrates, such as rice and rice straw, wheat bran, paper pulp, hay, bagasse, coconut coir, and so on, are used for microbial growth to produce biomolecules on a solid substrate in the absence or near absence of free water. The nutrients are slowly released during the process determining the possibility to use the same substrate for long periods of time. SSF represents an easy and low energy consumption process, requiring minimal waste pretreatment and generating less wastewater in comparison with submerged fermentation. Nevertheless, the main challenges of this technique are the microorganism selection, inoculum preparation for large-scale cultivation and the maintenance of biological, chemical and physical parameters. On the other hand, product recovery is both technically and economically feasible (Šelo et al., 2021).

In SmF which bacteria, requiring high moisture percentage, are the main microorganisms employed, the substrates are quickly used, therefore fresh nutrients need to be constantly supplemented. In this case, the most used substrates are soluble sugars, molasses, liquid media, sewage/wastewater, fruit and vegetable juices (Aruldass et al., 2015; Mehri et al., 2021; Zhang et al., 2022).

From an industrial point of view, submerged fermentation is the most widely employed practice, due to its substrate utilization easiness, recovery of targeted metabolites, and control of process parameters. In regard to this last factor, it is important to highlight that even if the fermentation parameters are easily controlled in bioreactors, the main challenge in submerged fermentation of agro-industrial waste is represented by process design, since nutrients concentration, pH, temperature, aeration, and dissolved oxygen, are the main parameters affecting microbial growth and their productivity (Kuma et al., 2022).

Depending on the agro-industrial waste composition and the microorganisms employed, it can be necessary to have a substrate pretreatment, mainly by physical, chemical, or enzymatic degradation, in order to make the substrate more suitable and metabolizable by microbes. In this scenario two main different fermentation modes can be carried out, defined as "Separate Hydrolysis and Fermentation" (SHF) and "Simultaneous Saccharification and Fermentation" (SSF).

In the SHF process, enzymatic hydrolysis is carried out previously and separately from fermentation and every step can be performed at optimal pH and temperature conditions. However, accumulation into the medium of the main hydrolysis end-product can inhibit the enzymatic activity. From the plant costs point of view the use, in this process of two separate reactors, increases the capital costs. This fermentation approach has been largely implemented for pigment production by *Monascus purpureus* by using whey as a fermentation substrate (Mehri et al., 2021).

SSF combines hydrolytic enzymes and fermenting microorganisms in the same vessel at the same time and using the same conditions. This method shows a further advantage in preventing the buildup of hydrolysis products, which can further hamper the hydrolysis rate. SSF also seems to reduce the enzyme inhibition, because the microorganisms employed can ferment, at the same time, the released hydrolysis products, so a lower enzyme amount is required. This increases the overall productivity, the concentration and the final product yield, reducing moreover the contamination risks and shortening the fermentation time.

Finally, microbial pigments, such as carotenoids, can be carried out both via batch and fed-batch fermentation. In batch fermentation, the initial phases of the process are carried out at a high concentration of substrate, while the final stages are characterized by a high concentration of the products. This causes microbial growth inhibition and pigment synthesis reduction because of the lack of nutrients when the stationary phase is reached. In order to overcome this issue, enhancing both biomass and pigment productions, fed-batch fermentation can be applied, in which nutrient reduction will be balanced by intermittent feeding of a new medium (Sundararajan & Ramasamy, 2024; Villegas-Mendez et al., 2023).

4 Representative Pigments Produced from Fermentation of Agro-Industrial Residues

Isolating colored microorganisms from their natural environment is easy, but developing processes that can be used on an industrial scale is much more complex. Many published reviews list microbial pigments that have already been isolated, some of them produced on a large scale. This book chapter is dedicated to microbial pigments produced from agro-industrial waste, rich in polysaccharides, proteins, and bioactive compounds, contributing to higher profitability, a sustainable economy, and minimizing environmental problems (de Medeiros et al., 2022). Agro-industrial residues encompass a wide range of substrates, including crop residues like rice straw, sugarcane bagasse, wheat shell, soybean hulls, oil palm empty fruit bunches (EFBs) and corn stalk, as well as animal residues such as feathers, whey, blood, livestock feces and urine, and agro-industrial byproducts like sugarcane bagasse, cassava bagasse, soy hulls, olive mill waste and sugar-beet pulp (Balbino et al., 2023). When using agro-residues, the global environmental impacts of the biocolorants produced should be considered as the critical estimation of sustainability aspects based on life cycle assessment.

Carotenoids are the most advanced microbial pigments, colorants, and dyes in terms of industrial development. Carotenoids have been recognized for their potential health benefits, including the prevention of diseases such as cancer and eye macular degradation, making them valuable for the pharmaceutical and food industries. Although they are currently produced from conventional nitrogen and carbon substrates, more and more research is focused on the use of agricultural co-products. Carotenoids from agro-food by-products can value low-cost renewable resources and reduce industrial waste, providing positive economic and environmental impacts. Green production of carotenoid colorants based on the fermentation of agro-waste should also emphasize the benefits linked with sustainable extraction methods and the use of nano-technological approaches to protect carotenoids from degradation (Raval et al., 2024). Moreover, most of the production processes for microbial pigments are developed and will be developed in the coming years and decades under a biorefinery framework. The development of a sustainable and effective microbial biorefinery is influenced by several factors, including an adequate selection of microbial strain as the main aspect. As an example, the oleaginous red yeast *Rhodotorula sp.* has been identified as one powerful source of high-added-value biomolecules, such as carotenoids, lipids, and enzymes. Even if known for over a century, the implementation of *Rhodotorula sp.* as a resource for valuable products has not been commercialized yet. Many authors investigated different species of this genus, with various residues: *Rhodotorula mucilaginosa* grown on molasses and cassava wastewater (Machado et al., 2022); *Rhodotorula toruloides* grown on sugarcane molasses (Ochoa-Viñals et al., 2024); *Rhodotorula glutinis* grown on olive mill wastewater (Hladnik et al., 2024). The accumulation of carotenoid reached the maximum yield when pretreated molasses at 40 g/L (715.4 ± 15.1 µg/g d.w) was used, which represents a low value compared to carotenoid production of strains belonging to *Blakeslea trispora* or genetically modified strains of *Yarrowia lipolytica* (up to 35–40 g of β-carotene or lycopene per liter of fermentation medium, 35–40 kg per m^3) (Ma et al., 2022). The second example chosen among microbial carotenoids is astaxanthin. This carotenoid, implemented as an antioxidant and food pigment, can be produced by several microorganisms (yeast *Xanthophyllomyces dendrorhous* (former name *Phaffia rhodozyma*), microalgae *Haematococcus pluvialis*, bacteria *Paracoccus carotinifaciens*) in auto- or heterotrophic conditions. Corn steep liquor, cotton husk, corncob, olive pomace, mesquite pods, residual brewery yeast, whey, wheat wastes and urea were tested for solid-state fermentation processes in order to evaluate the economic viability of industrial-scale facilities for producing powdered astaxanthin (PAX) and an astaxanthin oil mixture (AXO) (Dursun et al., 2020). These economic studies on biotechnological solutions are of crucial importance.

Microbial pigments have highly variable structures (Fig. 1), absorbing in a wide range of visible spectrum, and thus giving rise to a wide range of colors, from black, brown, green, blue, red, purple, violet, orange, yellow and pink. Hereafters are some typical examples, combining the production of these dyes with the use of agricultural waste as growth substrates.

Azaphilones (red, orange, and yellow) are already produced in Asia at ton-scale. Produced with *Monascus* filamentous fungi strains on steamed rice using solid fermentation in Asia and currently being developed in Europe, South America, and the USA with *Talaromyces/Penicillium* filamentous strains, the process could be easily switched to the use of agro-industrial residues. Fermented durian seed, potato pomace, rice flour, sugarcane molasses, corncob hydrolysate, cheese whey, sugarcane bagasse, and grape waste gave promising results. The pro of filamentous fungi is their wide range of enzymes, enabling them to easily break down many substrates.

Blue dyes are highly sought after by manufacturers. Researchers around the world are carrying out screenings to find new blue chemical structures and/or new blue microorganisms. The market is currently partly covered by phycocyanin, produced by the microalga *Arthrospira platensis* (former *Spirulina platensis*) or the microalga *Galdieria sulphuraria*. When growing these microalgae it is recommended to use agro-industrial liquid anaerobic digestates instead of the agro-industrial residues themselves (Leca et al., 2024).

Microorganisms also have the advantage of producing pigments with carbon structures that do not exist elsewhere in nature (plants, for example). Market access is undoubtedly a little more complicated (more advanced toxicological studies, discussions with regulatory bodies such as EFSA or FDA) but the potential gains (absence of competition) can be remarkable.

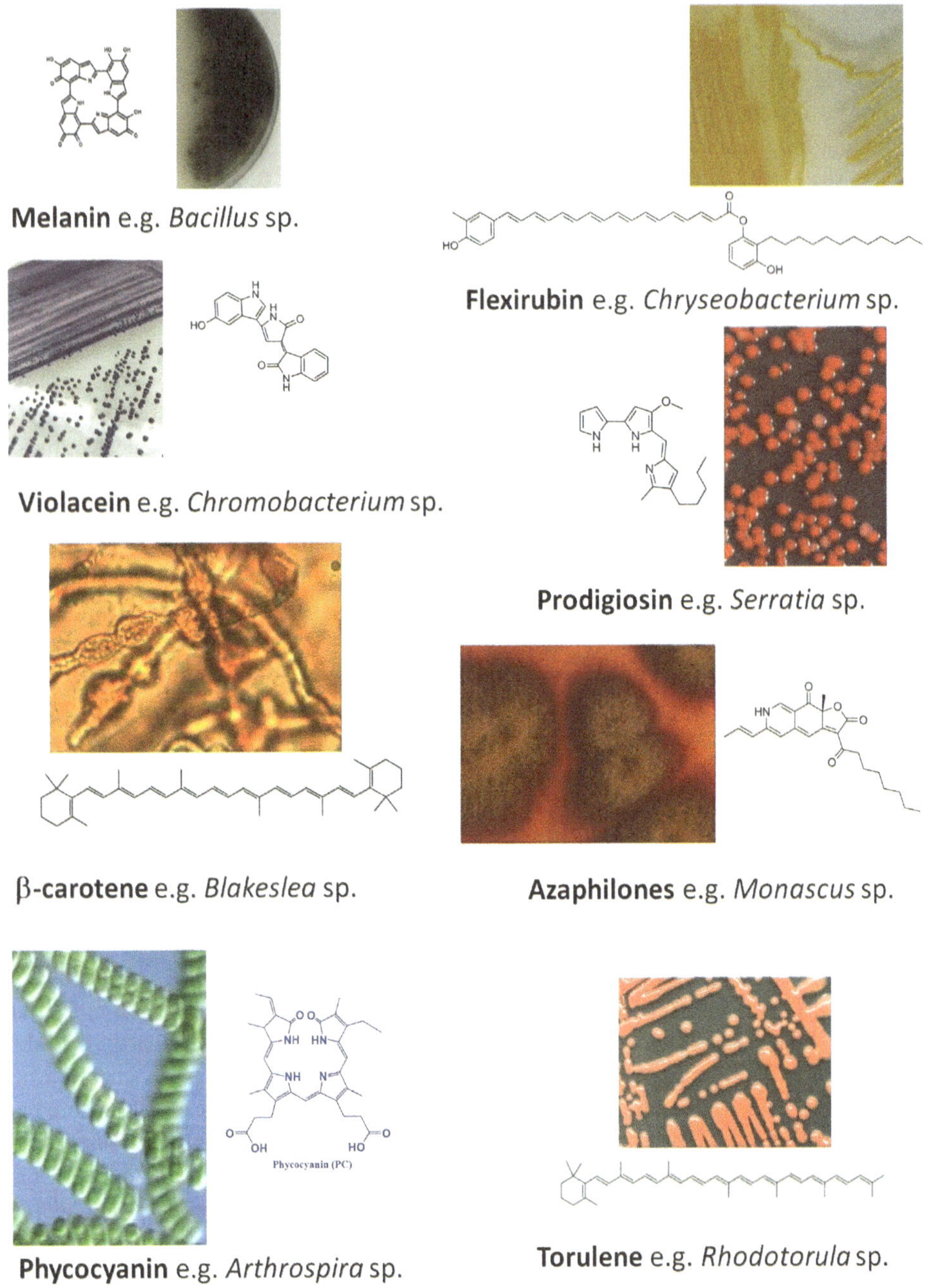

Fig. 1 Main microbial pigments structures and appearance

Violacein is a natural antimicrobial violet–purple pigment mainly produced by *Chromobacterium violaceum* and *Janthinobacterium lividum*. Violacein has also been found to have antifungal, antiviral, and antioxidant activities. Cassarini et al. (2022) studied the production of this molecule on protein-rich agro-industrial co-products, *Chromobacterium vaccinii* grew on gluten (wheat protein) but the production of pigment was boosted by the addition of the amino-acid tryptophan. Violacein now appears to be effectively produced by wild strains or genetically modified strains, however, the true market is not clear, pigment/colorant or antimicrobial? Prodigiosin is another bacterial pigment of interest, red-colored with versatile biological properties such as antimicrobial, apoptosis inducer, and antineoplastic agent. Reused brewing by-products, wheat bran, tilapia scale hydrolysate, and rice straw, were successfully used as nutrient sources for prodigiosin production (Miglani et al., 2023). Textile dyeing is another potential outlet for large-scale production of prodigiosin.

To summarize this part, engineers, biotechnologists involved in microbial pigment production from fermentation of agro-industrial residues should never forget (1) to be cost-effective (high biomass level, high pigment/colorant level in the biomass, easy and cheap extraction, purification of the pigment(s)/colorant(s)…); (2) to be safe (market = FOOD/FEED, no mycotoxin production if fungi is used, pigment is non-toxic itself; (market = TEXTILE, no allergenicity). It is also important that microbial pigment production does not generate more waste than the starting agro-residues (solid/dried agro-residues are in most cases part of a liquid fermentation medium, which means a dilution effect). The most elegant solution is to produce a few recoverable compounds of economic interest in a SINGLE fermentation followed by the final digestion of the remaining organic matter for energy production.

5 Pigment Extraction and Chemical Characterization

Biotechnological techniques, such as fermentation, anaerobic digestion, and composting, are processes able to convert abundant and low-cost waste biomasses into biorefinery products such as bioplastics, biofuels, biofertilizers, and secondary metabolites. Among those secondary chemicals can be included a wide array of important bioactive molecules, which belong to different chemical classes. Pigments are molecules characterized by having an extended system of conjugated double bonds, known as the chromophore, which can be present in different structural frameworks, like carotenoids, azaphylones, anthocyanins, betalains, and chlorophylls; interestingly, recently there is increasing interest in pigments because it is becoming clear that they are important not only as colorants but also as molecules showing beneficial health properties. The downstream stage of the biotechnological microbial pigments production is considered the pigments extraction from the biomass and their eventual chemical characterization. Among the available extraction methodologies, a selection must be done according to the nature of the microorganism because the different cellular walls composition can have different permeabilities to polar and apolar solvents and according to the polarity of the pigments to be extracted (Martínez et al., 2020). Conventional and emerging new technologies are available for pigment extraction from the different biomasses; among the conventional ones, Solid–liquid extraction, Maceration, and Hydro-distillation are usually applied, whereas among the emerging technologies, Ultrasound-assisted extraction, Microwave-assisted extraction, Pressurized-liquid extraction, Supercritical-fluid extraction, Enzyme-assisted extraction and fermentation technology are gaining applications. The development of novel extraction methods is aimed at obtaining higher pigment yield, lower solvent depletion, minimized environmental effects and the extraction convenience.

The use of microbial enzymes having the capability to disrupt cell walls could offer a valid novel option among the non-thermal pigments extraction methodologies. The main critical drawbacks of the techniques using common organic solvents are: (i) the toxic nature of the solvents used in the extraction, (ii) environmental pollution risks (iii) evaporation of large amounts of liquid for the solvent recovery and recycling requirement, as well as separation and purification of the bioactive compound; therefore, eco-friendly green solvents (ethanol, water, edible oils, inert gases), ionic liquids (ILs), and deep eutectic solvents (DES) have great potential for future applications (Mussagy et al., 2024a). The extracted pigments can eventually be chemically characterized by chromatography coupled or followed by various spectroscopic techniques, commonly UV–visible, mass spectroscopy and nuclear magnetic resonance; indeed high-performance liquid chromatography (HPLC) has provided the best pigments separations technique using different stationary phases, both normal phase and reversed phase, combined with online UV–vis and mass spectrometry detectors, having either APCI (atmospheric pressure chemical ionization) or ESI (electrospray ionization) ion sources.

Very commonly C18 and C30 stationary phases are used in HPLC systems for carotenoid separations; the C30 stationary phase allows also for the separations of carotenoid isomers. Azaphylone type pigments are commonly separated on the C18 stationary phase (Ruiz-Sánchez et al., 2023).

A comprehensive review of general aspects of carotenoids analysis in food and feed has recently been published in Meléndez-Martínez et al. (2022). Mussagy and Dufossé (2023) have recently reviewed astaxanthin production and recovery; also, very recently Mussagy et al. (2023 and 2024b), have reviewed on the production and recovery of the rare carotenoids rhodoxanthin, adonirubin and adonixanthin. An updated perspective of microbial carotenoid production *versus* synthetic pigments has very recently become available (Sundararajan & Ramasamy, 2024). Microbial pigments represent a sustainable source for pigment production thanks to their high economic value and eco-friendly nature; moreover, considering that consumers greatly demand natural pigments, efforts should be made to find microbial strains with higher pigment yields and novel eco-friendly pigment recovery techniques.

6 Microbial Pigments Applications

Currently, the implementation of microbial pigments is gaining more interest from the industrial point of view being applied in several sectors, mainly represented by feed, food, pharmaceutical, nutraceutical, textile, cosmetics, and energy (Fig. 2). These bioactive secondary metabolites, produced by using different substrates including agro-industrial waste, once extracted can be exploited for specific applications through biotechnological tools.

Microbial Pigments in Feed Industry

Fisheries and poultry husbandry are the blossoming sectors where microbial pigments have been employed for improving coloring and nutritional properties (Sundararajan & Ramasamy, 2024).

Supplementing feed with pigments like microbial carotenoids, allows for increase in the animal products quality, making them nutritionally improved and economically valuable, resulting in much more appreciated by the consumers and gainful for the producers. The pink/orange flesh color in crustaceans and farmed fishes as well as the yellow accentuated color of hen's skin and yolk egg, obtained by feed supplementation with astaxanthin, canthaxanthin, and zeaxanthin, are the main representative examples of the microbial pigments implementation in this sector.

Microbial Pigments in Food and Nutraceutical Industry

Foods with attractive colors besides nutritional and health-promoting properties have always gained interest in the food industry. At present, the pigments approved by the Food and Drug Administration (FDA) comprise carotene produced by *Blakeslea trispora* and *Monascus* pigments, riboflavin extracted from *Ashbya gossypii*, Arpink Red from *Penicillium oxalicum*, astaxanthin produced by *Xanthophyllomyces dendrorhous* and cantaxanthin obtained by *Bradyrhizobium sp.* Pigments produced by these microorganisms are largely used in several food products, such as cake icing, ice cream, processed seafoods, meat, ketchup, cheese, energy drinks, red rice, etc. These microbial pigments can be applied in order to improve food characteristics but also as preventing their deterioration by using them either as food-preserving agents or food packaging additives, because of their antioxidant property that hampers the oxidation process (Sundararajan & Ramasamy, 2024), as well as freshness indicators for intelligent packaging.

The implementation of food-based substances for health well-being and disease prevention is determining the current increase of the nutraceutical sectors. In this scenario, microbial dyes, such as melanin, violacein, prodigiosin, and carotenoids, due to their health-related benefits, are emerging as a new generation of nutraceuticals that can replace plant-based counterparts having with lower cost and lower toxicity.

Microbial Pigments in Pharmaceutical and Cosmetic Industries

Most of the microbial pigments, mainly represented by prodigiosin, carotenoids, azaphilones, quinones, melanins, violacein, and flavins, have been reported to have antioxidant, antimicrobial, antiviral, antiproliferative and many more biological activities such as gastroprotective effects, cytotoxic effects on carcinogenic cell line, antiviral activity, anti-inflammatory, anti-diabetic and anti-obesity effects, making them promising biomolecules for pharmaceuticals. Moreover, due to their potential as UV protection agents, antimicrobials and antioxidants, microbial pigments can find large resonance also in the cosmetic industry (Grewal et al., 2022).

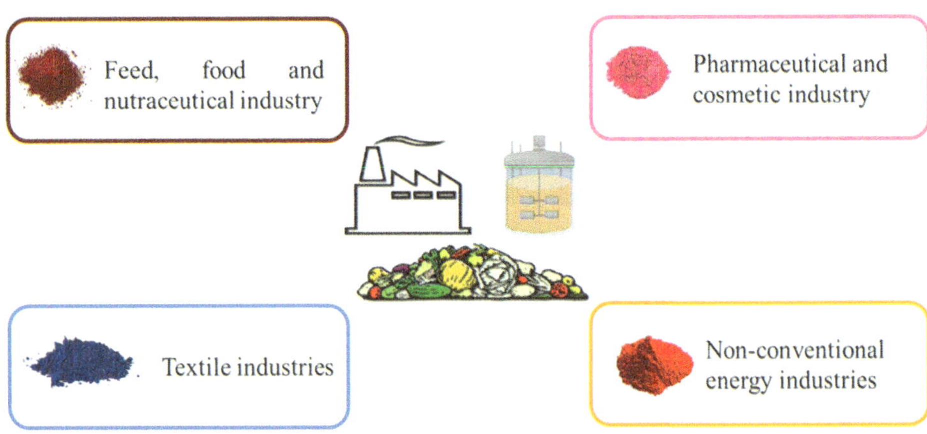

Fig. 2 Main microbial pigments industrial applications

Microbial Pigments in Textile and Energy Industries

The textile industry is the second most polluting industrial sector due to the employment of synthetic, non-biodegradable dyes and mordants. To overcome these issues, bacterial pigments have been suggested as an alternative in order to overcome the environmental pollution, health risks and allergenicity posed by synthetic dyes. Moreover, since several microbial dyes are characterized by antimicrobial activity, their implementation in the textile sector can allow the development of protective clothing (Venil et al., 2021), representing an interesting goal that, in the SARS-CoV-2 pandemic era, should represent a focus for both the scientific and industrial communities.

Finally, among the several industries sectors applying microbial dyes, the energy sector is recently gaining more attraction for obtaining innovative and low-cost photovoltaic devices. In this sense, the development of devices for sunlight conversion into electrical energy by the implementation of microbial dyes as sensitizers represents an interesting alternative (Spadaro et al., 2024; Tropea et al., 2024).

7 Circular Economy and Biorefineries

The modern model of economic production and consumption is not environmentally, economically, or socially sustainable. Currently, most human practices still follow the linear economy—a business model that focuses on taking, making, and disposing of resources. If this model continues, humankind will face increasing resource scarcity. To change this scenario, the circular economy offers an alternative model that proposes moving away from the traditional extract-produce-dispose framework to one that is sustainable and renewal-focused. By emphasizing the extension of a product's life cycle, the circular economy aims to maximize value and promote sustainability across environmental, economic, and social dimensions (Hamam et al., 2021). This approach involves innovative redesign and technological advancements to minimize waste (i.e., agro-industrial) through closed-loop systems, which support cradle-to-cradle processes and the concept of waste as a resource. In fact, agro-industrial wastes, rich in nutrients and organic matter, are excellent raw materials for producing various biomolecules, including natural pigments (Kaur et al., 2022). These wastes provide essential substrates for microbial growth and pigment synthesis. Their diverse composition allows to produce multiple pigments using bacteria, yeast, and microalgae, while their low cost enhances the economic process feasibility. Many microorganisms are highly capable of metabolizing complex organic matter, making these substances effective and efficient substrates for fermentation and pigment production because of their carbon and nitrogen contents. For instance, orange industrial processing waste has been successfully used as the exclusive nutrient source for the natural pigments production by *Monascus purpureus* and *Penicillium purpurogenum* through solid-state, semi-solid-state and submerged fermentations. The results indicated that *M. purpureus* was more efficient than *P. purpurogenum* in pigment production in solid-state fermentation, achieving 9 AU/$g_{biomass}$. In semi-solid-state fermentation, the presence of suspending waste orange peels in liquid media allowed a pigment production of 0.95 AU/mL. Submerged fermentations using boiled or hydrodistilled orange peel residues extracts as fermentation media demonstrated that pigment production could reach up to 0.58 AU mL (Kantifedaki et al., 2018). High-value pigments, such as carotenoids like torulene and torularhodin, have been successfully produced by the yeast *Rhodotorula mucilaginosa* using waste from the olive oil industry. This yeast demonstrated robust growth across various liquid fractions of the residue, metabolizing glucose and fructose to achieve a peak total carotenoid production of 7.3 mg/L (Ghilardi et al., 2020). Another species from the same genus, viz., *Rhodotorula glutinis* yeast, was employed to produce β-carotene using the same residue, olive mill wastewater, achieving a yield of 21 μg/g yeast (Hladnik et al., 2024). In another work, carotenoid production by fermentation reached 129.5 mg/L, using *Phaffia rhodozyma* yeast cultivated in a low-cost food waste (Lai et al., 2022).

While examples of waste valorization can be found in the literature, discussing the circular economy is incomplete without considering biorefineries, which play a crucial role in this framework. Biorefineries are essential for converting biomass, such as agricultural residues, into a diverse array of valuable products, including natural pigments. This process relies on microorganisms in central bioconversion operations, utilizing biochemical pathways to efficiently transform waste biomass into these valuable pigments (Mussagy et al., 2024c). Bio-based biorefineries integrate the management and valorization of waste biomass, enabling the efficient production of high-value products such as carotenoids, azaphylones, anthocyanins, betalains, chlorophylls, and others. In this context, various microorganisms have been investigated under biorefinery and/or circular economy principles to produce pigments. However, it is important to highlight that microbial production for a single biomolecule (i.e., produce only astaxanthin using yeast *P. rhodozyma*) is not as sustainable as desired. The challenges include substantial energy requirements and initial venture costs associated with establishing the entire industrial bioprocess as demonstrated by Mussagy et al. (2024c). Moreover, there are significant ongoing expenses linked to large water volumes, specific nutritional needs, and the broader operational conditions necessary for microbial growth which contributes to the high costs and energy intensity of pigment-focused biorefineries, rendering them less competitive compared to traditional industries based on chemical, electricity and extractive processes. Several researchers have applied the concept of multiproduct cascade biorefineries. For example, a biorefinery aimed at sustainable valorization of microalgae (*Arthrospira maxima*) biomass achieved a yield of 16.3 wt % phycocyanin pigment. Additionally, lipids and other pigments extracted from the phycocyanin-rich biomass yielded 4.43 wt % and 3.0 wt %, respectively. Furthermore, bioethanol was produced through continuous fermentation of the residual carbohydrate-rich biomass, achieving a volumetric ethanol productivity of 4.09 g/L·h in the process (Vignesh et al., 2023). Cyanobacterium *Synechocystis sp.* strain PCC6803 was also employed for producing several biomolecules, including carotenoids, proteins, and lipids by applying a cascade biorefinery approach (Imbimbo et al., 2023). Integrated yeast-based biorefinery for the simultaneous production of different value-added products with minimum waste production was recently suggested as a sustainable and cost-effective process, performed by utilizing *Rhodotorula pacifica* strain INDKK cultivated in lignocellulosic biomass to produce 210.4 mg/L of β-carotene. The yeast also produced 11.8 g/L of lipid production in bioreactors with 5% (v/v) molasses (Deeba et al., 2022).

As clearly observed, microbial-based biorefineries are crucial in the circular economy as they enable the extraction of valuable products from agricultural residues and other biomass sources. However, it is essential to prioritize economic, environmental, and sustainable considerations throughout the biorefinery process. Rather than solely focusing on extracting natural pigments, biorefineries should adopt a cascade approach to extract multiple products simultaneously. In the natural pigments production sector specifically, biorefineries play a crucial role by facilitating the concurrent production of multiple pigments and lipids. This simultaneous production reduces the overall cost of pigment manufacturing, allowing microbial biorefineries to compete favorably with synthetic pigments.

8 Conclusion and Final Remarks

Pigments have been wide implemented in several industrial sectors and the increasing "*chemophobia*" as a consequence of the several concerns for environmental pollution and human and animal health, is addressing the attention to natural microbial pigments implementation. Microbial cell factories for pigment production have constantly increased in the last years due to the possibility of controlling the growth conditions and the several advantages of plants and animals. Microbial pigments have been successfully implemented for several industrial applications and the progress in biotechnologies allowed microbial pigment production to reach the business scale, addressing mainly the interest in fermentation processes based on the implementation of agro-industrial residues for reducing the pigments production costs in various industries, promoting sustainability, addressing the transition to a circular bioeconomy and minimizing at the same time environmental contamination.

Nevertheless, the difficulty in predicting the amount of these residues generated per year, as well as their heterogeneity and the complex biochemical composition, are all factors that result in a difficult valorization of the process design. Moreover, microorganism screening and selection, by applying genetic engineering techniques, and the evolution of the best fermentative process to be carried out using agro-industrial residue as well as the extraction protocols end product recovery, are all competitive issues to be evaluated for commercial prospectives. In particular, downstream process scaling up, for pigment recovery still needs to be deeply faced from an industrial point of view due to the nature of microbial pigments, being secondary metabolites that can be accumulated either intracellularly or extracellularly, often characterized by environmental instability.

References

Aman, M. M., Ahangari, H., Mousazadeh, S., Hosseini, S. M., & Dufossé, L. (2021). Microbial pigments as an alternative to synthetic dyes and food additives: A brief review of recent studies. *Bioprocess and Biosystems Engineering, 2021*, 1–12.

Arashiro, L. T., Boto-Ordóñez, M., Van Hulle, S. W. H., Ferrer, I., Garfí, M., & Rousseau, D. P. L. (2020). Natural pigments from microalgae grown in industrial wastewater. *Bioresource Technology, 303*, 122894. https://doi.org/10.1016/j.biortech.2020.122894

Arikan, E. B., Canli, O., Caro, Y., Dufossé, L., & Dizge, N. (2020). Production of bio-based pigments from food processing industry by-products (apple, pomegranate, black carrot, red beet pulps) using Aspergillus carbonarius. *Journal of Fungi, 6*, 240. https://doi.org/10.3390/jof6040240

Aruldass, C. A., Chidambaram, R., Venil, K., & Ahmad, W. A. (2015). Violet pigment production from liquid pineapple waste by Chromobacterium violaceum UTM5 and evaluation of its bioactivity. *RSC Advances, 5*, 51524. https://doi.org/10.1039/c5ra05765e

Balbino, T. R., Sanchez-Munoz, S., Díaz-Ruíz, E., Rocha, T. M., Mier-Alba, E., Custodio, I. S., et al. (2023). Lignocellulosic biorefineries as a platform for the production of high-value yeast derived pigments—A review. *Bioresource Technology, 386*, 129549. https://doi.org/10.1016/j.biortech.2023.129549

Bardhan, P., Eldiehy, K. S. H., Daimary, N., Gohain, M., Goud, V. V., Deka, D., & Mandal, M. (2022). Structural characterization of mixed rice straw and deoiled algal cake-based substrate as a potential bioenergy feedstock for microbial lipids and carotenoid production. *Waste and Biomass Valorization, 13*, 195–212.https://doi.org/10.1007/s12649-021-01512-z

Barreto, J. V. D. O., Casanova, L. M., Junior, A. N., Reis-Mansur, M. C. P. P., & Vermelho, A. B. (2023). Microbial pigments:

Major groups and industrial applications. *Microorganisms, 11*, 2920.https://doi.org/10.3390/microorganisms11122920

Borba, C. M., das Neves, T. M., Costa, M. C., & de Medeiros, B. J. F. (2018). Carotenoid production by *Sporidiobolus pararoseus* in agroindustrial medium: optimization of culture conditions in shake flasks and scale-up in a stirred tank fermenter. *Brazilian Journal of Chemical Engineering, 35*(2), 509–520. https://doi.org/10.1590/0104-6632.20180352s20160545

Cassarini, M., Crônier, D., Besaury, L., et al. (2022). Protein-rich agro-industrial co-products are key substrates for growth of *Chromobacterium vaccinii* and its Violacein Bioproduction. *Waste Biomass Valor, 13*, 4459–4468. https://doi.org/10.1007/s12649-022-01798-7

Casullo de Araújo, H. W., Fukushima, K., & Takaki, C. G. M. (2010). Prodigiosin production by *Serratia marcescens* UCP 1549 using renewable-resources as a low cost substrate. *Molecules, 15*, 6931–6940. https://doi.org/10.3390/molecules15106931

Cipolatti, E. P., Remedi, R. D., dos Santos, S. C., Rodrigues, A. B., Gonçalves Ramos, J. M., Veiga Burkert, C. A., Furlong, E. B., & de Medeiros Burkert, J. F. (2019). Use of agroindustrial byproducts as substrate for production of carotenoids with antioxidant potential by wild yeasts. *Biocatalysis and Agricultural Biotechnology, 20*, 101208. https://doi.org/10.1016/j.bcab.2019.101208

Coutinho Rodrigues, O. H., Itokazu, A. G., Rorig, L., Maraschin, M., Gordo Correa, R., Pimentel-Almeida, W., & Moresco, R. (2021). Evaluation of astaxanthin biosynthesis by *Haematococcus pluvialis* grown in culture medium added of cassava wastewater. *International Biodeterioration and Biodegradation, 163*, 105269. https://doi.org/10.1016/j.ibiod.2021.105269

de Carvalho, J. C., Goyzueta-Mamani, L. D., Molina-Aulestia, D. T., Magalhães Júnior, A. I., Iwamoto, H., Ambati, R., Ravishankar, G. A., & Soccol, C. R. (2022). Microbial astaxanthin production from agro-industrial wastes—raw materials, processes, and quality. *Fermentation, 8*, 484. https://doi.org/10.3390/fermentation8100484

de Medeiros, T. D. M., Dufossé, L., & Bicas, J. L. (2022). Lignocellulosic substrates as starting materials for the production of bioactive biopigments. *Food Chemistry: X, 13*, 100223. https://doi.org/10.1016/j.fochx.2022.100223

Deeba, F., Kumar, K. K., Wani, S. A., Singh, A. K., Sharma, J., & Gaur, N. A. (2022). Enhanced biodiesel and β-carotene production in *Rhodotorula pacifica* INDKK using sugarcane bagasse and molasses by an integrated biorefinery framework. *Bioresource Technology, 351*, 127067. https://doi.org/10.1016/j.biortech.2022.127067

Dufossé, L., Fouillaud, M., Caro, Y., Mapari, S. A., & Sutthiwong, N. (2014). Filamentous fungi are large-scale producers of pigments and colorants for the food industry. *Current Opinion in Biotechnology, 26*, 56–61. https://doi.org/10.1016/j.copbio.2013.09.007

Dursun, D., & Dalgıç, A. C. (2016). Optimization of astaxanthin pigment bioprocessing by four different yeast species using wheat wastes. *Biocatalysis and Agricultural Biotechnology, 7*, 1–6. https://doi.org/10.1016/j.bcab.2016.04.006

Dursun, D., Koulouris, A., & Dalgıç, A. C. (2020). Process simulation and techno economic analysis of astaxanthin production from agro-industrial wastes. *Waste and Biomass Valorization, 11*(3), 943–954. https://doi.org/10.1007/s12649-018-0439-y

El-Fouly, M. Z., Sharaf, A. M., Shahin, A. A. M., El-Bialy Heba, A., & Omara, A. M. A. (2015). Biosynthesis of pyocyanin pigment by Pseudomonas aeruginosa. *Journal of Radiation Research and Applied Sciences, 8*(1), 36–48. https://doi.org/10.1016/j.jrras.2014.10.007

Ghilardi, C., Sanmartin Negrete, P., Carelli, A. A., & Borroni, V. (2020). Evaluation of olive mill waste as substrate for carotenoid production by *Rhodotorula mucilaginosa*. *Bioresources and Bioprocessing, 7*, 52. https://doi.org/10.1186/s40643-020-00341-7

Gonzalez, C. N. A., Gómez-García, R., & Kuddus, M. (2023). Food waste conversion, methods and protocols in food. *Science, 11*, 161–168. https://doi.org/10.1007/978-1-0716-3303-8_11

Goswami, G., Chaudhuri, S., & Dutta, D. (2015). Studies on the stability of a carotenoid produced by a novel isolate using low cost agro-industrial residue and its application in different model systems. *LWT-Food Science and Technology, 63*, 780–790. https://doi.org/10.1016/j.lwt.2015.03.017

Grewal, J., Wołacewicz, M., Pyter, W., Joshi, N., Drewniak, L., & Pranaw, K. (2022). Colorful treasure from agro-industrial wastes: A sustainable chassis for microbial pigment production. *Frontiers in Microbiology, 13*, 832918. https://doi.org/10.3389/fmicb.2022.832918

Hamam, M., Chinnici, G., Di Vita, G., Pappalardo, G., Pecorino, B., Maesano, G., & D'Amico, M. (2021). Circular economy models in agro-food systems: A review. *Sustainability, 13*, 3453. https://doi.org/10.3390/su13063453

Han, T., Lu, H., Zhao, Y., Xu, H., Zhang, Y., & Li, B. (2019). Two-step strategy for obtaining *Dunaliella* sp. biomass and β-carotene from anaerobically digested poultry litter wastewater. *International Biodeterioration and Biodegradation, 143*, 104714. https://doi.org/10.1016/j.ibiod.2019.06.002

Hladnik, L., Vicente, F. A., Grilc, M., & Likozar, B. (2024). β-Carotene production and extraction: A case study of olive mill wastewater bioremediation by *Rhodotorula glutinis* with simultaneous carotenoid production. *Biomass Conversion and Biorefinery, 14*(7), 8459–8467. https://doi.org/10.1007/s13399-022-03081-0

Imbimbo, P., D'Elia, L., Corrado, I., Alvarez-Rivera, G., Marzocchella, A., Ibáñez, E., Pezzella, C., Branco dos Santos, F., & Monti, D. M. (2023). An alternative exploitation of *Synechocystis* sp. PCC6803: A cascade approach for the recovery of high added-value products. *Molecules, 28*, 3144. https://doi.org/10.3390/molecules28073144

Kantifedaki, A., Kachrimanidou, V., Mallouchos, A., Papanikolaou, S., & Koutinas, A.A. (2018). Orange processing waste valorisation for the production of bio-based pigments using the fungal strains *Monascus purpureus* and *Penicillium purpurogenum*. *Journal of Cleaner Production, 185*, 882e890. https://doi.org/10.1016/j.jclepro.2018.03.032

Kaur, P., Ghoshal, G., & Jain, A. (2019). Bio-utilization of fruits and vegetables waste to produce b-carotene in solid-state fermentation: Characterization and antioxidant activity. *Process Biochemistry, 76*, 155–164. https://doi.org/10.1016/j.procbio.2018.10.007

Kaur, P., Singh, S., Ghoshal, G., Ramamurthy, P. C., Parihar, P., Singh, J., & Singh, A. (2022). Valorization of agri-food industry waste for the production of microbial pigments: An eco-friendly approach. In: Nayak, S. K., Baliyarsingh, B., Mannazzu, I., Singh, A., Mishra, B. B. (Eds.), *Advances in agricultural and industrial microbiology*. Springer. https://doi.org/10.1007/978-981-16-8918-5_8

Keskin, A., Ünlü, A. E., & Takaç, S. (2023). Utilization of olive mill wastewater for selective production of lipids and carotenoids by *Rhodotorula glutinis*. *Applied Microbiology and Biotechnology, 107*, 4973–4985. https://doi.org/10.1007/s00253-023-12625-x

Kuma, V., Sharma, N., Umesh, M., Selvaraj, M., Al-Shehri, B. M., Chakraborty, P., Duhan, L., Sharma, S., Pasrija, R., Awasthi, M. K., Lakkaboyana, S. R., Andler, R., Bhatnagar, A., & Maitra, S. S. (2022). Emerging challenges for the agro-industrial food waste utilization: A review on food waste biorefinery. *Bioresource Technology, 362*, 127790. https://doi.org/10.1016/j.biortech.2022.127790

Kumar, S. A., Vinay, N. D., Gokulakrishnan, M., Shraddha, P., Arivalagan, P., Thivaharan, V., Ramesh, V., & Raj, S. (2021). Production and extraction of red pigment by solid-state fermentation of broken rice using *Monascus sanguineus* NFCCI 2453. *Biocatalysis and Agricultural Biotechnology, 33*, 101964. https://doi.org/10.1016/j.bcab.2021.101964ù

Lai, J. X., Chen, X., Bu, J., Hu, B. B., & Zhu, M. J. (2022). Direct production of astaxanthin from food waste by *Phaffia rhodozyma*. *Process Biochemistry, 113*, 224–233. https://doi.org/10.1016/j.procbio.2022.01.003

Leca, M. A., Beigbeder, J.-B., Castel, L., Sambusiti, C., Le Guer, Y., & Monlau, F. (2024). Cultivation of *Arthrospira platensis* using different agro-industrial liquid anaerobic digestates diluted with geothermal water: A sustainable culture strategy. *Algal Research, 77*, 103348. https://doi.org/10.1016/j.algal.2023.103348

Liu, J., Sun, Z., Zhong, Y., Gerken, H., Huang, J., & Chen, F. (2013). Utilization of cane molasses towards cost-saving astaxanthin production by a *Chlorella zofingiensis* mutant. *Journal of Applied Phycology, 25*, 1447–1456. https://doi.org/10.1007/s10811-013-9974-x

Ma, Y., Liu, N., Greisen, P., Li, J., Qiao, K., Huang, S., & Stephanopoulos, G. (2022). Removal of lycopene substrate inhibition enables high carotenoid productivity in *Yarrowia lipolytica*. *Nature Communications, 13*(1), 572. https://doi.org/10.1038/s41467-022-28277-w

Machado, W. R. C., Murari, C. S., Duarte, A. L. F., & Bianchi, V. L. (2022). Optimization of agro-industrial coproducts (molasses and cassava wastewater) for the simultaneous production of lipids and carotenoids by *Rhodotorula mucilaginosa*. *Biocatalysis and Agricultural, 42*, 102342. https://doi.org/10.1016/j.bcab.2022.102342

Maglione, G., Zinno, P., Tropea, A., Mussagy, C. U., Dufossé, L., Giuffrida, D., & Mondello, A. (2024). Microbes' role in environmental pollution and remediation: A bioeconomy focus approach. *AIMS Microbiology, 10*(3), 723–755. https://doi.org/10.3934/microbiol.2024033

Maj, A., Dziewit, L., Drewniak, L., Garstka, M., Krucon, T., Piatkowska, K., Gieczewska, K., Czarnecki, J., Furmanczyk, E., Lasek, R., Baj, J., & Bartosik, D. (2020). In vivo creation of plasmid pCRT01 and its use for the construction of carotenoid-producing *Paracoccus* spp. strains that grow efficiently on industrial wastes. *Microbial Cell Factories, 19*, 141. https://doi.org/10.1186/s12934-020-01396-z

Martínez, J. M., Delso, C., Aguilar, D. E., Álvarez, I., & Raso, J. (2020). Organic-solvent-free extraction of carotenoids from yeast *Rhodotorula glutinis* by application of ultrasound under pressure. *Ultrasonics Sonochemistry, 61*, 104833. https://doi.org/10.1016/j.ultsonch.2019.104833

Mehri, D., Perendeci, N. A., & Goksungur, Y. (2021). Utilization of whey for red pigment production by *Monascus purpureus* in submerged fermentation. *Fermentation, 7*, 75.https://doi.org/10.3390/fermentation7020075

Meléndez-Martínez, A. J., Mandić, A. I., Bantis, F., Böhm, V., Borge, G. I. A., Brnčić, M., Bysted, A., Cano, M. P., Dias, M. G., Elgersma, A., Fikselová, M., García-Alonso, J., Giuffrida, D., Gonçalves, V. S. S., Hornero-Méndez, D., Kljak, K., Lavelli, V., Manganaris, G. A., Mapelli-Brahm, P., & O'Brien, N. (2022). A comprehensive review on carotenoids in foods and feeds: Status quo, applications, patents, and research needs. *Critical Reviews in Food Science and Nutrition, 62*(8), 1999–2049. https://doi.org/10.1080/10408398.2020.1867959

Miglani, K., Singh, S., Singh, D. P., & Krishania, M. (2023). Sustainable production of prodigiosin from rice straw derived xylose by using isolated *Serratia marcescens* (CMS 2): Statistical optimization, characterization, encapsulation & cost analysis. *Sustainable Food Technology, 1*(6), 837–849. https://doi.org/10.1039/d3fb00100h

Morales-Oyervides, L., Ruiz-Sánchez, J. P., Oliveira, J. C., Sousa-Gallagher, M. J., Morales-Martínez, T. K., Albergamo, A., Salvo, A., Giuffrida, D., Dufossé, L., & Montañez, J. (2020). Medium design from corncob hydrolyzate for pigment production by *Talaromyces atroroseus* GH2: Kinetics modeling and pigments characterization. *Biochemical Engineering Journal, 161*, 107698. https://doi.org/10.1016/j.bej.2020.107698

Müjdeci, G. N. (2022). Experimental modeling and optimization of melanin production by Aureobasidium Pullulans NBRC 100716 in carrot peel extract. *Environmental Progress and Sustainable Energy, 41*, e13919. https://doi.org/10.1002/ep.13919lof12

Mussagy, C. U., & Dufossé, L. (2023). A review of natural astaxanthin production in a circular bioeconomy context using Paracoccus carotinifaciens. *Bioresource Technology, 369*, 128499. https://doi.org/10.1016/j.biortech.2022.128499

Mussagy, C. U., Giuffrida, D., Meléndez-Martínez, A. J., & Dufossé, L. (2023). Rhodoxanthin: The new hit of the natural carotenoids market? *Trends in Food Science and Technology, 136*, 135–144. https://doi.org/10.1016/j.tifs.2023.04.016

Mussagy, C. U., Dias Ana, C. R. V., Santos-Ebinuma, V. C., Sadek, M., Shaaban, A. M., de Andrade, C. R., Haddad, F. F., dos Santos, J. L., Scarim, C. B., Pereira, J. F. B., Floriano, J. F., Herculano, R. D., & Mustafa, A. (2024a). Is the carotenoid production from *Phaffia rhodozyma* yeast genuinely sustainable? A comprehensive analysis of biocompatibility, environmental assessment, and techno-economic constraints. *Bioresource Technology, 397*, 130456. https://doi.org/10.1016/j.biortech.2024.130456

Mussagy, C. U., Farias, F. O., Tropea, A., Santi, L., Mondello, L., Giuffrida, D., Meléndez- Martínez, A. J., & Dufossé, L. (2024b). Ketocarotenoids adonirubin and adonixanthin: Properties, health benefits, current technologies, and emerging challenges. *Food Chemistry, 443*. https://doi.org/10.1016/j.foodchem.2024.138610

Mussagy, C. U., Hucke, H. U., Ramos, N. F., Ribeiro, H. F., Alves, M. B., Mustafa, A., Pereira, J. F. B., & Farias, F. O. (2024c). Tailor-made solventes for microbial carotenoids recovery. *Applied Microbiology and Biotechnology, 108*, 230. https://doi.org/10.1007/s00253-024-13049-x

Nguyen, T. H., Tran, T. H. T., Tran, T. N., Doan, C. T., Ngo, V. A., Ho, N. D., Do, V. C., Nguyen, A. D., & Nguyen, V. B. (2022). Utilization of by-product of groundnut oil processing for production of prodigiosin by microbial fermentation and its novel potent anti-nematodes effect. *Agronomy, 12*, 41. https://doi.org/10.3390/agronomy12010041

Ochoa-Viñals, N., Alonso-Estrada, D., Faife-Pérez, E., Chen, Z., Michelena-Alvarez, G., Martínez-Hernández, J. L., García-Cruz, A., & Ilina, A. (2024). β-Carotene production from sugarcane molasses by a newly isolated *Rhodotorula toruloides* L/24-26-1. *Archives of Microbiology, 206*(6), 245. https://doi.org/10.1007/s00203-024-03973-x

Pyter, W., Grewal, J., Bartosik, D., Drewniak, L., & Pranaw, K. (2022). Pigment production by *Paracoccus* spp. strains through submerged fermentation of valorized lignocellulosic wastes. *Fermentation, 8*, 440. https://doi.org/10.3390/fermentation8090440.

Rane, D. V., Pawar, P. P., Odaneth, A. A., & Lali, A. M. (2023). Microbial oil production by the oleaginous red yeast, *Rhodotorula glutinis* NCIM 3168, using corncob hydrolysate. *Biomass Conversion and Biorefinery, 13*, 1987–1997. https://doi.org/10.1007/s13399-021-01298-z

Raval, S. Y., Arya, P., Jain, M., Sosa, T., Trivedi, P., Dabhi, R., & Raval, V. H. (2024). Sustainable biosynthesis of β-carotene utilizing sugarcane bagasse: Depiction and biotechnological implications. *Biomass Conversion and Biorefinery*. https://doi.org/10.1007/s13399-024-05815-8

Ribeiro, J. E. S., Martini, M., Altomonte, I., Salari, F., Nardoni, S., Sorce, C., da Silva, F. L. H., & Andreucci, A. (2017). Production of *Chlorella protothecoides* biomass, chlorophyll and carotenoids using the dairy industry by-product scotta as a substrate. *Biocatalysis and Agricultural Biotechnology, 11*, 207–213. https://doi.org/10.1016/j.bcab.2017.07.007

Rishu, K., Suchitra, G., & Mayurika, G. (2021). Microalgae bioremediation: A perspective towards wastewater treatment along with industrial carotenoids production. *Journal of Water Process Engineering, 40*, 101794. https://doi.org/10.1016/j.jwpe.2020.101794

Rodrigues, D. B., Flores, É. M. M., Barin, J. S., Mercadante, A. Z., Jacob-Lopes, E., & Zepka, L. Q. (2014). Production of carotenoids from microalgae cultivated using agroindustrial wastes. *Food Research International, 65*, 144–148. https://doi.org/10.1016/j.foodres.2014.06.037

Rodrigues, T. A., Schueler, T. A., da Silva, A. J. R. E. F. C., & Oliveira, F. J. S. (2019). Valorization of solid wastes from the brewery and biodiesel industries for the bioproduction of natural dyes. *Brazilian Journal of Chemical Engineering, 36*(1), 99–107. https://doi.org/10.1590/0104-6632.20190361s20170608

Ruiz-Sánchez, J. P., Montañez, J. P., Giuffrida, D., Dufossé, L., & Morales-Oyervides, L. (2023). Production of pigments under submerged culture through repeated batch fermentation of immobilized *Talaromyces atroroseus* GH2. *Fermentation, 9*(2), 171. https://doi.org/10.3390/fermentation9020171

Sekatresna, W., Dharma, A., Zein, R., & Chaidir, Z. (2015). Isolation and characterization of microalgae isolated from palm oil mill effluent (POME) for biodiesel feed stocks with β-carotene as co-product. *Journal of Chemical and Pharmaceutical Research, 7*(9S), 222–231.

Šelo, G., Planinic, M., Tišma, M., Tomas, S., Koceva, K. D., & Bucic-Kojic, A. (2021). Comprehensive review on valorization of agro-food industrial residues by solid-state fermentation. *Foods, 10*, 927. https://doi.org/10.3390/foods10050927

Spadaro, D., Tropea, A., Citro, I., Trocino, S., Giuffrida, D., Rigano, F., Morales-Oyervides, L., Brinkhoff, T., Tiso, T., Dufossé, L., Calogero, G., & Mondello, L. (2024). Development of innovative dye sensitized solar cells (DSSCs) based on co-sensitization of natural microbial pigments. *Dyes and Pigments, 229*, 112311. https://doi.org/10.1016/j.dyepig.2024.112311

Su, A., Chi, S., Li, Y., Tan, S., Qiang, S., Chen, Z., & Meng, Y. (2018). Metabolic redesign of *Rhodobacter sphaeroides* for lycopene production. *Journal of Agriculture and Food Chemistry, 66*, 5879–5885. https://doi.org/10.1021/acs.jafc.8b00855

Sundararajan, P., & Ramasamy, S. P. (2024). Current perspectives on industrial application of microbial carotenoid as an alternative to synthetic pigments. *Sustainable Chemistry and Pharmacy, 37*, 101353. https://doi.org/10.1016/j.scp.2023.101353

Tarangini, K., & Mishra, S. (2014). Production of melanin by soil microbial isolate on fruit waste extract: Two step optimization of key parameters. *Biotechnology Reports, 4*, 139–146. https://doi.org/10.1016/j.btre.2014.10.0012

Tran, T. N., Tran, N.-T., Tran, T.-A., Pham, D.-C., Su, C.-H., Nguyen, H. C., Barrow, C. J., & Ngo, D.-N. (2023). Highly active astaxanthin production from waste molasses by mutated *Rhodosporidium toruloides* G17. *Fermentation, 9*, 148. https://doi.org/10.3390/fermentation9020148

Tropea, A., Spadaro, D., Trocino, S., Giuffrida, D., Salerno, T. M. G., Ruiz-Sanchez, J. P., Montañez, J., Morales-Oyervides, L., Dufossé, L., Mondello, L., & Calogero, G. (2024). Development of dye-sensitized solar cells using pigment extracts produced by *Talaromyces atroroseus* GH2. *Photochemical and Photobiological Sciences*. https://doi.org/10.1007/s43630-024-00566-x

Tropea, A. (2022a). Biofuels production and processing technology. *Fermentation, 8*, 319. https://doi.org/10.3390/fermentation8070319

Tropea, A. (2022b). Foodwaste. Valorization. *Fermentation, 8*, 168. https://doi.org/10.3390/fermentation8040168

Usmani, Z., Sharma, M., Sudheer, S., Gupta, V. K., Bhat, R. (2020). Engineered microbes for pigment production using waste biomass. *Current Genomics, 21*, 80–95. https://doi.org/10.2174/1389202921999200330152007

Venil, C. K., Aruldass, C. A., Dufossé, L., Zakaria, Z. A., & Ahmad, W. A. (2014). Current perspective on bacterial pigments: Emerging sustainable compounds with coloring and biological properties for the industry-an incisive evaluation. *RSC Advances, 4*, 39523–39529. https://doi.org/10.1039/c4ra06162d

Venil, C. K., Dufossé, L., Velmurugan, P., Malathi, M., & Lakshmanaperumalsamy, P. (2021). Extraction and application of pigment from *Serratia marcescens* SB08, an insect enteric gut bacterium, for textile dyeing. *Textiles, 1*, 21–36. https://doi.org/10.3390/textiles1010003

Vidyalakshmi, R., Paranthaman, R., Murugesh, S., & Singaravadivel, K. (2009). Microbial bioconversion of rice broken to food grade pigments. *Global Journal of Biotechnology and Biochemistry, 4*(2), 84–87.

Vignesh, K., Atchaya, R., Pavan, K. R. G., Shraddha, B., Jaffar, A. B. M., Emerson, A. S., Arun, P. R., & Donato, A. G. A. (2023). A cascaded biorefinery for the sustainable valorization of *Arthrospira maxima* biomass: A circular bioeconomy approach. *Bioresource Technology Reports, 23*, 101510. https://doi.org/10.1016/j.biteb.2023.101510

Villegas-Mendez, M. A., Montanez, J., Contreras-Esquivel, J. C., Salmeron, I., Koutinas, A. A., & Morales-Oyervides, L. (2023). Scale-up and fed-batch cultivation strategy for the enhanced co-production of microbial lipids and carotenoids using renewable waste feedstock. *Journal of Environmental Management, 339*, 117866. https://doi.org/10.1016/j.jenvman.2023.117866

Wang, X., Zhang, M. M., Liu, S. F., Xu, R. L., Mou, J. H., Qin, Z. H., Zhou, Z. G., Li, H. Y., Lin, C., Sze, K., & Sun, Z. (2021). Synergistic bioconversion of lipids and carotenoids from food waste by *Dunaliella salina* with fulvic acid via a two-stage cultivation strategy. *Energy Conversion and Management, 234*, 113908. https://doi.org/10.1016/j.enconman.2021.113908

Wu, K. C., Ho, K. C., Tang, C. C., & Yau, Y. H. (2018). The potential of foodwaste leachate as a phycoremediation substrate for microalgal CO_2 fixation and biodiesel production. *Environmental Science and Pollution Research*. https://doi.org/10.1007/s11356-018-1242-9

Zhang, S., Zhao, W., Nkechi, O., Lu, P., Bai, J., Lin, Q., & Liu, J. (2022). Utilization of low-cost agricultural by-product rice husk for *Monascus pigments* production via submerged batch-fermentation. *Journal of the Science of Food and Agriculture, 102*, 2454–2463. https://doi.org/10.1002/jsfa.11585

Zheng, X., Hu, R., Chen, D., Chen, J., He, W., Huang, L., Lin, C., Chen, H., Chen, Y., Zhu, J., Qi, F., & Xue, T. (2021). Lipid and carotenoid production by the Rhodosporidium toruloides mutant in cane molasses. *Bioresource Technology, 326*, 124816. https://doi.org/10.1016/j.biortech.2021.124816

Sustainable Production of Microbial Biopolymer Utilizing Waste

Chi-Ching Lee

Abstract

The globe has faced significant challenges in recent decades due to waste accumulation from the expansion of agriculture and industrialization. This could be caused by the rapid increase in the global population, which puts a strain on limited natural resources and requires the development of a food supply. Microbial biopolymers produced from various wastes are earning increased interest because of their biodegradability and compostability, biocompatibility, and biobased origin. Many studies have demonstrated the strong potential of biopolymers as biomaterials. Therefore, microbial biopolymers are commonly applied in the food, textile, agricultural, environmental, and medicinal industries. Several types of waste from the food and agricultural industry can be considered desirable feedstocks for sustainable production in microbial biopolymers. A number of microorganisms are involved in biopolymer production, such as polyhydroxyalkanoates (PHAs) and polyhydroxybutyrates (PHBs) from bacteria *Cupriavidus necator*, xanthan gum from bacterial *xanthomonas campestris*, pullulan from fungus *Aureobasidium pullulans* and β-Glucan from microalgae *Euglena gracilis*. This chapter will mention the different characteristics of microbial biopolymers in PHAs, polysaccharides, and polyamides. Various types of wastes and microorganisms in biopolymer production will be discussed. Besides, the key determinants and the challenges in the sustainable production of microbial biopolymers are also highlighted. The microbial biopolymer produced from waste utilization is a promising strategy for environmental sustainability.

Keywords

Microbial biopolymer · Sustainable production · Food and agricultural waste · Industrial waste · Polyhydroxyalkanoates · Bacterial cellulose · Alginate · Xanthan gum · β-glucan

1 Introduction

With the development of civilization and the advancement of science and technology, environmental problems have arisen continuously, although human beings live a convenient life. The need for environmentally friendly alternatives has increased due to rising concerns about environmental sustainability and the depletion of fossil fuel supplies. Various environmental issues have been recognized for decades, such as water pollution, soil pollution, waste and so on. Among them, pollutants in water include a variety of toxic derivatives, especially heavy metals, aromatic molecules, and dyes. The toxicity produced by these toxicants of environmental pollution has potential harm to living things (Salave et al., 2022). Biopolymers are an achievable strategy to lessen the environmental effects of traditional plastics because they are made from sustainable resources. Microorganism-produced biopolymers have drawn much interest because of their biodegradability, biocompatibility, and variety of functional characteristics. Numerous microbes produce microbial biopolymers as part of their metabolisms, including cellulose, polyhydroxyalkanoates (PHAs), and polysaccharides. These biopolymers are used in textiles, agriculture, medicine, packaging, and other manufacture. However, high production expenses

C.-C. Lee (✉)
Faculty of Engineering and Natural Sciences, Department of Food Engineering, Istanbul Sabahattin Zaim University, Istanbul, Türkiye
e-mail: chi.lee@izu.edu.tr

Faculty of Technologies, Department of Food Technology and Nutrition, Klaipeda State University of Applied Sciences, Klaipėda, Lithuania

K. Sharma (ed.), *Bio-prospecting of Novel Microbial Bioactive Compounds for Sustainable Development*, Sustainable Landscape Planning and Natural Resources Management, https://doi.org/10.1007/978-3-031-82178-3_7

and complicated production processes frequently limit the development of microbial biopolymers, emphasizing the need for economical and sustainable manufacturing techniques (Biswas et al., 2022). One inventive and environmentally friendly strategy to deal with these issues is to use waste as a substrate for microbial biopolymer production. Typical types of waste include food waste, industrial byproducts, and agricultural residues. Waste substances are plentiful, reasonably priced, and full of nutrients that can promote microbial growth. Aside from lowering manufacturing costs, converting these waste streams into valuable biopolymers additionally contributes to promoting a circular economy by alleviating waste disposal concerns (Reddy et al., 2021). Table 1 shows the overview of sustainable production in microbial biopolymers by fermentation and their industrial application. This chapter will explore the various types of waste that may be utilized to produce biopolymers, the microbes involved, and the fermentation procedures employed. It will also discuss the downstream processing strategies necessary for the extraction and purification of biopolymers and the economic and ecological impacts. The advantages of introducing waste usage into the synthesis of biopolymers and the possibility for commercial production will be demonstrated via industrial applications.

2 Microbial Biopolymers

Microbial biopolymers are naturally occurring and are assembled of monomers that originate from microorganisms, including algae, fungi, and bacteria, through the consumption of carbon sources during growth. They are appropriate for numerous industrial applications because of their extensive functions, biodegradability, and biocompatibility. Polymerase chain reactions naturally mediated by enzymes result in the formation of biopolymers during intricate cell metabolism. Protein-based polymers like collagen, gelatin, and albumin, as well as polysaccharides like chitosan, starch, and cellulose derivatives, are naturally biodegradable polymers. Extracellular polymeric substances (EPS) and polyhydroxyalkanoates (PHAs), which are generated by a wide variety of microbial species, are examples of microbial biopolymers that might be used as alternatives for facilitating waste bioremediation to perform more effectively. PHA and poly(3-hydroxybutyrate-co-3-hydroxyvalerate) (P(3HB-co-3HV), commonly known as PHBV) are biodegradable polymers produced by microbial fermentation (Jabeen et al., 2024). The functions of biopolymers include energy reserve materials as a source of carbon, protecting microorganisms from environmental stress, and supporting cellular activity. Among the various types of biopolymers, microbial EPSs are particularly involved and generally categorized into three primary groups, including polynucleotides and nucleic acids containing cell genetic components like DNA and RNA; polypeptides and proteins providing specific biological functions; polycarbohydrates, which are the most widely utilized biopolymers (Lee et al., 2024). The biopolymers of the polysaccharide type consist of chains of polymeric carbohydrates containing monosaccharide units. When exposed to water, the numerous surface hydroxyl groups in these chains produce viscous hydrogels (Berradi et al., 2023).

2.1 Characteristics of Microbial Biopolymers

Microbial biopolymers have a number of distinctive characteristics, leading to them being appropriate for various applications, but they also have certain drawbacks. Table 2 displays the characteristics of microbial biopolymers. Their capacity to decompose naturally into non-toxic components is one of their main advantages; this allows them to be an ecologically benign substitute for synthetic polymers while lowering pollution. Nevertheless, their biodegradation may occur under particular circumstances and result in a reduced shelf life. Another significant characteristic is biocompatibility, which permits the safe use of these biopolymers in pharmaceutical and medical applications without causing an unpleasant immune reaction. Microbial biopolymers are able to adjust to changing environmental conditions and change accordingly. However, compared to synthetic substitutes, they could occasionally provide comparatively lower mechanical durability and resilience (Lee & Demirci, 2023; Siracusa, 2019). In addition, microbial biopolymers have excessive mechanical qualities, ranging from strong and rigid to flexible and elastic, allowing applications in building, packaging, and textiles. Notwithstanding their adaptability, these biopolymers might need to be blended or reinforced in some applications in order to satisfy particular mechanical requirements. Many biological processes are catalyzed by microbial biopolymers, such as proteins or polypeptides. By using this in the chemical process, the activation energy is reduced. Thus, these catalysts allow microbes to break down nutrients and create necessary nutrients for biological processes (Du et al., 2019). Another crucial characteristic is thermal stability, which allows some biopolymers to retain their characteristics at high temperatures and be used in applications requiring resistance to heat (Lee & Taştemir, 2023). Certain ones, meanwhile, might not be as thermally stable as their synthetic equivalence. Microbial biopolymers function as energy storage factories, allowing bacteria to store excess energy for potential metabolic demands.

The water solubility and hydrophilicity of microbial biopolymers are the crucial properties to interact with water, dissolve, or swell are their. For microbial cells, the

Table 1 The overview of sustainable production in microbial biopolymers by fermentation and their industrial application

Microbial biopolymers	Microorganisms involved	Waste and substrates used	Production process	Industrial applications	References
Polyhydroxyalkanoates (PHAs)	*Cupriavidus necator*, *Pseudomonas putida*	Agricultural waste, glycerol waste, molasses	Aerobic fermentation	Biodegradable plastics, packaging materials	Park et al. (2024), Rebocho et al. (2020)
Bacterial Cellulose	*Gluconacetobacter xylinus*, *Komagataeibacter xylinus*	Fruit and vegetable waste, molasses	Static culture, agitated culture	Biodegradable composites, wound dressings	Du et al. (2020)
Alginate	*Azotobacter vinelandii*, *Pseudomonas spp.*	Sugarcane bagasse, industrial effluents	Fermentation, aerobic culture	Water treatment, biodegradable films	Bi et al. (2023)
Xanthan Gum	*Xanthomonas campestris*	Molasses, corn steep liquor	Submerged fermentation	Food stabilizer, biodegradable packaging	Asase and Glukhareva (2024), Crugeira et al. (2023)
Polyglutamic Acid (PGA)	*Bacillus subtilis*	Agricultural waste, soybean waste	Aerobic fermentation	Biodegradable superabsorbent materials, soil conditioner	Mohanraj et al. (2019)
Pullulan	*Aureobasidium pullulans*	Starch, corn syrup, hydrolyzed food waste	Submerged fermentation	Edible films, biodegradable coatings	Viveka et al. (2021)
Chitosan and Chitin	*Aspergillus niger*, *Rhizopus spp.*	Shrimp shells, fungal biomass	Solid-state fermentation, enzymatic extraction	Water treatment, biodegradable packaging	Terkula Iber et al. (2022)
Gellan Gum	*Sphingomonas elodea*	Sucrose, lactose, corn steep liquor	Submerged fermentation	Food gelling agent, biodegradable packaging	Abdl Aali and Al-Sahlany (2024)
Phycobiliproteins	*Spirulina platensis*	CO_2, light, wastewater nutrients	Photoautotrophic cultivation	Natural colorants, fluorescent labels	dos Santos Borges et al. (2023)

Table 2 Characteristics of microbial biopolymers

Characteristic	Description	Advantages	Disadvantages	Examples
Biodegradability	Can be broken down by natural processes into non-toxic components	Environmentally friendly; reduces pollution	May degrade under specific conditions only; potential for shorter shelf life	Polyhydroxyalkanoates (PHAs), polylactic acid (PLA), bacterial cellulose
Biocompatibility	Safe for use in medical and pharmaceutical applications	Suitable for medical devices, drug delivery systems, and tissue engineering; minimal immune response	Potential for limited mechanical strength and durability in some applications	Bacterial cellulose, alginates, chitosan
Mechanical Properties	Range from flexible and elastic to strong and rigid	Versatile applications in packaging, textiles, construction; customizable properties	May require blending or reinforcement for certain applications	Polyhydroxybutyrate (PHB) for rigid plastics, pullulan for flexible films
Thermal Stability	Maintains properties at elevated temperatures	Suitable for heat-resistant applications; enhanced processing and durability	Some biopolymers may have lower thermal stability compared to synthetic polymers	Polyhydroxybutyrate (PHB), xanthan gum
Water Solubility and Hydrophilicity	Can interact with water, dissolve, or swell	Useful in food thickeners, hydrogels, water-absorbent materials; enhanced bioavailability in drugs	May not be suitable for applications requiring water resistance or durability	Xanthan gum, pullulan, alginate
Film-Forming Ability	Can form transparent, flexible, and barrier films	Applications in packaging, coatings, and edible films; improves shelf life and protection of products	Mechanical properties of films may be inferior to synthetic alternatives	Pullulan, bacterial cellulose, gellan gum
Functional Groups and Chemical Modifiability	Contain functional groups for chemical modification	Enhanced functionality for specific applications; ability to create derivatives with improved performance	Chemical modifications can be complex and costly; potential for reduced biodegradability	Chitosan (amino groups), alginate (carboxyl groups)
Renewable and Sustainable Production	Produced from renewable resources using sustainable microbial processes	Reduces reliance on fossil fuels and petrochemicals; environmentally sustainable production processes	Production costs can be higher than petrochemical-based polymers; scalability issues	Polyhydroxyalkanoates (PHAs) from agricultural waste, bacterial cellulose from fruit waste

biopolymers serve as protective allies, defending them from external threats. Microbial biopolymers function as biofilms that adhere to cell surfaces, protecting them from external factors, including chemicals and physical stresses. This characteristic is valuable for improving bioavailability in medication delivery systems as well as for uses in food thickeners, hydrogels, and materials that absorb water. However, this characteristic can make them less useful in situations where durability or water resistance is needed. They may create films that are translucent, flexible, and barrier-free, which is advantageous for packaging, coatings, and edible films, even though the mechanical characteristics of these films may not always be equally effective as those of synthetic alternatives. Between bacteria and the environment they inhabit, microbial biopolymers serve as channels for transmission. To arrange their activities and their reaction to places that need alertness, the biopolymers facilitate the receive and delivery of signals for communications (Moradali & Rehm, 2020). Furthermore, microbial biopolymers have functional groups that enable chemical alterations, improving their usefulness for particular applications and allowing the synthesis of derivatives with better performance. The technical complexity and expense of these alterations and possible decreases in biodegradability may provide difficulties. Finally, by using sustainable microbiological techniques for producing these biopolymers from renewable resources, fossil fuel demand could decrease and promote ecologically friendly production. Notwithstanding these benefits, scaling up production is still difficult, and microbial biopolymers might have greater manufacturing costs than polymers based on petrochemicals (Di Bartolo et al., 2021).

2.2 Polyhydroxyalkanoates

The fundamental family of intracellular biopolyesters known as polyhydroxyalkanoates (PHAs) is produced from over three hundred different species of both Gram-positive and Gram-negative bacteria like *Cupriavidus necator* (formerly known as *Ralstonia eutropha* or *Alcaligenes eutrophus*), *Bacillus subtilis*, and *Pseudomonas* spp., as intracellular carbon and energy storage compounds (Zhou et al., 2023). PHAs are an ecologically beneficial substitute for petroleum-derived polymers (Abdl Aali & Al-Sahlany, 2024). Poly(3-hydroxybutyrate) (PHB) and poly(3-hydroxybutyrate-co-3-hydroxyvalerate) (P(3HB-co-3HV) or PHBV) are the most well-known members of the PHA class. PHAs may be manufactured into films, fibers, and other materials because of their biodegradable and biocompatible features. They resemble traditional polymers in their thermoplastic traits. Numerous medical tissue engineering, controlled administration of medication, orthopedic surgery (Taylor et al., 2021), wound healing (Oran et al., 2022), cardiovascular treatments (Râpă et al., 2022), and also food packaging materials (Urbina et al., 2019) have been developed using these natural polyesters. As a result of its in vivo degradation into 3-hydroxybutyric Acid, a metabolite present in human blood, P(3HB) exhibits exceptional biodegradable and biocompatible attributes. It also has more accessible processing, reminiscent of traditional thermoplastic polymers, and is more resistant to ultraviolet (UV) damage than polypropylene. The monomers of 3-hydroxybutyrate (3HB) and 3-hydroxyvalerate (3HV) comprise the copolymer known as PHBV. PHBV produced from the Gram-negative bacteria *Alcaligenes eutrophus* was commercialized as a biodegradable thermoplastic material in 1970 by a British chemical company, Imperial Chemical Industries Ltd., under the trade name Biopol™. Compared to PHB alone, PHBV has distinct characteristics due to the addition of 3HV components to the polymer chain in addition to 3HB. These properties include increased flexibility and decreased brittleness. Because PHBV is simpler to process in the mold and exhibits lower melting and crystallinity than PHB, it is also utilized to improve the qualities of PHB (Bossu et al., 2021).

Microbial fermentative production of PHVB involves cultivating bacteria *Cupriavidus necator* (previously known as *Ralstonia eutropha*) and *Pseudomonas putida* in the presence of carbon sources like glucose or waste substances. It is possible to modify the ratio of 3HB to 3HV in the biopolymer by adjusting the carbon sources and fermentation parameters. PHBV is completely biodegradable and disintegrates into carbon dioxide and water in natural environments such as soil and the ocean. Because of its biocompatibility, it may be used in pharmaceutical and medical applications, including drug delivery systems, tissue engineering frameworks, and sutures. The ability to be melted and reshaped gives PHBV its thermoplasticity, making it appropriate for various production techniques like extrusion, injection casting, and film blowing. Besides, By modulating the 3HB to 3HV ratio, PHBV may be designed to have certain mechanical characteristics such as impact resistance, tearing strength, and elasticity. Moreover, Flexibility is improved, and crystallinity is reduced with higher 3HV concentrations. PHBV's manufacturing scope and value are expanded since it has a lower melting point and glass transition temperature than PHB (Râpă et al., 2022).

2.3 Polysaccharides

Microorganisms produce long-chain carbohydrates called polysaccharides, such as xanthan gum, gellan gum, and bacterial cellulose. Agricultural waste industries are the usual source of these biopolymers, including cellulose, starch, pectin, chitin/chitosan, and other derivatives. Polysaccharides are the most prevalent natural biopolymers in plants, animals, and microbes. These biopolymers can be formed extracellularly or intracellularly, and they are biological macromolecules comprised of more than 10 simple sugar monomers that are sequentially connected by glycosidic linkages. Their diverse physical and chemical characteristics make them appropriate for many applications in the industry (Vaishnav & Choudhary, 2021). A significant biopolymer is xanthan gum, which is produced in around 30,000 tons annually globally. It is a hetero-polysaccharide produced from *Xanthomonas* spp. (Sworn, 2021). and used in food, pharmaceutical, and petrochemical industries as a thickening and stabilizing ingredient. It is extremely viscous, stable at several pH values and temperatures, and has gel-forming ability. Xanthan gum can be found in various food wastes, including discarded malt grains, cheese whey, and citrus wastes. The extracellular polysaccharide gellan gum is a naturally occurring polymer produced by the bacteria *Sphingomonas elodea* under a fermentation process in an industrial condition. It offers high acid stability, high resilience, translucent gels, and significant thermostability. Gellan gum demonstrates promising industrial applications as a gelling agent, a removing, and a highly viscous biogum (Demirci et al., 2020; Feketshane et al., 2022).

Bacterial cellulose is a naturally occurring biomaterial synthesized by *Acetobacter xylinum*, *Achromobacter*, *Aerobacter*, *Agrobacterium*, *Gluconacetobacter xylinus*, *Pseudomonas*, *Rhizobium*, *Sarcina*, *Alcaligenes* and *Zoogloea*. Bacterial cellulose displays exceptional water-holding ability because of its high surface area to mass ratio and strong hydrophilicity. In addition, it is relatively affordable to manufacture, has a high degree of crystallinity, and has remarkable mechanical strength (Arun et al., 2022a, 2022b). Its exceptional mechanical qualities, high water retention capacity, and exceptional suspension stability are all a result of its distinctive three-dimensional reticulated network structure, which is constructed from cellulose nanofibers with a diameter of 20–100 nm. Due to the water-holding capacity and unique nanofibrillar structure of cellulose, it can be found in biomedical applications such as wound dressings and tissue engineering, in the food industry for low-calorie desserts, and in paper and textile industries. The most plentiful renewable plant substance after cellulose is hemicellulose. They connect to lignin by covalent bonds and are related to cellulose moieties by hydrogen bonds. Coniferyl alcohol, synapyl alcohol, and p-hydroxyphenyl alcohol are the three main monomers of lignin. The cell wall is given mechanical elasticity and defense against microbial invasion by this biopolymer (Rao et al., 2023).

The polysaccharide water-soluble alginate can be produced from specific brown algal cell walls and bacteria, including *Ascophyllum nodosum*, *Azotobacter vinelandii* and *Pseudomonas* spp., and so on. β-D-mannuronic acid and 1–4 linked α-L-guluronic residues have been identified in alginate (Bi et al., 2023). Alginate is reported for its remarkable biocompatibility, biodegradability, and lack of antigenicity, which has led to its widespread usage in pharmaceutical and biomedical applications, such as wound dressings, drug delivery, and tissue engineering. Moreover, alginate reveals a high water-holding capacity and gelling in the presence of calcium ions for use as a thickening and stabilizer in the food industry. Pullulan is an exopolysaccharide produced by *Aureobasidium pullulans* under the fermentation process and is widely used in food, pharmaceutical and biomedical applications (Lee et al., 2016; Raychaudhuri et al., 2020). The bacterium obtains this polysaccharide to support its own mechanisms of cell adhesion, flocculation, and aggregation in various conditions. Pullulan exhibits excellent biodegradability, biocompatibility, fat resistance, water solubility, non-toxicity, and edible and antistatic films with low oxygen and moisture permeability. Pullulan is a desirable option for food coating and packaging because of these qualities (Singh et al., 2023). Curdlan, an exopolysaccharide, is comprised of glucose connected by β-(1,3)-glycosidic bonds and produced by a variety of bacteria, including *Agrobacterium* spp., *Paenibacillus* spp., *Rhizobium* spp., also yeast *Saccharomyces cerevisiae*, yeast *Candida* spp., and fungal *Aureobasidium pullulan*. Because of curdlan's prebiotic, viscosifying, and water-holding qualities, the food and pharmaceutical industries have consistently utilized it. In preclinical conditions, curdlan has demonstrated immunoregulatory and anticancer activity. It can be used as a gelling agent in foods, as a wound dressing in biomedicine, and as an enzyme immobilizer in biotechnology (Chaudhari et al., 2021; Lee et al., 2015).

2.4 Polyamides

Polyamides primarily comprise long chains of amino acids and other nitrogen-containing compounds. Biodegradability, biocompatibility, and potential applications across various industries have piqued interest in polyamides. Microbial polyamides are produced by natural metabolic processes and may be obtained from renewable resources, unlike manufactured polyamides like nylon. Cyanophycin is produced from Synechocystis, Anabaena, and other cyanobacteria. It consists of side chains of arginine attached to a poly-L-aspartic acid backbone. It is water-insoluble, biodegradable, has high nitrogen content, and acts as a carbon and nitrogen storage molecule in cyanobacteria. It may be used as a source of amino acids for dietary supplements, biodegradable polymers, and as a slow-releasing nitrogen fertilizer in agriculture (Sharon et al., 2023). Polyglutamic acid produced from *Bacillus subtilis* is composed of repeated units of glutamic acid. The application of polyglutamic acid may be found in drug delivery, as a thickening and flavor enhancer in food, and as a soil conditioner and fertilizer in agriculture due to its Water-soluble, biodegradable, and non-toxic. Elastin-like polypeptides and silk-like polymers are produced from engineered bacteria, typically *Escherichia coli*. Their structure is composed of repeating pentapeptide sequences and fibrous protein found in natural elastin and in silk fibroin produced by spiders and silkworms. The physical and chemical properties of polyamides include biodegradability, biocompatibility, mechanical structure, and high nitrogen content. Due to their amino acid composition, microbial polyamides have a high nitrogen content, which can be beneficial in certain agricultural applications (Freger & Ramon, 2021).

Polyamides like cyanophycin have a high nitrogen concentration, which makes them suitable for use as slow-release fertilizers. Through time, these fertilizers will give plants a consistent source of nitrogen. Moreover, polyamides may be converted into biodegradable polymers as biodegradable plastics, which provide a less polluting and environmentally friendly substitute for conventional plastics. By hydrolyzing cyanophycin, amino acids are released that can be added to feed or utilized as dietary supplements. Polyamides are biocompatible and may be utilized to construct scaffolds for tissue engineering, drug delivery systems, and medical devices (Tahouni et al., 2023). In contrast to traditional polyamides derived from petroleum, the environmental effect of producing microbial polyamides from waste materials and renewable resources can be greatly reduced. Because of their propensity to decompose naturally, they prevent plastic pollution from building up in the environment. A cost-effective way to add value to agricultural and industrial waste processes is by employing waste substrates for manufacturing (Lee, 2023a; Venoor et al., 2021).

3 Waste as a Substrate for Biopolymer Production

Utilizing waste as an initial substrate for microorganisms to produce biopolymers is an economical and environmentally friendly method that solves waste management issues while reducing biopolymer extraction expenses. Rich in organic substances, various waste types, including food, industrial, and agricultural waste, can be ideal substrates for the growth of different microorganisms and the synthesis of biopolymers. The growing global population forces pressure on limited natural resources and demands an increase in the food supply. The agriculture industry has generated a substantial amount of waste as a result of this change. Rapid industrialization and population shifts to metropolitan regions have often resulted in environmental contamination, raising worries about water pollution, waste discharge into the environment, and environmental deprivation. Given the chance to improve agricultural productivity, modern agriculture has been determined to be a beneficial response to the worldwide rise in population. However, it has also resulted in an increase in the world's carbon footprint, and farming produces waste that has an effect on both society and economic activity. Food loss and waste during farming and agricultural processes is estimated by the Food and Agricultural Organization (FAO) to be 1.3 billion tons annually (Blakeney, 2019). As agricultural waste is disposed of in landfills, which contributes to global warming, using these residues and turning them into biopolymers is an effective way to lower the global environmental impact. An unstable climate and extreme weather events like droughts and flooding are caused in part by food loss and waste, which accounts for 8–10% of global greenhouse gas emissions (FAO Publications Catalogue, 2023; Lee & Neio Demirci, 2023).

3.1 Waste Biorefinery

An establishment that transforms biomass—biological material from live or recently living organisms—into bio-based feedstocks is known as a biorefinery. It is a type of refinery where biomass is processed to produce energy and other valuable byproducts. The biorefinery demonstrated that sustainable biomass conversion to energy, chemicals, and materials is achieved via industrial facilities. The main motivations behind the development of biorefineries are the requirement for chemicals and materials and the need for environmentally friendly development to alleviate climate change and provide energy security (Mehariya et al., 2020). Improving waste stream management is demonstrated by the biological processing in the waste biorefinery to produce biopolymers and bioenergy, which not only responds to energy and ecological security concerns but also generates bioenergy. The production feedstock is inexpensive and sustainable, making it an eco-friendly substrate. Numerous waste materials have been effectively valorized into commercially viable and valuable bio-based items, including food waste, wastewater or sludge (Paul et al., 2020), lignocellulosic material, side streams from industries (such as those of paper and pulp, beer and wine, starch and juice, forest), agro-industrial byproducts, and agricultural waste (Leong et al., 2021). These waste materials present themselves as promising carbon and nitrogen sources with significant value-added potential for synthesizing biopolymers. The waste streams may be converted into biopolymers by varying their content and can be obtained from various industries. Various Microbes can be used to construct

Fig. 1 Sustainability of microbial biopolymers from waste streams to emergent application

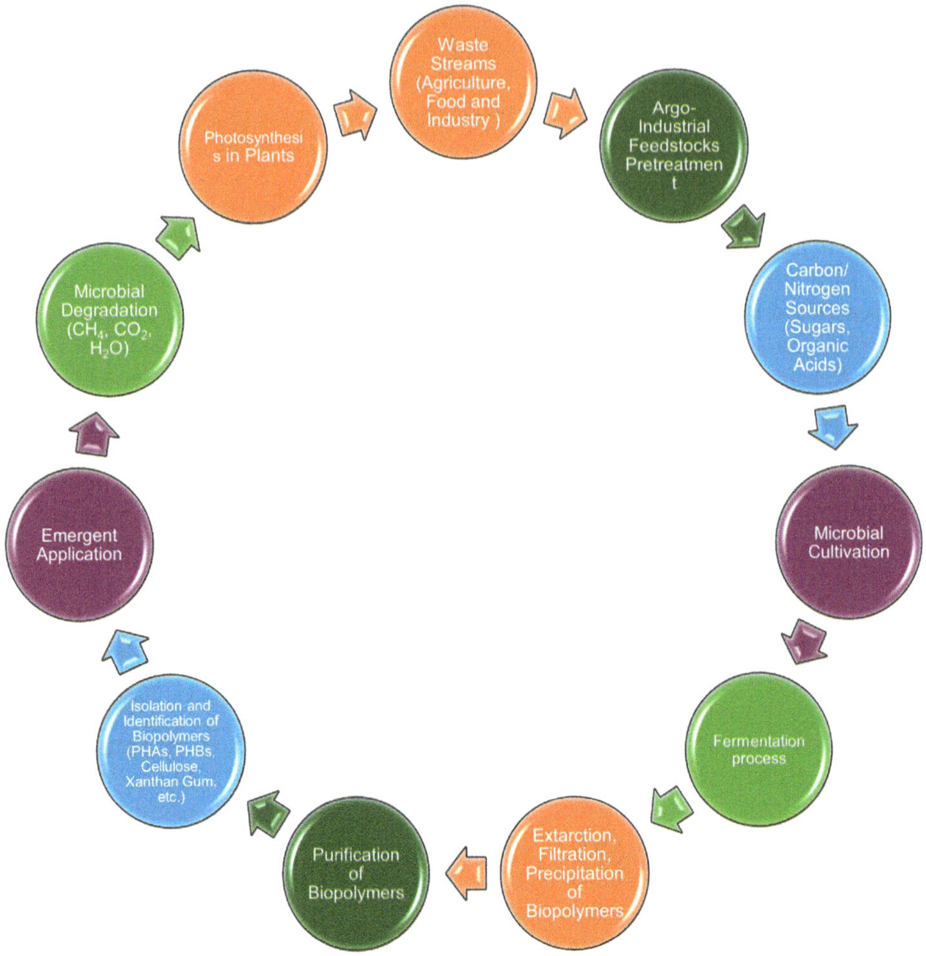

biopolymers, which can be divided into multiple categories. These biopolymers are substances stored in the cytoplasm of various microbe species. Several green biopolymers have drawn a lot of interest from the scientific community and industry due to their ability to replace traditional non-biodegradable and petroleum-based plastics. Due to their immense fundamental attributes, which include their non-toxicity, biodegradability, thermoplasticity, piezoelectricity, and insolubility in water, they have the potential to replace petroleum-derived plastics like polypropylene and polystyrene(Ansari et al., 2023, 2024; Venkata Mohan et al., 2016). Figure 1 presents a close loop for the sustainability of microbial biopolymers from waste streams to emergent applications.

3.2 Agricultural Waste

Agricultural waste refers to undesired garbage generated by agricultural operations. This category includes organic waste, such as manure and silage; plastic waste, such as mulch film and pesticide bottles; and other wastes, such as hazardous waste from the use of pesticides and fertilizers. In the agricultural process, the cultivation of crops and the supply of animal feed result in the production of organic wastes such as livestock manures, straw, and so on (Koul et al., 2022) Crop residues, fruit and vegetable waste, and livestock farming waste are three major examples of agricultural waste. The disposal of these wastes is sometimes problematic due to their abundance and poor management. Crop residues are rich in cellulose, hemicellulose, and lignin. They are capable of hydrolyzing into fermentable sugars for microbial development. The common crop residues are maize stover, wheat stalks, rice husks, and sugarcane bagasse. Fruit and vegetable wastes are usually high in carbohydrates like peels, pulps, and other byproducts from fruit and vegetable processing to be hydrolyzed by microbes or enzymes to produce biopolymers. Animal manure may be digested anaerobically to produce biogas because of nutrients and organic substances. Animal feces and urine make up the majority of manure, sometimes referred to as livestock manure. However, it typically contains plant material, most commonly straw or stalks, that has been used as animal bedding and has absorbed feces and urine. The remaining digestate can then be utilized as a nutrient-rich substrate for producing microbial biopolymers (Duque-Acevedo et al., 2020).

Typical agricultural non-natural wastes include pesticide containers, plastics, bags, sheets, tires, batteries, medical waste, old equipment, oil, packaging waste materials, and many more items. Approximately 1.5% of the total amount of plastic packaging in England's waste stream is composed of plastic packaging waste from agriculture. The numerous kinds of plastic wastes that arise can be large and unclean, resulting in their handling challenging. The UK produces about 135,500 tons of agricultural plastic trash annually. On the other hand, Slurries and dung are examples of common natural waste. 95% of all agricultural waste (43 million tons) was found to be manure and slurries in England. Manures and slurries are not considered waste in the UK and are not subject to waste management legislation if they are utilized as fertilizer for agricultural purposes. Certain limits on the spreading of manure and sludge to land are caused by the need to comply with additional rules, such as those pertaining to groundwater. Microorganisms can use agricultural waste as a substrate to produce biopolymers under a fermentative process. Bacteria like *Cupriavidus necator* use hydrolysates from agricultural wastes to produce PHA. Also, hydrolysates from fruit and vegetable waste are used by *Gluconacetobacter xylinus* to produce bacterial cellulose (He et al., 2020).

3.3 Food Waste

A significant substrate for the synthesis of biopolymers is food waste from restaurants, households, and food processing. Food waste may be converted into small compounds appropriate for microbial fermentation since it is usually abundant in proteins, lipids, and carbohydrates. Food waste can be observed at every stage of the manufacturing process, including on farms, in transportation, at stores, and ultimately at the consumer. Among the causes include cooking losses, deliberate food waste, and losses due to mildew, pests, or insufficient temperature management (Palaniveloo et al., 2020). Any portion of food that is not typically eaten or consumed, such as an eggshell, coffee grounds, banana peel, watermelon rind, or corn cob, is considered non-edible food waste. Kitchen waste includes expired products, general plate scrapings and leftovers, and peelings. Meat waste is out of date, as well as off cuts, small bones, and carcasses. Fruits and vegetables turn into waste because they are expired and damaged. Dairy waste is rich in lactose and proteins. Whey and other dairy byproducts are also ideal substrates for the microbial synthesis of biopolymers. The waste from seafood is commonly made up of bones, cartilage, and shells containing rich proteins and minerals like calcium, magnesium, and phosphorus. Therefore, food waster can PHAs were produced by *Haloferax mediterranei* using hydrolyzed food leftover. Pullulan was also produced by *Aureobasidium pullulans* using hydrolyzed brewery waste. Dairy waste was used for bacterial cellulose production (Kadier et al., 2021).

3.4 Industrial Waste

A significant amount of global industrial waste is produced annually. The disposal of these wastes on public property presents numerous environmental risks. Using industrial wastes for microbial biopolymer production is an environmentally friendly, economical, and productive approach to use them. Industrial waste comprises manufacturing waste from a variety of different processes, including sludges, product residues, kiln dust, slags, and ashes. The primary industrial wastes are metallurgy, non-metallurgy, and organic substances. Depending on the raw materials utilized, the manufacturing procedures, and the product outlets, waste may vary from industry to industry (Rajinipriya et al., 2018). Industrial waste is an all-encompassing phrase used to describe anything regarded to be no longer of use once a production process has been finished. Many areas of industrial manufacture generate waste, including mining, textile mills, and food and oil refineries. Biopolymer synthesis may also be achieved by using industrial waste, such as byproducts and effluents from different manufacturing operations. Numerous chemical substances and nutrients that might promote microbial growth are frequently present in high concentrations of these contaminants (Kadier et al., 2021). Biodiesel waste, or glycerol waste, is a byproduct from diesel production that contains much carbon and may be utilized as a substrate for the microbial manufacture of biopolymers. Molasses is condensed sugar residue left over after sugar refining to utilize for fermentation. Organic waste may be fermented into biopolymer by specific microorganisms, such as pulp and paper mill effluents containing rich cellulose and lignin. Biopolymer production from industrial waste has demonstrated that carbon sources such as waste glycerol are used in PHA production (Zhou et al., 2023). Xanthan Gum is prepared from molasses by the bacteria *Xanthomonas campestris*. Pulp and paper mill effluents are used to produce bacterial cellulose (Hussain et al., 2019).

4 Microorganisms Involved in Biopolymer Production

Biopolymers are produced by various microorganisms, such as algae, fungi, and bacteria. These microorganisms can generate a variety of biopolymers due to their unique metabolic pathways and capacities. Both external and intracellular biopolymers can be produced by different microorganisms. Polysaccharides, polyamides, polyesters, and polyphosphates are the starting points for metabolites in bacterial metabolic synthesis (Srisawat et al., 2022). Fungal biopolymers are valuable resources for a wide range of environmental applications, such as biosurfactant agents, soil emendation, enzyme immobilization, and heavy metal absorption. Algae are microorganisms that perform photosynthesis and do not only require pure water for growth. They may be grown on wastewater as well. From a sustainability perspective, polyesters and polysaccharides are the two most important biopolymers. Important microbial biopolymers and their environmental applications are discussed in the following section (Saharan et al., 2024).

4.1 Bacterial Biopolymer

The bacteria *Cupriavidus necator* produced by the PHAs was cultivated in the culture medium from non-alcoholic beer-producing brewery wastewater. This medium contains maltose as a primary carbon source, and the optimal maltose concentration is 40 g/L with a C/N/P ratio of 100:2:15 (Amini et al., 2020). The strain *Cupriavidus necator* LPB 1421 could produce polyhydroxybutyrates (PHBs) by combining hydrolyzed milled maize starchy fraction and crude glycerol as substrates. Following optimization phases, the productivity of 0.162 g/L per hour was reported, and a production yield of 11.64 g/L was revealed with a maximum accumulation of 62% of PHBs. A viable and profitable biorefinery might be established by combining the proposed bioplastic process with a unit that produces bioethanol (de Mello et al., 2023). It is possible to use mango pulp waste to produce a substantial amount of nanocellulose from *Komagataeibacter xylinus*. This approach reduces bacterial cellulose production costs while utilizing agro-food waste, which is widely generated in many countries (García-Sánchez et al., 2019). Rotten banana medium yielded the maximum yield of bacterial cellulose from *Komagataeibacter medellinensis*, which presented the highest tensile characteristics. With a maximum rate of degradation temperature of 368 °C, the cellulose produced from cheese whey medium exhibited the highest temperature. Regarding economics and environmental issues, food supply chain waste and agricultural waste byproducts are beneficial alternative substrates for producing bacterial cellulose with various characteristics and applications (Molina-Ramírez et al., 2018).

Gluconacetobacter xylinus KCCM 41,431 produced bacterial cellulose under static culture conditions by using pure and crude waste glycerol. Using both pure and crude glycerol, the maximum production of bacterial cellulose was 1.05 and 0.99 g per liter per day, respectively (Ho Jin et al., 2019). The ability of potato peels (*Solanum tuberosum*) to facilitate alginate production was displayed from *Azotobacter nigricans* through solid-state fermentation under the response surface methodology. Following optimization, 4 days of incubation at pH 7.5, 30 °C, 1 mL inoculum, and 0.1% corn steep liquor resulted in the maximum amount of alginate (Saeed et al., 2023). *Azotobacter chrococcum* has the potential to manufacture microbial alginate by using sweet whey during the fermentation process. With sweet whey medium adjusted to 45 g/L lactose content as a carbon source and under optimal fermentation conditions, the maximum alginate production with 12.56% and 0.08 g per liter per hour of productivity were achieved (Hendawy et al., 2019).

It was established what characteristics of xanthan gum may be obtained from lignocellulosic agroindustrial wastes generated by *Xanthomonas campestris* pv. campestris 1866 and 1867. When identifying compounds with rough surfaces, all gums showed strong thermal stability and distinct categories (da Silva et al., 2018). When adding moist olive pomace to the xanthan gum-producing process from *Xanthomonas campestris*, the stress induced by moist olive pomace caused a microbial reaction that increased the synthesis of xanthan gum, changed its structure and viscosity, and provided antioxidant activities. This result highlights a novel moist olive pomace application that supports the sustainability of the olive oil-producing chain (Crugeira et al., 2023; Liorančas & Lee, 2024). *Bacillus subtilis* ZC-5 was utilized to produce γ-polyglutamic Acid by using industrial wastes, including soybean residue, chicken feces, and monosodium glutamate waste liquor. The yield of γ-polyglutamic Acid climbed to 6.57%, surpassing the single factor experiments' highest value by 20.6% under the optimal condition containing 15.09 g of crude extract of glutamic Acid after isoelectric crystallization, 25.15 g of soybean residue, and 62.35 g chicken feces (Zhang et al., 2019). Table 3 exhibits recent studies on biopolymers derived from bacteria that utilize waste.

4.2 Fungal Biopolymer

Pullulan, a type of α-glucan, produced by *Aureobasidium pullulans* Y2092 was assessed on a laboratory scale by utilizing industrial waste residual brewery's yeast under the submerged fermentation culture. The media components for optimum pullulan production from *Aureobasidium pullulans* MTCC 1991 were discovered when cassava (*Manihot esculenta*) waste was used in solid-state fermentation. It was determined that 0.6 g per liter of yeast extract produced the highest pullulan yield with 6.45 g per liter of culture broth (Viveka et al., 2021). Pullulan production was

Table 3 Biopolymers derived from bacteria utilizing waste

Microbial biopolymers	Bacterial strains	Waste utilization	Significant results	References
Polyhydroxyalkanoates (PHAs)	*Cupriavidus necator*	Non-alcoholic beer-producing brewery wastewater	This medium contained maltose as a primary carbon source, and the optimal maltose concentration is 40 g/L with a C/N/P ratio of 100:2:15	Amini et al. (2020)
Polyhydroxybutyrates (PHB)	*Cupriavidus necator* LPB 1421	Hydrolyzed milled maize starchy fraction and crude glycerol	Following optimization phases, the productivity of 0.162 g/L per hour was reported, and a production yield of 11.64 g/L was revealed with a maximum accumulation of 62% of PHB	de Mello et al. (2023)
Bacterial Cellulose	*Komagataeibacter xylinus*	Mango pulp waste	This approach reduced bacterial cellulose production costs while utilizing agro-food waste, which is widely generated in many countries	García-Sánchez et al. (2019)
	Komagataeibacter xylinus	Food supply chain waste (rotten banana and cheese whey)	Rotten banana medium yielded the maximum yield of bacterial cellulose and presented the highest tensile characteristics. With a maximum rate of degradation temperature of 368 °C, the cellulose produced from cheese whey medium exhibited the highest temperature	Molina-Ramírez et al. (2018)
	Gluconacetobacter xylinus KCCM 41,431	Waste glycerol	The maximum production of bacterial cellulose was 1.05 and 0.99 g per liter per day, respectively under static culture	Ho Jin et al. (2019)
Alginate	*Azotobacter nigricans*	Potato peels (*Solanum tuberosum*)	Through solid-state fermentation under the response surface methodology. For optimization, 4 days of incubation at pH 7.5, 30 °C, 1 mL inoculum and 0.1% corn steep liquor resulted in the maximum amount of alginate	Saeed et al. (2023)
	Azotobacter chrococcum NCBI MH249629	Sweet whey	With sweet whey medium adjusted to 45 g/L lactose content as a carbon source and under optimal fermentation conditions, the maximum alginate production with 12.56% and 0.08 g/L per hour of productivity was achieved	Hendawy et al. (2019)
Xanthan Gum	*Xanthomonas campestris* pv. campestris 1866 and 1867	Lignocellulosic agroindustrial wastes	When it came to the identification of compounds with rough surfaces, all gums showed strong thermal stability and distinct categories	da Silva et al. (2018)
	Xanthomonas campestris	Moist olive pomace	The stress induced by moist olive pomace caused a microbial reaction that increased the synthesis of xanthan gum, changed its structure and viscosity, and provided antioxidant activities. This result highlights a novel moist olive pomace application that supports the sustainability of the olive oil-producing chain	Crugeira et al. (2023)
γ-polyglutamic acid	*Bacillus subtilis* ZC-5	Industrial wastes including monosodium glutamate waste liquor, chicken manure and soybean cake	The yield of γ-polyglutamic acid climbed to 6.57%, surpassing the single factor experiments' highest value by 20.6% under the optimal condition containing 15.09 g of crude extract of glutamic acid after isoelectric crystallization, 25.15 g of soybean residue, and 62.35 g chicken feces	Zhang et al. (2019)

evaluated by *Aureobasidium pullulans* MTCC 2195 using jackfruit seed powder. The results indicated that the highest concentrations of pullulan were 18.76 g/L under the optimal fermentative medium with 2% jackfruit seed powder and 0.3% yeast extract (Sharmila et al., 2013).

For the β-glucan production, the addition of deproteinated potato juice water was added into the fermentation medium, followed by the inoculation of the medium with the yeast strain Candida utilis ATCC 9950 with A 5 Liter bioreactor for yeast incubation for 72 h. Eighty-two percent of the β-glucan was extracted from yeast cells during the fermentation process (Bzducha-Wróbel et al., 2020). The isolation of β-glucans was performed with wet sediments from red and white wine lees. After clarification of white wine grape juice, the liquid supernatant was collected to use for the fermentation process to produce acetic Acid by using acetic acid bacteria naturally present in wine grape juice. The purified fraction of β-glucan from red wine lees (28.95%) was obtained with 2 M NaOH for 0.5 h, whereas white wine lees (43.37%) were obtained with 0.5 M NaOH for 1 h. Both of their fermentation processes were contacted at 90 °C with 1:5 sample to NaOH concentration. Using winery yeast waste lees can be a novel approach to producing β-glucan (Varelas et al., 2016).

Very high polarity sugar from sugarcane mills and soybean cake from biodiesel manufacturing processes have been used to assess the microbial synthesis of fumaric Acid by *Rhizopus arrhizus* NRRL 2582. Fumaric Acid is a precursor for biopolymers like polybutylene succinate and can be polymerized into various bioplastics. Using a two-stage bioprocess that involved the generation of crude enzymes by solid-state fermentations of *Aspergillus* or *Rhizopus arrhizus* grown on soybean cake, followed by enzymatic hydrolysis of the soybean cake, which can be transformed into a nutrient-rich hydrolysate (Papadaki et al., 2018). The fumaric acid production by 17 distinct *Rhizopus* species was evaluated by Utilizing the pentose-rich hydrolysate obtained from the diluted acid pretreatment of empty fruit bunches, an agricultural waste mostly generated in the palm oil industry. Fumaric Acid was produced in 96 h by *Rhizopus circicans* 1475 and *Rhizopus* 3271 at 30 °C and 120 rpm, yielding 5.65 g/L and 5.25 g/L, respectively (Pairazamán et al., 2024). Table 4 presents recent studies on biopolymers derived from fungi that utilize waste.

4.3 Microalgal Biopolymer

Microalgae can effectively treat waste streams with high levels of organic carbon and nitrogen. This approach can be cost-effective if valuable products are simultaneously generated from low-cost waste streams. Phycocyanin is a water-soluble protein-pigment complex with a strong blue color found in phycobiliproteins. The co-production of phycocyanin and PHB was studied in non-axenic *Arthrospira platensis* MUR126 with supplemental mixed-waste organic carbon like acetate, oxalate, and glycerol. The result revealed that a higher biomass yield of phycocyanin and PHB (3.6–7.8%) was found in all cultures than in photoautotrophic control. Phycocyanin content also increased two to three-fold over eight days (Shayesteh et al., 2023). The unique potential of the textile dye Direct Green 6 treatment may increase the synthesis of the biopolymer PHBs by *Spirulina platensis*. The maximum simultaneous co-production of PHBs (12.7%) occurred at a Direct Green 6 level of 60 mg/L. *Spirulina platensis* could utilize Direct Green 6 dye as a nitrogen and carbon source to accumulate important bioplastics and metabolites simultaneously (Lee, 2023b; Moradi et al., 2021). The effect of cheese whey as a viable substrate on phycobiliproteins production was determined by the mixotrophic culture of *Arthrospira platensis*. A higher amount of phycocyanin (3.52 mg/mL) was produced by *Arthrospira platensis* under mixotrophy than the yield of phycocyanin (2.55 mg/mL) in photoautotrophic conditions (Cavallini et al., 2024). The PHBs production via microbial fermentation was performed by the biomass of the Nordic microalgae, including *Chlorococcum* spp., *Chlorella vulgaris*, *Coelastrum astroideum*, and *Desmodesmus* spp. cultivated in synthetic wastewater. The results showed that hydrolysate of *Chlorococcum* sp. grown on BG-11 medium produced the most PHBs (27%), followed by hydrolysate from *Desmodesmus* spp. grown on synthetic wastewater. Nordic microalgal biomass produced on wastewater can thus serve as a low-cost feedstock for sustainable bioplastic manufacture (Mehariya et al., 2023).

An innovative bio-refinery approach was developed to efficiently utilize de-oiled algae cake as a bio-resource for biopolymer PHBs production, attempting to make the biorefinery process more cost-effective and environmentally friendly by entirely implementing the algal biomass without producing algal residues. The findings displayed that one gram of de-oiled cake yielded 0.41 g of PHBs. It supports algal refineries in terms of sustainability and the circular use of algae biomass for the development of numerous products (Arun et al., 2022a, 2022b). The production of β-glucan from potato liquor was reported. 75% of the algal mass was β-glucan, produced by alga *Euglena gracilis* under the cultivation of potato liquor (Šantek et al., 2010). Nonylphenol toxicity on algal cell growth was significantly higher than bisphenol A and 17α-ethinylestradiol. Chlorophyll degradation was induced by nonylphenol, while 17α-ethinylestradiol decreased production. Increased yields of colloidal polysaccharides and proteins responded to 17α-ethinylestradiol with nonylphenol undergoing abiotic degradation in alga-free water and 17α-ethinylestradiol biodegrading in alga-spiked water (Yang et al., 2021).

Table 4 Biopolymers derived from fungi utilizing waste

Microbial biopolymers	Fungal strains	Waste utilization	Production process	References
Pullulan	*Aureobasidium pullulans* Y2092	Residual brewery's yeast	A laboratory scale is utilized for industrial waste residual brewery's yeast under the submerged fermentation culture	Oliveira et al. (2023)
	Aureobasidium pullulans MTCC 1991	Cassava (*Manihot esculenta*) waste	It was determined that 0.6 g per liter of yeast extract produced the highest pullulan yield with 6.45 g per liter of culture broth under solid-state fermentation	Viveka et al. (2021)
	Aureobasidium pullulans MTCC 2195	Jackfruit seed powder	The results indicated that the highest concentrations of pullulan were 18.76 (g/L) under the optimal fermentative medium with 2% jackfruit seed powder and 0.3% yeast extract	Sharmila et al. (2013)
Bioplastics	*Rhizopus arrhizus* NRRL 2582 *Aspergillus* spp.	Very high polarity sugar from sugarcane mills soybean cake from biodiesel manufacturing	Using a two-stage bioprocess that involved the generation of crude enzymes by solid-state fermentations followed by enzymatic hydrolysis of the soybean cake which can be transformed into a nutrient-rich hydrolysate	Papadaki et al. (2018)
	17 distinct *Rhizopus* spp.	Empty fruit bunches–an agricultural waste from palm oil industry	Fumaric Acid was produced in 96 h by *Rhizopus circicans* 1475 and *Rhizopus* 3271 at 30 °C and 120 rpm, yielding 5.65 g/L and 5.25 g/L, respectively	Pairazamán et al. (2024)
Chitin, chitosan and Chitin-glucan	*Pleurotus sajor-caju*, *Lentinula edodes*, *Agaricus bisporus*, *Auricula-judae*, *Trametes versicolor*, *Armillaria mellea*, *Pleurotus ostreatus*, and *Pleurotus eryngii*	Agro industrial wastes	During submerged fermentation, these species produced yields of chitin and chitosan of around 85–196 and 10–40 mg/g of dried biomass, respectively	Philibert et al. (2017)
β-glucan	Candida utilis ATCC 9950	Deproteinated potato juice water	The addition of deproteinated potato juice water was added into the fermentation medium, followed by the inoculation in a 5-L bioreactor for yeast incubation for 72 h. 82% of the β-glucan was extracted from yeast cells during the fermentation process	Bzducha-Wróbel et al. (2020)
	Saccharomyces cerevisiae VIN 13	Winery spent yeast waste biomass with Red wine lees and white wine lees	The purified fraction of β-glucan from red wine lees (28.95%) was obtained with 2 M NaOH for 0.5 h, whereas white wine lees (43.37%) were obtained with 0.5 M NaOH for 1 h. Both of their fermentation processes were contacted at 90 °C with a 1:5 sample at NaOH concentration	Varelas et al. (2016)

Cadmium elimination from *Scenedesmus obliquus* biofilm with K3 carriers might be attributed to more rapid EPS production over 72 h compared to suspended *Scenedesmus obliquus*. Under cadmium (II) stress, the *Scenedesmus obliquus* biofilm displayed greater tolerance in pigment concentration, particularly chlorophyll content. On the third day of exposure to 10 mg/L of cadmium, the biofilm's extracellular polymeric substance grew dramatically to 34.91 mg/g, in comparison with 22.52 mg/g in suspended *Scenedesmus obliquus* (Ma et al., 2021). Table 5 demonstrates recent studies on biopolymers derived from microalgae that utilize waste.

5 Key Determinants in the Production of Microbial Biopolymers

Recently, microbial biopolymers have attracted much attention due to their potential to replace synthetic polymers. A number of different factors impact the production of microbial biopolymers, and each of them is important for determining productivity, efficiency, and profitability. Apart from essential factors like microorganism selection and substrate composition, some important variables are involved in the fermentation process, such as genetic engineering, metabolic pathway optimization, and downstream processing (Bernard et al., 2024; Lee & Küçükata, 2023). Producing biopolymers depends substantially on a meticulous choice of microbial strains. In order to efficiently produce microbial biopolymers, microorganism selection is essential. Owing to their distinct genetic and metabolic characteristics, microbes differ in their capacity for biopolymer synthesis. *Gluconacetobacter xylinus* is effective in bacterial cellulose production, and *Cupriavidus necator* is well-known for synthesizing PHAs. In the manufacturing process, the Food and Drug Administration granted approval for just a few commercially available strains (Paul et al., 2020). Along with possessing the genetic pathways required for biopolymer synthesis, the selected microbes must also be able to grow in a variety of substrates and environmental conditions. The selection of microbial strain and its metabolic pathways directly impact efficiency and productivity, as well as functional, mechanical, and rheological characteristics. Figure 2 shows different factors influencing the sustainable production of microbial biopolymers.

Microbial biopolymer synthesis is strongly influenced by the characteristics and availability of the substrate. The medium composition is one of the most important elements that influence effective microbial fermentation. The type and quantity of substrates given to the microbes are extremely significant. The most critical component for synthesizing microbial biopolymers in the culture medium is an appropriate amount of carbon and nitrogen. Common carbon sources include industrial byproducts, agricultural waste, and simple sugars. A sufficient nitrogen source is required for microbial development, but the most suitable nitrogen sources promote the development of biopolymers rather than biomass formation. Depending on the different types of biopolymers, microorganisms frequently utilize amino acids and ammonium salts as nitrogen sources and polysaccharides like glucose, sucrose, and fructose as carbon sources (Sharma et al., 2024). In addition, microbes need nutrients, minerals, and trace elements to synthesize microbial biopolymers. Minerals and vitamins are supplemental nutrients that might improve output even further. Sulfur, iron, phosphorus, yeast extract, beef extract, and other trace elements must be present properly for the microbial metabolic pathway to function. Glucose, molasses, glycerol, and other agricultural leftovers are commonly appropriate substrates. Any substances that can obstruct microbial activity must also be eliminated from the media (Bernard et al., 2024). Any variation triggers an interference in the metabolic pathway, which then affects the biopolymer manufacturing process. Other intrinsic factors, like pH, temperature, and oxygen availability, also need to be considered. The optimum situation for microbial growth and biopolymer synthesis requires adjusting the bioreactor's environmental parameters. The optimal temperature and pH ranges for manufacture depend on a number of variables. Each microbe has a specific temperature range that affects how it consumes energy. *Cupriavidus necator*, for instance, thrives at temperatures close to 30 °C. The optimum pH level for bacterial cellulose synthesis is pH 6–7. This range guarantees both microbial stability and enzyme activity. Deviations from the optimal conditions may hinder microbial growth rates and the production of biopolymers. Oxygen availability may be used to modify the production and properties of microbial biopolymers. For aerobic microbes to continue their metabolic processes, they need a sufficient supply of oxygen and enough aeration (Shaikh et al., 2023). Extraction, filtration, precipitation, and purification have a direct impact on the quality of biopolymers, and they also have an impact on the total yield of biopolymers. Downstream processing plays a critical role in the recovery of biopolymers, hence ultimately influencing manufacturing costs. To extract biopolymers from microbial cells to achieve the required purity levels, effective cell lysis procedures must be used. More output is obtained from the extraction of a biopolymer from commercially utilized ingredients than from agro-industrial waste. The cost of biopolymer is increased in the end by solvent extraction and extra stages in the downstream processing process. Filtration and rinsing for bacterial cellulose and solvent extraction for PHB are typical approaches (Aramvash et al., 2015).

Table 5 Biopolymers derived from microalgae utilizing waste

Microbial biopolymer	Microalgal strain	Waste utilization	Significant results	References
Phycocyanin and polyhydroxybutyrate (PHB)	*Arthrospira platensis* MUR126	Mixed-waste organic carbon with acetate, oxalate, and glycerol	Higher biomass yield of phycocyanin and PHB (3.6–7.8%) was found in all cultures than in photoautotrophic control. Phycocyanin content also increased two to three-fold over eight days	Shayesteh et al. (2023)
	Spirulina platensis	The textile dye Direct Green 6	The maximum simultaneous co-production of PHB (12.7%) occurred at a Direct Green 6 level of 60 mg/L. *Spirulina platensis* could utilize Direct Green 6 dye as a nitrogen and carbon source to accumulate important bioplastics and metabolites simultaneously	Cavallini et al. (2024), Moradi et al. (2021)
	Nordic microalgae including *Chlorococcum* spp., *Chlorella vulgaris*, *Coelastrum astroideum*, and *Desmodesmus* spp.	Synthetic wastewater	Hydrolysate of *Chlorococcum* sp. grown on BG-11 medium produced the most PHB (27%), followed by hydrolysate from *Desmodesmus* spp. grown on synthetic wastewater. Nordic microalgal biomass produced on wastewater can thus serve as a low-cost feedstock for sustainable bioplastic manufacture	Mehariya et al. (2023)
	Chlorella vulgaris	De-oiled algae cake	The finding displayed that 1 g of de-oiled cake yielded 0.41 g of PHB. It supports algal refineries in terms of sustainability and the circular use of algae biomass for the development of numerous products	Arun et al., (2022a, 2022b)
Phycocyanin	*Arthrospira platensis*	Cheese whey	A higher amount of phycocyanin (3.52 mg/mL) was produced by *Arthrospira platensis* under mixotrophy than the yield of phycocyanin (2.55 mg/mL) in photoautotrophic conditions	Cavallini et al. (2024)
β-Glucan	*Euglena gracilis*	Potato liquor	75% of the algal mass was β-glucan produced in the medium containing potato liquor	Šantek et al. (2010)
Extracellular polymeric substance (EPS)	*Tetraselmis* spp.	Contaminated estuarine sediments	Nonylphenol degraded abiotically, increasing toxicity to microalgae. Increased colloidal EPS allowed microalgae to biodegrade 17α-ethinylestradiol	Yang et al. (2021)
	Scenedesmus obliquus	Cadmium polluted water	Cadmium elimination from *Scenedesmus obliquus* biofilm might be attributed to more rapid EPS production over 72 h compared to suspended *Scenedesmus obliquus*. On the third day of exposure to 10 mg/L of cadmium, the biofilm's extracellular polymeric substance grew dramatically to 34.91 mg/g, in comparison with 22.52 mg/g in suspended *Scenedesmus obliquus*	Ma et al. (2021)

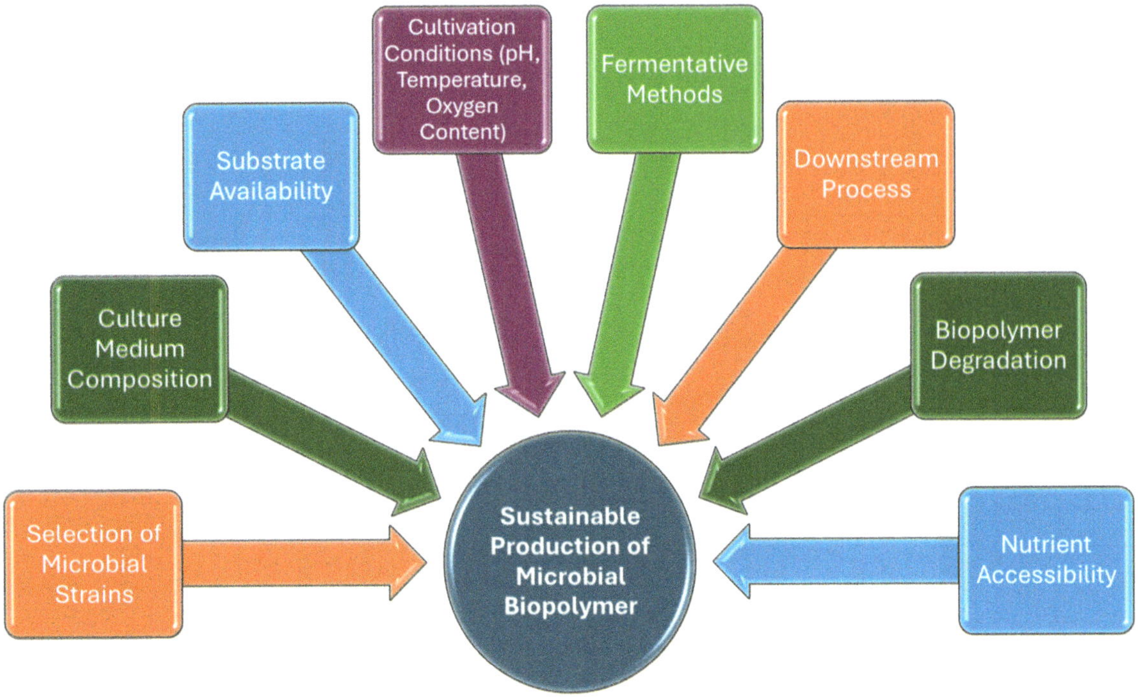

Fig. 2 Key determinants in the sustainable production of microbial biopolymers

The production costs of many stages in the PHA manufacturing process may be significantly reduced by downstream processing. Some biopolymers' characteristics degrade over time due to the temperature step required in downstream processing. Purification procedures must then be used to eliminate contaminants. For the production process to remain economically viable, it is imperative to balance the need for purity and economical processing. Various process parameters, including agitation techniques, the kind of bioreactor used, feeding methods, inoculum density, mass flow, enthalpy of stream, and processing time, influence the final properties of produced biopolymers. The fermentation and bioreactor types that may be used to produce microbial biopolymers include simple batch reactors, fed-batch systems, and semi-continuous systems. Although batch fermentation is simple and straightforward to manage, production may not be desirable. Fed-batch fermentation can increase output by offering a more constant growing environment and allowing for a regulated nutrient supply. Continuous fermentation allows for continuous production and can be especially effective for industrial-scale operations by preserving a steady-state condition. It is necessary to take into account the design, scalability, controllability, simplicity of inspection, and the specific requirements of the biopolymer production. Based on the particular production goals, each approach must be evaluated for its distinctive combination of benefits and limitations. Transferring lab-scale biopolymer manufacturing to an industrial system presents challenges. Maintaining consistent needs, preventing contamination, and enhancing the flow of nutrients and oxygen are essential elements for successful large-scale production (Das et al., 2023). Following production, the biopolymer undergoes changes in its rheological and structural characteristics. Biopolymer degradation has been found to have an influence on poly(3-hydroxybutyrate-co-3-hydroxyvalerate), alginate, chitosan, bacterial cellulose, and polylactic Acid. The intention of changes in physicochemical properties and crystallinity is to deteriorate biopolymers (Shaikh et al., 2024). There are several potentials to improve the synthesis of microbial biopolymers through genetic engineering and metabolic pathway modification. Microorganisms can be genetically modified to produce biopolymers by overexpressing essential enzymes involved in the process, rerouting metabolic pathways to promote biopolymer production, and improving tolerance to high concentrations of substrate or product. For instance, PHAs synthesis may rise in *Cupriavidus necator* when PHAs synthase is overexpressed. These genetic improvements can make The production method more commercially feasible by significantly increasing yield and efficiency.

6 The Challenges of Microbial Biopolymer Production from Waste

A desirable alternative for materials produced chemically is microbial polymers. One major environmental challenge is the waste management from the different industrial sectors. The agro-industrial waste is unrestrained accumulation and detrimental even though it is not a significant bioburden. Their valorization develops rapidly, particularly when these wastes may be utilized to produce other valuable goods like biopolymers. Table 6 indicates the challenges and their potential solutions in the production of microbial biopolymers utilizing wastes. The transition from a laboratory small-scale to an industrial large-scale manufacturing is exceedingly difficult. One of the main obstacles is increasing production to an industrial scale while keeping costs down. The accomplishment of biopolymer production may be scaled up, and their viability for industrial production depends on a number of factors, including the cost of substrates, the amount of biopolymer generated over feed rate, and the expenses associated with downstream processing (Chavan et al., 2021). Developing microbial biopolymers from waste is a significant problem in terms of economic feasibility. The total cost of manufacturing must be comparable with typical petrochemical-based polymers, even though implementing waste as a substrate can lower the cost of raw materials. This contains all the costs from microbial cultivation, fermentation, downstream processing, and purification. The first step in making the production of microbial biopolymers economically feasible is obtaining bacteria that can utilize readily available and affordable raw materials. Furthermore, maximizing the fermentation output through the utilization of genetically engineered strains and biotechnological approaches may significantly reduce the overall cost of production (Das et al., 2023).

One of the most significant operating costs in polymer recovery is the water removal of biomass after it has been developed. It is essential to enhance the generation of biopolymers from waste substrates by optimizing the fermentation process. To maintain the best possible microbial activity, variables like pH, temperature, oxygen levels, and the availability of nutrients must be carefully regulated. The simplest and least costly approach is gravitational settling. However, because microbial biopolymers have minuscule molecules, membrane filtration may also be used in conjunction with centrifugation. To prevent an increase in operating expenses, it must be precisely optimized. There are benefits and drawbacks to each type of fermentation process, including fed-batch, batch, and continuous fermentation. Commercial viability depends on creating fermentation systems that are scalable, affordable, and capable of handling enormous amounts of waste. The content of waste streams varies greatly, including food leftovers, industrial byproducts, and agricultural waste. The consistent quality and effectiveness of the manufacturing of biopolymers may be impacted by this variability. A variety of waste possesses differing concentrations of nutrients, pollutants, and inhibitors to affect the development of microorganisms and the generation of biopolymers. Production cannot be reliable or efficient until waste feedstocks are standardized and strong microbial strains capable of handling a variety of substrates have been established (Mohapatra et al., 2017).

Furthermore, it could be difficult to achieve sustainable development objectives due to the unpredictable nature of legislation and policies. The economic climate of environmental policy is impacted by uncertainty. The system also affects the microbial biopolymers' development, acceptance, and scalability. A network of roles includes business enterprises, academic scholars, lawmakers, and regulatory authorities. Moreover, establishing new supply chain-hosting players necessitates a drawn-out process for creating and using innovative economic models (Liguori et al., 2016). Designing a model to predict and adjust to changes in the future in order to achieve desired results requires a detailed understanding of the interplay between varied ecosystem complexity and environmental sustainability and biodiversity unpredictability. When biopolymers produced from waste are to be used in food items, pharmaceutical products, and medical treatment, they must adhere to strict regulations controlling safety, quality, and performance. Furthermore, biopolymers especially those originating from genetically modified organisms—are viewed and accepted by the general population, which might impact their popularity. Highlighting the economical and environmental advantages of biopolymers requires education and awareness initiatives (Paul et al., 2020).

7 Conclusion and Future Perspectives

This chapter has outlined many resources for these wastes as well as a variety of methods that might be used as a potential strategy to increase microbial biopolymer production and the value of the waste. Food and agricultural waste products, which would otherwise need to be artificially produced, are also in high demand in industry. Initiatives to combat global warming and prohibitions on plastic bags are expanding the market for bio-based polymers. To lower the cost of the final product, a lot of work has been done to add inexpensive nanoparticles to the biopolymers. The depth of knowledge on the sustainable production of microbial biopolymers required much more improvement in comparison to the information available on physical and chemical approaches. As a result, adequate knowledge

Table 6 The challenges and potential solutions in the production of microbial biopolymers utilizing wastes

Challenge	Description	Potential solutions
Variability in Waste Composition	Waste streams have inconsistent nutrient and contaminant levels, affecting microbial growth and biopolymer yield. Standardizing feedstocks is necessary for stable production	• Develop pretreatment processes to standardize waste composition • Use robust microbial strains capable of handling variable substrates • Implement monitoring and adjustment systems
Microbial Strain Optimization	Natural microbes may not efficiently convert waste to biopolymers. Genetic engineering and metabolic pathway optimization are required but can be complex and costly	• Employ advanced genetic engineering techniques, such as CRISPR and synthetic biology • Use adaptive laboratory evolution to improve microbial efficiency • Collaborate with biotech companies for strain development
Process Optimization	Requires precise control of pH, temperature, nutrients, and oxygen levels. Developing cost-effective, scalable processes is essential for commercial viability	• Use bioprocess modeling and simulation to optimize conditions • Implement automated control systems for real-time process adjustments • Scale-up experiments to identify and solve potential issues early
Economic Feasibility	High production costs must be competitive with petrochemical-based polymers. Costs include waste collection, preprocessing, fermentation, and purification	• Enhance microbial productivity and yield through strain improvement and process optimization • Develop cost-effective waste collection and preprocessing methods • Improve downstream processing efficiency
Downstream Processing and Purification	Extraction and purification of biopolymers from microbial cultures are complex and expensive. Ensuring high purity and performance standards while reducing costs is challenging	• Develop novel, low-cost purification techniques (e.g., membrane filtration, adsorption) • Integrate processes to minimize steps and reduce costs • Optimize recovery methods for higher efficiency
Regulatory and Market Acceptance	Must meet regulatory standards for safety and quality, especially for food, medical, and pharmaceutical uses. Public perception and acceptance of biopolymers also affect market success	• Conduct comprehensive safety and quality assessments to meet regulatory requirements • Engage in public outreach and education to improve the perception of biopolymers • Cooperate closely with regulatory bodies
Environmental Impact	Assessing the overall sustainability of the production process, including energy use, emissions, and resource consumption, is crucial to ensure environmental benefits	• Perform life cycle assessments (LCAs) to identify and mitigate environmental impacts • Develop energy-efficient processes and use renewable energy sources • Implement circular economy practices to reuse waste streams

is not available about the major biopolymer synthesis and recovery ecologically by utilizing waste. This revised assessment provides unique insight into the expanding interest in microbial biopolymer production with an environmentally-friendly approach. Sustainable production and utilization of biopolymers are useful strategies for circular bioeconomy and provide a chance to solve environmental problems. Moreover, the chasm between academics and industry remains considerably wide. Numerous investigations were limited to being carried out in small-scale production. In order to introduce the technology to the masses for commercialization, a comprehensive strategy must be established. The potential of biorefinery to produce various biopolymers would significantly lower greenhouse gas emissions. Because the costs of the fractional raw materials used in production are lower than those of fresh raw materials, the products derived from food and argo-industrial wastes also counterbalance the production expenses of these biopolymers. Biomedical businesses, food industries, automation products, agricultural commodities, wastewater treatments, cosmetic manufacturing, and biocatalyst casings are merely a few of the industries that utilize these biopolymers extensively. Thus, a significant strategy to reduce the amount of different waste dumped in landfills and river streams is to produce biopolymers from a variety of waste.

Acknowledgements The author would like to appreciate Istanbul Sabahattin Zaim University, Department of Food Engineering.

References

Abdl Aali, R., & Al-Sahlany, S. (2024). Gellan gum as a unique microbial polysaccharide: Its characteristics, synthesis, and current application trends. *Gels, 10*(3), 183. https://doi.org/10.3390/gels10030183

Amini, M., Yousefi-Massumabad, H., Younesi, H., Abyar, H., & Bahramifar, N. (2020). Production of the polyhydroxyalkanoate biopolymer by Cupriavidus necator using beer brewery wastewater containing maltose as a primary carbon source. *Journal of Environmental Chemical Engineering, 8*(1), 103588. https://doi.org/10.1016/j.jece.2019.103588

Ansari, F., Lee, C.-C., Rashidimehr, A., Eskandari, S., Ashaolu, T. J., Mirzakhani, E., Pourjafar, H., & Jafari, S. M. (2024). The role of probiotics in improving food safety; detoxification of heavy metals and chemicals. *Toxin Reviews, 43*(1), 63–91. https://doi.org/10.1080/15569543.2023.2283768

Ansari, F., Lee, C.-C., Rashidimehr, A., Eskandari, S., Joshua Ashaolu, T., Mirzakhani, E., Pourjafar, H., & Jafari, S. M. (2023). The role of probiotics in ımproving food safety: Inactivation of pathogens and biological toxins. *Current Pharmaceutical Biotechnology, 24*. https://doi.org/10.2174/1389201024666230601141627

Aramvash, A., Gholami-Banadkuki, N., Moazzeni-Zavareh, F., & Hajizadeh-Turchi, S. (2015). An environmentally friendly and efficient method for extraction of PHB biopolymer with non-halogenated solvents. *Journal of Microbiology and Biotechnology, 25*(11), 1936–1943. https://doi.org/10.4014/JMB.1505.05053

Arun, J., Vigneshwar, S. S., Swetha, A., Gopinath, K. P., Basha, S., Brindhadevi, K., & Pugazhendhi, A. (2022a). Bio-based algal (Chlorella vulgaris) refinery on de-oiled algae biomass cake: A study on biopolymer and biodiesel production. *Science of the Total Environment, 816*, 151579. https://doi.org/10.1016/j.scitotenv.2021.151579

Arun, R., Shruthy, R., Preetha, R., & Sreejit, V. (2022). Biodegradable nano composite reinforced with cellulose nano fiber from coconut industry waste for replacing synthetic plastic food packaging. *Chemosphere, 291*. https://doi.org/10.1016/J.CHEMOSPHERE.2021.132786

Asase, R. V., & Glukhareva, T. V. (2024). Production and application of xanthan gum—prospects in the dairy and plant-based milk food industry: A review. *Food Science and Biotechnology, 33*(4), 749–767. https://doi.org/10.1007/S10068-023-01442-7

Di Bartolo, A., Infurna, G., & Dintcheva, N. T. (2021). A review of bioplastics and their adoption in the circular economy. *Polymers, 13*(8). https://doi.org/10.3390/POLYM13081229

Bernard, K. N. M., Prakash, O., Juneja, C., Panchal, D., Sylvere, N. K., & Pal, S. (2024). Development and techno-economic analysis of Grewia biopolymer-based dual coagulant system for wastewater treatment at pilot scale. *Bioresource Technology, 397*. https://doi.org/10.1016/j.biortech.2024.130514

Berradi, A., Aziz, F., Achaby, M. El, Ouazzani, N., & Mandi, L. (2023). a comprehensive review of polysaccharide-based hydrogels as promising biomaterials. *Polymers, 15*(13), 2908. https://doi.org/10.3390/POLYM15132908

Bi, D., Yang, X., Lu, J., & Xu, X. (2023). Preparation and potential applications of alginate oligosaccharides. *Critical Reviews in Food Science and Nutrition, 63*(29), 10130–10147. https://doi.org/10.1080/10408398.2022.2067832

Biswas, M. C., Jony, B., Nandy, P. K., Chowdhury, R. A., Halder, S., Kumar, D., Ramakrishna, S., Hassan, M., Ahsan, M. A., Hoque, M. E., & Imam, M. A. (2022). Recent advancement of biopolymers and their potential biomedical applications. *Journal of Polymers and the Environment, 30*(1), 51–74. https://doi.org/10.1007/S10924-021-02199-Y

Blakeney, M. (2019). Food loss and waste and food security. In *Food loss and food waste* (pp. 1–26). Edward Elgar Publishing. https://doi.org/10.4337/9781788975391.00006

Bossu, J., Le Moigne, N., Dieudonné-George, P., Dumazert, L., Guillard, V., & Angellier-Coussy, H. (2021). Impact of the processing temperature on the crystallization behavior and mechanical properties of poly[R-3-hydroxybutyrate-co-(R-3-hydroxyvalerate)]. *Polymer, 229*. https://doi.org/10.1016/J.POLYMER.2021.123987

Bzducha-Wróbel, A., Koczoń, P., Błażejak, S., Kozera, J., & Kieliszek, M. (2020). Valorization of deproteinated potato juice water into β-Glucan preparation of C. utilis origin: Comparative study of preparations obtained by two ısolation methods. *Waste and Biomass Valorization, 11*(7), 3257–3271. https://doi.org/10.1007/S12649-019-00641-W/FIGURES/5

Cavallini, A., Torre, S., Usai, L., Casula, M., Fais, G., Nieri, P., Concas, A., & Lutzu, G. A. (2024). Effect of cheese whey on phycobiliproteins production and FAME profile by Arthrospira platensis (Spirulina): Promoting the concept of a circular bio-economy. *Sustainable Chemistry and Pharmacy, 40*, 101625. https://doi.org/10.1016/j.scp.2024.101625

Chaudhari, V., Buttar, H. S., Bagwe-Parab, S., Tuli, H. S., Vora, A., & Kaur, G. (2021). Therapeutic and industrial applications of curdlan with overview on its recent patents. *Frontiers in Nutrition, 8*, 646988. https://doi.org/10.3389/FNUT.2021.646988

Chavan, S., Yadav, B., Tyagi, R. D., & Drogui, P. (2021). A review on production of polyhydroxyalkanoate (PHA) biopolyesters by thermophilic microbes using waste feedstocks.

Bioresource Technology, 341, 125900. https://doi.org/10.1016/j.biortech.2021.125900

Crugeira, P. J. L., Almeida, H. H. S., Marcet, I., Rendueles, M., Pires, M. G., Rafael, H. M., Rodrigues, A. I. G., Santamaria-Echart, A., & Barreiro, M. F. (2023). Biosynthesis of antioxidant xanthan gum by Xanthomonas campestris using substrates added with moist olive pomace. *Food and Bioproducts Processing, 141*, 210–218. https://doi.org/10.1016/j.fbp.2023.08.008

Das, A., Ringu, T., Ghosh, S., & Pramanik, N. (2023). A comprehensive review on recent advances in preparation, physicochemical characterization, and bioengineering applications of biopolymers. *Polymer Bulletin, 80*(7), 7247–7312. https://doi.org/10.1007/S00289-022-04443-4/FIGURES/31

de Mello, A. F. M., Vandenberghe, L. P. de S., Machado, C. M. B., Valladares-Diestra, K. K., de Carvalho, J. C., & Soccol, C. R. (2023). Polyhydroxybutyrate production by Cupriavidus necator in a corn biorefinery concept. *Bioresource Technology, 370*, 128537. https://doi.org/10.1016/j.biortech.2022.128537

Demirci, M., Lee, C. C., Çavuş, M., & Çağlar, M. Y. (2020). Oleogels for food applications. *Biopolymer-Based Formulations: Biomedical and Food Applications*, 781–811. https://doi.org/10.1016/B978-0-12-816897-4.00031-X

dos Santos Borges, A. V., Andrade, B. B., Santana, J. S., de Almeida Medeiros, R. M., de Souza, C. O., de Jesus Assis, D., da Silva, J. B. A., Tavares, P. P. L. G., & Cardoso, L. G. (2023). Biopolymers synthesized by microalgae grown in wastewater: A technological survey. *BioEnergy Research, 17*(1), 73–86. https://doi.org/10.1007/s12155-023-10680-w

Du, R., Wang, Y., Zhao, F., Qiao, X., Song, Q., Li, S., Kim, R.-C., Pan, L., Han, Y., Xiao, H., & Zhou, Z. (2020). Production, optimization and partial characterization of bacterial cellulose from Gluconacetobacter xylinus TJU-D2. *Waste and Biomass Valorization, 11*(5), 1681–1690. https://doi.org/10.1007/s12649-018-0440-5

Du, J., Li, L., & Zhou, S. (2019). Microbial production of cyanophycin: from enzymes to biopolymers. *Biotechnology Advances, 37*(7). https://doi.org/10.1016/j.biotechadv.2019.05.006

Duque-Acevedo, M., Belmonte-Ureña, L. J., Cortés-García, F. J., & Camacho-Ferre, F. (2020). Agricultural waste: Review of the evolution, approaches and perspectives on alternative uses. *Global Ecology and Conservation, 22*, e00902. https://doi.org/10.1016/J.GECCO.2020.E00902

FAO publications catalogue 2023. (2023). *FAO Publications Catalogue 2023*. https://doi.org/10.4060/CC7285EN

Feketshane, Z., Alven, S., & Aderibigbe, B. A. (2022). Gellan gum in wound dressing scaffolds. *Polymers, 14*(19), 4098. https://doi.org/10.3390/POLYM14194098

Freger, V., & Ramon, G. Z. (2021). Polyamide desalination membranes: Formation, structure, and properties. *Progress in Polymer Science, 122*, 101451. https://doi.org/10.1016/J.PROGPOLYMSCI.2021.101451

García-Sánchez, M. E., Robledo-Ortiz, J. R., Jiménez-Palomar, I., González-Reynoso, O., & González-García, Y. (2019). Production of bacterial cellulose by Komagataeibacter xylinus using mango waste as alternative culture medium. *Revista Mexicana de Ingeniería Química, 19*(2), 851–865. https://doi.org/10.24275/rmiq/Bio743

He, F., Yang, H., Zeng, L., Hu, H., & Hu, C. (2020). Production and characterization of bacterial cellulose obtained by Gluconacetobacter xylinus utilizing the by-products from Baijiu production. *Bioprocess and Biosystems Engineering, 43*(5), 927–936. https://doi.org/10.1007/S00449-020-02289-6/TABLES/5

Hendawy, W., El Kanani, Y., El Batawy, O., & El mongy, T. (2019). Optimization the alginate production conditions in sweet whey medium by azotobacter chrococcum. *Arab Universities Journal of Agricultural Sciences, 27*(1), 479–489. https://doi.org/10.21608/ajs.2019.43586

Ho Jin, Y., Lee, T., Kim, J. R., Choi, Y.-E., & Park, C. (2019). Improved production of bacterial cellulose from waste glycerol through investigation of inhibitory effects of crude glycerol-derived compounds by Gluconacetobacter xylinus. *Journal of Industrial and Engineering Chemistry, 75*, 158–163. https://doi.org/10.1016/j.jiec.2019.03.017

Hussain, Z., Sajjad, W., Khan, T., & Wahid, F. (2019). Production of bacterial cellulose from industrial wastes: a review. *Cellulose, 26*(5), 2895–2911. https://doi.org/10.1007/S10570-019-02307-1

Jabeen, N., Atif, M. (2024). Polysaccharides based biopolymers for biomedical applications: A review. *Polymers for Advanced Technologies, 35*(1), e6203. https://doi.org/10.1002/PAT.6203

Kadier, A., Ilyas, R. A., Huzaifah, M. R. M., Harihastuti, N., Sapuan, S. M., Harussani, M. M., Azlin, M. N. M., Yuliasni, R., Ibrahim, R., Atikah, M. S. N., Wang, J., Chandrasekhar, K., Amirulislam, M., Sharma, S., Punia, S., Rajasekar, A., Asyraf, M. R. M., & Ishak, M. R. (2021). Use of industrial wastes as sustainable nutrient sources for bacterial cellulose (BC) production: Mechanism, advances, and future perspectives. *Polymers, 13*(19), 3365. https://doi.org/10.3390/POLYM13193365

Koul, B., Yakoob, M., & Shah, M. P. (2022). Agricultural waste management strategies for environmental sustainability. *Environmental Research, 206*, 112285. https://doi.org/10.1016/J.ENVRES.2021.112285

Lee, C. C., Chen, J., & Frank, J. F. (2015). Role of cellulose and colanic acid in attachment of shiga toxin-producing escherichia coli to lettuce and spinach in different water hardness environments. *Journal of Food Protection, 78*(8), 1461–1466. https://doi.org/10.4315/0362-028X.JFP-15-056

Lee, C. C., Chen, J., & Frank, J. F. (2016). Influence of extracellular cellulose and colanic acid production on the survival of shiga toxin-producing escherichia coli on spinach and lettuce after chlorine treatment. *Journal of Food Protection, 79*(4), 666–671. https://doi.org/10.4315/0362-028X.JFP-15-375

Lee, C. C., Suttikhana, I., & Ashaolu, T. J. (2024). Techno-functions and safety concerns of plant-based peptides in food matrices. *Journal of Agricultural and Food Chemistry*. https://doi.org/10.1021/ACS.JAFC.4C02464/ASSET/IMAGES/MEDIUM/JF4C02464_0002.GIF

Lee, C.-C., & Demirci, M. (2023). Zeaxanthin. In *Handbook of food bioactive ingredients* (pp. 1–34). Springer International Publishing. https://doi.org/10.1007/978-3-030-81404-5_53-1

Lee, C.-C., & Küçükata, Y. Ş. (2023). Microbial production of zeaxanthin. In *Microbial production of food bioactive compounds* (pp. 1–38). Springer International Publishing. https://doi.org/10.1007/978-3-030-81403-8_36-1

Lee, C.-C., & Neio Demirci, M. (2023). Different types of fryers for the food industry. In *High-temperature processing of food products* (pp. 293–326). Elsevier. https://doi.org/10.1016/B978-0-12-818618-3.00003-3

Lee, C.-C., & Taştemir, İ. A. (2023). Direct heat exchangers in the food industry. In *Thermal processing of food products by steam and hot water* (pp. 181–208). Elsevier. https://doi.org/10.1016/B978-0-12-818616-9.00008-0

Lee, C.-C. (2023a). Astaxanthin. In *Handbook of food bioactive ingredients* (pp. 687–727). Springer International Publishing. https://doi.org/10.1007/978-3-031-28109-9_54

Lee, C.-C. (2023b). Microbial production of astaxanthin. In *Microbial production of food bioactive compounds* (pp. 1–38). Springer International Publishing. https://doi.org/10.1007/978-3-030-81403-8_37-1

Leong, H. Y., Chang, C. K., Khoo, K. S., Chew, K. W., Chia, S. R., Lim, J. W., Chang, J. S., & Show, P. L. (2021). Waste biorefinery towards a sustainable circular bioeconomy: A solution to global issues. *Biotechnology for Biofuels, 14*(1), 1–15. https://doi.org/10.1186/S13068-021-01939-5/TABLES/3

Liguori, R., Ventorino, V., Pepe, O., & Faraco, V. (2016). Bioreactors for lignocellulose conversion into fermentable sugars for production of high added value products. *Applied Microbiology and Biotechnology, 100*(2), 597–611. https://doi.org/10.1007/s00253-015-7125-9

Lioranĉas, V., & Lee, C.-C. (2024). Antimicrobial role and sensory ınfluence of oregano essential oil on the shelf-life of atlantic cod loins. *Darnıos Aplınkos Vystymas, 21*(1), 121–126. https://doi.org/10.52320/dav.v21i1.300

Ma, X., Chen, Y., Liu, F., Zhang, S., & Wei, Q. (2021). Enhanced tolerance and resistance characteristics of Scenedesmus obliquus FACHB-12 with K3 carrier in cadmium polluted water. *Algal Research, 55*, 102267. https://doi.org/10.1016/j.algal.2021.102267

Mehariya, S., Plöhn, M., Jablonski, P., Stagge, S., Jönsson, L. J., & Funk, C. (2023). Biopolymer production from biomass produced by Nordic microalgae grown in wastewater. *Bioresource Technology, 376*, 128901. https://doi.org/10.1016/j.biortech.2023.128901

Mehariya, S., Marino, T., Casella, P., Iovine, A., Leone, G. P., Musmarra, D., & Molino, A. (2020). Biorefinery for agro-ındustrial waste ınto value-added biopolymers: Production and applications. In *Biorefineries: A step towards renewable and clean energy* (pp. 1–19). Springer. https://doi.org/10.1007/978-981-15-9593-6_1

Mohanraj, R., Gnanamangai, B. M., Ramesh, K., Priya, P., Srisunmathi, R., Poornima, S., Ponmurugan, P., & Robinson, J. P. (2019). Optimized production of gamma poly glutamic acid (γ-PGA) using sago. *Biocatalysis and Agricultural Biotechnology, 22*, 101413. https://doi.org/10.1016/j.bcab.2019.101413

Mohapatra, S., Maity, S., Dash, H. R., Das, S., Pattnaik, S., Rath, C. C., & Samantaray, D. (2017). Bacillus and biopolymer: Prospects and challenges. *Biochemistry and Biophysics Reports, 12*, 206–213. https://doi.org/10.1016/j.bbrep.2017.10.001

Molina-Ramírez, C., Castro, C., Zuluaga, R., & Gañán, P. (2018). Physical characterization of bacterial cellulose produced by Komagataeibacter medellinensis using food supply chain waste and agricultural by-products as alternative low-cost feedstocks. *Journal of Polymers and the Environment, 26*(2), 830–837. https://doi.org/10.1007/S10924-017-0993-6/TABLES/5

Moradali, M. F., & Rehm, B. H. A. (2020). Bacterial biopolymers: from pathogenesis to advanced materials. *Nature Reviews Microbiology, 18*(4), 195–210. https://doi.org/10.1038/s41579-019-0313-3

Moradi, Z., Madadkar Haghjou, M., Zarei, M., Colville, L., & Raza, A. (2021). Synergy of production of value-added bioplastic, astaxanthin and phycobilin co-products and direct green 6 textile dye remediation in Spirulina platensis. *Chemosphere, 280*, 130920. https://doi.org/10.1016/j.chemosphere.2021.130920

Oliveira, V. R. L., Coelho, V. V., Sérvulo, E. F. C., & Calado, V. M. de A. (2023). Study of the molecular weight of pullulan produced by aureobasidium pullulans from ındustrial Waste. *Materials Research, 26*(suppl 1). https://doi.org/10.1590/1980-5373-mr-2023-0060

Oran, D., Unal, S., & Gunduz, O. (2022). Evaluation of bacterial cellulose/quince seed mucilage composite scaffold for wound dressing. *Emergent Materials, 5*(2), 315–321. https://doi.org/10.1007/S42247-022-00352-4

Pairazamán, O. D., Woiciechowski, A. L., Zevallos, L. A., Tanobe, V. O. A., Zandona, A., & Soccol, C. R. (2024). Fumaric acid production by Rhizopus species from acid hydrolysate of oil palm empty fruit bunches. *Brazilian Journal of Microbiology, 55*(2), 1179–1187. https://doi.org/10.1007/s42770-024-01322-0

Palaniveloo, K., Amran, M. A., Norhashim, N. A., Mohamad-Fauzi, N., Peng-Hui, F., Hui-Wen, L., Kai-Lin, Y., Jiale, L., Chian-Yee, M. G., Jing-Yi, L., Gunasekaran, B., & Razak, S. A. (2020). Food waste composting and microbial community structure profiling. *Processes, 8*(6), 723. https://doi.org/10.3390/PR8060723

Papadaki, A., Papapostolou, H., Alexandri, M., Kopsahelis, N., Papanikolaou, S., de Castro, A. M., Freire, D. M. G., & Koutinas, A. A. (2018). Fumaric acid production using renewable resources from biodiesel and cane sugar production processes. *Environmental Science and Pollution Research, 25*(36), 35960–35970. https://doi.org/10.1007/s11356-018-1791-y

Park, S., Roh, S., Yoo, J., Ahn, J. H., Gong, G., Lee, S.-M., Um, Y., Han, S. O., & Ko, J. K. (2024). Tailored polyhydroxyalkanoate production from renewable non-fatty acid carbon sources using engineered Cupriavidus necator H16. *International Journal of Biological Macromolecules, 263*, 130360. https://doi.org/10.1016/j.ijbiomac.2024.130360

Paul, E., Bessière, Y., Dumas, C., & Girbal-Neuhauser, E. (2020). Biopolymers production from wastes and wastewaters by mixed microbial cultures: strategies for microbial selection. *Waste and Biomass Valorization, 12*(8), 4213–4237. https://doi.org/10.1007/S12649-020-01252-6

Philibert, T., Lee, B. H., & Fabien, N. (2017). Current status and new perspectives on chitin and chitosan as functional biopolymers. *Applied Biochemistry and Biotechnology, 181*(4), 1314–1337. https://doi.org/10.1007/s12010-016-2286-2

Rajinipriya, M., Nagalakshmaiah, M., Robert, M., & Elkoun, S. (2018). Importance of agricultural and industrial waste in the field of nanocellulose and recent industrial developments of wood based nanocellulose: A review. *ACS Sustainable Chemistry and Engineering, 6*(3), 2807–2828. https://doi.org/10.1021/ACSSUSCHEMENG.7B03437/ASSET/IMAGES/MEDIUM/SC-2017-03437S_0021.GIF

Rao, J., Lv, Z., Chen, G., & Peng, F. (2023). Hemicellulose: Structure, chemical modification, and application. *Progress in Polymer Science, 140*, 101675. https://doi.org/10.1016/J.PROGPOLYMSCI.2023.101675

Râpă, M., Stefan, L. M., Seciu-Grama, A. M., Gaspar-Pintiliescu, A., Matei, E., Zaharia, C., Stănescu, P. O., & Predescu, C. (2022). Poly(3-hydroxybutyrate-co-3-hydroxyvalerate) (P(3HB-co-3HV))/Bacterial Cellulose (BC) biocomposites for potential use in biomedical applications. *Polymers, 14*(24), 5544. https://doi.org/10.3390/POLYM14245544

Raychaudhuri, R., Naik, S., Shreya, A. B., Kandpal, N., Pandey, A., Kalthur, G., & Mutalik, S. (2020). Pullulan based stimuli responsive and sub cellular targeted nanoplatforms for biomedical application: Synthesis, nanoformulations and toxicological perspective. *International Journal of Biological Macromolecules, 161*, 1189–1205. https://doi.org/10.1016/J.IJBIOMAC.2020.05.262

Rebocho, A. T., Pereira, J. R., Neves, L. A., Alves, V. D., Sevrin, C., Grandfils, C., Freitas, F., & Reis, M. A. M. (2020). Preparation and characterization of films based on a natural P(3HB)/mcl-PHA blend obtained through the co-culture of cupriavidus necator and pseudomonas citronellolis in apple pulp waste. *Bioengineering, 7*(2), 34. https://doi.org/10.3390/bioengineering7020034

Reddy, M. S. B., Ponnamma, D., Choudhary, R., & Sadasivuni, K. K. (2021). A comparative review of natural and synthetic biopolymer composite scaffolds. *Polymers, 13*(7). https://doi.org/10.3390/POLYM13071105

Saeed, S., Tayyab, M., Mehmood, T., Awan, A. R., Firyal, S., Nadeem, F., & Irfan, M. (2023). Valorization of potato peel for production of alginate and optimization of the process through response surface methodology (RSM) by using Azotobacter nigricans. *Biomass Conversion and Biorefinery, 13*(5), 3893–3901. https://doi.org/10.1007/s13399-021-01357-5

Saharan, B. S., Kamal, N., Badoni, P., Kumar, R., Saini, M., Kumar, D., Sharma, D., Tyagi, S., Ranga, P., Parshad, J., Goyal, C., Kumar, R., Nehra, M., Seth, C. S., Duhan, J. S., & Mandal, N. K. (2024). Biopolymer and polymer precursor production by microorganisms: Applications and future prospects. *Journal of Chemical Technology and Biotechnology, 99*(1), 17–30. https://doi.org/10.1002/jctb.7524

Salave, S., Rana, D., Sharma, A., Bharathi, K., Gupta, R., Khode, S., Benival, D., & Kommineni, N. (2022). Polysaccharide based implantable drug delivery: Development strategies, regulatory requirements, and future perspectives. *Polysaccharides, 3*(3), 625–654. https://doi.org/10.3390/POLYSACCHARIDES3030037

Šantek, B., Felski, M., Friehs, K., Lotz, M., & Flaschel, E. (2010). Production of paramylon, a β-1,3-glucan, by heterotrophic cultivation of Euglena gracilis on potato liquor. *Engineering in Life Sciences, 10*(2), 165–170. https://doi.org/10.1002/elsc.200900077

Shaikh, S., Rashid, N., Onwusogh, U., McKay, G., & Mackey, H. R. (2023). Effect of nutrients deficiency on biofilm formation and single cell protein production with a purple non-sulphur bacteria enriched culture. *Biofilm, 5*. https://doi.org/10.1016/j.bioflm.2022.100098

Shaikh, S., Rashid, N., McKay, G., & Mackey, H. R. (2024). Resource recovery through bioremediation of fuel-synthesis wastewater in a biofilm photobioreactor using purple non-sulfur bacteria: A circular bioeconomy approach. *Chemical Engineering Journal Advances, 19*. https://doi.org/10.1016/j.ceja.2024.100614

Sharma, M., Tellili, N., Kacem, I., & Rouissi, T. (2024). Microbial biopolymers: from production to environmental applications—a review. *Applied Sciences, 14*(12), 5081. https://doi.org/10.3390/APP14125081

Sharmila, G., Muthukumaran, C., Nayan, G., & Nidhi, B. (2013). Extracellular biopolymer production by aureobasidium pullulans MTCC 2195 using jackfruit seed powder. *Journal of Polymers and the Environment, 21*(2), 487–494. https://doi.org/10.1007/s10924-012-0459-9

Sharon, I., Hilvert, D., & Schmeing, T. M. (2023). Cyanophycin and its biosynthesis: Not hot but very cool. *Natural Product Reports, 40*(9), 1479–1497. https://doi.org/10.1039/D2NP00092J

Shayesteh, H., Laird, D. W., Hughes, L. J., Nematollahi, M. A., Kakhki, A. M., & Moheimani, N. R. (2023). Co-producing phycocyanin and bioplastic in arthrospira platensis using carbon-rich wastewater. *Biotech, 12*(3), 49. https://doi.org/10.3390/biotech12030049

da Silva, J. A., Cardoso, L. G., de Jesus Assis, D., Gomes, G. V. P., Oliveira, M. B. P. P., de Souza, C. O., & Druzian, J. I. (2018). Xanthan gum production by Xanthomonas campestris pv. campestris IBSBF 1866 and 1867 from lignocellulosic agroindustrial wastes. *Applied Biochemistry and Biotechnology, 186*(3), 750–763. https://doi.org/10.1007/S12010-018-2765-8/FIGURES/5

Singh, R. S., Kaur, N., Singh, D., Purewal, S. S., & Kennedy, J. F. (2023). Pullulan in pharmaceutical and cosmeceutical formulations: A review. *International Journal of Biological Macromolecules, 231*, 123353. https://doi.org/10.1016/J.IJBIOMAC.2023.123353

Siracusa, V. (2019). Microbial degradation of synthetic biopolymers waste. *Polymers, 11*(6), 1066. https://doi.org/10.3390/polym11061066

Srisawat, P., Higuchi-Takeuchi, M., & Numata, K. (2022). Microbial autotrophic biorefineries: Perspectives for biopolymer production. *Polymer Journal, 54*(10), 1139–1151. https://doi.org/10.1038/s41428-022-00675-3

Sworn, G. (2021). Xanthan gum. In *Handbook of hydrocolloids* (pp. 833–853). https://doi.org/10.1016/B978-0-12-820104-6.00004-8

Tahouni, Y., Cheng, T., Lajewski, S., Benz, J., Bonten, C., Wood, D., & Menges, A. (2023). Codesign of biobased cellulose-filled filaments and mesostructures for 4D printing humidity responsive smart structures. *3D Printing and Additive Manufacturing, 10*(1), 1–14. https://doi.org/10.1089/3DP.2022.0061

Taylor, C. S., Illangakoon, U., Dawson, J. I., Kanczler, J. M., Behbehani, M., Humphrey, E., Majid, Q., Lukasiewicz, B., Nigmatullin, R., Heseltine, P., Oreffo, R. O. C., Haycock, J. W., Terracciano, C., Harding, S. E., Edirisinghe, M., & Roy, I. (2021). Harnessing polyhydroxyalkanoates and pressurized gyration for hard and soft tissue engineering. *ACS Applied Materials and Interfaces, 13*(28), 32624–32639. https://doi.org/10.1021/ACSAMI.0C19689

Terkula Iber, B., Azman Kasan, N., Torsabo, D., & Wese Omuwa, J. (2022). A review of various sources of chitin and chitosan in nature. *Journal of Renewable Materials, 10*(4), 1097–1123. https://doi.org/10.32604/jrm.2022.018142

Urbina, L., Eceiza, A., Gabilondo, N., Corcuera, M. Á., & Retegi, A. (2019). Valorization of apple waste for active packaging: multicomponent polyhydroxyalkanoate coated nanopapers with improved hydrophobicity and antioxidant capacity. *Food Packaging and Shelf Life, 21*. https://doi.org/10.1016/J.FPSL.2019.100356

Vaishnav, A., & Choudhary, D. K. (2021). Microbial polymers: Applications and ecological perspectives. *Microbial Polymers: Applications and Ecological Perspectives*, 1–710. https://doi.org/10.1007/978-981-16-0045-6

Varelas, V., Tataridis, P., Liouni, M., & Nerantzis, E. T. (2016). Valorization of winery spent yeast waste biomass as a new source for the production of β-Glucan. *Waste and Biomass Valorization, 7*(4), 807–817. https://doi.org/10.1007/s12649-016-9530-4

Venkata Mohan, S., Modestra, J. A., Amulya, K., Butti, S. K., & Velvizhi, G. (2016). A circular bioeconomy with biobased products from CO_2 sequestration. *Trends in Biotechnology, 34*(6), 506–519. https://doi.org/10.1016/j.tibtech.2016.02.012

Venoor, V., Park, J. H., Kazmer, D. O., & Sobkowicz, M. J. (2021). Understanding the effect of water in polyamides: A review. *Polymer Reviews, 61*(3), 598–645. https://doi.org/10.1080/15583724.2020.1855196

Viveka, R., Varjani, S., & Ekambaram, N. (2021). Valorization of cassava waste for pullulan production by Aureobasidium pullulans MTCC 1991. *Energy and Environment, 32*(6), 1086–1102. https://doi.org/10.1177/0958305X20908065

Yang, Q., Xu, W., Luan, T., Pan, T., Yang, L., & Lin, L. (2021). Comparative responses of cell growth and related extracellular polymeric substances in Tetraselmis sp. to nonylphenol, bisphenol A and 17α-ethinylestradiol. *Environmental Pollution, 274*, 116605. https://doi.org/10.1016/j.envpol.2021.116605

Zhang, C., Wu, D., & Ren, H. (2019). Economical production of agricultural γ-polyglutamic acid using industrial wastes by Bacillus subtilis. *Biochemical Engineering Journal, 146*, 117–123. https://doi.org/10.1016/j.bej.2019.03.013

Zhou, W., Bergsma, S., Colpa, D. I., Euverink, G.-J.W., & Krooneman, J. (2023). Polyhydroxyalkanoates (PHAs) synthesis and degradation by microbes and applications towards a circular economy. *Journal of Environmental Management, 341*, 118033. https://doi.org/10.1016/j.jenvman.2023.118033

Microbiological Technology to Extract Bioactive Molecules from Agricultural and Industrial Waste

Muyideen Olaitan Bamidele, Michael Bola Bamikale, Motolani Adepeju Bamidele, Eliseo Cardenas Hernandez, Guillermo Castillo Olvera, Jose Sandoval Cortes, and Cristobal Noe Aguilar

Abstract

This study explores the innovative use of microbiological technology to extract bioactive molecules from agricultural and industrial waste. Agricultural and industrial by-products, often regarded as waste, are rich in bioactive compounds with significant potential for pharmaceutical, nutraceutical, and industrial applications. Traditional extraction methods have limitations in efficiency and sustainability. Therefore, we investigated the use of biological approaches, including enzyme-assisted extraction and microbial fermentation, to enhance the recovery of these valuable compounds. Our findings demonstrate that specific microorganisms can effectively utilize agricultural wastes as substrates, facilitating the extraction and recovery of bioactive compounds. The application of these biological methods not only improves extraction efficiency but also promotes the sustainable use of waste materials, contributing to waste reduction and environmental protection. This comprehensive review highlights microbiological technologies' potential in valorizing agro-industrial waste, presenting a promising alternative to conventional extraction techniques and paving the way for future research and industrial applications.

Keywords

Waste · Bioactive compound · Microbe · Extraction

1 Introduction

Among all extractive methodologies to recover phytochemicals and bioactive molecules from agricultural and industrial wastes, microbiological technology plays a crucial role in extracting procedures due to several features, including environmental sustainability, economic growth, and the advancement of biotechnological research (Lemes et al., 2022).

Microbiological technology is relevant for waste utilization because in addition to allowing the recovery of compounds of interest, it helps reduce the overall waste burden on the environment, promoting a circular economy where waste materials are converted into valuable products, reducing the reliance on new raw materials and minimizing environmental impact, and the negative impacts associated with waste disposal, such as soil and water contamination (Krakowska-Sieprawska et al., 2022).

For this reason, microbiological technology can provide economic benefits by creating new revenue streams from waste materials and reducing disposal costs. It also promotes the development of new industries and job opportunities related to waste processing and bioactive compound production. It also drives innovation in biotechnology and microbiology, developing new methods and applications for waste-derived bioactive molecules (Lemes et al., 2022).

Our research group is a pioneer in developing bioprocesses based on microbiological technology for the recovery of bioactive molecules, particularly polyphenolic compounds (Martins et al., 2011). However, this technology can extract proteins and enzymes, hormones, vitamins, antioxidants and antibiotics, among others. All these molecules are often used in pharmaceuticals, nutraceuticals, and functional foods due to their health-promoting properties. Microbiological extraction technology involves using microorganisms such as bacteria, fungi, and yeasts to extract bioactive molecules from various sources, including agricultural and industrial waste.

M. Olaitan Bamidele · M. Bola Bamikale · M. Adepeju Bamidele · E. Cardenas Hernandez · G. Castillo Olvera · J. Sandoval Cortes · C. Noe Aguilar (✉)
Bioprocesses and Bioproducts Research Group, Food Research Department, School of Chemistry, Universidad Autónoma de Coahuila, Saltillo, México
e-mail: cristobal.aguilar@uadec.edu.mx

K. Sharma (ed.), *Bio-prospecting of Novel Microbial Bioactive Compounds for Sustainable Development*, Sustainable Landscape Planning and Natural Resources Management, https://doi.org/10.1007/978-3-031-82178-3_8

This technology leverages the metabolic capabilities of microorganisms to convert waste materials into valuable compounds. Bacteria are often used for their rapid growth and diverse metabolic pathways; fungi can produce a wide range of enzymes and secondary metabolites while yeasts are industrially utilized for fermentation processes and production of specific bioactive compounds.

It is important to consider that several agricultural wastes can be used, including crop residues, fruit and vegetable peels, and other plant materials. In contrast, industrial waste comprises by-products from food processing, breweries, and other industrial processes. Sometimes a pre-treatment of Waste is required, such as physical, chemical, or enzymatic treatment, to break down complex substrates into simpler forms that microorganisms can quickly metabolize.

The key process of microbiological technology is the fact of the growth of microorganisms releasing and/or producing bioactive molecules. This can be done in different types of bioreactors, including submerged fermentation, where microorganisms are grown in a liquid medium, or solid-state fermentation, where microorganisms are grown on solid substrates in the absence or near-absence of free water.

To maximize the response of phytochemicals separation, the optimization of conditions is required, particularly controlling factors such as pH, temperature, oxygen levels, and nutrient supply to maximize the yield of bioactive molecules. After fermentation, the bioactive molecules are extracted from the microbial culture using various techniques such as solvent extraction using solvents to dissolve and separate the desired compounds; filtration and centrifugation to remove microbial cells and other solids; chromatography for purifying and concentrating the bioactive molecules. Further purification, formulation, and packaging of the bioactive molecules for their intended use is required for several applications, including pharmaceuticals production of antibiotics, enzymes, and other therapeutic agents, food industry where preservatives, flavor enhancers, and nutritional supplements are highly demanded. In agriculture, for the development of biopesticides and growth-promoting agents and environment where bioremediation agents for cleaning up pollutants and waste treatment are needed.

For this reason, microbiological extraction technology is a versatile and sustainable approach for harnessing bioactive molecules from waste materials, with wide-ranging applications across various industries, utilizing agricultural and industrial wastes for the recovery of bioactive molecules is significant for several reasons, encompassing environmental, economic, and societal benefits. The complex nature of waste utilization promotes collaboration across disciplines, including microbiology, chemical engineering, environmental science, and material science.

In brief, utilizing agricultural and industrial wastes to recover bioactive molecules presents significant environmental, economic, and societal benefits. This approach not only addresses waste management challenges but also contributes to sustainable development, innovation, and the overall well-being of society.

2 Microorganisms Used in Bioactive Molecule Extraction

The extraction of bioactive molecules from natural sources has become a cornerstone of modern biotechnology, with microorganisms playing a pivotal role due to their diverse metabolic capabilities. Among these microorganisms, bacteria, fungi, and other microbial entities have shown remarkable potential in producing a wide array of bioactive compounds with applications in pharmaceuticals, agriculture, and environmental management.

2.1 The Role of Bacteria, Fungi, and Other Microorganisms

Microorganisms play a pivotal role in the extraction of bioactive molecules due to their ability to produce a diverse array of secondary metabolites with significant biological activities. The primary microorganisms involved in this process include bacteria, fungi, and other microorganisms like actinomycetes and microalgae.

- **Bacteria**: Bacteria are prolific producers of bioactive compounds, including antibiotics, enzymes, and other secondary metabolites. For instance, the genus Streptomyces is renowned for producing over two-thirds of the clinically useful antibiotics of natural origin (Cuozzo et al., 2023). These bacteria utilize agro-industrial waste as a substrate, which can enhance the recovery of bioactive molecules. The fermentation processes carried out by bacteria are highly efficient and can be optimized for large-scale production.
- **Fungi**: Fungi, particularly endophytic fungi, are another crucial group of microorganisms involved in the extraction of bioactive compounds. They produce a wide range of secondary metabolites with antimicrobial, antiviral, and anticancer properties (Hashem et al., 2023). Fungi, such as *Penicillium* and *Aspergillus*, have been extensively studied for their ability to produce bioactive molecules. These organisms can be isolated from various plant tissues and induce the production of novel bioactive compounds when cultured under specific conditions (Bhaskar et al., 2024).

- **Other Microorganisms**: Other microorganisms, including actinomycetes and microalgae, also significantly contribute to the production of bioactive compounds. Actinomycetes, a group of filamentous bacteria, are notable for their ability to produce a vast array of bioactive molecules including antibiotics and antitumor agents (Diarra et al., 2024). Microalgae are known for their production of pigments, antioxidants, and other bioactive compounds, with potential applications in pharmaceuticals and nutraceuticals (Menon et al., 2023).

2.2 Microbial Enzymes

Microorganisms are essential players in the extraction of bioactive molecules primarily because of their ability to produce a wide range of enzymes. These microbial enzymes serve as metabolic catalysts that facilitate the conversion of complex substrates to valuable bioactive compounds. The diversity and functionality of microbial enzymes are crucial in various biotechnological applications, including the recovery of bioactive molecules from agricultural and industrial waste (Alloun & Calvio, 2024).

2.2.1 Diversity of Microbial Enzymes

Microbial enzymes are derived from a plethora of microorganisms, including bacteria, fungi, actinomycetes, and microalgae. Each microorganism group produces unique enzymes that have specific biochemical and catalytic properties. For instance, bacteria such as *Bacillus* and *Streptomyces* generate a range of enzymes, including amylases, proteases, and lipases, which are widely utilized in industrial processes (Vojnovic et al., 2024). Fungi, particularly species of *Aspergillus* and *Penicillium*, produce cellulases and xylanases crucial for the breakdown of plant biomass (Tong et al., 2024).

2.2.2 Functions of Microbial Enzymes

The primary function of microbial enzymes in bioactive molecule extraction is to catalyze the breakdown of complex organic materials into simpler bioactive compounds. Various enzymatic activities facilitate this process, including hydrolysis, oxidation, and reduction. For example, cellulases break down cellulose into glucose units, which can then be fermented into bioactive molecules such as ethanol and organic acids (Fan et al., 2024). Microbial enzymes also play a pivotal role in biotransformation processes, modifying substrates to produce new bioactive compounds with enhanced biological activity. Enzymes, such as oxidoreductases and transferases, are particularly important in this context, enabling the synthesis of complex molecules that are otherwise difficult to produce through chemical synthesis (González-Granda et al., 2023). The robust nature and multifunctionality of microbial enzymes make them invaluable for various biotechnological applications. They are used in the food industry to enhance product quality and shelf life, as pharmaceuticals for drug synthesis, and in environmental biotechnology for waste management and bioremediation (Liu & Kokare, 2023). The use of microbial enzymes in the fermentation of agricultural and industrial wastes not only aids in the extraction of bioactive compounds but also promotes sustainability by reducing waste and minimizing environmental impact (Kaur et al., 2023).

2.3 Microbial Fermentation and Its Relevance

Microbial fermentation is a critical biotechnological process for the extraction and production of bioactive molecules. This process harnesses the metabolic capabilities of microorganisms such as bacteria, fungi, and yeast to convert organic substrates into valuable compounds. The relevance of microbial fermentation for bioactive molecule extraction lies in its efficiency, sustainability, and versatility.

- **Microbial Fermentation Process**: Microbial fermentation involves using microorganisms to break down complex organic materials under anaerobic and aerobic conditions. This process can be divided into two main stages: primary fermentation, which produces energy and primary metabolites (e.g., ethanol and lactic acid), and secondary fermentation, which generates bioactive secondary metabolites. These secondary metabolites include polyphenols, polysaccharides, amino acids, and vitamins, all of which have significant applications in the pharmaceutical, nutraceutical, and food industries (Elshafie et al., 2023).

2.4 Types of Microorganisms Used

Various microorganisms are employed in fermentation, each of which contributes to unique enzymatic activities and metabolic pathways. Commonly used microorganisms include the following:

- **Bacteria**: *Lactobacillus* and *Streptomyces* are widely used for their ability to produce lactic acid, antibiotics, and other bioactive compounds.
- **Fungi**: *Aspergillus* and *Penicillium* species are known for producing enzymes such as amylases and cellulases, which are essential for the breakdown of starch and cellulose.
- **Yeast**: *Saccharomyces cerevisiae* is a primary organism used in ethanol production and for the synthesis of vitamins and other bioactive compounds.

2.5 Benefits of Microbial Fermentation

The benefits of microbial fermentation for bioactive molecule extraction include the following:

- **Efficiency**: Microorganisms can rapidly convert substrates into the desired products under controlled conditions, making the process highly efficient.
- **Sustainability**: Fermentation processes often use renewable resources and can be conducted with minimal environmental impacts. For instance, agro-industrial wastes can be utilized as substrates, thereby reducing waste and promoting sustainability.
- **Versatility**: The adaptability of microorganisms allows them to produce a wide range of bioactive compounds that can be tailored to specific needs through genetic and metabolic engineering.
- **Technological Advancements**: Advancements in biotechnology have enhanced the efficiency and scope of microbial fermentation. Techniques such as metagenomics and synthetic biology have enabled the discovery and optimization of novel microbial strains and enzymes, thereby expanding the repertoire of bioactive compounds that can be produced. Additionally, bioreactor design and process optimization have improved the scalability and economic viability of fermentation processes.

3 Enzymatic Extraction Processes

Enzymatic extraction processes have become essential in biotechnology to efficiently retrieve bioactive compounds from various natural sources. This method leverages the specificity and efficiency of enzymes to break down cell walls and release target molecules, thereby improving yield and purity. Enzyme-assisted extraction (EAE) uses specific enzymes to hydrolyze cell wall components, thereby facilitating the release of bioactive compounds. Commonly used enzymes include cellulases, hemicellulases, pectinases, and proteases. These enzymes target polysaccharides and proteins, break them into smaller units, and free the desired molecules (Kleekayai et al., 2023).

3.1 Mechanisms of the Enzymatic Breakdown of Waste Materials

Enzymes are biocatalysts that accelerate chemical reactions without being consumed. In waste management, enzymes, such as cellulases, proteases, lipases, and amylases, degrade organic materials. These enzymes function by hydrolyzing chemical bonds within waste polymers, converting them into smaller molecules that are more manageable and less harmful to the environment (Ashraf et al., 2023).

3.2 Mechanisms of Enzymatic Degradation of Plastics

Plastic waste, particularly synthetic polymers such as polyethylene terephthalate (PET), poses significant environmental challenges owing to their durability and resistance to degradation. Enzymatic degradation of plastics occurs via a two-step mechanism.

- **Adsorption**: The enzymes are first absorbed onto the polymer surface. This step is critical because it brings the catalytic sites of the enzymes close to the polymer chains.
- **Hydrolysis**: The adsorbed enzymes catalyze the hydrolytic cleavage of polymer chains. For instance, PETases hydrolyze the ester bonds in PET, breaking down the polymer into its monomers, such as terephthalic acid and ethylene glycol. Microorganisms can further metabolize these monomers, facilitating complete biodegradation.

3.3 Microbial Enzymes in Waste Degradation

Microorganisms such as bacteria and fungi produce a diverse array of enzymes that can degrade various waste materials. For example.

- **Ligninolytic Enzymes**: These include laccases, peroxidases, and manganese peroxidases, which break down lignin, a complex polymer found in the plant biomass. Ligninolytic enzymes are pivotal in the decomposition of lignocellulosic waste, converting it into simpler, valuable products such as biofuels and biochemicals (Ahmad et al., 2023).
- **Cellulases and Hemicellulases**: These enzymes degrade cellulose and hemicellulose into fermentable sugars. This process is crucial for the conversion of agricultural residues into bioethanol, an alternative renewable energy source (Jayakumar et al., 2023).

3.4 Types of Enzymes Used

Enzymatic extraction processes leverage the catalytic properties of enzymes to break down complex molecules into simpler ones, thereby facilitating the extraction of valuable compounds from various materials. Different types of enzymes are utilized in these processes, each tailored to specific substrates.

- **Cellulases**: Cellulases are enzymes that catalyze the hydrolysis of cellulose into glucose. They are crucial for the breakdown of plant biomass and are widely used in the biofuel industry to convert lignocellulosic materials into fermentable sugars. Cellulases also play a significant role in the extraction of bioactive compounds from plant materials (Sanjeewa et al., 2023).
- **Proteases**: Proteases hydrolyze peptide bonds in proteins and break them into amino acids and smaller peptides. These enzymes are used in various industries, including food processing, pharmaceuticals, and waste management. In oil extraction, proteases can break down cell wall proteins and oleosins, thereby improving the yield and quality of the extracted oil (Song et al., 2023).
- **Lipases**: Lipases catalyze the hydrolysis of fats (lipids) into glycerol and free fatty acids. These enzymes are extensively used in the food industry for the modification of fats and oils, biodiesel production, and the pharmaceutical industry for the synthesis of enantiopure compounds. Lipases are also crucial for degrading lipid-rich waste materials (Enespa et al., 2023).
- **Amylases**: Amylases hydrolyze starch into sugars such as maltose and glucose. These enzymes are essential to the food and beverage industry for producing syrups, alcohol, and other products. They also contribute to the bioconversion of agricultural waste into biofuels (Abolore et al., 2023).
- **Laccases**: Laccases are enzymes that catalyze the oxidation of phenolic and non-phenolic substrates. They are involved in the degradation of lignin in plant cell walls, making them valuable in the paper and pulp industries and in the bioremediation of phenolic pollutants (Jeyabalan et al., 2023).
- **Pectinases**: Pectinase breaks down pectin, a polysaccharide in plant cell walls. These enzymes are widely used in the fruit juice and wine industries to clarify juices and wines and extract essential oils and other valuable compounds from plant materials (Cosgrove, 2024).

3.5 Optimization Strategies for Enzymatic Extraction

Enzymatic extraction processes have gained significant attention because of their efficiency and eco-friendliness in extracting valuable compounds from various raw materials. Optimization of these processes is crucial for enhancing yield, purity, and cost-effectiveness. Several strategies can be employed to optimize enzymatic extraction.

- **Selection of Appropriate Enzymes**: Choosing an appropriate type of enzyme is fundamental. Enzymes, such as cellulases, proteases, and lipases, have specific substrates upon which they act. The selection was based on the nature of the raw material and the desired product. For instance, cellulases are used as plant materials rich in cellulose, whereas proteases are ideal proteinaceous substrates (Sarwan et al., 2024).
- **Enzyme Concentration**: The enzyme concentration can significantly affect the efficiency of the extraction process. Higher enzyme concentrations can increase the hydrolysis rate, but this must be balanced against cost considerations. Optimizing enzyme concentration involves determining the minimal effective dose that achieves the maximum yield (Sun et al., 2023).
- **pH and Temperature Control**: Enzymes function optimally in specific pH and temperature ranges. Maintaining these conditions is critical for maximizing enzymatic activity. For example, most cellulases operate best at a pH of approximately 5.0 and temperatures between 50 and 60 °C. Deviations from these optimal conditions can result in reduced enzyme efficiency and stability (Kabir & Ju, 2023).
- **Substrate Pretreatment**: Pretreating the substrates can enhance the accessibility of enzymes to their target sites. Physical methods, such as grinding or milling, and chemical treatments, such as acid or alkali treatments, can help break down complex structures, making the substrate more amenable to enzymatic action. Pretreatment can significantly increase the yield of the desired product (Hu et al., 2023).
- **Enzyme Synergy**: The use of a combination of enzymes may be more effective than using a single enzyme. A mixture of cellulases, hemicellulases, and pectinases can break down plant cell walls more efficiently than any one of these enzymes alone. Synergistic effects can lead to more complete hydrolysis of the substrate, improving the overall extraction efficiency (Tong et al., 2023).
- **Reaction Time**: The duration of the enzymatic reaction affected the yield and quality of the extracted product. Optimization involves finding the balance between sufficient reaction time for complete hydrolysis and practical considerations, such as processing speed and cost. Prolonged reaction times can lead to degradation of sensitive compounds; therefore, careful monitoring is essential (Singla et al., 2023).
- **Use of Response Surface Methodology (RSM)**: RSM is a statistical technique used to optimize the conditions of enzymatic extraction processes. This involves designing experiments to evaluate the effects of multiple variables and their interactions. RSM helps identify the optimal conditions for maximum yield and efficiency with minimal experimental runs, thereby saving time and resources (Setyani et al., 2023).

4 Fermentation-Based Extraction (Micheal)

4.1 Fermentation

Fermentation can be defined as a catabolic process in which an organism breaks down carbohydrates, such as starch or sugar, into an alcohol or an acid. In other words, it is a biological process in which microorganisms convert organic compounds into simpler molecules, generating energy such as adenosine triphosphate (ATP) (Boantza, 2020). Fermentation is often used for making wine, mead, cheese, and beer and is a natural way to extract bioactive compounds. It does not involve high heat or radiation, which can be harmful. Instead, it uses enzymes to break down plant cell walls, releasing, beneficial compounds like phenolics. This process also removes toxins and changes phenolic compounds into other helpful ones (Vidhu, 2020). In fermentation, lactic acid bacteria such as *Enterococcus, Streptococcus, Leuconostoc, Lactobacillus,* and *Pediococcus,* yeasts and molds such as *Debaryomyces, Kluyveromyces, Saccharomyces, Geotrichium, Mucor, Penicillium*, and *Rhizopus* species are majorly employed (Sharma et al., 2020). Fermentation can be classified into lactic acid and alcohol fermentation.

4.1.1 Lactic Acid Fermentation

Lactic acid fermentation involves the conversion of sugars into lactic acid by bacteria, fungi, and other microorganisms in anaerobic conditions. This process is widely utilized in producing fermented foods and beverages like yogurt, sauerkraut, kimchi, and sourdough bread (Emkani et al., 2022). During lactic acid fermentation, as described in Fig. 1, the enzyme lactate dehydrogenase converts pyruvate, a glycolysis byproduct, into lactic acid, enabling glycolysis to proceed without oxygen. Microorganisms such as *Lactobacillus, Streptococcus,* and *Leuconostoc*, commonly found in dairy products and fermented vegetables, are employed in lactic acid fermentation (de Oliveira Junqueira et al., 2023).

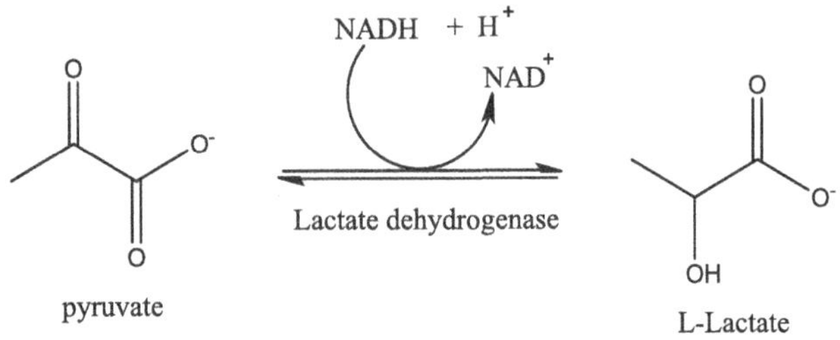

Fig. 1 Lactic acid fermentation

Lactic acid production contributes to the flavor and texture of fermented foods. They act as preservatives by lowering pH and inhibiting harmful bacteria growth and enhance the digestibility and nutritional value of certain foods by breaking down complex carbohydrates and proteins into simpler forms (Aguirre-Garcia et al., 2024). Furthermore, nicotinamide adenine dinucleotide (NADH) transfers its electrons directly to pyruvate, producing lactate as a byproduct. Lactic acid bacteria play a crucial role in preserving and producing wholesome foods. This process is cost-effective and environmentally sustainable, making it important in the food and pharmaceutical industries (Aguirre-Garcia et al., 2024).

4.1.2 Alcohol Fermentation

Alcohol fermentation is another type of two-step process. It involves the conversion of pyruvate to ethanol with the aid of yeast. At first, the pyruvate loses a carboxyl group, forming acetaldehyde and releasing CO_2. Secondly, NADH donates electrons to acetaldehyde, producing ethanol and regenerating NAD^+. This process described in Fig. 2, is essential in producing alcoholic beverages and industrial ethanol. Ethanol fermentation sustains glycolysis, promoting energy production. Its significance spans biological and industrial spheres, contributing to flavor enhancement and energy sustainability (Vamvakas & Kapolos, 2020).

4.2 Principle of Microbial Fermentation

Microbial fermentation involves converting organic compounds into simpler molecules like ethanol or lactic acid through the metabolic activities of microorganisms like yeast and bacteria (Shanmugam & Ingram, 2022). This principle is mostly used in numerous industrial processes, including the production of fermented foods, alcoholic

Fig. 2 Alcohol fermentation

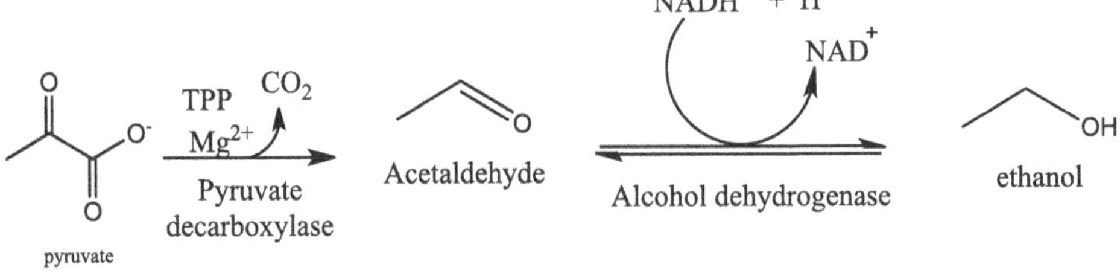

beverages, biofuels, and biopharmaceuticals. The end products vary depending on the type of microorganism involved and the specific metabolic pathways employed (Shanmugam & Ingram, 2022).

A critical step in this process is glycolysis, where microorganisms break down glucose and other sugars into pyruvate, generating ATP for their own energy needs. Anaerobic respiration occurs when pyruvate is converted into byproducts like ethanol or lactic acid, allowing microbes to generate ATP anaerobically (Watanabe, 2024). The specific byproducts formed depend on the type of microorganism and environmental conditions, which can be harvested for various industrial, pharmaceutical, and food applications (Sadh et al., 2018). The fundamental principle underlying this versatile technology is the ability of microbes to thrive in anaerobic conditions and convert organic compounds into valuable byproducts. Typical products from fermentation processes include organic acids, alcohols, secondary metabolites and gases (Sun et al., 2022).

4.3 Fermentation Processes

Fermentation processes are classified into two parts: solid state and sub-merged or liquid fermentation. The kind of substrate, the intended result, process economics, and scalability are some variables that influence the decision between different fermentation processes (Sadh et al., 2018).

4.3.1 Solid-State Fermentation

Solid-state fermentation is a process in which microorganisms adhere to the surface of a solid substrate and develop there with little to no water, mostly involving the substrate-particle interface. Fermented foods, animal feed additives, enzymes, bioactive compounds, and other biotechnological products are produced using this approach (Yadav et al., 2022). Solid-state fermentation is sustainable for agricultural wastes, as it requires less water. It produces high product concentrations and purity due to the absence of extra water. It also reduces running costs due to the absence of a liquid phase and contributes to environmental sustainability by transforming organic leftovers into valuable products (Chilakamarry et al., 2022). Solid-state fermentation produces fermented foods like tempeh, koji, and miso, as well as enzymes, probiotics, and bioactive compounds like antibiotics, antioxidants, and antimicrobial agents (Yafetto, 2022).

Solid-state fermentation has contamination risks, slow growth, and tricky monitoring like every other method. It is also harder to scale up and control than other methods. Plus, getting the product out can be tough, and the nutrients might not be easy for the microbes to reach in dense substrates. These issues can lower product quality, increase costs, and make recovery more complicated. Unwanted microbes can hang out in the solid substrate, messing up the product's consistency and yield (Mattedi et al., 2023).

Research has shown that solid-state fermentation of *Aspergillus niger's* in wineries, olive mills, and breweries produced lignocellulolytic enzymes and antioxidant compounds (Leite et al., 2021). Pineapple peel and core processed via solid-state fermentation using *Aspergillus niger* GH1 yielded free phenols and other bioactive compounds (Paz-Arteaga et al., 2023). Solid-state fermentation of avocado seeds using *Aspergillus niger* helped to recover bioactive compounds. These bioactive compounds have been reported to exhibit several biological activities that are salient in medical and healthcare sectors (Yepes-Betancur et al., 2021).

4.3.2 Sub-Merged or Liquid Fermentation

Submerged fermentation is when enzymes and other reactive compounds are submerged in a liquid like alcohol, oil, or nutrient broth to create biomolecules. In this method, microorganisms grow in nutrient-rich broth with a 95% water content, and various products are formed with better control over process variables like pH and temperature, using defined culture media for consistent reproducibility across experiments (Martínez-Medina et al., 2019).

Submerged fermentation offers several advantages due to its high water content, nutrient-rich broth, and ability to produce a wide range of products. It promotes microbial growth and metabolism, enhancing nutrient absorption. This method is useful for producing industrial products such as enzymes, organic acids, antibiotics, biofuels, and pharmaceuticals (Dudekula et al., 2020). This method needs proper monitoring in the area of mass and energy transfer, scaling, and maintaining continuous operation (Hegab et al., 2013).

Several research has employed this method in recovering bioactive compounds from waste sources. Submerged fermentation of the edible mushroom *Pleurotus ostreatus* led to the production of fatty acids, phenolic metabolites, nucleotides and alkaloids (Jiamworanunkul, 2019). Liquid-submerged fermentation of onion skin using *Lactiplantibacillus plantarum* TB 11–32 produced bioactive compounds, including phenolic compounds (Ramires et al., 2023).

Both the solid-state and the submerged fermentation methods have been applied industrially and scientifically; however, one yield is better than the other due to differences in the metabolisms carried out by the microorganisms.

4.4 Fermentation of Different Waste Sources

Various waste sources can undergo fermentation to produce biofuels, bioplastics, and organic fertilizers. Such sources encompass agricultural waste, food waste, municipal solid waste, sewage sludge, industrial waste, forest residues, algae, aquatic plants, animal by-products, and green waste. Selections of the fermentation process are product-specific (Ong et al., 2020). Table 1 below shows the waste sources for fermentation, the organisms required for fermentation, and the recovered bioactive compounds.

4.5 Bioactive Compounds from Fermentation-Based Extraction

4.5.1 Bioactivities Compounds Obtained from Fermentation

Several bioactive compounds have been recovered from various sources via fermentation. These compounds were extracted utilizing a meticulous selection of microorganisms and precise control of fermentation conditions (Siddiqui et al., 2023).

Antibiotics, such as penicillin, streptomycin, erythromycin, and tetracycline, are examples of fermentation-derived compounds that combat bacterial infections by inhibiting bacterial growth (Sadh et al., 2018). Polyphenols, flavonoids, and carotenoids play a crucial role in neutralizing harmful free radicals, thereby mitigating the risk of chronic diseases like cancer, cardiovascular disorders, skin diseases, and neurodegenerative diseases (Lemes et al., 2022). Probiotics are compounds known for their ability to promote gut health and aid digestion, they are recovered via fermentation, and they help in the production of probiotic-rich foods like yogurt, sauerkraut, and kombucha. Enzymes such as amylase, protease, lipase, and cellulase are pivotal in various industries, including food processing, detergent manufacturing, biofuel production, and pharmaceuticals (Yi et al., 2020).

Vitamins such as vitamin B12, riboflavin, and vitamin K2 are important compounds from fermentation processes. They are essential for myriad physiological functions within the body. Organic acids like acetic acid, lactic acid, citric acid, and gluconic acid exhibit antimicrobial properties and contribute to food preservation (Thompson et al., 2020).

Bioactive peptides, derived from proteins through enzymatic hydrolysis during fermentation, exhibit remarkable health-promoting properties. Meanwhile, polyunsaturated fatty acids synthesized by selected microorganisms during fermentation play pivotal roles in brain function, inflammation regulation, and cardiovascular health (Chai et al., 2020).

It is evident that through fermentation, bioactive compounds are recovered from waste sources, and these compounds not only enrich our diets but also enhance our well-being, underscoring the immense value of this transformative bioprocess.

5 Bioprospecting and Identification of Novel Microbial Strains

Bioprospecting involves systematically searching for bioactive compounds from natural sources, including microorganisms, to discover novel drugs, enzymes, and other biologically significant molecules. This process is critical for developing new therapeutics and their industrial applications. Methods for bioprospecting and isolation of novel microbes, screening techniques for identifying high-yielding strains, and examples of novel bioactive molecules are highlighted below.

5.1 Methods for Bioprospecting and Isolation of Novel Microbes

Bioprospecting and isolation of novel microbes can be achieved through various methods tailored to specific goals and environments. The following methods are used in the bioprospecting and isolation of microbes.

- **Isolation of Pure Microbial Strains**: This method focused on obtaining individual microbial species from environmental samples. It often involves culturing microbes on selective media to isolate the colonies of interest. This approach is useful for studying specific organisms and their properties (Lemke & DeSalle, 2023).
- **Symbiotic Communities**: Instead of isolating a single species, this method investigates the microbial communities that work together. This approach is particularly relevant for understanding complex ecosystem interactions and synergistic effects (García et al., 2023).
- **Culture-Independent Methods**: These methods involve direct analysis of environmental DNA or RNA to identify and study microbes without culturing them. Metagenomics, metatranscriptomics, and metaproteomics have been employed to explore microbial diversity and function in situ (Tessler et al., 2023).
- **Screening for Bioactive Compounds**: This method involves isolating and screening microbes for the production of novel bioactive compounds. Traditional bioprospecting often relies on culturing potential producers

Table 1 Bioactive compounds from fermentative extraction of different waste sources

Waste sources	Organisms for fermentation	Bioactive compounds	References
Cereal bran	*Bacillus subtilis*	Phenolic compounds, flavonoids, glucans, pigments	Hadj Saadoun et al. (2021), Lemes et al. (2022)
Fruits and vegetable waste	*Lactobacillus planetarium, Aspergillus niger*	Phenolic compounds,	
Fish waste	*Chryseobacterium* species and *Bacillus* species	Omega 3	
Milk	*Lactobacillus species; Rhizopus oryzae MTCC 8784*	Peptides, Lactic acid production	
Apple Pomace	*Aspergillus niger, Rhizopus oryzae 1526*	Epicatechin, catechins, chlorogenic acid, hydroxycinnamates, phloretin glycosides, quercitin glycosides, procyanidins, anthocyanins, fumaric acid	Sadh et al. (2018)
Wheat bran, Coffee pulp	*Aspergillus niger*	Polygalactouronase	
Rice bran	*Rhizopus oryzae*	Phenolic compounds	
Corn husk	*Bacillus cereus*	Cellulose	
Peels of citrus fruits, soybean meal, tofu dreg, Brewer's spent grain	*Rheinheimera sp., Lysinibacillus sp., Trametes versicolor*	Laccase	Yin et al. (2022)
Vegetable waste	*Bacillus subtilis, Rhizopus oligosporus, Fusarium flocciferum, Aspergillus, Yarrowia, and Trichoderma species*	Functional carbohydrates and glycosidases	Sabater et al. (2020)
Grape Must, Corn *Stover,* Sugar Beet Molasses, Sugarcane Bagasse, Cassava Wastewater,	*Bacillus subtilis Aspergillus species*	2-Phenylethanol, Indole-3-acetic acid	Mitri et al. (2022)
Buriti pulp	*Lacticaseibacillus casei* SJRP38	β-carotene, lycopene	Borgonovi et al. (2021)
Yellow mustard flour and milk whey	Lactic acid bacteria	Total polyphenols, DL-3-phenyllactic acid (PLA), lactic acid	Escrivá et al. (2021)
Tomato waste	*Aspergillus niger*	Lycopene	Sadh et al. (2018)

followed by bioassays to detect bioactivity (Ajar Nath, 2021).

- **Exploring Extreme Environments**: Actinobacteria, for example, are often isolated from unique and extreme environments, such as saline sediments. This approach helps discover microbes with unique adaptations and potential biotechnological applications (Guan et al., 2020).

5.2 Screening Techniques for Identifying High-Yielding Strains

- **Bioassay-Guided Screening**: Involves testing microbial extracts against target organisms or cells to identify bioactive compounds. For example, antimicrobial assays can identify strains producing antibiotics, while cytotoxicity assays can discover anticancer agents.
- **Chemical Screening**: Analytical techniques such as mass spectrometry (MS) and nuclear magnetic resonance (NMR) spectroscopy are used to analyze microbial extracts for the presence of novel compounds. High-resolution MS can identify compounds based on their mass-to-charge ratio, while NMR provides structural information.
- **Genomic and Metabolomic Screening**: Genomic approaches involve screening microbial genomes for gene clusters associated with the biosynthesis of bioactive compounds, such as polyketides and non-ribosomal peptides. Metabolomic screening involves analyzing the metabolic profiles of microbial cultures to identify novel metabolites.
- **High-Content Screening (HCS)**: HCS combines automated microscopy and image analysis to screen large libraries of microbial extracts or compounds for bioactivity. This technique is beneficial for discovering compounds with specific cellular effects, such as the modulation of signaling pathways.

5.3 Examples of Novel Bioactive Molecules Discovered

Antibiotics:

- **Teixobactin**: Discovered from soil bacteria, teixobactin is a novel antibiotic that shows activity against gram-positive bacteria, including antibiotic-resistant strains like multi-drug resistant *Staphylococcus aureus*.

Anticancer Agents:

- **Salinosporamide A**: Isolated from the marine bacterium *Salinispora tropica*, this compound has potent anticancer activity and is being developed to treat multiple myeloma.
- **Enzymes**: Lipases have been discovered from various microbes, including Pseudomonas species, and are used in industrial applications such as biocatalysis and biodiesel production.
- **Probiotics**: Novel strains of probiotics have been isolated from human milk, such as strains with potential health benefits for gut health and immune modulation.

6 Genetic Engineering and Biotechnological Advances

The constant search for techniques to improve processes has led to tools from different fields being used in different areas. For example, genetic engineering and biotechnology are novel tools that can have applications in obtaining bioactive molecules from agricultural and industrial waste. These tools are commonly used to obtain genetically modified microorganisms, which may have applications in fermentation processes for obtaining polyphenols, antioxidants, carotenoids, polysaccharides, and bioactive peptides from waste. Gene editing processes can improve the capacity of microorganisms to produce enzymes that degrade plant material, which can improve bioprocesses for the extraction of bioactive molecules (Nayak et al., 2022).

6.1 Genetic Modification of Microbial Strains for Enhanced Extraction

Green extraction technologies have used microbial strains such as fungi, bacteria, yeasts, and enzymes to obtain bioactive molecules from waste. Nogales-Bueno et al. (2020) used cellulase, glucosidase and pectinase enzymes to extract anthocyanins and flavonols from grape skins, obtaining an increase in phenolic extraction. Amorim et al. (2019) obtained prebiotic arabinoxylooligosaccharides from beer bagasse. However, studies with genetically modified strains to improve waste degradation have been limited.

Genetic and metabolic engineering modification and genomic editing using techniques such as CRISPR-Cas9 allow the identification and modification of genes to improve the degradation and biosynthesis capacity of microbial strains (Sarkar et al., 2023). Doria et al. (2022) used the bacterium *Bacillus subtilis* genetically modified for enhanced production of cellulases and xylanases to extract bioactive compounds from cauliflower discarded from a food processing industry. In fact, bacteria of the genus *Bacillus* have been widely investigated as biotechnological tools for obtaining bioactive molecules (Herrmann et al., 2024). Another study has used *Yarrowia lipolytica* Po1g, a strain genetically modified to obtain higher lipid production when grown in a sugarcane bagasse hydrolysate (Tsigie et al., 2011). Additionally, fungi of the genus *Aspergillus* spp. have been widely studied as promising tools for obtaining high-value-added compounds through genetic modifications (Jin et al., 2021). Genetic modifications can help in higher enzyme production, reduced toxic microbial waste, improved nutrient utilization, and improved ability to use cellular energy, increasing the yield and efficiency of extraction bioprocesses (Kosariya et al., 2023). Figure 3 illustrates the process of genetic modification to improve the capabilities of a microorganism to obtain bioactive molecules.

6.2 Biotechnological Tools for Process Optimization

Currently, there are a large number of biotechnological tools that allow the improvement of bioprocesses, for example, metabolic engineering, omics technologies, biosensors, and synthetic biology, which aim to improve microbial strains to obtain more bioactive molecules, as well as adequate control of bioprocesses (Florez-Orrego et al., 2022). These tools allow the optimization of fermentation processes in bioreactors and obtain more and better microbial enzymes.

7 Integrated Approaches and Process Optimization

Integrated approaches refer to the use of multiple tools for process optimization, for example, the use of mathematical models, simulations and bioinformatics. Integrated approaches can help predict the behavior of the production systems, allowing better operational control and the ability to optimize designs, which helps reduce energy consumption, raw material use, and waste generation. In addition, the use of agricultural and industrial waste has a positive impact on the environment and operating costs (Madhavan et al., 2022).

7.1 Integration of Microbiology with Bioprocess Engineering

Bioprocess engineering depends mostly on microbiology to obtain bioactive molecules. Microbiology helps to understand the metabolism of microorganisms and their nutritional requirements and applications, which allows the design and development of biotechnological processes and

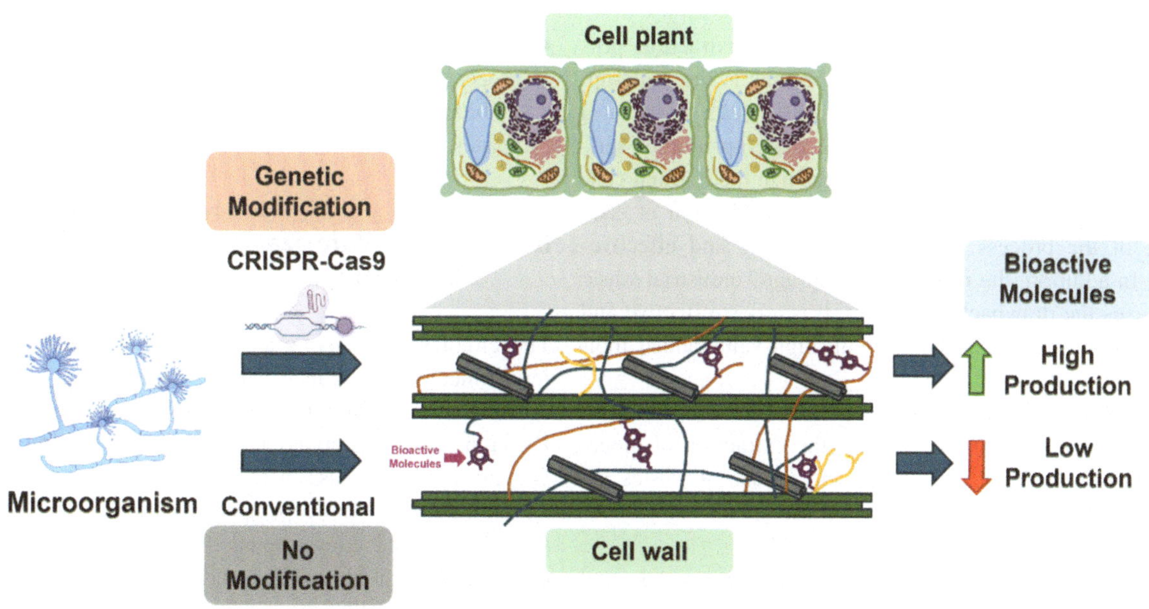

Fig. 3 General diagram of the genetic modification of microorganisms

their optimization. In addition, the integration of microbiology has allowed the development of a wide range of fermentation processes, either for the use of the enzymatic machinery of the microorganism or for the individual use of the enzyme complex to obtain bioactive molecules (Afzaal et al., 2022). This integration is essential for the advancement of microbiological technology.

7.2 Use of Bioinformatics for Data Analysis and Optimization

Bioinformatics is a tool that uses computational and statistical technologies to analyze data obtained from biological processes. Bioinformatics allows the study of the genetic composition of microorganisms used in bioprocesses by analyzing genomic sequences and identifying genes and metabolic pathways that allow optimization (Helleckes et al., 2023). One of the most widely used bioinformatics tools is the modeling of metabolic networks, which allows the prediction of the behavior of modified microorganisms (Paliwal & Dutta, 2023). The use of Big Data and Machine Learning also helps in the analysis of large amounts of data that cannot be analyzed using conventional techniques, allowing the identification of behavioral patterns and correlations that allow the optimization of biotechnological processes, achieving efficient, profitable and sustainable processes (Helleckes et al., 2023).

7.3 Strategies for Scaling up the Extraction Processes

One of the significant challenges in the extraction processes using microorganisms and enzymes is the scale-up. Generally, fermentations can behave in a certain way at a laboratory scale; however, industrial scale-up implies serious challenges when trying to homogenize the optimal conditions (Xia et al., 2021). Therefore, optimization at each stage of the process ensures its efficiency and effectiveness. In addition, the incorporation of agro-industrial waste increases the drawbacks; however, it is a cost-effective and novel strategy that can help obtain high-value-added molecules (Ranjitha et al., 2022).

In the following, strategies that can be considered for the scaling up of bioprocesses for the extraction of bioactive molecules from agricultural and industrial wastes are mentioned:

- **Microorganisms**: They can include bacteria, yeasts, and fungi that have the capacity to produce enzymes that degrade the plant wall of agricultural wastes and achieve the release of bioactive molecules. In addition, strains with greater extraction capacity can be obtained through genetic engineering, which increases the utilization of the material (Oyewole et al., 2022).
- **Fermentation**: There are different types of fermentation, for example, in solid, semi-solid, or submerged states, that can be optimized depending on the characteristics of the microorganism. From the type of fermentation, it is possible to optimize the conditions of temperature, pH, and agitation, among others. In addition, the composition of the medium can be improved by analyzing the nutrients and their appropriate concentrations (Kosariya et al., 2023).
- **Quality control**: control and monitoring are essential; for example, through the use of sensors, the critical parameters of the system can be determined, and through automated strategies, they can be controlled at optimal levels (Phillips, 2020).
- **Modeling and Simulation**: Through mathematical models, it is possible to study microbial growth and the production of bioactive molecules, which allows predicting the prediction of the system's behavior. In addition, using simulation programs allows for studying different scenarios and conditions that allow for optimizing the parameters of the bioprocess (Acorsi et al., 2022).
- **Viability of the process**: The composition of the waste is important for the correct extraction of bioactive molecules, so the material selection is important. In addition, it is necessary to implement a circular economy approach where the by-products of the process, e.g., biomass, are reused. This will allow the optimization and viability of process scale-up (Xia et al., 2021). Finally, it is necessary to consider regulations, standards, and safety in the extraction processes.
- Implementing these strategies, a robust and efficient bioprocess can be developed for the production/extraction of bioactive molecules from agricultural and industrial wastes.

8 Waste Valorization and Circular Economy

The circular economy is a broad approach that uses different methods to reduce waste production, defining economic growth separately from the use of natural resources. Different countries apply this concept, promoting a cyclical and responsible use of resources to support sustainability. From an economic perspective, the circular economy seeks to replace conventional linear models, the flow of energy and material resources, addressing problems such as environmental pollution, social equity, and long-term economic sustainability (Fidelis et al., 2019). According to the United Nations Sustainable Development Goals, a circular economy is the key to improving and promoting social equity and economic growth; it also reduces the extraction of raw materials by closing material cycles. In particular, Goal 9 (Industry, Innovation, and Infrastructure) and Goal 12 (Responsible Consumption and Production) highlight the importance of technological progress to achieve environmental goals of greater energy and resource efficiency. Therefore, by reducing the use of resources, the welfare gains obtained from economic activities can increase, reducing degradation and pollution along the entire cycle of a food supply chain and, as a result increasing the quality of life (Fidelis et al., 2019).

The high demand for raw materials, mainly for human and animal consumption and energy production in the world has led to the expansion of the industry, however, it generates large amounts of waste. These residues are the losses of up to 50% of raw materials, which occur during harvesting, post-harvest, transportation, processing, storage and consumption. These losses are estimated at 680 billion dollars per year, corresponding to 25–35% of the total food production in the world. These losses correspond to 1300 million tons per year, since this material is neither consumed nor transformed using the appropriate processes (Lemes et al., 2022; Sabater et al., 2020). Agroindustrial wastes can be of vegetable or animal origin. Vegetable waste includes leaves, straws, stems, bark, seeds, and shells generated during cleaning, processing, cooking, packaging, and storage. In the case of cereals, waste accounts for 35% of the total production, including wastewater, rice effluents, corn grits, and beer grain. As for animal wastes, they correspond to skins, hooves, feathers, hides, bones, blood, fat, and whey, and the meat sector reports losses of up to 23%, while the dairy industry generates 4 to 11 million wastes per year. In general, these residues have a high content of bioactive compounds such as phenols, polyphenols, cellulose, hemicellulose, lignin, fibers, pigments, oligosaccharides, phytosterols, organic acids, proteins, lipids, minerals, and vitamins (Chikezie Ogbu & Nnaemeka Okey, 2023).

Therefore, the industry has sought better waste management to mitigate the negative environmental impact, which is an important step for the transition to a circular economy. Therefore, new strategies have been implemented to valorize these substrates, seeking to take advantage of waste and by-products and convert them into high-value-added products such as in the formulation of medicines, bioactive ingredients, functional foods, or natural addictive products (Fig. 4) (Lemes et al., 2022; Sabater et al., 2020). The valorization of agro-industrial wastes is of great importance since the market value of the products derived from these wastes is estimated at 633.3, 48 billion, and 46 billion dollars in the beverage, medical, food, and consumer goods sectors, respectively (Khaksar et al., 2022).

Therefore, the recovery of bioactive compounds from agro-industrial waste is of great interest to the industry, and novel methods have been proposed to obtain bioactive compounds, such as bioprocesses (fermentation or enzymatic),

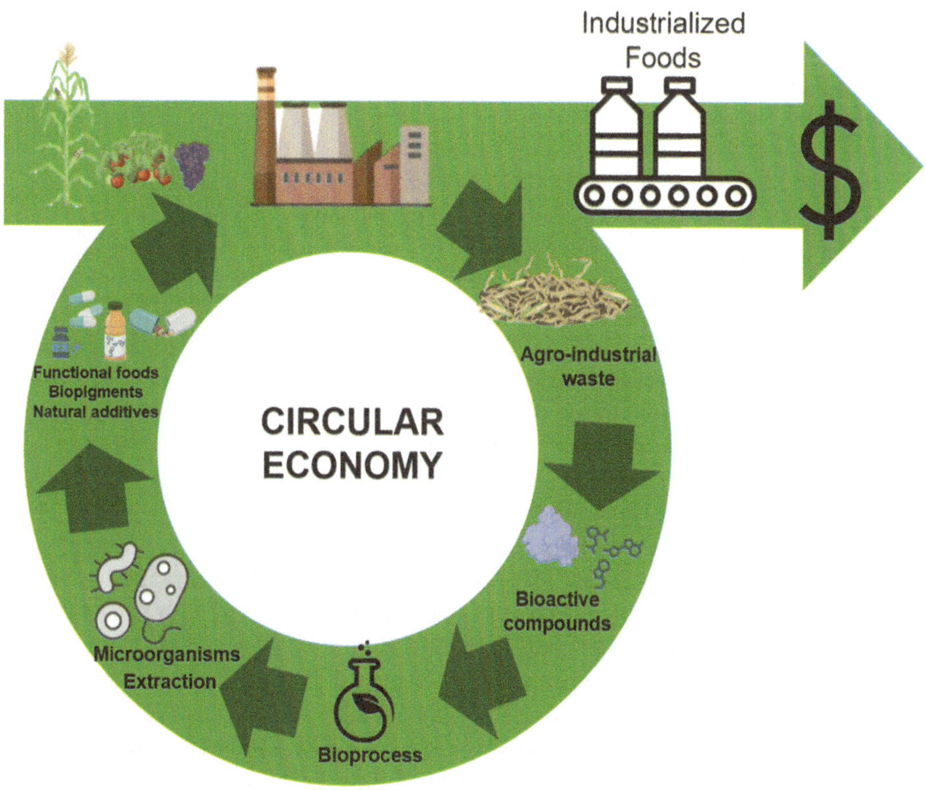

Fig. 4 Circular economy in agroindustry: waste revalorization through bioprocesses

as described in this chapter. The use of microorganisms to obtain value-added biomolecules, such as bio pigments, biofuels, and biodegradable materials, plays an important role in promoting a circular economy (Dahiya et al., 2022).

Bioprocesses applied to agro-industrial wastes using bacteria, fungi, and yeasts have proven profitable. Recent studies show a return on investment and an internal rate of return of over 10%. In addition, the break-even point of the capital investment is estimated at approximately seven years. These financial indicators highlight the economic viability and growth potential of the valorization of agro-industrial waste through bioprocesses, thus contributing to a sustainable and profitable business model (Sabater et al., 2020). In addition, the use of bioprocesses has proven to be a profitable investment but also has advantages such as improved extraction time, low solvent use, and increased quality and purity of the final product obtained at low costs compared to conventional extraction processes. Other advantages on an industrial scale include cost reduction, as the enzymes produced by microorganisms act as catalysts that provide greater savings, resulting in greater efficiency and sustainability in obtaining compounds through bioprocesses (Lemes et al., 2022).

- The application of bioprocesses in the production of xylooligosaccharides, poly-γ-glutamates, acids, and arabinoxylooligosaccharides using agro-industrial residues, such as grape pomace, soybean hulls, and grain residues, in beer production has proven to be highly profitable because of the high selling price of these ingredients. In comparison, producing compounds such as carotenoids, glycosidases, and isomaltose from soybean residues and cane molasses is less profitable because of the long fermentation times and lower market prices of these products compared to those of other functional foods. In addition, the production of oligosaccharides and biopolymers can act as prebiotics and be a highly lucrative strategy to add value to agro-industrial waste, regardless of the microorganism used (Sabater et al., 2020).

9 Challenges and Prospects

Owing to the high demand for food, the linear food industry generates millions of tons of waste, which accumulates and causes economic and environmental challenges. However, this waste is recognized as a valuable raw material with diverse applications, as it is rich in bioactive compounds. Therefore, their valorization will provide many economic, environmental, and social opportunities. However, further development is required to integrate a fully evolving valorization system as a sustainable and efficient industrial tool (Khaksar et al., 2022; Yadav et al., 2024). In recent years, consumers have shown interest in products with bioactive ingredients, which are free of synthetic additives. Therefore bioactive compounds obtained through a bioprocess are related to health as they can be applied in a variety of foods as functional, nutraceuticals, supplements or used as food additives not only have application in food but also in the pharmaceutical industry as the development of drugs such as vaccines or drug precursors offering a natural and sustainable alternative to synthetic products (Yadav et al., 2024). However, the bioactive compounds obtained present certain disadvantages, such as the susceptibility of thermosensitive compounds, variable stability and loss of activity, especially those already incorporated in food, and controlled release of compounds, this is because most compounds are evaluated under controlled conditions. These factors indicate the importance of developing new technologies in industrial environments as well as the optimization of processes, use of artificial intelligence, or model-based predictive control for improving the efficiency, productivity, and sustainability of bioprocesses (Pai et al., 2022; Rathore et al., 2021). In addition, more research is needed on the application and exploration of microorganisms for use in waste materials generated annually in various sectors worldwide and thus contribute to the circular economy at the commercial level (Dahiya et al., 2022).

- One of the biggest challenges is always the opinion of the consumer; it is necessary to convince a certain population that food or products obtained by microorganisms are safe and beneficial to human health. However, sustainable food production may not be a challenge, but trying to convince policymakers to make sustainable decisions and be willing to accept new technologies for optimizing and improving products (Ramírez Rojas et al., 2022).

10 Conclusion

Traditional methods have been employed in recovering bioactive compounds from various agro-industrial wastes, and they have been characterized by low efficiency and sustainability. -assisted fermentation for extraction has helped recover these bioactive compounds with much efficiency and sustainability. This is because microorganisms take most agro-industrial wastes as substrates. As such, they give promising results regarding efficiency and sustainability in recovering bioactive compounds, promoting the circular economy model.

Future research is imperative in determining the specific biochemical pathways through which the microorganisms solubilize bioactive compounds.

References

Abolore, R. S., Tsegaye, B., Jaiswal, S., & Jaiswal, A. K. (2023). An overview of industrial enzymes in beverage production and processing. In *Value-addition in beverages through enzyme technology* (pp. 1–26). Elsevier. https://doi.org/10.1016/B978-0-323-85683-6.00013-2

Acorsi, R. L., De Giovanni, M. Y. G., Olivo, J. E., & Andrade, C. M. G. (2022). Modeling and simulation of batch sugarcane alcoholic fermentation using the metabolic model. *Fermentation, 8*(2), 82. https://doi.org/10.3390/fermentation8020082

Afzaal, M., Saeed, F., Ahmad, A., Saeed, M., Usman, I., & Nouman, M. (2022). Microbial processing for valorization of waste and application. In *Environmental microbiology: Advanced research and multidisciplinary applications* (pp. 188–210). Bentham Science Publishers. https://doi.org/10.2174/9781681089584122010012

Aguirre-Garcia, Y. L., Nery-Flores, S. D., Campos-Muzquiz, L. G., Flores-Gallegos, A. C., Palomo-Ligas, L., Ascacio-Valdés, J. A., Sepúlveda-Torres, L., & Rodríguez-Herrera, R. (2024). Lactic acid fermentation in the food industry and bio-preservation of food. *Fermentation, 10*(3), 168. https://doi.org/10.3390/fermentation10030168

Ahmad, N., Aslam, S., Hussain, N., Bilal, M., & Iqbal, H. M. N. (2023). Transforming lignin biomass to value: Interplay between ligninolytic enzymes and lignocellulose depolymerization. *BioEnergy Research, 16*(3), 1246–1263. https://doi.org/10.1007/s12155-022-10541-y

Ajar Nath, Y. (2021). Microbial biotechnology for bio-prospecting of microbial bioactive compounds and secondary metabolites. *Journal of Applied Biology and Biotechnology*. https://doi.org/10.7324/JABB.2021.92ed

Alloun, W., & Calvio, C. (2024). Bio-driven sustainable extraction and ai-optimized recovery of functional compounds from plant waste: A comprehensive review. *Fermentation, 10*(3), 126. https://doi.org/10.3390/fermentation10030126

Amorim, C., Silvério, S. C., & Rodrigues, L. R. (2019). One-step process for producing prebiotic arabino-xylooligosaccharides from brewer's spent grain employing Trichoderma species. *Food Chemistry, 270*, 86–94. https://doi.org/10.1016/j.foodchem.2018.07.080

Ashraf, M., Ullah, N., Khan, I., Tremel, W., Ahmad, S., & Tahir, M. N. (2023). Photoreforming of waste polymers for sustainable hydrogen fuel and chemicals feedstock: Waste to energy. *Chemical Reviews, 123*(8), 4443–4509. https://doi.org/10.1021/acs.chemrev.2c00602

Bhaskar, P., Jain, D., & Srivastava, R. (2024). *Untapped bioactive compounds from endophytic fungi with potential antioxidant activity* (pp. 229–257). https://doi.org/10.1007/978-3-031-49112-2_11

Boantza, V. D. (2020). Fermentation. In *Encyclopedia of early modern philosophy and the sciences* (pp. 1–16). Springer International Publishing. https://doi.org/10.1007/978-3-319-20791-9_478-1

Borgonovi, T. F., Casarotti, S. N., & Penna, A. L. B. (2021). Lacticaseibacillus casei SJRP38 and buriti pulp increased bioactive compounds and probiotic potential of fermented milk. *LWT, 143*, 111124. https://doi.org/10.1016/j.lwt.2021.111124

Chai, K. F., Voo, A. Y. H., & Chen, W. N. (2020). Bioactive peptides from food fermentation: A comprehensive review of their sources, bioactivities, applications, and future development. *Comprehensive Reviews in Food Science and Food Safety, 19*(6), 3825–3885. https://doi.org/10.1111/1541-4337.12651

Chikezie Ogbu, C., & Nnaemeka Okey, S. (2023). Agro-industrial waste management: The circular and bioeconomic perspective. In *Agricultural waste—new insights*. IntechOpen. https://doi.org/10.5772/intechopen.109181

Chilakamarry, C. R., Mimi Sakinah, A. M., Zularisam, A. W., Sirohi, R., Khilji, I. A., Ahmad, N., & Pandey, A. (2022). Advances in solid-state fermentation for bioconversion of agricultural wastes to value-added products: Opportunities and challenges. *Bioresource Technology, 343*, 126065. https://doi.org/10.1016/j.biortech.2021.126065

Cosgrove, D. J. (2024). Structure and growth of plant cell walls. *Nature Reviews Molecular Cell Biology, 25*(5), 340–358. https://doi.org/10.1038/s41580-023-00691-y

Cuozzo, S., de Moreno de LeBlanc, A., LeBlanc, J. G., Hoffmann, N., & Tortella, G. R. (2023). Streptomyces genus as a source of probiotics and its potential for its use in health. *Microbiological Research, 266*, 127248. https://doi.org/10.1016/j.micres.2022.127248

Dahiya, D., Sharma, H., Rai, A. K., & Nigam, P. S. (2022). Application of biological systems and processes employing microbes and algae to reduce, recycle, reuse (3Rs) for the sustainability of circular bioeconomy. *AIMS Microbiology, 8*(1), 83–102. https://doi.org/10.3934/microbiol.2022008

de Oliveira Junqueira, A. C., Moreira Melo, N. T., Skorupa Parachin, N., & Costa Paes, H. (2023). Loss of a functional mitochondrial pyruvate carrier in Komagataella phaffii does not improve lactic acid production from glycerol in aerobic cultivation. *Microorganisms, 11*(2), 483.

Diarra, U., Osborne-Naikatini, T., & Subramani, R. (2024). Actinomycetes associated with hymenopteran insects: a promising source of bioactive natural products. *Frontiers in Microbiology, 15*. https://doi.org/10.3389/fmicb.2024.1303010

Doria, E., Buonocore, D., Marra, A., Bontà, V., Gazzola, A., Dossena, M., Verri, M., & Calvio, C. (2022). Bacterial-assisted extraction of bioactive compounds from cauliflower. *Plants, 11*(6), 816. https://doi.org/10.3390/plants11060816

Dudekula, U. T., Doriya, K., & Devarai, S. K. (2020). A critical review on submerged production of mushroom and their bioactive metabolites. *3 Biotech, 10*(8), 337. https://doi.org/10.1007/s13205-020-02333-y

Elshafie, H. S., Camele, I., & Mohamed, A. A. (2023). A comprehensive review on the biological, agricultural and pharmaceutical properties of secondary metabolites based-plant origin. *International Journal of Molecular Sciences, 24*(4), 3266. https://doi.org/10.3390/ijms24043266

Emkani, M., Oliete, B., & Saurel, R. (2022). Effect of lactic acid fermentation on legume protein properties, a review. *Fermentation, 8*(6), 244. https://doi.org/10.3390/fermentation8060244

Enespa, C. P., & Singh, D. P. (2023). Sources, purification, immobilization and industrial applications of microbial lipases: An overview. *Critical Reviews in Food Science and Nutrition, 63*(24), 6653–6686. https://doi.org/10.1080/10408398.2022.2038076

Escrivá, L., Manyes, L., Vila-Donat, P., Font, G., Meca, G., & Lozano, M. (2021). Bioaccessibility and bioavailability of bioactive compounds from yellow mustard flour and milk whey fermented with lactic acid bacteria. *Food and Function, 12*(22), 11250–11261. https://doi.org/10.1039/D1FO02059E

Fan, L., Ma, S., Li, L., & Huang, J. (2024). Fermentation biotechnology applied to wheat bran for the degradation of cell wall fiber and its potential health benefits: A review. *International Journal of Biological Macromolecules, 275*, 133529. https://doi.org/10.1016/j.ijbiomac.2024.133529

Fidelis, M., de Moura, C., Kabbas Junior, T., Pap, N., Mattila, P., Mäkinen, S., Putnik, P., Bursać Kovačević, D., Tian, Y., Yang, B., & Granato, D. (2019). Fruit seeds as sources of bioactive compounds: Sustainable production of high value-added ingredients from by-products within circular economy. *Molecules, 24*(21), 3854. https://doi.org/10.3390/molecules24213854

Florez-Orrego, D., Sharma, S., & Navabi, S. (2022). Editorial: Integration and optimization in the chemical process industry.

Frontiers in Chemical Engineering, 4. https://doi.org/10.3389/fceng.2022.961022

García, F. C., Clegg, T., O'Neill, D. B., Warfield, R., Pawar, S., & Yvon-Durocher, G. (2023). The temperature dependence of microbial community respiration is amplified by changes in species interactions. *Nature Microbiology, 8*(2), 272–283. https://doi.org/10.1038/s41564-022-01283-w

González-Granda, S., Albarrán-Velo, J., Lavandera, I., & Gotor-Fernández, V. (2023). Expanding the synthetic toolbox through metal-enzyme cascade reactions. *Chemical Reviews, 123*(9), 5297–5346. https://doi.org/10.1021/acs.chemrev.2c00454

Guan, T.-W., Lin, Y.-J., Ou, M.-Y., & Chen, K.-B. (2020). Isolation and diversity of sediment bacteria in the hypersaline aiding lake, China. *PLOS ONE, 15*(7), e0236006. https://doi.org/10.1371/journal.pone.0236006

Hadj Saadoun, J., Bertani, G., Levante, A., Vezzosi, F., Ricci, A., Bernini, V., & Lazzi, C. (2021). Fermentation of agri-food waste: A promising route for the production of aroma compounds. *Foods, 10*(4), 707. https://doi.org/10.3390/foods10040707

Hashem, A. H., Attia, M. S., Kandil, E. K., Fawzi, M. M., Abdelrahman, A. S., Khader, M. S., Khodaira, M. A., Emam, A. E., Goma, M. A., & Abdelaziz, A. M. (2023). Bioactive compounds and biomedical applications of endophytic fungi: A recent review. *Microbial Cell Factories, 22*(1), 107. https://doi.org/10.1186/s12934-023-02118-x

Hegab, H. M., ElMekawy, A., & Stakenborg, T. (2013). Review of microfluidic microbioreactor technology for high-throughput submerged microbiological cultivation. *Biomicrofluidics, 7*(2). https://doi.org/10.1063/1.4799966

Helleckes, L. M., Hemmerich, J., Wiechert, W., von Lieres, E., & Grünberger, A. (2023). Machine learning in bioprocess development: From promise to practice. *Trends in Biotechnology, 41*(6), 817–835. https://doi.org/10.1016/j.tibtech.2022.10.010

Herrmann, L. W., Letti, L. A. J., Penha, R. de O., Soccol, V. T., Rodrigues, C., & Soccol, C. R. (2024). Bacillus genus industrial applications and innovation: First steps towards a circular bioeconomy. *Biotechnology Advances, 70*, 108300. https://doi.org/10.1016/j.biotechadv.2023.108300

Hu, Y., Priya, A., Chen, C., Liang, C., Wang, W., Wang, Q., Lin, C. S. K., & Qi, W. (2023). Recent advances in substrate-enzyme interactions facilitating efficient biodegradation of lignocellulosic biomass: A review. *International Biodeterioration and Biodegradation, 180*, 105594. https://doi.org/10.1016/j.ibiod.2023.105594

Jayakumar, M., Gindaba, G. T., Gebeyehu, K. B., Periyasamy, S., Jabesa, A., Baskar, G., John, B. I., & Pugazhendhi, A. (2023). Bioethanol production from agricultural residues as lignocellulosic biomass feedstock's waste valorization approach: A comprehensive review. *Science of the Total Environment, 879*, 163158. https://doi.org/10.1016/j.scitotenv.2023.163158

Jeyabalan, J., Veluchamy, A., Kumar, A., Chandrasekar, R., & Narayanasamy, S. (2023). A review on the laccase assisted decolourization of dyes: Recent trends and research progress. *Journal of the Taiwan Institute of Chemical Engineers, 151*, 105081. https://doi.org/10.1016/j.jtice.2023.105081

Jiamworanunkul, S. (2019). Effective antioxidant production through submerged fermentation of edible mushrooms. *Thai Journal of Pharmaceutical Sciences, 43*(4), 213–218.

Jin, F.-J., Hu, S., Wang, B.-T., & Jin, L. (2021). Advances in genetic engineering technology and its application in the industrial fungus Aspergillus oryzae. *Frontiers in Microbiology, 12*. https://doi.org/10.3389/fmicb.2021.644404

Kabir, M. F., & Ju, L.-K. (2023). On optimization of enzymatic processes: Temperature effects on activity and long-term deactivation kinetics. *Process Biochemistry, 130*, 734–746. https://doi.org/10.1016/j.procbio.2023.05.031

Kaur, M., Singh, A. K., & Singh, A. (2023). Bioconversion of food industry waste to value added products: Current technological trends and prospects. *Food Bioscience, 55*, 102935. https://doi.org/10.1016/j.fbio.2023.102935

Khaksar, G., Sirijan, M., Suntichaikamolkul, N., & Sirikantaramas, S. (2022). Metabolomics for agricultural waste valorization: shifting toward a sustainable bioeconomy. *Frontiers in Plant Science, 13*. https://doi.org/10.3389/fpls.2022.938480

Kleekayai, T., Khalesi, M., Amigo-Benavent, M., Cermeño, M., Harnedy-Rothwell, P., & FitzGerald, R. J. (2023). Enzyme-assisted extraction of plant proteins. In *Green protein processing technologies from plants* (pp. 131–178). Springer International Publishing. https://doi.org/10.1007/978-3-031-16968-7_6

Kosariya, A., Kumar, D., Pant, K. K., Bhatt, R., & Chandra, V. (2023). Strategies for strain improvement of economically important microorganisms. In *Industrial microbiology and biotechnology* (pp. 695–727). Springer Nature. https://doi.org/10.1007/978-981-99-2816-3_24

Krakowska-Sieprawska, A., Kiełbasa, A., Rafińska, K., Ligor, M., & Buszewski, B. (2022). Modern methods of pre-treatment of plant material for the extraction of bioactive compounds. *Molecules, 27*(3), 730. https://doi.org/10.3390/molecules27030730

Leite, P., Belo, I., & Salgado, J. M. (2021). Co-management of agro-industrial wastes by solid-state fermentation for the production of bioactive compounds. *Industrial Crops and Products, 172*, 113990. https://doi.org/10.1016/j.indcrop.2021.113990

Lemes, A. C., Egea, M. B., Oliveira Filho, J. G. de, Gautério, G. V., Ribeiro, B. D., & Coelho, M. A. Z. (2022). Biological approaches for extraction of bioactive compounds from agro-industrial by-products: A review. *Frontiers in Bioengineering and Biotechnology, 9*. https://doi.org/10.3389/fbioe.2021.802543

Lemke, M., & DeSalle, R. (2023). The next generation of microbial ecology and its importance in environmental sustainability. *Microbial Ecology, 85*(3), 781–795. https://doi.org/10.1007/s00248-023-02185-y

Liu, X., & Kokare, C. (2023). Microbial enzymes of use in industry. In *Biotechnology of microbial enzymes* (pp. 405–444). Academic Press.

Madhavan, A., Arun, K., Sindhu, R., Alphonsa Jose, A., Pugazhendhi, A., Binod, P., Sirohi, R., Reshmy, R., & Kumar Awasthi, M. (2022). Engineering interventions in industrial filamentous fungal cell factories for biomass valorization. *Bioresource Technology, 344*, 126209. https://doi.org/10.1016/j.biortech.2021.126209

Martínez-Medina, G. A., Barragán, A. P., Ruiz, H. A., Ilyina, A., Martínez Hernández, J. L., Rodríguez-Jasso, R. M., Hoyos-Concha, J. L., & Aguilar-González, C. N. (2019). Fungal proteases and production of bioactive peptides for the food industry. In *Enzymes in food biotechnology* (pp. 221–246). Elsevier. https://doi.org/10.1016/B978-0-12-813280-7.00014-1

Martins, S., Mussatto, S. I., Martínez-Avila, G., Montañez-Saenz, J., Aguilar, C. N., & Teixeira, J. A. (2011). Bioactive phenolic compounds: Production and extraction by solid-state fermentation. A review. *Biotechnology Advances, 29*(3), 365–373.

Mattedi, A., Sabbi, E., Farda, B., Djebaili, R., Mitra, D., Ercole, C., Cacchio, P., Del Gallo, M., & Pellegrini, M. (2023). Solid-state fermentation: Applications and future perspectives for biostimulant and biopesticides production. *Microorganisms, 11*(6), 1408. https://doi.org/10.3390/microorganisms11061408

Menon, R., Thomas, R., Sood, N., & Gokhale, T. (2023). Marine phytoplankton: Bioactive compounds and their applications in medicine. In *Marine bioactive molecules for biomedical and pharmacotherapeutic applications* (pp. 251–282). Springer Nature. https://doi.org/10.1007/978-981-99-6770-4_13

Mitri, S., Koubaa, M., Maroun, R. G., Rossignol, T., Nicaud, J.-M., & Louka, N. (2022). Bioproduction of 2-Phenylethanol through yeast fermentation on synthetic media and on agro-industrial waste and by-products: A review. *Foods, 11*(1), 109. https://doi.org/10.3390/foods11010109

Nayak, A., Kisku, K., Dhara, M., & Naik, U. C. (2022b). Genetically modified microbes as an effective tool for sustainable solid waste management. In *Relationship between microbes and the environment for sustainable ecosystem services* (Vol. 2, pp. 255–271). Elsevier. https://doi.org/10.1016/B978-0-323-89937-6.00008-5

Nogales-Bueno, J., Baca-Bocanegra, B., Heredia, F. J., & Hernández-Hierro, J. M. (2020). Phenolic compounds extraction in enzymatic macerations of grape skins identified as low-level extractable total anthocyanin content. *Journal of Food Science, 85*(2), 324–331. https://doi.org/10.1111/1750-3841.15006

Ong, K. L., Fickers, P., Pang, K. W., Dharma, P. R., Luk, H. S., Uisan, K., & Lin, C. S. K. (2020). Fermentation of fruit and vegetable wastes for biobased products. In *Food industry wastes* (pp. 255–273). Elsevier. https://doi.org/10.1016/B978-0-12-817121-9.00012-7

Oyewole, O. A., Idris, A. D., Bello, A. B., Yakubu, J. G., & Saidu, M. M. (2022). *Microbial enzymes in the recycling of wastes* (pp. 189–213). https://doi.org/10.1007/978-981-19-0155-3_10

Pai, S., Hebbar, A., & Selvaraj, S. (2022). A critical look at challenges and future scopes of bioactive compounds and their incorporations in the food, energy, and pharmaceutical sector. *Environmental Science and Pollution Research, 29*(24), 35518–35541. https://doi.org/10.1007/s11356-022-19423-4

Paliwal, A., & Dutta, H. (2023). Mathematical modeling in fermentation processes. In *Advances in food process engineering* (pp. 217–236). Apple Academic Press. https://doi.org/10.1201/9781003303848-11

Paz-Arteaga, S. L., Ascacio-Valdés, J. A., Aguilar, C. N., Cadena-Chamorro, E., Serna-Cock, L., Aguilar-González, M. A., Ramírez-Guzmán, N., & Torres-León, C. (2023). Bioprocessing of pineapple waste for sustainable production of bioactive compounds using solid-state fermentation. *Innovative Food Science and Emerging Technologies, 85*, 103313. https://doi.org/10.1016/j.ifset.2023.103313

Phillips, J. A. (2020). "Novel" sensors for the monitoring of fermentation processes. In *Computer control of fermentation processes* (pp. 15–72). CRC Press. https://doi.org/10.1201/9781003068372-2

Ramires, F. A., Bavaro, A. R., D'Antuono, I., Linsalata, V., D'Amico, L., Baruzzi, F., Pinto, L., Tarantini, A., Garbetta, A., Cardinali, A., & Bleve, G. (2023). Liquid submerged fermentation by selected microbial strains for onion skins valorization and its effects on polyphenols. *World Journal of Microbiology and Biotechnology, 39*(10), 258. https://doi.org/10.1007/s11274-023-03708-y

Ramírez Rojas, A. A., Swidah, R., & Schindler, D. (2022). Microbes of traditional fermentation processes as synthetic biology chassis to tackle future food challenges. *Frontiers in Bioengineering and Biotechnology, 10*. https://doi.org/10.3389/fbioe.2022.982975

Ranjitha, K., Reddy, V. R., & Oberoi, H. S. (2022b). Microbial valorization of food industry wastes for production of nutraceutical molecules. In *Sustainable microbial technologies for valorization of agro-industrial wastes* (pp. 115–134). CRC Press. https://doi.org/10.1201/9781003191247-6

Rathore, A. S., Mishra, S., Nikita, S., & Priyanka, P. (2021). Bioprocess control: Current progress and future perspectives. *Life, 11*(6), 557. https://doi.org/10.3390/life11060557

Sabater, C., Ruiz, L., Delgado, S., Ruas-Madiedo, P., & Margolles, A. (2020). Valorization of VEGETABLE FOOD WASTE AND BY-PRODUCTS THROUGH FERMENTATION PROCESSes. *Frontiers in Microbiology, 11*. https://doi.org/10.3389/fmicb.2020.581997

Sadh, P., Kumar, S., Chawla, P., & Duhan, J. (2018). Fermentation: A boon for production of bioactive compounds by processing of food industries wastes (By-Products). *Molecules, 23*(10), 2560. https://doi.org/10.3390/molecules23102560

Sanjeewa, K. K. A., Herath, K. H. I. N. M., Kim, Y.-S., Jeon, Y.-J., & Kim, S.-K. (2023). Enzyme-assisted extraction of bioactive compounds from seaweeds and microalgae. *TrAC Trends in Analytical Chemistry, 167*, 117266. https://doi.org/10.1016/j.trac.2023.117266

Sarkar, S., Yadav, M., & Kumar, A. (2023). CRISPR/Cas9 system: An advanced approach for the improvement of industrially important microorganisms. In *Industrial microbiology and biotechnology* (pp. 69–97). Springer Nature. https://doi.org/10.1007/978-981-99-2816-3_4

Sarwan, J., Uddin, N., Sahani, S., Mittal, K., Maurya, D. D. K. P., & Jagadeesh Chandra Bose, K. (2024). Isolation and characterization of novel Bacillus with dual enzymes, biodegradation potential: a special reference to cellulolytic and proteolytic activities. *Biomass Conversion and Biorefinery*. https://doi.org/10.1007/s13399-024-05329-3

Setyani, W., Murwanti, R., Sulaiman, T. N. S., & Hertiani, T. (2023). Application of response surface methodology (RSM) for the optimization of ultrasound-assisted extraction (UAE) of Moringa oleifera: Extraction yield, content of bioactive compounds, and biological effects in vitro. *Plants, 12*(13), 2455. https://doi.org/10.3390/plants12132455

Shanmugam, K. T., & Ingram, L. O. (2022). Principles and practice of designing microbial biocatalysts for fuel and chemical production. *Journal of Industrial Microbiology and Biotechnology, 49*(2). https://doi.org/10.1093/jimb/kuab016

Sharma, R., Garg, P., Kumar, P., Bhatia, S. K., & Kulshrestha, S. (2020). Microbial fermentation and its role in quality improvement of fermented foods. *Fermentation, 6*(4), 106. https://doi.org/10.3390/fermentation6040106

Siddiqui, S. A., Erol, Z., Rugji, J., Taşçı, F., Kahraman, H. A., Toppi, V., Musa, L., Di Giacinto, G., Bahmid, N. A., Mehdizadeh, M., & Castro-Muñoz, R. (2023). An overview of fermentation in the food industry—looking back from a new perspective. *Bioresources and Bioprocessing, 10*(1), 85. https://doi.org/10.1186/s40643-023-00702-y

Singla, M., Singh, A., & Sit, N. (2023). Effect of microwave and enzymatic pretreatment and type of solvent on kinetics of ultrasound assisted extraction of bioactive compounds from ripe papaya peel. *Journal of Food Process Engineering, 46*(6). https://doi.org/10.1111/jfpe.14119

Song, P., Zhang, X., Wang, S., Xu, W., Wang, F., Fu, R., & Wei, F. (2023). Microbial proteases and their applications. *Frontiers in Microbiology, 14*. https://doi.org/10.3389/fmicb.2023.1236368

Sun, C., Meng, X., Sun, F., Zhang, J., Tu, M., Chang, J.-S., Reungsang, A., Xia, A., & Ragauskas, A. J. (2023). Advances and perspectives on mass transfer and enzymatic hydrolysis in the enzyme-mediated lignocellulosic biorefinery: A review. *Biotechnology Advances, 62*, 108059. https://doi.org/10.1016/j.biotechadv.2022.108059

Sun, W., Shahrajabian, M. H., & Lin, M. (2022). Research progress of fermented functional foods and protein factory-microbial fermentation technology. *Fermentation, 8*(12), 688. https://doi.org/10.3390/fermentation8120688

Tessler, M., Cunningham, S. W., Ingala, M. R., Warring, S. D., & Brugler, M. R. (2023). An environmental DNA primer for microbial and restoration ecology. *Microbial Ecology, 85*(3), 796–808. https://doi.org/10.1007/s00248-022-02168-5

Thompson, H. O., Önning, G., Holmgren, K., Strandler, H. S., & Hultberg, M. (2020). Fermentation of Cauliflower and white beans with Lactobacillus plantarum—Impact on levels of riboflavin, folate, Vitamin B12, and amino acid composition. *Plant Foods for Human Nutrition, 75*(2), 236–242. https://doi.org/10.1007/s11130-020-00806-2

Tong, L., Li, Y., Lou, X., Wang, B., Jin, C., & Fang, W. (2024). Powerful cell wall biomass degradation enzymatic system from saprotrophic Aspergillus fumigatus. *The Cell Surface, 11*, 100126. https://doi.org/10.1016/j.tcsw.2024.100126

Tong, X., He, Z., Zheng, L., Pande, H., & Ni, Y. (2023). Enzymatic treatment processes for the production of cellulose nanomaterials: A review. *Carbohydrate Polymers, 299*, 120199. https://doi.org/10.1016/j.carbpol.2022.120199

Tsigie, Y. A., Wang, C.-Y., Truong, C.-T., & Ju, Y.-H. (2011). Lipid production from Yarrowia lipolytica Po1g grown in sugarcane bagasse hydrolysate. *Bioresource Technology, 102*(19), 9216–9222. https://doi.org/10.1016/j.biortech.2011.06.047

Vamvakas, S.-S., & Kapolos, J. (2020). Factors affecting yeast ethanol tolerance and fermentation efficiency. *World Journal of Microbiology and Biotechnology, 36*(8), 114. https://doi.org/10.1007/s11274-020-02881-8

Vidhu, A. (2020). Fermentation: A green herbal extraction process for polyphenols. *Biomedical Journal of Scientific and Technical Research, 31*(4). https://doi.org/10.26717/BJSTR.2020.31.005137

Vojnovic, S., Aleksic, I., Ilic-Tomic, T., Stevanovic, M., & Nikodinovic-Runic, J. (2024). Bacillus and Streptomyces spp. as hosts for production of industrially relevant enzymes. *Applied Microbiology and Biotechnology, 108*(1), 185. https://doi.org/10.1007/s00253-023-12900-x

Watanabe, D. (2024). Sake yeast symbiosis with lactic acid bacteria and alcoholic fermentation. *Bioscience, Biotechnology, and Biochemistry, 88*(3), 237–241. https://doi.org/10.1093/bbb/zbad167

Xia, J., Wang, G., Fan, M., Chen, M., Wang, Z., & Zhuang, Y. (2021). Understanding the scale-up of fermentation processes from the viewpoint of the flow field in bioreactors and the physiological response of strains. *Chinese Journal of Chemical Engineering, 30*, 178–184. https://doi.org/10.1016/j.cjche.2020.12.004

Yadav, P., Chauhan, A. K., Singh, R. B., Khan, S., & Halabi, G. (2022). Organic acids: microbial sources, production, and applications. In *Functional foods and nutraceuticals in metabolic and non-communicable diseases* (pp. 325–337). Elsevier. https://doi.org/10.1016/B978-0-12-819815-5.00053-7

Yadav, S., Malik, K., Moore, J. M., Kamboj, B. R., Malik, S., Malik, V. K., Arya, S., Singh, K., Mahanta, S., & Bishnoi, D. K. (2024). Valorisation of agri-food waste for bioactive compounds: Recent trends and future sustainable challenges. *Molecules, 29*(9), 2055. https://doi.org/10.3390/molecules29092055

Yafetto, L. (2022). Application of solid-state fermentation by microbial biotechnology for bioprocessing of agro-industrial wastes from 1970 to 2020: a review and bibliometric analysis. *Heliyon, 8*(3), e09173. https://doi.org/10.1016/j.heliyon.2022.e09173

Yepes-Betancur, D. P., Márquez-Cardozo, C. J., Cadena-Chamorro, E. M., Martinez-Saldarriaga, J., Torres-León, C., Ascacio-Valdes, A., & Aguilar, C. N. (2021). Solid-state fermentation—assisted extraction of bioactive compounds from hass avocado seeds. *Food and Bioproducts Processing, 126*, 155–163. https://doi.org/10.1016/j.fbp.2020.10.012

Yi, R., Pan, Y., Long, X., Tan, F., & Zhao, X. (2020). Enzyme producing activity of probiotics and preparation of compound enzyme. *Journal of Chemistry, 2020*, 1–8. https://doi.org/10.1155/2020/9140281

Yin, Y., Hu, Y., & Wang, J. (2022). Co-fermentation of sewage sludge and lignocellulosic biomass for production of medium-chain fatty acids. *Bioresource Technology, 361*, 127665. https://doi.org/10.1016/j.biortech.2022.127665

Microbial Production of Lipopeptides

M. Magdalena Rivera-Salas, J. Pablo Manjarrez-Quintero, Octavio Valdez-Baro, Jorge M. Heredia-Bátiz, Raymundo S. García-Estrada, and J. Basilio Heredia

Abstract

Lipopeptides are amphiphilic compounds since their molecular structure is formed by a fatty acid, which gives them hydrophobic characteristics. They are covalently linked to a hydrophilic peptide sequence that has between 7 and 10 amino acid residues. Lipopeptides are bioactive agents synthesized by the nonribosomal pathway in different species of bacteria and fungi. These biosurfactant compounds present a high biodiversity since they can present variations in the size and saturation of the fatty acid chain and the sequence, composition, and size of the peptide fraction. The high biodiversity of lipopeptides induces a wide range of properties such as surfactants, immunomodulators, antitumors, bioremediators, and mainly antimicrobials. The antimicrobial potential of lipopeptides is related to their ability to bind to the lipid molecules of cell membranes, where, depending on the concentration, they can cause alterations in the structure and permeability of the membranes, which can trigger cell death. The structural and functional diversity of lipopeptides has applications in different sectors, such as the food industry, where they can be used as emulsifiers and antimicrobials; in the recovery of crude oil from contaminated marine environments; in the agricultural sector as biocontrol products to reduce or eliminate diseases caused by plant pathogenic microorganisms, among others. Genetic engineering, metabolic engineering, and synthetic biology tools are used to increase the biodiversity, performance and effectiveness of lipopeptides to expand and improve their application in industrial activity.

Keywords

Lipopeptides · Biosurfactants · *Bacillus* · *Pseudomonas* · *Streptomyces*

1 Introduction

Lipopeptides are a heterogeneous group of amphiphilic compounds of microbial origin. As the definition suggests, lipopeptides combine a peptide and a fatty acid, where both the peptide and the fatty acid vary in size depending on the lipopeptide type. The structural and functional diversity of lipopeptides is due to differences in the sequence of amino acids that can be found in the L or D form, as well as in the length of the peptide portion, coupled with differences in isomerism, length, and saturation of the lipid portion. The high diversity of lipopeptides induces a wide range of biological activities that can have applications in different sectors, such as antimicrobial agents, immunomodulators, biosurfactants, and preservatives, among others. The synthesis of lipopeptides is considered part of the secondary metabolism of various microorganisms (bacteria, fungi, yeasts, and actinomycetes), which synthesize these molecules for different functions depending on their environmental conditions. Factors such as carbon and nitrogen sources, dissolved oxygen, pH, metal ions, temperature, agitation, and light are of great importance for the microbial production of lipopeptides, resulting in variations in the quantity and type of lipopeptides produced. Lipopeptides are biosynthesized by enzyme complexes called nonribosomal peptide synthetases. These mega enzymes are divided into multiple modules, and each module is subdivided into several catalytic domains. Together, these domains are responsible for the synthesis of a lipopeptide. Recent studies have focused on altering these multimodular enzymes to increase the yield and diversity of lipopeptides produced in fermentation processes (Biniarz et al., 2018; Pilz et al., 2023).

M. M. Rivera-Salas · J. P. Manjarrez-Quintero · O. Valdez-Baro · J. M. Heredia-Bátiz · R. S. García-Estrada · J. B. Heredia (✉)
Centro de Investigación en Alimentación y Desarrollo A. C., Culiacán, Sinaloa, México
e-mail: jbheredia@ciad.mx

K. Sharma (ed.), *Bio-prospecting of Novel Microbial Bioactive Compounds for Sustainable Development*, Sustainable Landscape Planning and Natural Resources Management, https://doi.org/10.1007/978-3-031-82178-3_9

2 Lipopeptide Biosynthesis

Due to the union of the two components, a peptide and a fatty acid, the biosynthesis of these precursors and their assembly is required. In the case of fatty acids, they follow the malonate pathway, where it is converted into malonyl-CoA, which is the precursor for all fatty acids. The number of enzymatic reactions depends on the type of fatty acid, mainly determined by the chain size. This is because the process repeats with the smaller-sized intermediate product, adding two carbons to the chain each time. Enzymes such as synthetases, reductases, isomerases, dehydratases, hydrolases, synthases, transferases, and esterases, among others, participate in this process (KEGG, 2024). The synthesis of the peptide part of lipopeptides differs from the synthesis of conventional peptides, which are synthesized in ribosomes through translation, where tRNAs are responsible for synthesizing the peptide chain depending on the genetic code. The peptides that make up lipopeptides follow the nonribosomal pathway, as tRNA synthetases are not involved. At least three proteins are required that behave analogously to their ribosomal counterparts but differ at the structural level. Another difference between ribosomal and nonribosomal peptides is that ribosomal ones result in greater precision as they are part of primary metabolism; however, they are limited to assembling peptides with the 20 essential amino acids.

In contrast, nonribosomal peptides involve hundreds of known non-essential or non-natural amino acids participating in peptide formation, resulting in greater structural diversity. The assembly of peptides is carried out by a set of enzymes called nonribosomal peptide synthetases, which operate in stages: initiation stage (A-domain), covalent loading (CP-domain), and peptide bond formation (C-domain). Different proteins perform these initiation and elongation reactions perform specific reactions such as the Adenylation-Domain, the Peptidyl Carrier Protein, the Condensation-Domain, the Thioesterase-Domain, and Tailoring Domains (Pilz et al., 2023).

3 Lipopeptide Classification

Lipopeptides are produced as mixtures of structural analogs. Each lipopeptide family can be composed of isoforms and homologs. Lipopeptides are classified into isoforms based on the number of carbons present in the peptide portion. In contrast, the sub-classification of isoforms into homologs is based on the number of carbons of the fatty acid chain. Different species of fungi and bacteria have been identified as producers of lipopeptides, highlighting *Bacillus* and *Pseudomonas* as the main producing genera (Biniarz et al., 2018; de Souza et al., 2018).

3.1 Lipopeptides Produced by *Bacillus* Spp.

Lipopeptides are the most widely studied secondary metabolites in numerous *Bacillus* species (*B. subtilis, B. amyloliquefaciens, B. licheniformis, B. megaterium, B. mycoides, and B. thuringiensis*). Lipopeptides produced by Bacillus bacteria are classified into three groups or families: surfactin, iturin, and fengycin. Surfactins are cyclic heptapeptides linked to a fatty acid (β-hydroxy) of 12–16 carbon atoms. Surfactin is characterized by its antibacterial, insecticidal, and antiviral activity. Lipopeptides of the surfactin family can interact with cell membranes and consequently cause alterations in lipid layers. The iturin family comprises iturin A, C, D, E, bacillomycin D, F, L, and mycosubtilin. The lipopeptides of the iturin family are cyclic heptapeptides linked to a fatty acid (β-amino) of between 14 and 17 carbon atoms. Members of the iturin family have antifungal and hemolytic activities, with iturin A being the lipopeptide with the most potent antifungal activity. Its mechanism of action is based on the formation of ion-conducting pores, which cause osmotic disturbances that can trigger cell death. Lipopeptides of the fengycin family are decapeptides linked to a fatty acid (β-hydroxy) with a chain length of between 14 and 18 carbon atoms. Members of the fengycin family, such as iturin A and B, have strong activity against filamentous fungi. Fengycins also interact with the cell membrane, causing alterations in its structure and permeability (Carolin et al., 2021; Fira et al., 2018; Yan et al., 2020).

3.2 Lipopeptides Produced by *Pseudomonas* Spp.

Lipopeptides produced by *Pseudomonas* species are mainly classified as orfamide, viscosin, amphisin, tolaasin, and syringomycin-like compounds. Orfamides are cyclic lipopeptides (CLPs) produced by *Pseudomonas* species, known for their biosurfactant properties. They consist of a cyclic oligopeptide lactone ring coupled to a fatty acid tail and are synthesized by nonribosomal peptide synthases (NRPSs). The primary orfamide compound produced by *P. protegens* is orfamide A, whereas *Pseudomonas sp.* CMR5c and CMR12a produce orfamide B. Additionally, *Pseudomonas sp.* CMR5c was found to make two new homologs, orfamide F and orfamide G. Orfamide A and B differ by a single amino acid, while orfamide B and G differ in the length of the fatty acid part. Orfamide A, B, and

G showed similar antifungal activities. The slight structural differences among orfamides (such as amino acid substitution and fatty acid chain length) do not significantly impact their broad-spectrum biological activities. This suggests a level of redundancy in the biological activity of these compounds. The identification of new orfamide homologs expands the understanding of the structural diversity and potential functional roles of these cyclic lipopeptides. The regulation of orfamide production by luxR-type regulators underscores the complex regulatory networks that control secondary metabolite production in *Pseudomonas* species. Understanding these regulatory mechanisms can aid in manipulating and optimizing biocontrol strains for enhanced agricultural applications (Ma et al., 2016).

Ciurko et al. (2023), studied the production of lipopeptide biosurfactant using a strain of *Pseudomonas antarctica* 28E. The biosurfactant produced by *P. antarctica* 28E, identified as viscosin, exhibited excellent surface-active properties, significantly reducing surface tension and improving emulsification. The critical micelle concentration (CMC) of viscosin was determined to be 20 mg/L. The viscosin biosurfactant reduced the contact angle on hydrophobic surfaces such as silicone and polystyrene, indicating enhanced wettability. No significant reduction was observed on hydrophilic surfaces like glass. The low CMC of viscosin indicates its high efficiency and potential for applications in various industries, including pharmaceuticals, food, and bioremediation.

Varnava et al. (2024), carried out a study focused on the characterization, optimal production, and ecotoxicity evaluation of a lipopeptide biosurfactant (BSF) produced by *Pseudomonas citronellolis* 620 C using oily wastewater. This BSF was identified as a cyclic lipopeptide (CLP) based on various analytical techniques. The analytical methods confirmed the BSF as a novel lipopeptide with a unique combination of fatty acids and amino acids, similar to known lipopeptides but with distinct differences, particularly in amino acid composition. The BSF showed no estrogenic activity and lower toxicity than other biosurfactants like surfactin and synthetic surfactants such as sodium dodecyl sulfate (SDS). The BSF produced by *P. citronellolis* 620 C exhibited lower toxicity than many synthetic surfactants, making it a promising candidate for environmental and industrial applications, including bioremediation and agriculture.

3.3 Lipopeptides Produced by *Streptomyces* Spp.

Streptomyces is one of the most important bacteria capable of producing lipopeptides. Due to the great variety of species within this genus of bacteria, the array of lipopeptides it produces is also wide. Lipopeptides are compounds with antibiotic potential that are mostly used in the agricultural and pharmaceutical industries. Below, we present some of the many lipopeptides produced by *Streptomyces* and how their production has evolved. The usage of lipopeptides as antibiotics was pioneered over 60 years ago when amphomycin was isolated and characterized from *Streptomyces canus*. This opened the way to the discovery of more antibiotic lipopeptides, such as albopeptins isolated from *Streptomyces albofaciens* JC-82–120 in 1986, which demonstrated to be effective against plant pathogenic diseases (Oikawa et al., 2022). Afterward, this class of compounds expanded in search of new lipopeptides, focusing on *Bacillus*, *Pseudomonas,* and *Streptomyces* as their production sources, and a great advance for the cause was made in 2003 when the lipopeptide daptomycin produced by *Streptomyces roseosporus* was approved by the Food and Drug Administration (FDA) for the treatment of skin and tissue infections, and shortly after for the treatment of bacteremia caused by *Staphylococcus aureus* (Clements-Decker et al., 2022).

A big interest in antibiotics has recently been generated because of multidrug-resistant pathogens. In 2018, a sample generated by *Streptomyces griseorubens* contained as major components the lipopeptides aspartocin B and aspartocin C, which had considerable activity against five MRSA strains (Tyurin et al., 2018). *Streptomyces sp.* KCB14A132 was studied due to the high interest of lipopeptides biosynthesized by microorganisms. It was found that this actinobacterium produces enamodin, enamodin B, and enamodin C, three cyclic lipopeptides with antibacterial activity against *Staphylococcus aureus*, which means they could be used for therapeutic treatment for bacterial infections (Son et al., 2018). Later, in 2019, the production of antibiotics by *Streptomyces parvulus* K-R1 was evaluated, the endophytic bacteria was fermented and a crude extract was obtained which showed great activity against *Aspergillus niger* MTCC-872, *Candida albicans* MTCC-183, *Proteus vulgaris* JNMC-7, *Pseudomonas aeruginosa* MTCC-741, *Klebsiella pneumoniae* JNMC-6, *Klebsiella pneumoniae* MTCC-3384, *Staphylococcus epidermis* JNMC-4 and *Staphylococcus aureus* JNMC-3. The extract was analyzed, and among the compounds present, it was found the lipopeptide fengycin (Chandrakar & Gupta, 2019).

Various lipopeptides have been described from *Streptomyces* species, like the most mentioned in the literature, daptomycin, and many others such as enduracidin, A54145 and CDA. Examples of this variety of compounds are rotihibins C and D, which can be produced by *Streptomyces scabies*; these lipopeptides have shown herbicidal activity against *Lemna minor* and *Arabidopsis thaliana*, and are variants of the lipopeptides rotihibins A and B, discovered from *Streptomyces graminofaciens*, which are plant growth inhibitors. The lipopeptide 5812-A/C produced by *Streptomyces roseoflavus* INA-Ac-5812 has inhibitory activity against Gram-positive pathogenic bacteria, similar to daptomycin (Planckaert et al., 2021). Daptomycin could be the most cited lipopeptide antibiotic in the literature; it is produced by *Streptomyces roseosporus* and has proven its potential, especially against multidrug-resistant bacteria. Recently, a new lipopeptide isolated from *Streptomyces sp.* RerS4 was called methyl (10-methylundecanoyl) threonylglycylalaninate ($C_{22}H_{41}N_3O_6$) and showed positive results against *Escherichia coli* CGMCC 1.1521 and *Bacillus subtilis* ATCC 21,332 (He et al., 2023).

In sum to these new antibiotics, the biosynthesis of new lipopeptides has gained interest; such is the case of this study where *Streptomyces nodosus* was used to produce a new skyllamycin termed oxy-skyllamycin. This study paves the way for generating new bioactive compounds that collaborate with the need for new drugs and antibiotics (Song et al., 2024). In the same sense of supplying the high demand for new antibiotics that also help prevent the development of resistance in bacteria, researchers in this area have taken on the task of finding new sources of antibiotic lipopeptides in recent years. In this extensive search, species of *Streptomyces* isolated from marine soils have been investigated, such as *Streptomyces parvulus*, which has shown that the lipopeptide produced by this species has great activity against Gram-negative and Gram-positive bacteria, and therefore, it can be used for the development of novel drugs (Kadaikunnan et al., 2024). Another example of *Streptomyces* found in marine soil is *Streptomyces* strain G222; it was found that this microorganism produces cyclic and linear variants of lichenysins and surfactins, lipopeptides that showed anti-biofilm and antibacterial activity and therefore contributed to solving the high demand for new antibiotics (Castaldi et al., 2024).

3.4 Lipopeptides Produced by Fungi

In addition to bacteria, it has been found that some fungi can also produce different lipopeptides. Lipopeptide antibiotics derived from fungi can be classified as cyclic depsipeptides, peptaibiotics, non-depsipeptide cyclic lipopeptides, and non-peptaibiotic linear lipopeptides.

Beginning with cyclic depsipeptides containing multiple fatty acid residues, we found beauvericin from *Beauveria bassiana*, enniatin G from *Halosarpheia ratnagiriensis*, beauvenniatin A from *Acremonium* spp., all with antimycobacterial activity and pleofungins A, C, and D from *Phoma sp.* with antifungal activity. Then, with cyclic β-depsipeptides, there are isaridin A from *Isaria* sp. with antimalarial activity and fusaristanin A from *Fusarium* sp. and *Fusarium decemcellulare* with antifungal activity. Continuing with cyclic depsipeptides containing piperazic or aromatic acids, they mention clavariospins A and B from *Clavariosis aquatica* with antifungal activity and paecilodepsipeptide A/gliotide from *Gliocladium* sp. and *Paecilomyces cinnamomeus* and SCH 218,157 and SCH 217,048 from *Pleosporaceae* with antimalarial activity.

To add other cyclic depsipeptides, we can refer to verlamelin A from *Simplicillium sp.* with antifungal activity; simplicilliumtide J from *Simplicillium* sp. with antifungal and antiviral activity; isariin C from *Isaria felina* with antimalarial activity; isariins C_2 and iso-isariins B from *Isaria sp.* and *B. felina* with insecticidal activity; glomosporin from *Glomospora sp.* with antifungal activity. Now, to mention some non-depsipeptide cyclic lipopeptides, cycloaspeptide D and cycloaspeptide E from *Penicillium* spp., and *Tricothecium* have antimalarial and insecticidal activity; ASP2397, AS2488053, AS2488059, and AS2529132 from *Acremonium persicinum* have antifungal activity; FR901379, FR 227,673 and FR190293 from *Chalara sp.* and *Tolypocladium parasiticum* have antifungal activity; FR209602, FR209603, FR209604, FR220897 and FR220899 from *Coleophoma crateriformis* and *C. empetri* have antifungal activity, and JNIR-15, sclerotiotides A, B, F and I from *A. sclerotium* have antifungal activity.

Following with peptaibiotic products (linear lipopeptides), we can find atroviridins A-C and neoatroviridins A-D from *Trichoderma atroviride* with antibacterial and antifungal activity; chalciporin A from *Sepedonium chalcipori* with antibacterial activity; longibrachins A-II-b, B-II, and B-III from *Trichoderma longibrachiatum* with antibacterial, antifungal and antimycoplasmal activity. Now, about non-peptaibol peptaibiotics, standing out, we find leucinostatins A and B from *Paecilomyces* spp. with antitrypanosomal and antimalarial activity; trichoderins A, B, and A1 from *Trichoderma sp.* with antimycobacterial activity; trichopolyn VI from *Trichoderma brevicompactum* with insecticidal activity and SCH 466,457, SCH 466,456 and SCH 643,432 from *Paecilomyces varioti* with antifungal activity. Finally, we come to non-peptaibiotic linear lipopeptides, where we can highlight simplicilliumtide B from *Simplicillium* spp. with antimycobacterial activity (Zhao et al., 2019).

4 Lipopeptide Applications

Lipopeptides have been used for different purposes, and their applications have evolved alongside science. In the beginning, lipopeptides were used in the very first link of the food industry, in the production sector. In the early 2000s, the potential use of lipopeptides in the agricultural industry was already known, and around the world, different sources and applications of these compounds continued

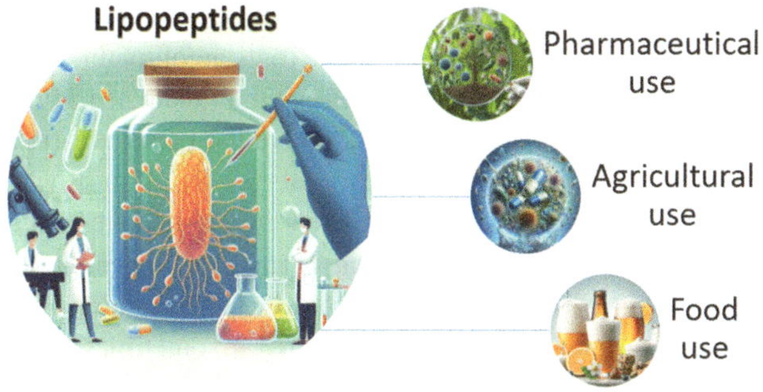

Fig. 1 The potential use of lipopeptides. Made with Microsoft Copilot

to be investigated. Lipopeptides can be used in pharmaceutical and agrochemical companies, so it is worth mentioning that some lipopeptides and their chemically synthesized derivatives have already been developed into commercial products and/or entered into clinical trials. Some examples of these cases are cilofungin, caspofungin, micafungin, anidulafungin, and rezafungin as antifungal agents; emodepside as an anthelmintic agent; fusafungine and a mixture of enniatins for the treatment of rhinosinusitis as a nasal spray; destruxins as insecticidal agents (Zhao et al., 2019) (Fig. 1).

4.1 Pharmaceutical and Nutraceutical Uses

There is plenty of information about possible uses of lipopeptides that go from the food industry to the bioremediation of heavy metals. Still, this section will focus on their pharmaceutical and nutraceutical uses. It is worth noting that every pharmacological product has to be approved and tested by the FDA to ensure its safety for living organisms and the environment. Lipopeptides can possess antibiotic properties and can be applied orally, injected, or topically. Still, their discovery and production with this purpose have decreased, given that in the last 60 years, only a few novel classes like daptomycin were brought onto the market. Ultra-short lipopeptides have proved effective against gram-positive bacteria in treating multi-resistant strains and against *Staphylococcus aureus*. Daptomycin is one of the better-known lipopeptides with antibiotic application. Tsushimycin is an analog of daptomycin that also shows antimicrobial activity. *Staphylococcus epidermis* LP01 could be used as a skin antimicrobial due to the lipopeptides it secretes.

Unfortunately, fungi, like bacteria, also generate resistance against conventional treatments, but fortunately, pelgipeptins secreted by *Paenibacillus elgii* could be used as antifungal treatment, having proven its potential against *Candida albicans*. Another lipopeptide with antifungal properties is WH1fungin from *Bacillus amyloquefaciens*. Echinocandin-based drugs are great against yeast and fungi like *Candida sp.* and *Aspergillus sp.* Caspofungin was the first echinocandin that reached the market. The lipopeptide pneumocandin B0 produced by *Glarea lozoyensis* was used as a base to develop that drug. Another case of success is anidulafungin, also based on echinocandin. After these two, the third antimycotic drug is called micafungin, where the hexapeptide FR901370 produced by *Coleophoma empetri* is used as a base.

More recently, it was developed rezafungin, one more echinocandin-related antimycotic. Another important pharmaceutical use of lipopeptides is against viruses and in developing vaccines. For example, lipopeptides can be used in the development of vaccines against viral infections like SARS-CoV-2, MERS-CoV, Influenza, Ebola, HIV, and HPV, or even vaccines for immunotherapy and cancer treatment (Pilz et al., 2023). The importance of cancer and the developing alternatives that help with the high cost of treatments and improve success rates, much research has been carried out throughout the years looking for other options for this disease. Some lipopeptides produced by *Bacillus subtilis* are mentioned as alternatives to this problem; lipopeptides like iturin, surfactin, and fengycin have the potential to inhibit a wide range of tumor cell lines such as breast cancer by influencing the cell checkpoints, regulatory proteins, and signaling pathways (Kumar et al., 2021).

Another very important disease is diabetes, and one of the key points in the treatment of this disease is the use of permeability improvers to increase insulin transport through the gastrointestinal tract; it has been proven that the use of some lipopeptides enhanced the passage of insulin through a Caco-2/HT-29 model (Garcia et al., 2018). For the treatment of HIV, some therapies greatly improve the survival of patients, but this brings with it adverse effects, thus leading to the need to develop new drugs. Some novel lipopeptides have shown positive effects with anti-HIV activity that could result in long-term therapy (Chong et al., 2019). Obesity is another disease that should not be underestimated; it has been shown that lipopeptides produced by *Bacillus velezensis* strain FJAT-52631 can present anti-obesogenic effects. Among these compounds, iturins, fengycins, and surfactins possess great lipase inhibitory activity (Chen et al., 2020).

4.2 Food Use

Lipopeptides and biosurfactants have emerged as promising natural preservatives in food science, offering potential alternatives to synthetic additives.

4.2.1 Bread and Cereals Products

With the knowledge that lipopeptides are also good surfactants, the food industry began to incorporate these biosurfactants in food processing. They were included as additives in different food matrices to improve or change their physicochemical properties. They emerged as more natural and ecological alternatives for the commercial surfactants available in the market. Thus, the incorporation of purified lipopeptide powder obtained from *Bacillus subtilis* SPB1 as an additive in dough formulation improves the hardness, springiness, cohesion, adhesion, and gas retention capacity in comparison with the commercial surfactant soy lecithin. So, SPB1 has shown to be very useful in the breadmaking industry (Mnif et al., 2013).

More recently, the usage of lipopeptides for the protection and preservation of food began to be evaluated, and thanks to multiple related research it has been shown that *Bacillus cereus* XZ30-2 could be used in the storage stage as a biological control agent against *Aspergillus*, an important pathogen attacking stored cereals (Wang et al., 2023).

4.2.2 Meat and Fish Products

Jemil et al. (2020), investigated the antioxidant properties of lipopeptides from *Bacillus methylotrophicus* DCS1 in sunflower oil-in-water emulsions and raw beef patties, demonstrating their efficacy in delaying lipid oxidation, offering a natural alternative to synthetic antioxidants like BHA, which is advantageous for clean-label food products. Wang et al. (2019), examined the preservative effects of lipopeptides, particularly bacillomycin D, from *Bacillus megaterium*, which substantially improved the shelf life in large yellow croaker by inhibiting spoilage microorganisms. Mouafo et al. (2020), focused on biosurfactants from *Lactobacillus* strains to enhance the shelf life of raw ground goat meat. The biosurfactants significantly inhibited microbial growth, lipid oxidation, and spoilage-related volatile compound production, extending the shelf life to 15 d at 4 °C. Collectively, these studies suggest that lipopeptides can be a promising natural alternative for extending the shelf life of meat and fish products.

4.2.3 Beverages

The role of adding lipopeptides to beverages such as juice, milk, beer, soda, and yogurt is to provide a preservation effect, as it is mainly used for the eradication of pathogenic microorganisms that can decompose food when proper preservation processes are not in place (Shi et al., 2018; Wu et al., 2019).

These lipopeptides have been shown to prolong the shelf life of beverages, as contamination is primarily derived from fungi and yeasts. The incorporation of 100 µg/mL is sufficient to reduce the microbial load to minimal levels, achieving good permeation into fungal cells, thereby exerting its fungicidal effect. This is due to its amphipathic structure; the fatty acid chain facilitates membrane permeability, allowing it to penetrate the cell, where the peptide part can exert its fungicidal effect (Thery et al., 2018).

Typically, milk has one of the shortest shelf lives due to the high initial load of fermentative microorganisms it contains. Even with proper pasteurization, seeking alternatives for controlling these microorganisms is sometimes necessary. The addition of lipopeptides to milk, such as brevibacillin V at 64 µg/mL, results in a reduction to minimal levels of bacteria such as *S. aureus* and *L. monocytogenes*, thereby prolonging shelf life, as without these contaminating agents, there is no decomposition of the milk (Wu et al., 2019).

In addition to the effect on microorganisms, it is also necessary to consider that incorporating lipopeptides such as iturin can affect the physicochemical characteristics. While there is a fungicidal effect on *S. cerevisiae* at 0.76 mg/mL of iturin in orange juice, it is interesting that it prolongs shelf life. After 4 d of storage, the yeast is no longer present, and the effect lasts for more than 10 d, similar to natamycin, which is used for beverage preservation against yeasts and fungi. However, acidity content and total soluble solids (ºBrix) are unaffected regarding physicochemical parameters. Still, vitamin C content is greatly reduced, along with a slight increase in browning reactions, resulting in some loss of visual appeal (Shi et al., 2018).

In summary, incorporating lipopeptides is beneficial for extending the shelf life of beverages. However, it is important to consider the quantity and type of lipopeptide, as well as the beverage it will be incorporated into, since it must be biocompatible, not alter the physicochemical characteristics, and evaluate the sensory quality so that it is not displeasing to consumers and can be a viable option for the preservation of these beverages.

4.3 Agricultural Use

Natural biocontrol agents, such as lipopeptides, are emerging as promising alternatives to chemical pesticides in agriculture. The structural differences of lipopeptides are strongly related to their antifungal and antibacterial activities (Fig. 2).

4.3.1 Antibacterial Activity

Lipopeptides have made significant progress in demonstrating their bactericidal effect; however, most of this evidence is in vitro, leaving room for opportunity in crops. Formulations based on spores and cells of lipopeptide-producing bacteria, such as those from the genera *Bacillus*

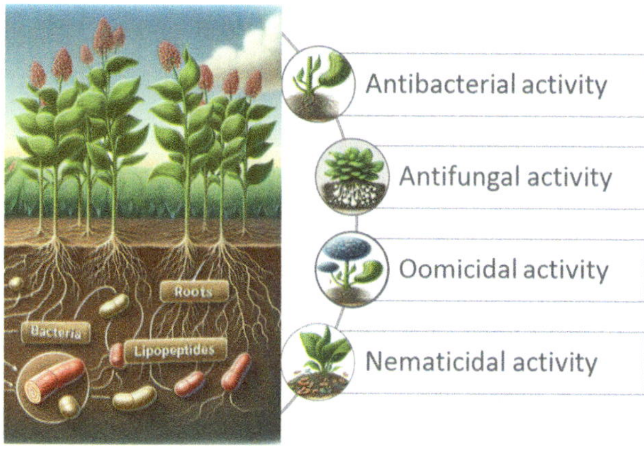

Fig. 2 Use of lipopeptides for the control of plant pathogenic microorganisms. Made with Microsoft Copilot

and *Pseudomonas,* have been tested. However, the efficacy of these formulations is little to none due to the viability of the cells under inadequate storage conditions. Therefore, lipopeptides in formulations have taken particular interest (Chen et al., 2019).

Some studies have been conducted on crops such as rice, tomato, and banana, using different species of *Bacillus* spp. that produce lipopeptides like surfactin, iturin, and fengycin under different culture conditions. All these factors affect the effectiveness of the lipopeptide tested in various crops and against different bacterial diseases (Table 1).

As we can see (Table 1), most of the tests conducted on crops involve lipopeptides from *B. amyloliquefaciens*, which have been tested against various plant pathogenic bacteria such as *R. solanacearum* and *A. tumefaciens*. The main attraction for using this bacterium is the production of surfactins, iturins, and fengycins under optimal temperature conditions of 30 °C, varying the time and agitation. This variation in time and agitation conditions and pathosystems results in different lipopeptide effectiveness, depending on the lipopeptide profile and concentration administered.

4.3.2 Antifungal Activity

The search for alternatives to chemical fungicides has increased, needing new sources of lipopeptides. It was found that *Bacillus amyloliquefaciens* FZB42 produces fengycin, which is useful in protecting cereal crops from the pathogenic fungus *Fusarium graminearum* (Hanif et al., 2019). In the same way, the lipopeptides gageopeptides A-D and gageotetrin B from marine *Bacillus subtilis* strain 109GGC020 showed positive results against *Magnaporthe oryzae* pathotype *Triticum* (MoT), the cause of Wheat blast (Chakraborty et al., 2020). *Streptomyces* can be used in agriculture because they produce lipopeptides with biological activity against plant pathogens. In this sense, a study was conducted to study the potential of *Streptomyces bikiniensis* HD-087 against *Magnaporthe oryzae*, the cause of rice blast. Surfactins, iturins, and fengycins were obtained from *S. bikiniensis,* and they showed noticeable effects against *M. oryzae*, proving these compounds would be the right alternative to chemical pesticides (Liu et al., 2022). Also, *Fusarium graminearum* and *Fusarium verticillioides* could be controlled with the help of mycosubtilin, a lipopeptide purified from *Bacillus subtilis* ATCC6633, to protect crops from these pathogens (Yu et al., 2021).

Surfactin A, the main component of a lipopeptide extract from *Bacillus halotolerans* LDFZ001, showed great activity against *Rhizoctonia solani* and, therefore, a great potential in plant protection (Feng et al., 2022). *Streptomyces* can be used in agriculture because they produce lipopeptides with biological activity against plant pathogenic fungi. In this sense, a study was conducted to study the potential of *Streptomyces bikiniensis* HD-087 against *Magnaporthe oryzae*, the cause of rice blast. Surfactins, iturins, and fengycins were obtained from *S. bikiniensis,* and they showed noticeable effects against *M. oryzae*, proving these compounds would be the right alternative to chemical pesticides (Liu et al., 2022).

Several studies have reported the antifungal activity of lipopeptides produce by *Bacillus* spp. against various plant pathogenic fungi: *Botrytis cinerea, Alternaria alternata, Colletotrichum gloeosporioides, Fusarium oxysporum, Aspergillus niger,* and *Mucor* sp (Table 2).

As can be seen (Table 2), the lipopeptides with antifungal activity against plant pathogenic fungi are mainly lipopeptides such as bacillomycin, iturin, and fengycin alone or in a mixture.

4.3.3 Oomicidal Activity

Li et al. (2019) reported that *Bacillus subtilis* GLB191 reduced grapevine downy mildew (*Plasmopara viticola*) by inhibiting pathogen sporulation and inducing plant defense responses, with lipopeptides surfactin and fengycin playing crucial roles. Sheng et al. (2024) highlighted the biocontrol potential of putisolvin from *Pseudomonas capeferrum* HN2-3 against *Phytophthora* blight in cucumbers by lysing zoospores, reducing disease symptoms in plants. The strain HN2-3, isolated from the rhizosphere soil of *Setaria viridis* plants, was identified as *Pseudomonas capeferrum* through genome analysis. Additionally, Peng et al. (2017), carried out a study that demonstrates the potential of *Bacillus mycoides* strains BM02 and NP02 as biocontrol agents against cucumber damping-off caused by *Pythium aphanidermatum* through the production of biosurfactants like surfactin A, which inhibited zoospore release and lysis. In

Table 1 Bactericidal activity in crops of various lipopeptides produced by bacteria

Microorganism	Lipopeptides	Culture conditions	Plant pathogenic bacteria	Results	References
B. licheniformis HN-5	Bacitracin A	28 °C 48 h 180 rpm	*Pantoea ananatis* in rice plants	Effectiveness: 49.94 ± 1.56%	Jin et al. (2021)
B. amyloliquefaciens 32a	Surfactin, fengycin and iturin	30 °C 18 h 200 rpm	*A. tumefaciens* C58 and B6 in tomato plants	At 0.4 mg wound^{-1} Biocontrol index and Gall weight reduction: 100%	Abdallah et al. (2018)
B. amyloliquefaciens FJAT-2349	Surfactin, fengycin, and iturin	30 °C 50 mL of growth medium in a flask of 250 mL 48 h 170 rpm	*R. solanacearum* FJAT-91 in tomato plants	At 12 days postinoculation Severity and incidence: 2–3% Effectiveness: 97.6% At 30 days postinoculation Severity and incidence: 61.4% Effectiveness: 38.6%	Chen et al. (2019)
B. amyloliquefaciens EA-CB0959	Surfactin, fengycin, and iturin	30 °C 180 mL of growth medium in an Erlenmeyer flask of 500 mL 5 d 140 rpm	*R. solanacearum* EAP-009 in banana plants	Incidence: 61.1% Severity: 5.9 ± 0.8	Villegas-Escobar et al. (2018)

Table 2 Antifungal activity of lipopeptides produced by *Bacillus* spp. against various plant pathogenic fungi

Microorganisms	Lipopeptides	Culture conditions	Plant pathogenic fungi	Results	References
B. methylotrophicus XT1 CECT 8661	Surfactin, bacillomycin, and fengycin	30 °C, 72 h, 200 rpm	*Botrytis cinerea*	High antifungal activity and trigger the antioxidant activity in fruit	Toral et al. (2018)
B. velezensis SL-6	Fengycins	30 °C, 24 h, 200 rpm	*Penicillium expansum, Alternaria alternate,* and *Botrytis cinerea in apple and pear fruits*	Inhibition of the radial growth of *B. cinerea* (72.67%), *A. alternate* (46.3), whereas no inhibition was detected for *P. expansum*	Cozzolino et al. (2020)
B. subtilis KLP15	Surfactin-type	37 °C, 48 h, 200 rpm	*Mucor sp., Aspergillus niger*	Antifungal activity: 75.1% (*Mucor sp.*) y 41.9% (*A. niger*)	Meena et al. (2020)
B. amyloliquefaciens MG3	Iturin A	32 °C, 60 h, 170 rpm	*Colletotrichum gloeosporioides*	Inhibit mycelia growth and spore germination	Yan et al. (2020)
B. subtilis IBFCBF-4	Iturin A, bacillomycin, fengycin, surfactin, bacilysin, mersacidin	30 °C, 48 h, 180 rpm	*Fusarium oxisporum*	Reduce the incidence of *Fusarium* wilt of watermelon (51.1%)	Zhu et al. (2020)

another study by Omoboye et al. (2019), chemical characterization and genetic analysis identified four novel banan-amide derivatives (D-G) produced by *Pseudomonas sp.* COW3. These compounds differ from previously described cyclic lipopeptides (CLPs) in their amino acid composition and fatty acid tail structures. The whole cells of COW3 showed antagonistic activity against *Pythium myriotylum*. These findings underscore the effectiveness of biocontrol agents in disease suppression and their role in sustainable agriculture.

4.3.4 Nematicidal Activity

Particularly lipopeptides from *Bacillus* species, offer a promising alternative to chemical nematicides in agriculture. Zheng et al. (2018), investigated the nematicidal activity of thumolycin from *Bacillus thuringiensis* BMB171 against *Caenorhabditis elegans,* demonstrating significant nematicidal effects with higher concentrations leading to increased nematode mortality. Chavarria-Quicaño et al. (2023) compared crude lipopeptide extracts from *Bacillus paralicheniformis* TB197 and *Bacillus subtilis* ATCC 21,332, finding that TB197 exhibited higher nematicidal activity due to its diverse lipopeptide production, including fengycins and lichenysins. Hussain et al. (2020), evaluated lipopeptides from *Bacillus subtilis* Hussain T-AMU against the root-knot nematode, *Meloidogyne incognita*, showing a significant reduction in juvenile mortality and egg hatching, highlighting its potential as a bio-nematicide. These studies underscore the efficacy of *Bacillus*-derived lipopeptides as biocontrol agents against nematodes, providing a sustainable and environmentally friendly alternative to chemical nematicides. Future research should focus on optimizing production and formulation for field applications, understanding detailed modes of action, and ensuring the safety and effectiveness of these biocontrol agents in diverse agricultural environments.

5 Techniques to Improve the Production of Lipopeptides

Various efficient microorganisms are used for lipopeptide production, which can be affected by multiple factors. These include the type of strain, the development stage of the microorganism, growth conditions (temperature, dissolved oxygen, pH, agitation, substrates, ions, among others), as well as genetic machinery. It is important to understand the genes and their regulation so that the biosynthesis of lipopeptides can be manipulated to favor a higher production of lipopeptides, potentially through genetic mutations (Luo et al., 2023; Pilz et al., 2023).

Research continues on the regulators and signal transduction involved in the biosynthesis of lipopeptides. It is possible to favor the production of lipopeptides by suppressing some genes and overexpressing others. Therefore, mutations that encourage this behavior must be chosen (Luo et al., 2023).

Performing genetic mutations in lipopeptide-producing bacteria can result in mutant strains capable of producing larger quantities of specific lipopeptides and not synthesizing others. In summary, knowledge of the regulatory mechanisms of genes involved in lipopeptide production can be useful for the genetic improvement of strains that prioritize maximizing lipopeptide yield, whether in a mixture or favoring a specific one. Genetic engineering information paves the way for pilot and industrial-scale development using metabolic engineering to produce cell factories, making lipopeptide production economically viable (Pilz et al., 2023).

The production of lipopeptides is expensive, although some research is dedicated to improving it. For example, in some *Bacillus* species, it is possible to increase lipopeptide yield by integrating genetic engineering and optimizing fermentation conditions. Fermentation time can influence the type and concentration of lipopeptide produced. Lipopeptides with antimicrobial activity, such as surfactins, are made during the exponential bacterial growth phase. In contrast, lipopeptides with antifungal activity, such as iturins and fengycins, are produced during the stationary phase. Although lipopeptide biosynthesis can occur at temperatures between 25 and 45 °C, the fermentation temperature also affects the type of lipopeptide produced. Agitation and the culture medium's initial pH value can also greatly influence lipopeptide production. Generally, high stirring speeds and a neutral pH favor optimal production of lipopeptides, so controlling different factors during fermentation is essential to maintaining optimal production (Pilz et al., 2023; Xu et al., 2020).

6 Conclusion

Fungi and bacteria have proven to be a natural source for producing lipopeptides. These microbial compounds attract attention for their low toxicity, high efficiency, and good degradability, so they have enormous potential to be used in different fields due to their wide range of biological applications, essentially as biosurfactants and antimicrobials. In recent years, research has focused on manipulating fermentation conditions and genetic machinery to promote greater production of lipopeptides in microbial factories.

References

Abdallah, D. B., Tounsi, S., Gharsallah, H., Hammami, A., & Frikha-Gargouri, O. (2018). Lipopeptides from Bacillus amyloliquefaciens strain 32a as promising biocontrol compounds against the plant pathogen Agrobacterium tumefaciens. *Environmental Science and Pollution Research, 25*(36), 36518–36529. https://doi.org/10.1007/s11356-018-3570-1

Biniarz, P., Coutte, F., Gancel, F., & Łukaszewicz, M. (2018). High-throughput optimization of medium components and culture conditions for the efficient production of a lipopeptide pseudofactin by Pseudomonas fluorescens BD5. *Microbial Cell Factories, 17*(1). https://doi.org/10.1186/s12934-018-0968-x

Carolin, C. F., Kumar, P. S., & Ngueagni, P. T. (2021). A review on new aspects of lipopeptide biosurfactant: Types, production, properties and its application in the bioremediation process. *Journal of Hazardous Materials, 407*, 124827. https://doi.org/10.1016/j.jhazmat.2020.124827

Castaldi, A., Truong, B. N., Vu, Q. T., Le, T. H. M., Marie, A., Le Pennec, G., Bourguet-Kondracki, M. L. (2024). Computational methods reveal a series of cyclic and linear Lichenysins and Surfactins from the Vietnamese Marine sediment-derived streptomyces strain G222. *Molecules, 29*(7). https://doi.org/10.3390/molecules29071458

Chakraborty, M., Mahmud, N. U., Gupta, D. R., Tareq, F. S., Shin, H. J., & Islam, T. (2020). Inhibitory effects of linear lipopeptides from a marine bacillus subtilis on the wheat blast fungus magnaporthe oryzae triticum. *Frontiers in Microbiology, 11*. https://doi.org/10.3389/fmicb.2020.00665

Chandrakar, S., & Gupta, A. K. (2019). Studies on the production of broad spectrum antimicrobial compound polypeptide (Actinomycins) and Lipopeptide (Fengycin) from Streptomyces sp. K-R1 associated with root of abutilon indicum against multidrug resistant human pathogens. *International Journal of Peptide Research and Therapeutics, 25*(2), 779–798. https://doi.org/10.1007/s10989-018-9727-4

Chavarria-Quicaño, E., De la Torre-González, F., González-Riojas, M., Rodríguez-González, J., & Asaff-Torres, A. (2023). Nematicidal lipopeptides from Bacillus paralicheniformis and Bacillus subtilis: A comparative study. *Applied Microbiology and Biotechnology, 107*(5–6), 1537–1549. https://doi.org/10.1007/s00253-023-12391-w

Chen, M., Wang, J., Liu, B., Zhu, Y., Xiao, R., Yang, W., & Chen, Z. (2020). Biocontrol of tomato bacterial wilt by the new strain Bacillus velezensis FJAT-46737 and its lipopeptides. *BMC Microbiology, 20*(1), 160. https://doi.org/10.1186/s12866-020-01851-2

Chen, M. C., Wang, J. P., Zhu, Y. J., Liu, B., Yang, W. J., & Ruan, C. Q. (2019). Antibacterial activity against Ralstonia solanacearum of the lipopeptides secreted from the Bacillus amyloliquefaciens strain FJAT-2349. *Journal of Applied Microbiology, 126*(5), 1519–1529. https://doi.org/10.1111/jam.14213%JJournalofAppliedMicrobiology

Chong, H. H., Xue, J., Zhu, Y. M., Cong, Z., Chen, T., Wei, Q., He, Y. X. (2019). Monotherapy with a low-dose lipopeptide HIV fusion inhibitor maintains long-term viral suppression in rhesus macaques. *Plos Pathogens, 15*(2). https://doi.org/10.1371/journal.ppat.1007552

Ciurko, D., Chebbi, A., Kruszelnicki, M., Czapor-Irzabek, H., Urbanek, A. K., Polowczyk, I., & Janek, T. (2023). Production and characterization of lipopeptide biosurfactant from a new strain of Pseudomonas antarctica 28E using crude glycerol as a carbon source. *RSC Advances, 13*(34), 24129–24139. https://doi.org/10.1039/d3ra03408a

Clements-Decker, T., Kode, M., Khan, S., & Khan, W. (2022). Underexplored bacteria as reservoirs of novel antimicrobial lipopeptides. *Frontiers in Chemistry, 10*. https://doi.org/10.3389/fchem.2022.1025979

Cozzolino, M. E., Distel, J. S., García, P. A., Mascotti, M. L., Ayub, M. J., Benazzi, L. M., Silva, P. G. (2020). Control of postharvest fungal pathogens in pome fruits by lipopeptides from a Bacillus sp. isolate SL-6. *Scientia Horticulturae, 261*, 108957. https://doi.org/10.1016/j.scienta.2019.108957

de Souza, C. G., Martins, F. I. C. C., Zocolo, G. J., Figueiredo, J. E. F., Canuto, K. M., & de Brito, E. S. (2018). Simultaneous quantification of lipopeptide isoforms by UPLC-MS in the fermentation broth from Bacillus subtilis CNPMS22. *Analytical and Bioanalytical Chemistry, 410*(26), 6827–6836. https://doi.org/10.1007/s00216-018-1281-6

Feng, Z. B., Xu, M. Z., Yang, J., Zhang, R. H., Geng, Z. G., Mao, T. T., & Zhang, H. X. (2022). Molecular characterization of a novel strain of <i>Bacillus halotolerans protecting wheat from sheath blight disease caused by Rhizoctonia solani Kuhn. *Frontiers in Plant Science, 13*. https://doi.org/10.3389/fpls.2022.1019512

Fira, D., Dimkić, I., Berić, T., Lozo, J., & Stanković, S. (2018). Biological control of plant pathogens by Bacillus species. *Journal of Biotechnology, 285*, 44–55. https://doi.org/10.1016/j.jbiotec.2018.07.044

Garcia, J., Fernández-Blanco, A., Teixidó, M., Sánchez-Navarro, M., & Giralt, E. (2018). D-polyarginine lipopeptides as intestinal permeation enhancers. *ChemMedChem, 13*(19), 2045–2052. https://doi.org/10.1002/cmdc.201800428

Hanif, A., Zhang, F., Li, P. P., Li, C. C., Xu, Y. J., Zubair, M., Gao, X. W. (2019). Fengycin produced by bacillus amyloliquefaciens FZB42 inhibits fusarium graminearum growth and mycotoxins biosynthesis. *Toxins, 11*(5). https://doi.org/10.3390/toxins11050295

He, H. R., Huang, J. R., Zhao, Z. Z., Du, P. Q., Li, J. S., Xin, J. L., Zheng, X. K. (2023). Whole genome analysis of Streptomyces sp. RerS4, a Rehmannia glutinosa rhizosphere microbe producing a new lipopeptide. *Heliyon, 9*(9). https://doi.org/10.1016/j.heliyon.2023.e19543

Hussain, T., Haris, M., Shakeel, A., Ahmad, G., Ahmad Khan, A., & Khan, M. A. (2020). Bio-nematicidal activities by culture filtrate of Bacillus subtilis HussainT-AMU: New promising biosurfactant bioagent for the management of Root Galling caused by Meloidogyne incognita. *Vegetos, 33*(2), 229–238. https://doi.org/10.1007/s42535-020-00099-5

Jemil, N., Ouerfelli, M., Almajano, M. P., Elloumi-Mseddi, J., Nasri, M., & Hmidet, N. (2020). The conservative effects of lipopeptides from Bacillus methylotrophicus DCS1 on sunflower oil-in-water emulsion and raw beef patties quality. *Food Chemistry, 303*, 125364. https://doi.org/10.1016/j.foodchem.2019.125364

Jin, P., Tan, Z., Wang, H., Liu, W., & Miao, W. (2021). Antimicrobial effect of Bacillus licheniformis HN-5 bacitracin A on rice pathogen Pantoea ananatis. *BioControl, 66*(2), 249–257. https://doi.org/10.1007/s10526-020-10052-9

Kadaikunnan, S., Vijayaragavan, P., Rathi, M. A., Balamuralikrishnan, B., Alharbi, N. S., Gopalakrishnan, V. K., & Sivanesan, R. (2024). Antibacterial and biofilm disruptive nonribosomal lipopeptides from Streptomyces parvulus against multidrug-resistant bacterial infections. *Journal of Infection and Public Health, 17*(3), 450–456. https://doi.org/10.1016/j.jiph.2024.01.001

KEGG. (2024). Fatty acid biosynthesis—Reference pathway. https://www.genome.jp/pathway/map00061+C00383

Kumar, A., Rabha, J., & Jha, D. K. (2021). Antagonistic activity of lipopeptide-biosurfactant producing Bacillus subtilis AKP, against Colletotrichum capsici, the causal organism of anthracnose disease of chilli. *Biocatalysis and Agricultural Biotechnology, 36*. https://doi.org/10.1016/j.bcab.2021.102133

Li, Y., Héloir, M. C., Zhang, X., Geissler, M., Trouvelot, S., Jacquens, L., & Adrian, M. (2019). Surfactin and fengycin contribute to the protection of a Bacillus subtilis strain against grape downy mildew by both direct effect and defense stimulation. *Molecular Plant Pathology, 20*(8), 1037–1050. https://doi.org/10.1111/mpp.12809

Liu, W., Wang, J. W., Li, S., Zhang, H. Q., Meng, L., Liu, L. P., Du, C. M. (2022). Genomic and biocontrol potential of the crude lipopeptide by streptomyces bikiniensis HD-087 against magnaporthe oryzae. *Frontiers in Microbiology, 13*. https://doi.org/10.3389/fmicb.2022.888645

Luo, K. C., Chen, Y. W., Qian, X. Y., Zhong, H. J., Onchari, M. M., Liu, X. H., Luo, C. P. (2023). Enhancing surfactin production in B. velezensis Bs916 combined cumulative mutagenesis and expression key enzymes. *Applied Microbiology and Biotechnology, 107*(13), 4233–4244. https://doi.org/10.1007/s00253-023-12590-5

Ma, Z., Geudens, N., Kieu, N. P., Sinnaeve, D., Ongena, M., Martins, J. C., & Höfte, M. (2016). Biosynthesis, chemical structure, and structure-activity relationship of orfamide lipopeptides produced by pseudomonas protegens and related species. *Frontiers in Microbiology, 7*. https://doi.org/10.3389/fmicb.2016.00382

Meena, K. R., Sharma, A., Kumar, R., & Kanwar, S. S. (2020). Two factor at a time approach by response surface methodology to aggrandize the Bacillus subtilis KLP2015 surfactin lipopeptide to use as antifungal agent. *Journal of King Saud University—Science, 32*(1), 337–348. https://doi.org/10.1016/j.jksus.2018.05.025

Mnif, I., Besbes, S., Ellouze-Ghorbel, R., Ellouze-Chaabouni, S., & Ghribi, D. (2013). Improvement of bread dough quality by Bacillus subtilis SPB1 biosurfactant addition: Optimized extraction using response surface methodology. *Journal of the Science of Food and Agriculture, 93*(12), 3055–3064. https://doi.org/10.1002/jsfa.6139

Mouafo, H. T., Mbawala, A., Tanaji, K., Somashekar, D., & Ndjouenkeu, R. (2020). Improvement of the shelf life of raw ground goat meat by using biosurfactants produced by lactobacilli strains as biopreservatives. *Lwt, 133*, 110071. https://doi.org/10.1016/j.lwt.2020.110071

Oikawa, H., Mizunoue, Y., Nakamura, T., Fukushi, E., Jiang, Y. L., Ozaki, T., & Minami, A. (2022). Structure and biosynthesis of the ribosomal lipopeptide antibiotic albopeptins. *Bioscience Biotechnology and Biochemistry, 86*(6), 717–723. https://doi.org/10.1093/bbb/zbac039

Omoboye, O. O., Geudens, N., Duban, M., Chevalier, M., Flahaut, C., Martins, J. C., Höfte, M. (2019). Pseudomonas sp. COW3 produces new bananamide-type cyclic lipopeptides with antimicrobial activity against pythium myriotylum and pyricularia oryzae. *Molecules, 24*(22), 4170. https://doi.org/10.3390/molecules24224170

Peng, Y. H., Chou, Y. J., Liu, Y. C., Jen, J. F., Chung, K. R., & Huang, J. W. (2017). Inhibition of cucumber Pythium damping-off pathogen with zoosporicidal biosurfactants produced by Bacillus mycoides. *Journal of Plant Diseases and Protection, 124*(5), 481–491. https://doi.org/10.1007/s41348-017-0110-z

Pilz, M., Cavelius, P., Qoura, F., Awad, D., & Brück, T. (2023). Lipopeptides development in cosmetics and pharmaceutical applications: A comprehensive review. *Biotechnology Advances, 67*. https://doi.org/10.1016/j.biotechadv.2023.108210

Planckaert, S., Deflandre, B., De Vries, A. M., Ameye, M., Martins, J. C., Audenaert, K., Devreese, B. (2021). Identification of novel rotihibin analogues in streptomyces scabies, including discovery of its biosynthetic gene cluster. *Microbiology Spectrum, 9*(1). https://doi.org/10.1128/Spectrum.00571-21

Sheng, J., Qin, X., Yang, X., Liu, Q., & Ma, Z. (2024). The biocontrol roles of cyclic lipopeptide putisolvin produced from Pseudomonas capeferrum HN2-3 on the Phytophthora blight disease in cucumbers. *Journal of Plant Diseases and Protection, 131*(2), 423–432. https://doi.org/10.1007/s41348-024-00874-5

Shi, J., Zhu, X., Lu, Y., Zhao, H., Lu, F., & Lu, Z. (2018). Improving iturin a production of bacillus amyloliquefaciens by genome shuffling and its inhibition against saccharomyces cerevisiae in orange juice. *9*. https://doi.org/10.3389/fmicb.2018.02683

Son, S., Ko, S. K., Kim, S. M., Kim, E., Kim, G. S., Lee, B., & Ahn, J. S. (2018). Antibacterial cyclic lipopeptide enamidonins with an enamide-linked acyl chain from a streptomyces species. *Journal of Natural Products, 81*(11), 2462–2469. https://doi.org/10.1021/acs.jnatprod.8b00497

Song, Y. H., Amaya, J. A., Murarka, V. C., Mendez, H., Hogan, M., Muldoon, J., & Caffrey, P. (2024). Biosynthesis of a new skyllamycin in Streptomyces nodosus: A cytochrome P450 forms an epoxide in the cinnamoyl chain. *Organic and Biomolecular Chemistry, 22*(14), 2835–2843. https://doi.org/10.1039/d4ob00178h

Thery, T., O'Callaghan, Y., O'Brien, N., & Arendt, E. K. (2018). Optimization of the antifungal potency of the amidated peptide H-Orn-Orn-Trp-Trp-NH_2 against food contaminants. *International Journal of Food Microbiology, 265*, 40–48. https://doi.org/10.1016/j.ijfoodmicro.2017.10.024

Toral, L., Rodríguez, M., Béjar, V., & Sampedro, I. (2018). Antifungal activity of lipopeptides from bacillus XT1 CECT 8661 against botrytis cinerea. *Frontiers in Microbiology, 9*. https://doi.org/10.3389/fmicb.2018.01315

Tyurin, A. P., Alferova, V. A., Paramonov, A. S., Shuvalov, M. V., Malanicheva, I. A., Grammatikova, N. E., & Korshun, V. A. (2018). Crystallomycin revisited after 60 years: Aspartocins B and C. *MEDCHEMCOMM, 9*(4), 667–675. https://doi.org/10.1039/c8md00002f

Varnava, C. K., Grenni, P., Mariani, L., Caracciolo, A. B., Hadjipakkou, H., Lefkaritis, G., & Tsipa, A. (2024). Characterization, production optimization and ecotoxicity of a lipopeptide biosurfactant by Pseudomonas citronellolis using oily wastewater. *Biochemical Engineering Journal, 205*, 109257. https://doi.org/10.1016/j.bej.2024.109257

Villegas-Escobar, V., González-Jaramillo, L. M., Ramírez, M., Moncada, R. N., Sierra-Zapata, L., Orduz, S., & Romero-Tabarez, M. (2018). Lipopeptides from Bacillus sp. EA-CB0959: Active metabolites responsible for in vitro and in vivo control of Ralstonia solanacearum. *Biological Control, 125*, 20–28. https://doi.org/10.1016/j.biocontrol.2018.06.005

Wang, M., Ma, Y., Mou, H., & Kong, Q. (2019). Bacillomycin D lipopeptides from marine Bacillus megaterium as antimicrobial and preservative agents for large yellow croaker, Larimichthys crocea. *Journal of Food Safety, 39*(4). https://doi.org/10.1111/jfs.12652

Wang, X. X., Du, Z. H., Chen, C. X., Guo, S., Mao, Q. Z., Wu, W., Yu, X. M. (2023). Antifungal effects and biocontrol potential of lipopeptide-producing Streptomyces against banana Fusarium wilt fungus Fusarium oxysporum f. sp. cubense. *Frontiers in Microbiology, 14*. https://doi.org/10.3389/fmicb.2023.1177393

Wu, Y., Zhou, L., Lu, F., Bie, X., Zhao, H., Zhang, C., & Lu, Y. (2019). Discovery of a Novel Antimicrobial Lipopeptide, Brevibacillin V, from Brevibacillus laterosporus fmb70 and its application on the preservation of skim milk. *Journal of Agricultural and Food Chemistry, 67*(45), 12452–12460. https://doi.org/10.1021/acs.jafc.9b04113

Xu, Y., Cai, D., Zhang, H., Gao, L., Yang, Y., Gao, J., & Chen, S. (2020). Enhanced production of iturin A in Bacillus amyloliquefaciens by genetic engineering and medium optimization. *Process Biochemistry, 90*, 50–57. https://doi.org/10.1016/j.procbio.2019.11.017

Yan, F., Li, C., Ye, X., Lian, Y., Wu, Y., & Wang, X. (2020). Antifungal activity of lipopeptides from Bacillus amyloliquefaciens MG3 against Colletotrichum gloeosporioides in loquat fruits. *Biological Control, 146*, 104281. https://doi.org/10.1016/j.biocontrol.2020.104281

Yu, C. J., Liu, X., Zhang, X. Y., Zhang, M. X., Gu, Y. Y., Ali, Q., Gu, Q. (2021). Mycosubtilin produced by bacillus subtilis ATCC6633 inhibits growth and mycotoxin biosynthesis of fusarium graminearum and fusarium verticillioides. *Toxins, 13*(11). https://doi.org/10.3390/toxins13110791

Zhao, P. C., Xue, Y., Li, J. H., Li, X., Zu, X. Y., Zhao, Z. Q., & Feng, S. X. (2019). Non-lipopeptide fungi-derived peptide antibiotics developed since 2000. *Biotechnology Letters, 41*(6–7), 651–673. https://doi.org/10.1007/s10529-019-02677-3

Zheng, D., Zeng, Z., Xue, B., Deng, Y., Sun, M., Tang, Y.-J., & Ruan, L. (2018). Bacillus thuringiensis produces the lipopeptide thumolycin to antagonize microbes and nematodes. *Microbiological Research, 215*, 22–28. https://doi.org/10.1016/j.micres.2018.06.004

Zhu, J., Tan, T., Shen, A., Yang, X., Yu, Y., Gao, C., & Zeng, L. (2020). Biocontrol potential of Bacillus subtilis IBFCBF-4 against Fusarium wilt of watermelon. *Journal of Plant Pathology, 102*(2), 433–441. https://doi.org/10.1007/s42161-019-00457-6

Role of Algae in Microbial Fuel Cell and the Production of Value Added Products

Megha Duggal, Manisha Bisht, and Manoj Pal

Abstract

Microalgae biomass is one of the sustainable and renewable sources, mainly used in biorefineries for the production of biofuels. Photosynthetic microbial fuel cell (PMFC) could be an alternative method for microalgae biomass production as it can be run using wastewater as a feeding source. Different configurations of PMFC were already explored to optimize the growth and biomass production. Microalgae remove the total carbon, N, and P from the wastewater and convert it into biomass. The production of O_2 in the cathodic chamber of PMFC also reduces the cost of aeration. Harvested microalgae biomass from the cathodic chamber is used to extract different metabolites such as fatty acids, essential oil, vitamins, polysaccharides, polyphenolic compounds. These metabolites are used in different industrial sectors such as food, agriculture feed, cosmetics, biorefineries, and cosmetics. Mass culturing of microalgae in PMFC rendered the process of biomass production cost-effective as simultaneously helped to treat wastewater and produce bio-electricity. However, biomass production at the pilot scale in PMFC still remains challenging and needs a lot of optimizations.

Keywords

Bioelectricity · Microalgae biomass · Polysaccharides · Polyphenolic compounds · Photosynthetic microbial fuel cell

M. Duggal · M. Pal (✉)
Department of Microbiology, Graphic Era (Deemed to Be University), Dehradun, India
e-mail: manojpal.bt@geu.ac.in

M. Bisht
Department of Chemistry, Motiram Baburam, Government Post Graduate College (M.B.G. PG College), Haldwani (Nainital), Uttarakhand, India

1 Introduction

The increasing population of the world and depleting natural energy sources are slowly pressing alarming conditions, where we need to think critically about alternative sources of energy. Microbial fuel cell (MFCs) is a type of fuel cell system that utilizes microbes, especially bacteria as a biocatalyst in the anodic chamber to produce electrical energy. Bacteria usually feed on biodegradable waste to generate energy. The generation of electrical energy with the degradation of bio-waste could be the best option for waste management. Since the discovery of MFC in 1911 by Potter and Waller, continuous efforts have been made to improve the efficiency of MFC and to bring the technology to a commercial level. In progressive efforts, Allen and Bennetto used various types of bacterial species and electron mediators to improve the efficiency of electron transfer. *Geobacter* and *Shewanella* are two most commonly used bacterial species in MFC. Both are known as electroactive bacteria or exoelectrogens as capable of transferring electrons out of the cell surface directly to an anodic electrode. However other bacterial species were also explored such as *Ruminococcaceae spp., Desulfobulbus spp., Bacillus, Geothrix, Pseudomonas, and Acidobacteria.*

The ability of microalgae to convert carbon dioxide (CO_2) into biomass and to produce oxygen (O_2) is an exemplary property to be utilised in the cathodic chamber of the microbial fuel cell (MFC) system. The bacterial species oxidise organic substrates in the anodic chamber and generate CO_2 that is utilised by algal species in the cathodic chamber by photosynthetic process in the presence of light. The photosynthetic process generates O_2 that is further reduced at the cathodic electrode. Due to the fixation of CO_2 into biomass and generation of O_2 and subsequent reduction of it at the cathodic electrode, the whole MFC system is considered as photosynthetic MFC (PMFCs). The application of microalgae in a cathodic chamber could be one of the options to be used in PMFC.

K. Sharma (ed.), *Bio-prospecting of Novel Microbial Bioactive Compounds for Sustainable Development*, Sustainable Landscape Planning and Natural Resources Management, https://doi.org/10.1007/978-3-031-82178-3_10

The following are the general biochemical reactions representing the functioning of PMFC.

(1) At anode (Oxidation): Organics → $CO_2 + H^+ + e^-$
(2) At cathode (Reduction): $2O_2 + 4H^+ + 4e^- \rightarrow 4H_2O$
(3) Overall biochemical reaction:
Organics + $O_2 \rightarrow + CO_2 + H_2O$ + external power

In PMFC, algal species not only fix CO_2 into biomass but also help to remove nitrogen, phosphorous, and ammonia from the waste. The surplus supply of O_2 from the algal partner helps to reduce the extra cost of aeration in the cathodic chamber.

2 Different Factors Controlling the Performance of PMFC

A number of factors affect the PMFC's activity such as the type of carbon source (simple to complex substrate) used in the cathodic chamber, geometry of the cathodic electrode (brush or planar), distance between electrodes, type of the microalgae used, geometry of MFC: tubular, circular, single chamber, double chamber, coupled PMFCs, sediment PMFCs, type of chemical and biological electron mediators, pH, biofouling, catalyst poisoning, CO_2 concentration, light intensity.

Usually, an increase in the pH of the cathodic chamber after a long time of operation of the PMFC hampered the overall performance by inhibiting the growth of algae. Similarly, the feeding of high carbon content rich wastewater in an anodic chamber leads to the generation of high CO_2 levels that may inhibit the growth of algae in a cathodic chamber. The presence of high mineral and ionic content in wastewater may also cause unbounded growth of the algae that results in the development of multilayers of algal growth on the surface of the cathodic electrode. The development of multiple layers inhibits the transport of electrons to the surface of the electrode. The crossover of the molecules and substrate from the anodic chamber to the cathodic chamber leads to a decrease in the cathodic potential by altering the surface of the electrode called poisoning. The presence of polyvalent cations in the medium inhibits the algal growth due to their adsorption on the cell surface which finally affects the efficiency of MFC. Decrease in the light intensity reaching the algal cells growing in depth of MFC or photobioreactor may also affect the performance of the fuel cell system. The interruption of light intensity mainly depends on the geometry and size of the fuel cell system.

3 Algae in PMFC

To optimize the efficiency of PMFC, a number of types of algae were explored. The CO_2 fixation efficiency depends upon the concentration of CO_2. Different types of algae show variable types of CO_2 fixation patterns at the same concentration of CO_2. Even the variability in cell shape, size, growth rate and nutrient requirements make them suitable to work in different environmental conditions, especially when the different types of carbon sources are used. Table 1 indicates the type of algae commonly used in PMFC.

3.1 Chlorella Vulgaris

It is freshwater, spherical, unicellular microalgae. The presence of a thick cell wall (100–200 nm) is the main identification characteristic of this algae that makes it resistant to the heavy metals and other toxic ions present in the wastewater.

3.2 Chlamydomonas Reinhardtii

Chlamydomonas reinhardtii is a green algae having approximately 10 µm size. It is widely distributed in temperate, tropical, subtropical and soil and water samples. Being its ability to use different waste materials for growth, it is the most extensively explored microalgae used in bioremediation of the soil and in wastewater treatment.

Table 1 Commonly used algae in PMFC

S. no.	Algae
1	*Chlorella vulgaris*
2	*Scenedismus obliquus*
3	*Desmodesmus sp.*
4	*Blue green Phormidium*
5	*Laminaria saccharina*
6	*Chlamydomonas reinhardtii*
7	*Spirulina platensis*
8	*Anabaena cylindrical*
9	*Chlorella pyrenoidosa*
10	*Synechococcus leopoliensis*

3.3 Scenedismus Obliquus

Scenedesmus spp. is a small green algae arranged in flat colonies. They are known for their ability to produce hydrogen gas (H_2), although the H_2 production ability is less than compared to *Chlamydomonas reinhardtii*. They produce H_2 gas in the presence of light and in anaerobic conditions. Due to their high tolerance power to environmental changes and high biomass and lipid synthesis, these are widely explored for industrial wastewater treatment.

3.4 Chlorella Pyrenoidosa

It is known for its higher growth rate on settled sludge and excellent efficiency for the removal of phosphorous and nitrogen from wastewater.

3.5 Desmodesmus Sp.

It is a type of green algae known for its higher growth rate and high biomass and lipid production. These are also suitable algae for the removal of pollutants from wastewater and extensively explored in wastewater treatment.

3.6 Laminaria Saccharina

It is a type of brown algae and they are known for their excellent metal removal efficiency from wastewater.

3.7 Spirulina Platensis

It is a multicellular blue-green microalgae having a length of 50–500 μm and width 3–4 μm belonging to the family cyanophyceae. Morphologically it is characterised by filamentous and cylindrical trichome. It has high COD, BOD, nitrogen and phosphorous removal efficiency.

3.8 Synechococcus Leopoliensis

Synechococcus leopoliensis are coccoid in shape and belong to the Cyanophyceae family. These are planktonic in nature and found in temperate and tropical oceans. They have been explored in wastewater treatment with other algal species.

3.9 Phormidium Spp.

These are the long filamentous algae, generally filament attached to the base. These are known for their high arsenic tolerance. These are known for their high nitrogen and phosphorous removal efficiency.

3.10 Anabaena Cylindrical

Cells are spherical, cylindrical and may be ovoidal in morphology, arranged in filaments known as trichome. Trichome again straight, curved and spiral. Heterocyst is present in between the cells or terminally at the top of trichome. They have good COD, nitrogen, phosphorous and heavy metal removal efficiency. In PMFC, they are generally used in mixed populations with other algal species.

4 Configuration of PMFC

Generally, three major configurations; (I) single chamber (II) double chamber (III) double chamber with attached photobioreactor are widely used in PMFC setup.

(I) **Single chamber**

In a single chamber, bacteria and algae are grown in the same chamber without using any cation-selective membrane (Fig. 1a). Single chamber PMFC is easy to operate, and cost-effective method (Chandra et al., 2012), where CO_2 produced by bacterial population is consumed by an algal partner. However, O_2 produced by algae is reduced to H_2O at the air cathode surface. Algae needs sunlight or photons to grow, therefore a source of light is always required nearby to the chamber.

(II) **Double Chambered**

Double-chambered PMFC is a little bit more advanced compared to single chambered as anodic and cathodic chambers are separated by a cation (H^+) exchange membrane (Fig. 1b) (Angioni et al., 2018; Saba et al., 2017). The bacterial population grows inside the anodic chamber with organic substrate and algae inside the cathodic chamber. Degradation of organic substrate by bacterial population produces electrons and protons (H^+). Electrons move in the outer electronic circuit (anode to cathode), whereas, protons (H^+) selectively pass through the membrane and react with O_2 produced by algae inside the cathodic chamber and forming the H_2O at the cathode surface (Yongjin et al., 2009).

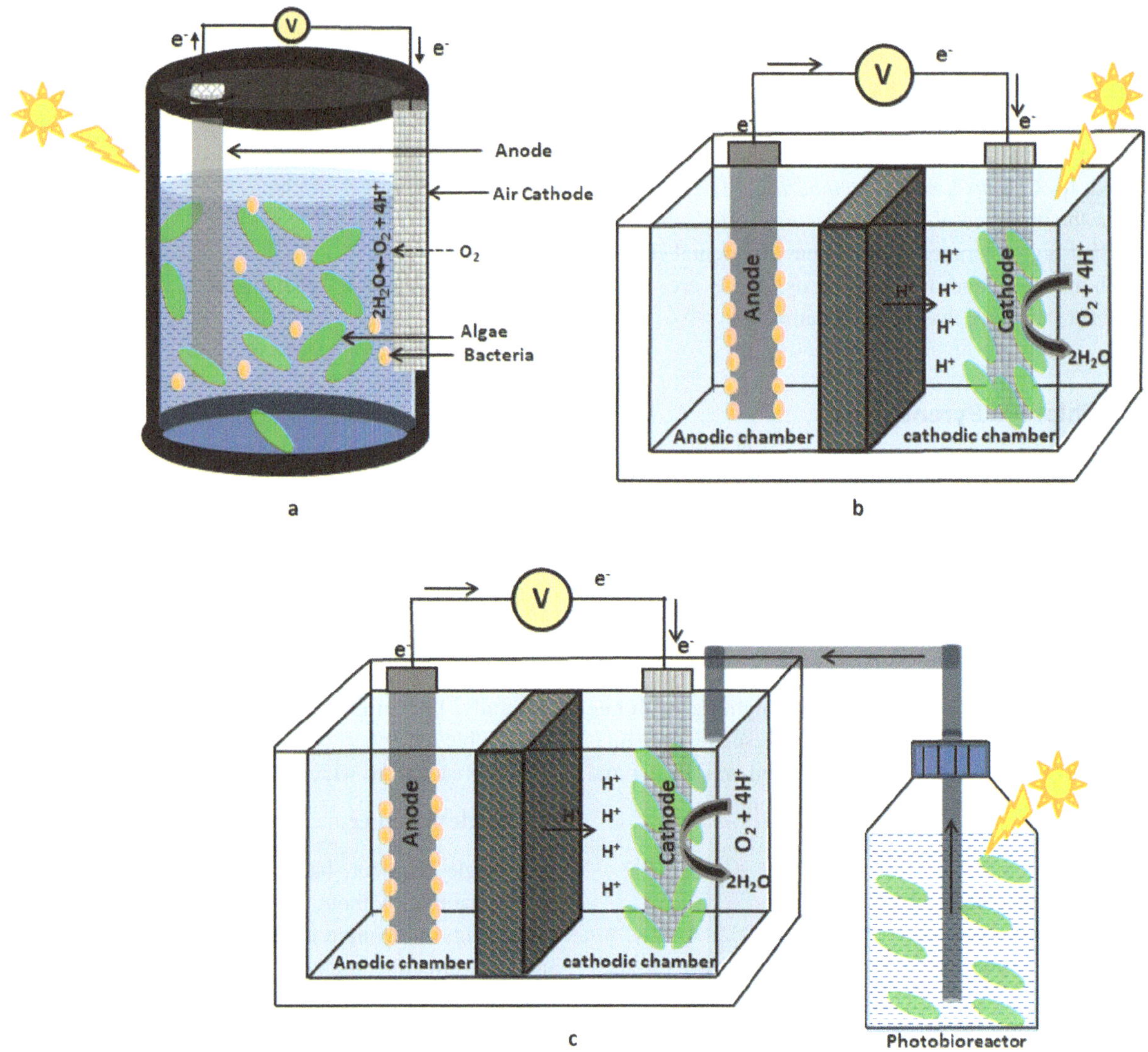

Fig. 1 Different configurations of MFCs, **a** single chambered with air cathode, **b** double chambered with cation-selective membrane at the centre, and **c** double chambered connected with photobioreactor

(III) **Double Chambered with Photobioreactor**

In another kind of a double-chambered PMFC, a photobioreactor is attached with a cathodic chamber for the continuous supply of algae (Fig. 1c). A light source is provided both to the photobioreactor and to the cathodic chamber. Continuous supply of algae to the cathodic chamber helps to keep the steady state that results in the increase in the power density (Colombo et al., 2017; Kannan & Donnellan, 2021). Further, different modifications were made in the set-up of PMFCs to enhance the power generation such as the use of multiple anode stacks, connecting anaerobic digestor, providing external aerated source (Hou et al., 2016; Schievano et al., 2018), etc.

5 Extraction of Value-Added Products from Algae

Generally, various types of waste materials are used as a carbon source in PMFC for waste treatment and concurrently to produce algal biomass with electrical energy. Recently, major emphasis was given to valorisation of biomass into useful value-added products besides cutting the cost and making the process energy efficient (González-López et al., 2020). Algal biofuel is one of the major products obtained from algal biomass, both liquid (biodiesel, bioethanol) and gaseous (biomethane, biogas) biofuels are extracted from algal biomass (Saad et al., 2019). Microalgae are the major source of various types of bioactive compounds such as polysaccharides, carotenoids, lipids, fatty acids of dietary value, proteins, amino acids, and auxins.

5.1 Polysaccharides

Many microalgae specifically, marine microalgae are rich sources of polysaccharides. Algae such as; *Porphyridium sp. Chlorella sp.*, *Aphanothece halophytica*, *Arthrospira platensis*, *Anbaena*, *Nostoc*, etc. are widely used for the extraction of exopolysaccharides and sulphated exopolysaccharides. Calcium-Spirulan extracted from *A. platensis* is a potent inhibitor of several viruses. In 2011, Radonic and coworkers found that a sulphated polysaccharide extracted from *A. platensis* and *P. purpureum* exhibit antiviral activities against two strains of *Vaccinia virus* and an *Ectromelia virus* (Radonić et al., 2011).

Similarly, high molecular weight and sulphated exopolysaccharide isolated from *Porphyridium* inhibited the growth of neoplastic mammalian cells (Geresh et al., 2002). In a likewise study, Shopen-Katz and coworkers showed the anticolon cancer effect of biomass extracted from the same algae (Shopen-Katz et al., 2000). Gardeva and coworkers studied the strong anti-tumour activity of sulphated polysaccharide isolated *P. cruentum* and suggested a good anticancer agent (Gardeva et al., 2009).

5.2 Proteins and Amino Acids

Microalgae are rich source of proteins and essential amino acids constituting 40–60% of total biomass (Koyande et al., 2019). Therefore algae could be a good source of protein in the human diet and food supplement (Caporgno & Mathys, 2018). Many microalgae such as *Chlorella sp.*, *Arthrospira sp.*, *Aphanizomenon sp.*, and *Dunaliella sp.*, having a protein content of 50% and more were already utilized in the human protein diet (Koyande et al., 2019). Many others microalgae were characterized as animal feed such as *Chlamydomonas sp.*, *Nannochloropsis sp.*, *Tetraselmis sp.*, *Dunaliella sp.*, (Koyande et al., 2019; Madeira et al., 2017). Some microalgae are highly rich in protein content such as *Arthrospira maxima* (70–75%), *Spirulina sp.*, (45–65%), and *Synechococcus sp.*, (60–65%) (Phillips & Williams, 2011). Microalgae are rich in many essential amino acids such as tryptophan, valine, lysine, and leucine. Many *spirulina sp.*, rich in protein content were characterized and graded as GRAS (Generally Recognised as Safe) by the FDA (Costa & Morais, 2013). Therefore algal biomass grown in PMFC could be used to meet the protein demand in food and feed.

5.3 Fatty Acids and Essential Oils

Microalgae being rich in lipid content are considered promising sustainable sources of third-generation biofuel (Behera et al., 2015). However, the presence of unique fatty acids and their derivatives make them be used as alternatives in food and supplements. The wider distribution of algae in different habitats is responsible for the varied composition of membrane lipids. Many microalgae are used as a source of different kinds of essential oils rich in omega-3 fatty acids, DHA, EPA, n-3 PUFA (Poly Unsaturated Fatty Acid) along with a lesser amount of monounsaturated (MUFA) and saturated fatty acids (SFA) (Remize et al., 2021). A number of dietary supplements rich in omega-3 fatty acids, DHA, and EPA, are available in the market and there is always a necessity of sustainable sources to meet the industry demand. For example, DHASCO is a commercially available DHA isolated from *Schizochytrium sp.* A number of algae were explored for the extraction of PUFA such as *Chlorella sp.*, *Schizochytrium sp.*, *Ulkenia sp.*, *Spirulina platensis.*, *Nannochloropsis sp.*, *Porphyridium cruentum*, *Parietochloris incise*, *Odontella aurita*, *Diacronema vlkianum*, *Phaedactylum tricornutum*, and *Isochrysis galbana*. Microalgae are also rich in different kinds of sterols such as ergosterol, cholesterol, and phytosterol.

5.4 Vitamins

Due to the limited quantity of vitamins in vegetables, algae could be a good alternative. Microalgae are rich in different vitamins such as vitamins D, K, B_{12}, B_9, B_6 with other vitamins such as A, C, and E (Del Mondo et al., 2020). For example, the algae, *Dunaliella* is rich in almost all kinds of vitamin B such as B_2, B_{12}, B_9, and B_3. Some important algae rich in vitamins are shown in Table 2.

Table 2 Different types of microalgae and vitamins

Name of the algae	Rich in vitamin	References
Arthrospira, Chlorella, Dunaliella, Tetraselmis, Chaetoceros	Vit. K	Aaronson et al. (1977), Santiago-Morales et al. (2018)
Chlamydomonas, Chlorella, Dunaliella, Nannochloris, Chaetoceros, Skeletonema, Isochrysis galbana, Tetraselmis, and *Nannochloropsis*	Vit. C	Aaronson et al. (1977), Brown and Farmer (1994), Fabregas and Herrero (1990)
Anabaena, Arthrospira, Synechococcus, Chlamydomonas, Chlorella, Dunaliella, Scenedesmus, Porphyridium, Chaetoceros, Nannochloropsis, Euglena	Vit. E	Aaronson et al. (1977), Brown and Farmer (1994), Fabregas and Herrero (1990), Fujita et al. (2008), Santiago-Morales et al. (2018)
Anabaena, Aphanizomenon, Arthrospira, Chlorella, Dunaliella, Tetraselmis, Isochrysis, Nannochloropsis	Vit. B	Aaronson et al. (1977), Fabregas and Herrero (1990)
Tetraselmis sp. CS-362, Tetraselmis suecica, Skeletonema costatum, Isochrysis galbana, Pavlova lutheri	Vit. D	De Roeck-Holtzhauer et al. (1991)
Anabaena cylindrical, Arthrospira, Tetraselmis suecica, Skeletonema costatum, Isochrysis galbana, Pavlova lutheri, Tetraselmis suecica, Skeletonema costatum	Vit. K	De Roeck-Holtzhauer et al. (1991), Tarento et al. (2018)

5.5 Polyphenolic Compounds

Different microalgae are rich in polyphenolic compounds having anticancer, anti-diabetic, antioxidant, and antimicrobial properties. Many microalgae are rich in polyphenolic compounds such as *Nodularia*, *Spirogyra sp.*, *Arthrospira sp.*, *Nostoc sp.*, *Euglena sp.*, *Caespitella pascheri*, and *spumigena* (Jerez-Martel et al., 2017; Tarento et al., 2018).

6 Conclusion

With the fastest growth and limited requirements, microalgae are diversified in different habitats. Being their ability to grow on organic waste materials such as wastewater from municipal and industrial sources, these are considered as excellent biomass for energy production. A number of methods are used for the mass production of microalgae but still a lot of optimisations are required to scale up the process. On the other side, biomass production of microalgae in MFCs could be a better alternative, simultaneously with wastewater treatment and electricity production. The use of algae in the cathode chamber of MFCs made the system O_2 rich which helps to replace the outer aeration source and leads to the reduction in the cost of PMFC operation. Microalgae biomass is not only used for energy production but also for the extraction of a number of value-added products. Nowadays, microalgae biomass is used in different sectors such as food industries, animal feeds, biofuel refineries, and cosmetics.

References

Aaronson, S., Dhawale, S. W., Patni, N. J., Deangelis, B., Frank, O., & Baker, H. (1977). The cell content and secretion of water-soluble vitamins by several freshwater algae. *Archives of Microbiology, 112*, 57–59. https://doi.org/10.1007/BF00446654

Angioni, S., Millia, L., Mustarelli, P., Doria, E., Temporiti, M. E., Mannucci, B., Corana, F., & Quartarone, E. (2018). Photosynthetic microbial fuel cell with polybenzimidazole membrane: synergy between bacteria and algae for wastewater removal and biorefinery. *Heliyon, 4*, e00560. https://doi.org/10.1016/j.heliyon.2018.e00560

Behera, S., Singh, R., Arora, R., Sharma, N. K., Shukla, M., & Kumar, S. (2015). Scope of algae as third generation biofuels. *Frontiers in Bioengineering and Biotechnology*.

Brown, M. R., & Farmer, C. L. (1994). Riboflavin content of six species of microalgae used in mariculture. *Journal of Applied Phycology, 6*, 61–65. https://doi.org/10.1007/BF02185905

Caporgno, M. P., & Mathys, A. (2018). Trends in microalgae incorporation into innovative food products with potential health benefits. *Frontiers in Nutrition*.

Chandra, R., Venkata Subhash, G., & Venkata Mohan, S. (2012). Mixotrophic operation of photo-bioelectrocatalytic fuel cell under anoxygenic microenvironment enhances the light dependent bioelectrogenic activity. *Bioresource Technology, 109*, 46–56. https://doi.org/10.1016/j.biortech.2011.12.135

Colombo, A., Marzorati, S., Lucchini, G., Cristiani, P., Pant, D., & Schievano, A. (2017). Assisting cultivation of photosynthetic microorganisms by microbial fuel cells to enhance nutrients recovery from wastewater. *Bioresource Technology, 237*, 240–248. https://doi.org/10.1016/J.BIORTECH.2017.03.038

Costa, J., & Morais, M. (2013). Microalgae for food production (pp. 405–428). https://doi.org/10.1201/b14070-17

De Roeck-Holtzhauer, Y., Quere, I., & Claire, C. (1991). Vitamin analysis of five planktonic microalgae and one macroalga. *Journal of Applied Phycology, 3*, 259–264. https://doi.org/10.1007/BF00003584

Del Mondo, A., Smerilli, A., Sané, E., Sansone, C., & Brunet, C. (2020). Challenging microalgal vitamins for human health. *Microbial Cell Factories, 19*, 201. https://doi.org/10.1186/s12934-020-01459-1

Fabregas, J., & Herrero, C. (1990). Vitamin content of four marine microalgae. Potential use as source of vitamins in nutrition. *Journal of Industrial Microbiology, 5*, 259–263. https://doi.org/10.1007/BF01569683

Fujita, T., Aoyagi, H., Ogbonna, J. C., & Tanaka, H. (2008). Effect of mixed organic substrate on α-tocopherol production by Euglena gracilis in photoheterotrophic culture. *Applied Microbiology and Biotechnology, 79*, 371–378. https://doi.org/10.1007/s00253-008-1443-0

Gardeva, E., Toshkova, R., Minkova, K., & Gigova, L. (2009). Cancer protective action of polysaccharide, derived from red microalga porphyridium cruentum—A biological background. *Biotechnology and Biotechnological Equipment, 23*. https://doi.org/10.1080/13102818.2009.10818540

Geresh, S., Mamontov, A., & Weinstein, J. (2002). Sulfation of extracellular polysaccharides of red microalgae: preparation, characterization and properties. *Journal of Biochemical and Biophysical Methods, 50*, 179–187. https://doi.org/10.1016/S0165-022X(01)00185-3

V. González-López, C., García-Cuadra, F., Jawiarczyk, N., M. Fernández-Sevilla, J., G. Acién-Fernández, F. (2020). Valorization of microalgae and energy resources, In *Sustainable mobility*. https://doi.org/10.5772/intechopen.90072

Hou, Q., Nie, C., Pei, H., Hu, W., Jiang, L., & Yang, Z. (2016). The effect of algae species on the bioelectricity and biodiesel generation through open-air cathode microbial fuel cell with kitchen waste anaerobically digested effluent as substrate. *Bioresource Technology, 218*. https://doi.org/10.1016/j.biortech.2016.07.035

Jerez-Martel, I., García-Poza, S., Rodríguez-Martel, G., Rico, M., Afonso-Olivares, C., & Gómez-Pinchetti, J. L. (2017). Phenolic profile and antioxidant activity of crude extracts from microalgae and cyanobacteria strains. *Journal of Food Quality, 2017*, 2924508. https://doi.org/10.1155/2017/2924508

Kannan, N., & Donnellan, P. (2021). Algae-assisted microbial fuel cells: A practical overview. *Bioresource Technology Reports, 15*, 100747. https://doi.org/10.1016/J.BITEB.2021.100747

Koyande, A. K., Chew, K. W., Rambabu, K., Tao, Y., Chu, D.-T., & Show, P.-L. (2019). Microalgae: A potential alternative to health supplementation for humans. *Food Science and Human Wellness, 8*, 16–24. https://doi.org/10.1016/j.fshw.2019.03.001

Madeira, M. S., Cardoso, C., Lopes, P. A., Coelho, D., Afonso, C., Bandarra, N. M., & Prates, J. A. M. (2017). Microalgae as feed ingredients for livestock production and meat quality: A review. *Livestock Science, 205*, 111–121. https://doi.org/10.1016/j.livsci.2017.09.020

Phillips, G. O., & Williams, P. A. (2011). Handbook of food proteins. *Handbook of Food Proteins*. https://doi.org/10.1533/9780857093639

Radonić, A., Thulke, S., Achenbach, J., Kurth, A., Vreemann, A., König, T., Walter, C., Possinger, K., & Nitsche, A. (2011). Anionic polysaccharides from phototrophic microorganisms exhibit antiviral activities to vaccinia virus. https://doi.org/10.4172/jaa.100002

Remize, M., Brunel, Y., Silva, J. L., Berthon, J. Y., & Filaire, E. (2021). Microalgae n-3 PUFAs production and use in food and feed industries. *Marine Drugs*. https://doi.org/10.3390/MD19020113

Saad, M. G., Dosoky, N. S., Zoromba, M. S., & Shafik, H. M. (2019). Algal biofuels: Current status and key challenges. *Energies, 12*. https://doi.org/10.3390/en12101920

Saba, B., Christy, A.D., Yu, Z., & Co, A. C. (2017). Sustainable power generation from bacterio-algal microbial fuel cells (MFCs): An overview. *Renewable and Sustainable Energy Reviews*. https://doi.org/10.1016/j.rser.2017.01.115

Santiago-Morales, I. S., Trujillo-Valle, L., Márquez-Rocha, F. J., & López Hernández, J. F. (20180. Tocopherols, phycocyanin and superoxide dismutase from microalgae: As potential food antioxidants. *Applied Food Biotechnology, 5*, 19–27. https://doi.org/10.22037/afb.v5i1.17884

Schievano, A., Colombo, A., Cossettini, A., Goglio, A., D'Ardes, V., Trasatti, S., & Cristiani, P. (2018). Single-chamber microbial fuel cells as on-line shock-sensors for volatile fatty acids in anaerobic digesters. *Waste Management, 71*. https://doi.org/10.1016/j.wasman.2017.06.012

Shopen-Katz, O., Ling, E., Himelfarb, Y., Lamprecht, S., Arad, S., & Shany, S. (2000). The effect of Porphyridium sp., biomass and of its polysaccharide in prevention and inhibition of human colon cancer. *Proceedings of the Era of Biotechnology*

Tarento, T. D. C., McClure, D. D., Vasiljevski, E., Schindeler, A., Dehghani, F., Kavanagh, J. M. (2018). Microalgae as a source of vitamin K1. *Algal Research, 36*, 77–87. https://doi.org/10.1016/j.algal.2018.10.008

Yongjin, Z., Pisciotta, J., Billmyre, R. B., & Baskakov, I. V. (2009). Photosynthetic microbial fuel cells with positive light response. *Biotechnology and Bioengineering, 104*. https://doi.org/10.1002/bit.22466

Harnessing Microbes: Eco-Friendly Solutions for Waste Management

Binu Gogoi, Arun Chettri, and Arun Kumar Rai

Abstract

Rapid development and urbanisation have increased land, water, and soil pollution levels. Many new technologies have been established to aid in the recycling and treatment of such harmful pollutants, but such modern procedures have drawbacks such as high prices and the generation of resistant toxic chemicals. With such negative environmental impacts, a safe and convenient biological approach in the form of microbial strains enters the picture as these microorganisms are known to exhibit a variety of biosynthetic activities. This characteristic of microorganisms aids in the resolution of several problems, including the treatment of oil spills that have damaged the environment, industrial effluent management, and waste management. Microbes contribute to ecological equilibrium by participating in biogeochemical cycles. Microbes play a significant role in nutrient cycling, which benefits soil and plant health. Furthermore, a variety of microorganisms break down harmful substances that have entered the soil as a result of the widespread use of chemical pesticides and fertilisers. DGGE and other molecular techniques have been widely used in waste treatment. This chapter focuses on the different ways that microbes are used in molecular-based environmental maintenance techniques, heavy metal detoxification, bioremediation, and cleanup of contaminated areas.

Keywords

Microbes · Environment · Bioremediation · Microbial enzymes

B. Gogoi · A. Chettri · A. K. Rai (✉)
Department of Botany, Sikkim University, Gangtok, Sikkim, India
e-mail: akrai@cus.ac.in

Abbreviations

AtzA	Atrazine chlorohydrolase
BPA	Bisphenol A
COD	Chemical oxygen demand
EPS	Extracellular polymeric substances
FISH	Fluorescence in situ hybridization
MIPs	Molecular imprinted printers
PCBs	Polychlorinated biphenyls
PCR	Polymerase chain reaction
PCR-DGGE	Polymerase chain reaction denaturant gradient gel electrophoresis
PCR-TGGE	Temperature gradient gel electrophoresis
POPs	Persistent Organic Pollutants
RISA	Ribosomal intergenic spacer analysis
RT-PCR	Reverse transcription polymerase chain reaction
T-RFLP	Terminal restriction length fragment polymorphism
TrzN	Triazine hydrolase

1 Introduction

The increase in a country's economic growth also correlates with the amount of destruction of natural sources, piling up of wastes, release of more hazardous substances, and use of more pesticides to increase yield. To cope with such environmental degradation and issues, a sustainable and eco-friendly approach should be considered that not only helps in tackling such issues but also doesn't leave any ill-effect afterward. In the present time, many pollutants and contaminants have increased massively and are discarded without treatment in the open. Such pollutants include wastes from industries, pesticides from agricultural lands, sewage and landfills. They not only pollute the environment but pose a major threat to all life forms. There are physical, chemical and biological methods for dealing with such hazardous

K. Sharma (ed.), *Bio-prospecting of Novel Microbial Bioactive Compounds for Sustainable Development*,
Sustainable Landscape Planning and Natural Resources Management, https://doi.org/10.1007/978-3-031-82178-3_11

chemicals, agriculture wastes and industrial effluents. Amongst all the three methods biological method comes as an ecofriendly approach as this involves the use of safe, practices and mainly by utilizing certain microorganisms that cannot only degrade such contaminants but also prove to be beneficial for plants and soil health. Many microorganisms are not only capable of bioremediation but also are a part of biogeochemical cycles which help in balancing the environment.

Microbes are present almost everywhere and have a diversified metabolic capability. Microbes help in maintaining the ecological balance of the environment. They not only degrade certain wastes but have also adapted themselves to degrade certain chemical and industrial wastes. They possess a diverse range of metabolic activities and have an excellent ability to break down toxic compounds (Kumar & Sachan, 2021).

Many microorganisms have become tolerant to various heavy metals and thus can carry out bioremediation (Yin et al., 2019). The microbial bioremediation process has certain advantages as compared to other remediation strategies. Microbial bioremediation is a natural process, cost-effective, requires less manpower, has no harmful effects, and completely degrades all the targeted pollutants (Abatenh et al., 2017). In several incidents of oil spills in aquatic bodies, specific microbial species were used to clean up the oil. In such circumstances, using other remedial procedures would have been ineffective and would have resulted in additional concerns shortly. Microbes can produce certain kinds of enzymes that help in the degradation of complex molecules into simple ones (Wiltschi et al., 2020).

Biological remediation of heavy metal-contaminated soil consists of four types: Bioremediation, Phytoremediation, Cyanoremediation and Mycoremediation. In bioremediation, microbes produce certain kinds of enzymes that are used for the degradation of pollutants. Methods like biosorption, bioaccumulation, biomineralization, bioleaching and microbe-metal interaction methods are utilized. In phytoremediation plants are used for the removal of heavy metals (Yaashikaa et al., 2022).

2 Microbes in Nutrient Cycling

Many microbes are known for their role in nutrient cycling. Most of them are soil microbes that help in nutrient cycling. Microorganisms can break down dead plants and animal matter which is then transformed into nutrients. The most important example is the nitrogen and carbon cycle. Nitrogen, which is abundantly present as atmospheric dinitrogen, cannot be utilized by plants. It needs to be converted into a form that can be taken up by plants. Some bacterial species can fix dinitrogen gas and carry out denitrification at the same time. In this process of nitrogen conversion, many enzymes are utilized. There are also other microorganisms involved in biogeochemical cycles like carbon, oxygen, phosphorous, and sulfur. Another important role that microbes play is in arsenic cycling by maintaining concentrations in the atmosphere, lithosphere and hydrosphere. Another group of microorganisms that help in increasing oxygen concentration in the atmosphere is cyanobacteria. Bacteria carry out nitrogen fixation with the help of the enzyme nitrogenase which combines nitrogen N_2 with hydrogen and produces a useful form of nitrogen such as ammonia. The importance of nutrient cycling helps in the production of biomolecules such as DNA, RNA, proteins and amino acids. Certain soil bacteria and fungi carry out mineral weathering which in turn aids in cycling and plant nutrition (Aislabie et al., 2013).

2.1 Carbon Cycling

There are certain amounts of biological mechanisms through which the microbes can carry out the carbon cycle. Numerous ways are taken up by soil microbes to influence land–atmosphere carbon exchanges. In this process, atmospheric CO_2 is converted into organic material by certain microbes. The carbon cycle occurs when carbon-fixing autotrophic organisms and chemoautotrophic microbes transfer carbon to the atmosphere and thus synthesize carbon dioxide into organic material. The carbon is then sent back to the atmosphere by various other pathways. Another way of carbon fixation is done by organic carbon-consuming heterotrophic microorganisms that include the decomposition of organic materials, and they utilise carbon from various sources like plants, animals or microbes. In this process, some amount of carbon is retained, and the rest of the CO_2 goes back to the atmosphere. Microbes use various amounts of carbon as various carbon and energy sources and for their survival via reproduction soil microbes transfer carbon to the atmosphere (Gougoulias et al., 2014).

2.2 Nitrogen Cycling

Microorganisms carry out denitrification mainly by various methods. The process of nitrification involves the conversion of inert nitrogen to forms that are assimilated by plants and is mediated by soil micro-organisms. Plants require nitrogen as nitrogen forms a component of chlorophyll and is a major important plant macronutrient. The availability of nitrogen in soil is dependent on microbes that convert inert N into N compounds that are assimilated by plants. The plant in turn helps in shaping the activity and composition of soil microbial communities. Plants take up various dissolved forms of N forms, varying from low to high molecular weight compounds that include amino acids, oligopeptides, nucleotides, and urea proteins (Moreau et al., 2019).

The way through which plants obtain nitrogen using microorganisms is by mutualism and multitrophic interaction. Reports suggest that dark septate endophyte (DSE) can facilitate the transfer of N to plants. DSE are facultative biotrophs associated with the roots of various plant species. DSE can break down organic nitrogen. When the C: N ratio of organic matter grows and the supply of N is limited, DSEs outrun arbuscular mycorrhizal fungi, which induces root nutrient uptake (Huusko et al., 2017). In addition to this many bacteria species can convert atmospheric N_2 to ammonia. Many plants thus form mutualistic interaction and release signals by secondary metabolites to nitrogen-fixing bacteria and then enter roots. This association forms nodules. Inside the plant root nodules, the plant gives photosynthetically fixed C to the bacteria which in turn provides fixed N to the host plant (Moreau et al., 2019). As the main structural element of amino acids, proteins, nucleic acids and other molecules. Microbial nitrogen converts gaseous dinitrogen to ammonia for assimilation. Microorganisms like archaebacteria to proteobacteria, and gram-positive eubacteria to fungi are responsible for carrying out such processes. The conversion of nitrite to nitric oxide involves two types of dissimilatory reductase i.e., cytochrome cd1 nitrite reductase and copper-containing nitrite reductase. Studies report that methanotrophic bacteria are also known to consume and produce nitric oxide. These organisms perform assimilatory nitrate reduction to nitrite which further gets reduced to nitric oxide and nitrous oxide. All these processes occur in the absence of oxygen and through enzymatic activities of nitrite and nitric oxide reductases (Rick & Thomas, 2001).

3 Metal Detoxification by Microbes

There is a humongous amount of metal pollution in the environment. Because heavy metals are cytotoxic and carcinogenic, their contamination has a detrimental impact on the ecosystem. Although many metals are necessary for life, a certain amount or level of metals are harmful to all life forms. Microbes are capable of converting metal waste into a lesser toxic form and then utilize the product as their energy source. Microorganisms such as bacteria, fungi, and algae can grow quickly, producing large biomasses. This biological solution is not only readily available and practicable, but it is also an effective technique for absorbing heavy metals from polluted areas. These microorganisms can break down heavy metals and with the help of certain enzymes, they induce the end products as a part of their metabolism (Abdu et al., 2017).

Heavy metal like chromium, cadmium, nickel, lead and copper causes damage by inducing stress to plants affecting the agricultural sector. Many different strategies such as physical, chemical and biological methods have been applied in the remediation of such metals. Biological methods are much more feasible and eco-friendly steps in comparison to physical and chemical methods. The hazardous effects of such a high concentration of heavy metals shorten plant growth and hamper other physiological processes. Certain microbes can produce enzymes that reduce the toxins to a less or reduced form. Another method by which the heavy metals are removed is cyanoremediation where algal species like *Spirulina, Spirogyra, Chlorella and Oedogonium* help in remediation. Fungal species like *Aspergillus, Cladosporium, Candida, Mortierella, Gliocladium, Fusarium, Ganoderma, Emericella and Aureobasidium* carry mycoremediation. Another fungal species *Trichoderma* is used in the remediation of agricultural wastes (Yaashikaa et al., 2022).

The genetic makeup of microbes makes it possible for them to absorb heavy metals and also possess the ability to convert heavy metals into less form (Abdu et al., 2017). Bioaugmentation is a method where genetically engineered microorganisms are brought into use for scaling up the existing microbes. This process helps in increasing the rate of metal degradation and helps in mitigating harmful pollutants (Jayaram et al., 2022). Studies have shown some fungal species such as *Aspergillus sp.* and *Penicillium sp.* have developed tolerance against heavy metals (Ezzouhri et al., 2009).

Bacteria are the most prevalent microbes on Earth, and their small size and rapid growth rate have been employed to remove heavy metal contaminants from the environment. Bacteria species like *Acidithiobacillus ferrooxidans, Leptospirillum ferrooxidans, Sulfolobus sp., Acidianus brierleyi* and *Sulfobacillus thermosulfidooxidans* can oxidize ferrous iron enzymically. Another important activity microbes carry out is bioleaching where metals such as copper, gold, uranium and ores are extracted with the help of chemolithoautotrophic microbes such as *A. ferrooxidans* and *A. thiooxidans*. Biosorption is a technique that includes the removal of metals, and radionuclides by utilizing biological methods mainly by bacteria, microalgae and fungi. Experimental studies have shown that ericoid mycorrhizal and ectomycorrhizal fungi can solubilize zinc, cadmium, copper phosphate and lead chlorophosphate by releasing phosphate and metals (Fomina et al., 2004; Gadd, 2010).

Photosynthetic organisms such as algae have high metal absorption capacity. Algae can be utilised to treat heavy metal concentrations of less than 10 mg/L. The heavy metal ions are accumulated in the algal mass surface. The algal phyla Phaeophyta, Chlorophyta, Rhodophyta, and Cyanophyta have the highest ability for metal adsorption, although Bacillariophyta has a somewhat lower capacity than other taxa (Lin et al., 2020). In phytoremediation, the soil microbes act in the plant's rhizosphere area and thereby influence chelated or complexed metals into soluble forms making it fit for plants to intake. Soil microbes have developed mechanisms to limit the effect of heavy metal toxicity for their survival (Abdu et al., 2017).

Fungi are capable of producing more stable enzymes even when the conditions are not favourable. Microbes can transform metalloids. Microbial genera, such as *Acidiothiobacillus*, can oxidize Sulphur compounds to sulfuric acid, which affects metals and alloys (Gadd, 2010). Another technique to deal with heavy metal pollutants is biosparging. In this technique air or oxygen and nutrients are applied with pressure for heavy metal degradation. In this method, bacteria can carry out biosparging activity given their population diversity and their internal structure (Choudhury & Chatterjee, 2022).

4 Microbial Enzymes Contributing to Bioremediation

Microbial enzymes like oxidoreductases, hydrolases, chromium reductase, Chromium reductase, atrazine dechlorinase, triazine hydrolase, alkane hydroxylases, laccase, carboxylesterases, peroxidases, phytase, horseradish peroxidase are the enzymes secreted by microorganisms (Table 1). The enzyme oxidoreductases help in the degradation of natural and artificial pollutants. *Bacillus safensis* a gram-positive bacteria helps in the degradation of petroleum compounds. Bacteria and fungi generate the enzyme oxidoreductase. This enzyme contributes to the detoxification of chemicals through oxidative coupling, which involves the oxidation of compounds by transferring electrons from reductants to oxidants, resulting in the release of chloride ions, CO_2, and methanol (Medina et al., 2018).

The enzyme Glyphosate oxidase is secreted by the bacterium *Pseudomonas sp.* which helps in bioremediating pesticides (Scott et al., 2010, yhn). White rot fungi can degrade

Table 1 Different enzymes secreted by microorganisms and their bioremediation activity

Enzyme	Microorganism	Contaminant	References
Atrazine dechlorinase	*Nocardioides sp.*	Triazine herbicides	Scott et al. (2010)
Hydrolase	*Achromobacter, Pseudomonas, Flavobacterium, Nocardia,* and *Bacillus cereus*	Carbofuran, carbaryl or parathion, diazinon, and coumaphos	Sutherland et al. (2002)
Alkane hydroxylases	*Arthrobacter, Burkholderia, Mycobacterium, Pseudomonas, Sphingomonas, Rhodococcus*	Hydrocarbons	Das and Chandran (2011)
Laccase	*Pycnoporous sanguineus*	Removal of pollutants like bromophenol blue and malachite green	Mayer and Staples (2002)
Protease	*Bacillus sp., Aspergillus sp., and Amycolatopsis sp.*	Removal of industrial wastes, treatment of wastewater	Kumar and Sharma (2019)
Dehalogenase	*Pseudomonas sp.*	Degradation of halogen acid	Zu et al. (2012)
Protease	*Bacillus licheniformis*	Marine crustacean wastes	Jellouli et al. (2011)
DMO	*Pseudomonas maltophilia*	Dicamba	Herman et al. (2005)
LinA	*Sphingobium sp., Sphingomoas sp.*	Hexachlorocyclohexane	Raina and Hall (2008)
TrzN	*Nocardioides sp.*	Herbicicdes chloro-s-triazine	Mulbry et al. (2002)
P450	*Pseudomonas putida*	Hexachlorobenzene and pentachlorobenzene	Chen et al. (2002)
OpdA (organophosphate hydrolase)	Agrobacterium radiobacter; *Pseudomonas diminuta; Flavobacterium*	Phosphotriester insecticides	Wijerathna-Yapa and Pathirana (2022)
SsoPox (phosphotriesterase-like lactonases)	*Sulfolobus solfataricus*	Degradation of pesticide paraoxon	Guerrero Ramírez et al. (2023)
OPAA (prolidases)	*Alteromonas sp.*	Compounds with cleaving P–F, P–O, P–CN, and P–S bond	Guerrero Ramírez et al. (2023)
Chromium reductase Atrazine dechlorinase	*Pseudomonas, Bacillus, Enterobacter, Rhodococcus erythropolis*	Transforming chromium	Thatoi et al. (2014)

PCBs. Many microorganisms like *Rhodococcus, Gordonia, Mycobacterium* and *Pseudomonas* contain catabolic genes that help in removing POPs. Many aerobic bacteria such as *Pseudomonas, Alcaligens, Sphingomonas, Rhodococcus* and *Mycobacterium* are helpful in the degradation of pesticides and hydrocarbons. On the other hand, anaerobic bacteria can bioremediation of polychlorinated biphenyls (PCBs), and dechlorination of chloroform and trichloroethylene (TCE) (Sharma, 2012).

Microbial enzymes like laccases, hydrolases, dehalogenases, dehydrogenases, proteases and lipases carry out the degradation of different pollutants by methods like oxidation, reduction and elimination. The enzyme protease helps in the elimination of keratinous wastes and converts them into useful products. The production of enzymes in the bioremediation of hydrocarbons and oil spills is attributed to numerous bacterial species (Bhandari et al., 2021). Many microorganisms generate laccase, an enzyme capable of oxidising phenolic and methoxyphenolic acids and depolymerizing lignin (Kim et al., 2002).

5 Microbial Remediation of Contaminated Soil

Many microbial species are capable of producing enzymes that convert the pollutants into less toxic forms (Table 2). The bioremediation method consists of ex-situ and in-situ bioremediation. In situ bioremediation the oxygen and nutrients are applied where the oxygen and nutrients are supplied in the soil which allows the selected bacteria sp. to grow and degrade contaminants. This method is beneficial as it doesn't impact the groundwater and soil. Another advantage of this method is cost-effectiveness, less pollution, and no disruption of nearby areas. Many fungal species like *Aspergillus, Cladosporium, Candida, Mortierella, Gliocladium, Fusarium, Ganoderma, Emericella* and *Aureobasidium* contain stable enzymes that help in mycoremediation (Mani & Kumar, 2014). Cyanoremediation is another method of bioremediation where algae species are used for the removal of heavy metals (Yaashikaa et al., 2022). Bioremediation of contaminated soil that contains heavy metals, a process called

Table 2 Microorganisms involved in various metal remediation

	Microorganism	Metal remediation	Method	References
Bacteria	*Candida lipolytica*	Fe, Zn	Soil washing	Luna et al. (2016)
	Citrobacter freundii	Cu, Cr	Precipitation	Gomaa and El-Meihy (2019)
	Burkholderia sp.	Pb, Mn, Cu, Zn, Cd, As	Bioleaching and flocculation	Yang et al. (2018)
	Bacillus circulans	Pb, Cd	Ce-precipitation	Das et al. (2009)
	Bacillus sp.	Cr	Reduction	Gnanamani et al. (2010)
	Pseudomonas sp.	Cr, Pb	Biosorption	Huang and Liu, (2013)
Fungi	*Aspergillus flavus, Fusarium sp., Cladosporium sp.*	Cr	Adsorption	Chandrakar et al. (2014)
	Aspergillus sp.	Ni, Cr	Bioaccumulation	Congeevaram et al. (2007)
	Trichoderma brevicompactum QYCD-6	Pb	Bioaccumulation	Zhang et al. (2020)
	Sporosarcina saromensis (M52)	Cr	Bioremediation	Ran et al. (2016)
	Cellulosmicrobium sp. (KX710177)	Pb	Bioremediation	Bhargava and Mishra (2018)
	Candidapara psilosis	Hg	Bioremediation	Muneer et al. (2013)
Algae	*Spirogyra sp.* *Spirulina sp.*	Cr	Bioremediation	Mane and Bhosle (2012)
	Nostoc sp.	Pb	Bioremediation	Kumaran et al. (2011)
	Spirogyra sp.	Cu	Bioremediation	Kumaran et al. (2011)

bioleaching is applied. In bioleaching, the treated soils were in good condition and also the heavy metals no longer had a negative effect on the environment.

6 Microbial Approaches for Wastewater Treatment and Industrial Effluents

The effluents from industries and the deposition of sewage water pose a major threat to the environment as well as human health. The untreated wastes from industries deplete the environment and are also a health hazard. These wastes are directly discarded in the nearby water bodies which kills all the life forms thriving on it. People thriving on such water sources face major issues. The negative impact of such discarded effluents from the industries can be calculated by seeing that the flowing water bodies have now become stagnant because of the huge amount of discarded materials, the level of impurities is so high that the water sources are no less than sewage canals. Only a few industries and factories treat their waste before discarding it. Also, the wastes from chemical and pharmaceutical companies pose a detrimental impact on the surrounding areas. The discarded materials within a few periods are discarded in the open but their negative impact keeps on affecting lives forever.

There are many ways to treat such wastes, There are different physical ways of wastewater treatment which are mentioned here. Recently microbial fuel cells have been for treating waste water where the heavy metals are removed. In addition to this microorganisms such as *Rhodoferax ferrireducens, Shewanella putrefaciens, Aeromonas hydrophila, Geobacter metallireducens and G. sulfurreducens* are capable of generating energy by sending electrons to the anode immediately (Kunwar et al., 2024; Malik et al., 2023). Greensand filtration is a process of wastewater treatment where glauconite is used for removing iron, hydrogen sulphide and manganese. The multimedia filtration technique uses three different layers of filtration i.e., anthracite, sand and garnet. In this method of filtration, about 10 to 25 micron-sized particles are removed. Another type of wastewater treatment includes microfiltration uses a barrier method for filtration. In ultrafiltration a particular amount of pressure is applied that separates solids from water via a barrier membrane. Here 0.005 to 0.01 size particles are removed. In nanofiltration, a semi-permeable membrane having a small pore size is used (Nahiun et al., 2021).

Molecular imprinted printers are synthetic polymers that consist of specific cavities and are made by copolymerization of cross-linking agents especially made for target molecules. MIPs are used in wastewater treatment. MIP-magnetic particles (Fe_3O_4) show considerable interest in separation, catalysis, and bioremediation. This process of wastewater treatment includes combining magnetic particles along with the molecular imprinting technique. Magnetic particles when combined with MIPs help in identifying the target molecules and also help to remove them from the complex matrix along with an external magnetic field (Huang et al., 2015). This process of combining magnetic MIPs was accomplished in the selective adsorption and separation of BPA from aqueous solution (Guo et al., 2011). Microbial proteases help in the solubilization of proteinaceous wastes and are known to lower the biological oxygen demand of aquatic systems. The protease produced by *Bacillus subtilis* was used in poultry farms for feathers and other keratin wastes (Ichida et al., 2001).

The microbial method consists of a process of denitrification where denitrifying bacteria genera like *Alcaligenes, Pseudomonas, Methylobacterium, Bacillus, Paracoccus, Hyphomicrobium* are capable of converting the products of nitrification into gaseous nitrogen compounds which helps them in removing from sewage (Wagner et al., 2002). A microbial consortium consisting of microorganisms *Proterobacteria, Bacterioides, Nitrospirales, Cyanobacteria and Bacillus sp.* was used for municipal wastewater treatment by PCR-DGGE technique (Guo et al., 2011) (Fig. 1).

Some bacterial species such as *Enterobacter cloacear, Gordonia alkanivorans, Halomonas eurihalina* can produce EPS that contains emulsifiers. These emulsifiers are in the form of polysaccharides, proteins and lipids which degrade

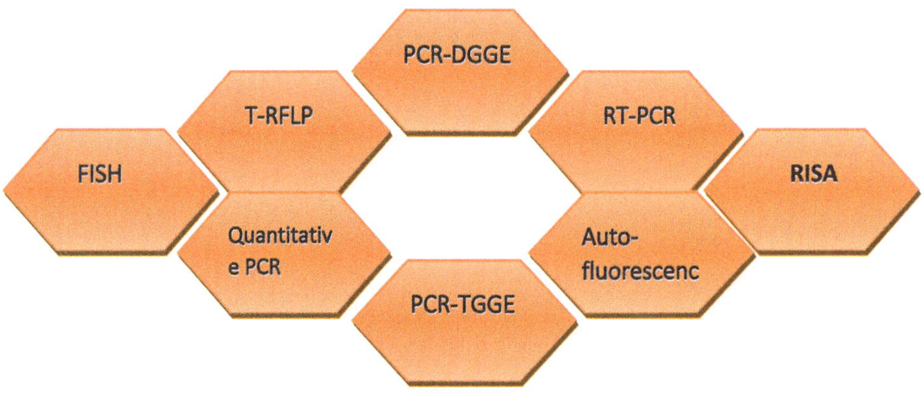

Fig. 1 Monitoring techniques of bioaugmentation

hydrocarbons (Ta-Chen et al., 2008). Bacteria consortia that include *Bacillus cereus, Bacillus myocoides, Bacillus subtilis, Bacillus sp., Micrococcus sp. and Pseudomonas sp.* can degrade azo dyes released from the textile industries (Mahmood et al., 2015). Studies have shown that white-rot fungi are capable of producing enzymes like laccase, Mn peroxidase, and lignin peroxidase. These enzymes can degrade xenobiotic compounds and dyes (Christian et al., 2005). Apart from the degradation of such compounds and dyes, certain fungal species like *Aspergillus niger* have high resistance to heavy metals like mercury, cadmium, and chromate that are present in water bodies released by industries. Many filamentous soil fungi like *Aspergillus terreus, Cadosporium cladosporioides, Fusarium oxysporum, Gliocladium roseum, Penicillium sp. and Trichoderma koningii* have been used for removing Cadmium (Massaccesi et al., 2002).

7 Conclusion

Bioremediation is only applicable to the removal of biodegradable contaminants leaving the non-biodegradable pollutants like plastic persisting in the environment and posing a risk. Since microorganisms require an optimal condition for their growth and survival only a few microorganisms can carry out such remediation strategies in unfavorable conditions. More approaches must be taken to deal with these non-biodegradable pollutants. The world of microbes is not yet completely known to us, more new microbial species are being discovered every day thus opening a wide diverse amount of metabolic activities that these microbes show which will help in addressing the issues of rising contaminants and pollutants in the environment. There is immense potential when it comes to how different microbes respond to various pollutants' degradation. More studies regarding how microbial enzymes function in various degradation processes will help in studying how effective these enzymes are in improving bioremediation. This approach of microbial remediation for wastes has more advantages than disadvantages but the success depends only when it is applied in the field to mitigate hazardous wastes and pollutants.

Acknowledgement and Contributions The authors gratefully recognise the support and facilities offered by the Department of Botany, Sikkim University, Gangtok, Sikkim, throughout the completion of this book chapter. The first author (BG) wrote the paper, while the second author (AKR) checked and modified the current manuscript.

References

Abatenh, E., Gizaw, B., Tsegaye, Z., & Wassie, M. (2017). The role of microorganisms in bioremediation-A review. *Open Journal of Environmental Biology, 2*(1), 038–046.

Abdu, N., Abdullahi, A. A., & Abdulkadir, A. (2017). Heavy metals and soil microbes. *Environmental Chemistry Letters, 15*(1), 6584. Springer.

Aislabie, J., Deslippe, J. R., & Dymond, J. (2013). Soil microbes and their contribution to soil services. *Ecosystem Services in New Zealand–Conditions and Trends. Manaaki Whenua Press, Lincoln, New Zealand, 1*(12), 143–161.

Bhandari, S., Poudel, D. K., Marahatha, R., Dawadi, S., Khadayat, K., Phuyal, S., Shrestha, S., Gaire, S., Basnet, K., Khadka, U., & Parajuli, N. (2021). Microbial enzymes used in bioremediation. *Journal of Chemistry, 2021*. Hindawi Limited.

Bharagava, R. N., & Mishra, S. (2018). Hexavalent chromium reduction potential of *Cellulosimicrobium sp.* isolated from common effluent treatment plant of tannery industries. *Ecotoxicology and Environmental Safety, 147*, 102–109.

Chandrakar, V., Sahu, S., Khare, J., & Sathpathy, S. (2014). Removal of pb and cr by fungi in municipal sewage water. *Indian Journal of Scientefic Research, 4*, 214–217.

Chen, X., Christopher, A., Jones, J. P., Bell, S. G., Guo, Q., Xu, F., Roa, Z., & Wong, L. L. (2002). Crystal structure of the F87W/Y96F/V247L mutant of cytochrome P-450cam with 1,3,5-trichlorobenzene bound and further protein engineering for the oxidation of pentachlorobenzene and hexachlorobenezene. *Journal of Biological Chemistry, 277*, 37519–37526.

Choudhury, S., & Chatterjee, A. (2022). Microbial application in remediation of heavy metals: an overview. *Archives of Microbiology, 204*(5). Springer Science and Business Media Deutschland GmbH.

Christian, V., Shrivastava, R., Shukla, D., Modi, H. A., & Vyas, B. R. M. (2005). Degradation of xenobiotic compounds by lignin-degrading white-rot fungi: enzymology and mechanisms involved. *Indian Journal of Experimental Biology.*

Congeevaram, S., Dhanarani, S., Park, J., Dexilin, M., & Thamaraiselvi, K. (2007). Biosorption of chromium and nickel by heavy metal resistant fungal and bacterial isolates. *Journal of Hazardous Materials, 146*(1–2), 270–277.

Das, P., Mukherjee, S., & Sen, R. (2009). Biosurfactant of marine origin exhibiting heavy metal remediation properties. *Bioresource Technology, 100*(20), 4887–4890.

Das, N., & Chandran, P. (2011). Microbial degradation of petroleum hydrocarbon contaminants: An overview. *Biotechnology Research International.*

Ezzouhri, L., Castro, E., Moya, M., Espinola, F., & Lairini, K. (2009). Heavy metal tolerance of filamentous fungi isolated from polluted sites in Tangier, Morocco. *African Journal of Microbiology Research, 3*(2), 35–48.

Fomina, M., & Gadd, G. M. (2003). Metal sorption by biomass of melanin-producing fungi grown in clay-containing medium. *Journal of Chemical Technology and Biotechnology: International Research in Process, Environmental and Clean Technology, 78*(1), 23–34.

Gadd, G. M. (2010). Metals, minerals and microbes: Geomicrobiology and bioremediation. *Microbiology, 156*(3), 609–643.

Gnanamani, A., Kavitha, V., Radhakrishnan, N., Rajakumar, G. S., Sekaran, G., & Mandal, A. B. (2010). Microbial products (biosurfactant and extracellular chromate reductase) of marine microorganism are the potential agents reduce the oxidative stress induced by toxic heavy metals. *Colloids and Surfaces B: Biointerfaces, 79*(2), 334–339.

Gomaa, E. Z., & El-Meihy, R. M. (2019). Bacterial biosurfactant from Citrobacter freundii MG812314. 1 as a bioremoval tool of heavy metals from wastewater. *Bulletin of the National Research Centre, 43*, 1–14.

Gougoulias, C., Clark, J. M., & Shaw, L. J. (2014). The role of soil microbes in the global carbon cycle: Tracking the below-ground microbial processing of plant-derived carbon for manipulating carbon dynamics in agricultural systems. *Journal of the Science of Food and Agriculture, 94*(12), 2362–2371. Wiley.

Guerrero Ramírez, J. R., Ibarra Muñoz, L. A., Balagurusamy, N., Frías Ramírez, J. E., Alfaro Hernández, L., & Carrillo Campos, J. (2023). Microbiology and Biochemistry of Pesticides Biodegradation. *International Journal of Molecular Sciences, 24*(21), 15969.

Guo, W., Hu, W., Pan, J., Zhou, H., Guan, W., Wang, X., et al. (2011). Selective adsorption and separation of BPA from aqueous solution using novel molecularly imprinted polymers based on kaolinite/Fe_3O_4 composites. *Chemical Engineering Journal, 171*(2), 603–611.

Herman, P. L., Behrens, M., Chakraborty, S., Chrastil, B. M., Barycki, J., & Weeks, D. P. (2005). A three-component dicamba O-demethylase from *Pseudomonas maltophilia*, strain DI-6: Gene isolation, characterization, and heterologous expression. *Journal of Biological Chemistry, 280*(26), 24759–24767.

Huang, D. L., Wang, R. Z., Liu, Y. G., Zeng, G. M., Lai, C., Xu, P., et al. (2015). Application of molecularly imprinted polymers in wastewater treatment: a review. *Environmental Science and Pollution Research, 22*, 963–977.

Huang, W., & Liu, Z. M. (2013). Biosorption of Cd (II)/Pb (II) from aqueous solution by biosurfactant-producing bacteria: Isotherm kinetic characteristic and mechanism studies. *Colloids and Surfaces B: Biointerfaces, 105*, 113–119.

Huusko, K., Ruotsalainen, A. L., & Markkola, A. M. (2017). A shift from arbuscular mycorrhizal to dark septate endophytic colonization in *Deschampsia flexuosa* roots occurs along primary successional gradient. *Mycorrhiza, 27*(2), 129–138.

Ichida, J. M., Krizova, L., LeFevre, C. A., Keener, H. M., Elwell, D. L., & Burtt, E. H., Jr. (2001). Bacterial inoculum enhances keratin degradation and biofilm formation in poultry compost. *Journal of Microbiological Methods, 47*(2), 199–208.

Jayaram, S., Ayyasamy, P. M., Aiswarya, K. P., Devi, M. P., & Rajakumar, S. (2022). Mechanism of microbial detoxification of heavy metals: A review. *Journal of Pure and Applied Microbiology, 16*(3), 1562–1574.

Jellouli, K., Ghorbel-Bellaaj, O., Ayed, H. B., Manni, L., Agrebi, R., & Nasri, M. (2011). Alkaline-protease from *Bacillus licheniformis* MP1: Purification, characterization and potential application as a detergent additive and for shrimp waste deproteinization. *Process Biochemistry, 46*(6), 1248–1256.

Kim, J. S., Park, J. W., Lee, S. E., & Kim, J. E. (2002). Formation of bound residues of 8-hydroxybentazon by oxidoreductive catalysts in soil. *Journal of Agricultural and Food Chemistry, 50*(12), 3507–3511.

Kumar, P., & Sachan, S. G. (2021). Exploring microbes as bioremediation tools for the degradation of pesticides. In *Advanced oxidation processes for effluent treatment plants* (pp. 51–67). Elsevier.

Kumar, A., & Sharma, S. (2019). *Microbes and enzymes in soil health and bioremediation* (pp. 353–366). Springer.

Kumaran, N., Sundaramanickam, A., & Bragadeeswaran, S. (2011). Absorption studies on heavy metals by isolated bacterial strain (*Pseudomonas sp.*) from uppanar estuarine water, southeast coast of India. *Journal of Applied Sciences in Environmental Sanitation, 6*(4).

Kunwar, S., Pandey, N., Bhatnagar, P., Chadha, G., Rawat, N., Joshi, N. C., et al. (2024). A concise review on wastewater treatment through microbial fuel cell: sustainable and holistic approach. *Environmental Science and Pollution Research, 31*(5), 6723–6737.

Lin, Z., Li, J., Luan, Y., & Dai, W. (2020). Application of algae for heavy metal adsorption: A 20-year meta-analysis. *Ecotoxicology and Environmental Safety, 190*. https://doi.org/10.1016/j.ecoenv.2019.110089

Luna, J. M., Rufino, R. D., & Sarubbo, L. A. (2016). Biosurfactant from *Candida sphaerica* UCP0995 exhibiting heavy metal remediation properties. *Process Safety and Environmental Protection, 102*, 558–566.

Mahmood, R., Sharif, F., Ali, S., Hayyat, M. U. (2015). *The Scientific World Journal*.

Malik, S., Kishore, S., Dhasmana, A., Kumari, P., Mitra, T., Chaudhary, V., & Rajput, V. D. (2023). A perspective review on microbial fuel cells in treatment and product recovery from wastewater. *Water, 15*(2), 316.

Mane, P. C., & Bhosle, A. B. (2012). Bioremoval of some metals by living algae *Spirogyra sp.* and *Spirullina sp.* from aqueous solution.

Mani, D., & Kumar, C. (2014). Biotechnological advances in bioremediation of heavy metals contaminated ecosystems: An overview with special reference to phytoremediation. *International Journal of Environmental Science and Technology, 11*, 843–872.

Massaccesi, G., Romero, M. C., Cazau, M. C., & Bucsinszky, A. M. (2002). Cadmium removal capacities of filamentous soil fungi isolated from industrially polluted sediments, in La Plata (Argentina). *World Journal of Microbiology and Biotechnology, 18*, 817–820.

Mayer, A. M., & Staples, R. C. (2002). Laccase: New functions for an old enzyme. *Phytochemistry, 60*(6), 551–565.

Medina, R., Gara, P. M. D., Fernández-González, A. J., Rosso, J. A., & Del Panno, M. T. (2018). Remediation of a soil chronically contaminated with hydrocarbons through persulfate oxidation and bioremediation. *Science of the Total Environment, 618*, 518–530.

Moreau, D., Bardgett, R. D., Finlay, R. D., Jones, D. L., & Philippot, L. (2019). A plant perspective on nitrogen cycling in the rhizosphere. *Functional Ecology, 33*(4), 540–552. Blackwell Publishing Ltd.

Mulbry, W. W., Zhu, H., Nour, S. M., & Topp, E. (2002). The triazine hydrolase gene trz N from *Nocardioides sp.* strain C190: cloning and construction of gene-specific primers. *FEMS Microbiology Letters, 206*(1), 75–79.

Muneer, B., JIqbal, M., Shakoori, F. R., & Shakoori, A. R. (2013). Tolerance and biosorption of mercury by microbial consortia: potential use in bioremediation of wastewater. *Pakistan Journal of Zoology, 45*(1).

Nahiun, K. M., Sarker, B., Keya, K. N., Mahir, F. I., Shahida, S., & Khan, R. A. (2021). A review on the methods of industrial waste water treatment. *Scientific Review, 7*(3), 20–31.

Raina, R., & Hall, P. (2008). Comparison of gas chromatography-mass spectrometry and gas chromatography-tandem mass spectrometry with electron ionization and negative-ion chemical ionization for analyses of pesticides at trace levels in atmospheric samples. *Analytical Chemistry Insights, 3*, ACI-S1005.

Ran, Z. H. A. O., Bi, W. A. N. G., Cai, Q. T., Li, X. X., Min, L. I. U., Dong, H. U., et al. (2016). Bioremediation of hexavalent chromium pollution by *Sporosarcina saromensis* M52 isolated from offshore sediments in Xiamen, China. *Biomedical and Environmental Sciences, 29*(2), 127–136.

Rick, W. Y., & Thomas, S. M. (2001). Microbial nitrogen cycles: Physiology, genomics and applications. *Current Opinion in Microbiology, 4*(3), 307–312.

Scott, C., Lewis, S. E., Milla, R., Taylor, M. C., Rodgers, A. J., Dumsday, G., et al. (2010). A free-enzyme catalyst for the bioremediation of environmental atrazine contamination. *Journal of Environmental Management, 91*(10), 2075–2078.

Sharma, S. (2012). Bioremediation: Features, strategies and applications. *Asian Journal of Pharmacy and Life Science, 2231*, 4423.

Sutherland, T., Russell, R., & Selleck, M. (2002). Using enzymes to clean up pesticide residues. *Pesticide Outlook, 13*(4), 149–151.

Ta-Chen, L., Chang, J. S., & Young, C. C. (2008). Exopolysaccharides produced by Gordonia alkanivorans enhance bacterial degradation activity for diesel. *Biotechnology Letters, 30*(7), 1201–1206.

Thatoi, H., Das, S., Mishra, J., Rath, B. P., & Das, N. (2014). Bacterial chromate reductase, a potential enzyme for bioremediation of hexavalent chromium: A review. *Journal of Environmental Management, 146*, 383–399.

Wagner, M., Loy, A., Nogueira, R., Purkhold, U., Lee, N., & Daims, H. (2002). Microbial community composition and function in wastewater treatment plants. *Antonie Van Leeuwenhoek, 81*, 665–680.

Wijerathna-Yapa, A., & Pathirana, R. (2022). Sustainable agro-food systems for addressing climate change and food security. *Agriculture, 12*(10), 1554.

Wiltschi, B., Cernava, T., Dennig, A., Casas, M. G., Geier, M., Gruber, S., et al. (2020). Enzymes revolutionize the bioproduction of value-added compounds: From enzyme discovery to special applications. *Biotechnology Advances, 40*, 107520.

Yaashikaa, P. R., Kumar, P. S., Jeevanantham, S., & Saravanan, R. (2022). A review on bioremediation approach for heavy metal detoxification and accumulation in plants. *Environmental Pollution, 301*.

Yang, Z., Shi, W., Yang, W., Liang, L., Yao, W., Chai, L., et al. (2018). Combination of bioleaching by gross bacterial biosurfactants and flocculation: A potential remediation for the heavy metal contaminated soils. *Chemosphere, 206*, 83–91.

Yin, K., Wang, Q., Lv, M., & Chen, L. (2019). Microorganism remediation strategies towards heavy metals. *Chemical Engineering Journal, 360*, 1553–1563.

Zhang, D., Yin, C., Abbas, N., Mao, Z., & Zhang, Y. (2020). Multiple heavy metal tolerance and removal by an earthworm gut fungus *Trichoderma brevicompactum* QYCD-6. *Scientific Reports, 10*(1), 6940.

Zu, L., Li, G., An, T., & Wong, P. K. (2012). Biodegradation kinetics and mechanism of 2, 4, 6-tribromophenol by *Bacillus sp.* GZT: A phenomenon of xenobiotic methylation during debromination. *Bioresource Technology, 110*, 153–159.

Microbial Secondary Metabolites Used Against Diabetes and Cancer—A Critical Appraisal

Sheetal Gupta, Wazen Longkumer, and A. Stephen

Abstract

Almost all living organisms synthesize secondary metabolites which allow them to survive under various conditions. To mankind, these secondary metabolites have proven to be pharmaceutically important. They possess various biological activities such as anticancer, antidiabetic, antioxidant and antimicrobial. In this review, we have discussed various details on microbes and their secondary metabolites. The mode of action of some of the antidiabetic secondary metabolites like Botryorhodine E, Aspergones, Neoechinulin, β-Caryophyllene, Carbazomycin B, (+)-Geodin and Aquastatin A and some anticancerous metabolites like sorbicillinoids, bleomycins, pluramycins, doxorubicin, romidepsin (dactinomycin) have been discussed in this chapter.

Keywords

Secondary metabolites · Anticancer · Antidiabetic · Actinomycetes · Biological activity

1 Introduction

The term microorganism includes a wide range of organisms, including eubacteria, archaebacteria, fungi, mycoplasmas, myxobacteria, viruses, etc. Microorganisms have existed on Earth for billions of years and have undergone several evolutionary changes to adapt to the ever-changing environmental conditions for survival. During this course of evolution, they have adapted various metabolic pathways to survive in varied conditions and compete with the other biotic communities (Bérdy, 2012).

Secondary metabolites are a group of low molecular mass compounds synthesized by living organisms that do not take part in primary metabolism and are not essential for the growth and development of the organism, but they help the plant to survive under various conditions by providing protection against biotic and abiotic stress. These secondary metabolites are used by mankind for industrial, agricultural, food and pharmacological purposes (Stephen & Kumaresan, 2020; Yadav, 2021). Microorganisms produce a wide array of these secondary metabolites, including some antibiotics, anti-tumor agents, anti-oxidants, anti-inflammatory compounds, hypocholesterolemic, pigments, toxins, immunomodulating agents, herbicides and pesticides (Ramírez-Rendon et al., 2022). There are nearly over 13,000 pharmaceutically important naturally occurring compounds that are of bacterial origins (Mohan et al., 2022). Some of the classical examples of secondary metabolites obtained from the microbial groups like *Streptomycetes* which synthesize a wide range of antibiotics like neomycins, erythromycins, streptomycins, tetracyclines, amphotericins, cypemycin, grisemycin, bottromycins and chloramphenicol (Bibb, 2013). These metabolites are extremely used to make drugs and cure some of the pertinent illness that occur in human beings. Diabetes and cancer share many common risk factors and by 2030, cancer will be the most common cause of mortality for those who have diabetes (Wang et al., 2023).

These authors contributed equally to this work—Sheetal Gupta and Wazen Longkumer.

S. Gupta · W. Longkumer · A. Stephen (✉)
Department of Botany, School of Life Sciences, St. Joseph's University, Bengaluru, India
e-mail: stephen@sju.edu.in

K. Sharma (ed.), *Bio-prospecting of Novel Microbial Bioactive Compounds for Sustainable Development*, Sustainable Landscape Planning and Natural Resources Management, https://doi.org/10.1007/978-3-031-82178-3_12

1.1 Diabetes

It is caused due to insulin not being produced by the pancreas or insulin sensitivity being lowered which makes the hormone less effective in reducing blood sugar levels (Marx, 2002). Blurred vision, frequent urination and fatigue are symptoms of a chronic illness known as diabetes. Enzymes necessary for sugar metabolism, such as α-amylase and β-glucosidase, serve as screening chemicals to determine whether substances have anti-diabetic properties (Bidon-Chanal et al., 2013).

Type 1 diabetes is said to be an autoimmune disease caused by the destruction of T-Cell mediated beta cells of the pancreas (Wang et al., 2016). It is believed to be caused due to genetic or environmental factors. Insulin deficiency caused by beta cell dysfunction and an increase in insulin resistance are the characteristics of type 2 diabetes (Chatterjee et al., 2017). Oxidative stress is vital in beta cell neogenesis, proliferation and survival. The beta cells are susceptible to oxidative stress caused by the insufficient expression of antioxidative enzymes and the production of reactive oxygen species (Wang & Wang, 2017). In accordance with this thought, antioxidative activity in metabolites can effectively reduce the deterioration of the beta cells of the pancreas.

1.2 Cancer

Cancer is the second most common cause of death worldwide, the first one being cardiovascular diseases. As a result of modern lifestyle changes, people are more exposed to carcinogens, which is resulting in a steady increase in cancer incidence worldwide (Stephen et al., 2017). In a broader sense, cancer refers to a group of diseases in which the body's normal cells lose their ability to contact inhibition and multiply uncontrollably, resulting in the formation of tumors. These tumors frequently metastasize and spread to different parts of the body, where they proliferate and affect various organs, impairing normal bodily functions and ultimately killing the patient.

This loss of the ability of contact inhibition is due to the DNA damage caused by various physical, biological and chemical factors. Physical factors include radiation, biological factors include oncogenic viruses like HPV (human papillomavirus), HBV (hepatitis B virus), HCV (hepatitis C virus), HIV-1 (human immunodeficiency virus-1), EBV (Epstein-Barr virus), Kaposi's sarcoma virus (Chen et al., 2014) and chemical factors include compounds like polycyclic aromatic hydrocarbons, arsenic, N- nitrosamines, dioxins, DDT, etc. (Soliman, 2018).

2 Microbial Secondary Metabolites as Therapeutic Agents

Microbes possess biologically active compounds that have a variety of functions that can help treat a multitude of diseases. Therapeutic properties of a compound can be deduced by studying the nature of various secondary metabolites and their action against biological systems. This can be in the form of inhibition of enzymes to prevent biological activity, increasing sensitivity to hormones, having cytotoxic effects on detrimental cells, disturbing membrane activity, arresting cell cycle and many other functions that will lead to prevention and control of diseases (Giurini et al., 2024).

2.1 Secondary Metabolites for the Treatment of Diabetes

Aquastatin A is derived from *Cosmospora sp.* have antidiabetic properties in the form of inhibition of protein tyrosine phosphatase 1B enzyme, which negatively regulates insulin signaling transduction (Table 1) (Fig. 1) (Seo et al., 2009), Other metabolites that act as PTP1B inhibitors are fumosorinone A, 6-O-methylalaternin, nordivaricatic acid and the divarinyl divarate (Deshmukh et al., 2023). Ethyl acetate extract of *Aspergillus terreus* MC751 was purified and isolated to obtain butyrolactone I, where the compound exhibited antidiabetic and antioxidant activity. It showed inhibition of α-glucosidase with an IC50 of 22.86 μg/mL. Endophyte *Aspergillus awamori* was isolated from *Acacia nilotica* and screened to obtain proteinaceous inhibitors for α-amylase and glucosidase (Singh & Kaur, 2016). 16 alkaloids were isolated and identified from *Aspergillus sp.*, a mangrove endophytic fungus. Out of 16, 6 are new alkaloids and named Aspergiamide A to F. Aspergiamide A & brevianamide K exhibited significant α-glucosidase inhibition activity (Ye et al., 2021). Catechol has antihyperglycemic activity and helps in carbohydrate metabolism. *Penicillium janthinellum* N29, a mangrove endophytic fungus, was studied for its diabetic activity against streptozotocin-induced mice. The exopolysaccharide from the fermented broth results in a reduction of blood glucose levels and improved glucose tolerance in the test organism. Moreover, an increase in insulin sensitivity and amelioration of insulin resistance was also observed (Shao et al., 2023). The biotransformation of catechol from catechin via microbes helps in the synthesis of this important compound. Biotransformation of catechol from catechin in bacteria is by *Lactobacillus hilgardii* (Alberto et al., 2004). *Chaetomium cupreum* (Sambandam & Mahadevan, 1993)

Table 1 List of antidiabetic microbial secondary metabolites and their source organism

Compound	Microorganism	Diabetes type	References
Aquastatin A	*Cosmospora sp.*	Diabetes type II	Seo et al. (2009)
Eurothiocins A and B	*Eurotium rubrum*	Diabetes type II	Liu et al. (2014)
Benthophoenin	*Streptomyces prunicolor*	Diabetes type 1	Shin-ya et al. (1993b)
Benthocyanins B and C	*Streptomyces prunicolor*	Diabetes type 1	Shin-ya et al. (1993a)
Carbazomycin B	*Streptomyces chromofuscus*	Diabetes type 1	Kato et al. (1993)
Exopolysaccharide	*Bacillus coagulans*	Diabetes type 1	Kodali and Sen (2008)
Ortho-coumaric acid	*Penicillium chrsogenum* var. *chrysogenum*	Diabetes type 1	Canturk et al. (2013)
Protocatechic acid	*Penicillium sp.*	Diabetes type 2	Kawakubo et al. (1993)
β-Caryophyllene	*Gilmaniella sp., Atractylodes lancea*	Diabetes type 2	Chen et al. (2016a, 2016b)
Zingiberene	*Gilmaniella sp., Atractylodes lancea*	Diabetes type 2	Chen et al. (2016a, 2016b)
β-eudesmol	*Gilmaniella sp., Atractylodes lancea*	Diabetes type 2	Chen et al. (2016a, 2016b)
4′,7,8-trihydroxyisoflavone	*Streptomyces sp.*	Diabetes type 1	Komiyama et al. (1989)
3′,4′,7-trihydroxyisoflavone	*Streptomyces sp.*	Diabetes type 1	Komiyama et al. (1989)
8-chloro-3′,4′,5,7-tetrahydroxyisoflavone	*Streptomyces sp.*	Diabetes type 1	Komiyama et al. (1989)
Exopolysaccharide	*Bacillus amyloliquefaciens*	Diabetes type 1	Chen et al. (2020)
Bioactive peptides, casein	*Bacillus sp.*	Diabetes type 1	Megrous et al. (2020)
Aspergiamide A	*Aspergillus sp.*	Diabetes type 2	Ye et al. (2021)
Neoechinulin	*Aspergillus repens* MA0197	Diabetes type 1	Yagi and Doi (1999)
Brevianamides	*Aspergillus sp.* *Aspergillus versicolor*	Diabetes type 2	Ye et al. (2021)
Benzomalvin A	*Penicillium spathulatum*	Diabetes type 2	Hussain et al. (2021)
Asperphenamate	*Penicillium spathulatum*	Diabetes type 2	Hussain et al. (2021)
Quinolactacins A1	*Penicillium spathulatum*	Diabetes type 2	Hussain et al. (2021)
Expansolide	*Penicillium expansum*	Diabetes type 2	Hussain et al. (2021)
Penicidone C	*Penicillium sp.* IFB-E022	Diabetes type 2	Hussain et al. (2021)
3,300-dihydroxy-60-O-desmethylterPhenyllin	*Penicillium chermesinum*	Diabetes type 2	Huang et al. (2011) Hussain et al. (2021)
Aspergones	*Aspergillus sp.* OCMD-1583	Diabetes type 2	Kong et al. (2015)
Botryorhodine E	*Meyerozyma guilliermondii, Kandelia obovata*	Diabetes type 2	Hussain et al. (2021)
Eucalyptol	*Nodulisporium sp. GS4d2II1a*	Diabetes type 1	Kim et al. (2020)
(+)-Geodin	*Penicillium chrysogenum SCSIO 41,001, Penicillium glabrum AJ117540*	Diabetes type 2	Hussain et al. (2021)

and *Aspergillus fumigatus* (Wang et al., 2014) are used for the biotransformation of tea polyphenols to bioactive theabrownins, a compound that can reverse insulin resistance (Liu et al., 2023).

Compounds JBIR-94 and JBIR-125 were isolated from fermented broths of *Streptomyces sp.* R56-07, a novel species of *Streptomyces*. The IC50 value of DPPH Radical-Scavenging Activity was 11.4 and 35.1 μM, respectively. This shows that the compounds have high antioxidative activity. Compounds such as benthocyanins A, B, C, and benthophoenin which are isolated from *Streptomyces prunicolor* also exhibit antioxidative activity. GancidinW was extracted from *Streptomyces paradoxus* VITALK03 and showed α-amylase and α-glucosidase inhibitory activity. It also prevented hemoglobin glycosylation and increased glucose uptake in yeast cells (Ravi et al., 2017). *Streptomyces* strain PM0324667 produced the simple aromatic compound NFAT-133. It induced glucose uptake in L6 skeletal muscle cells and reduced systemic glucose levels in diabetic animal models (Kulkarni-Almeida et al., 2011).

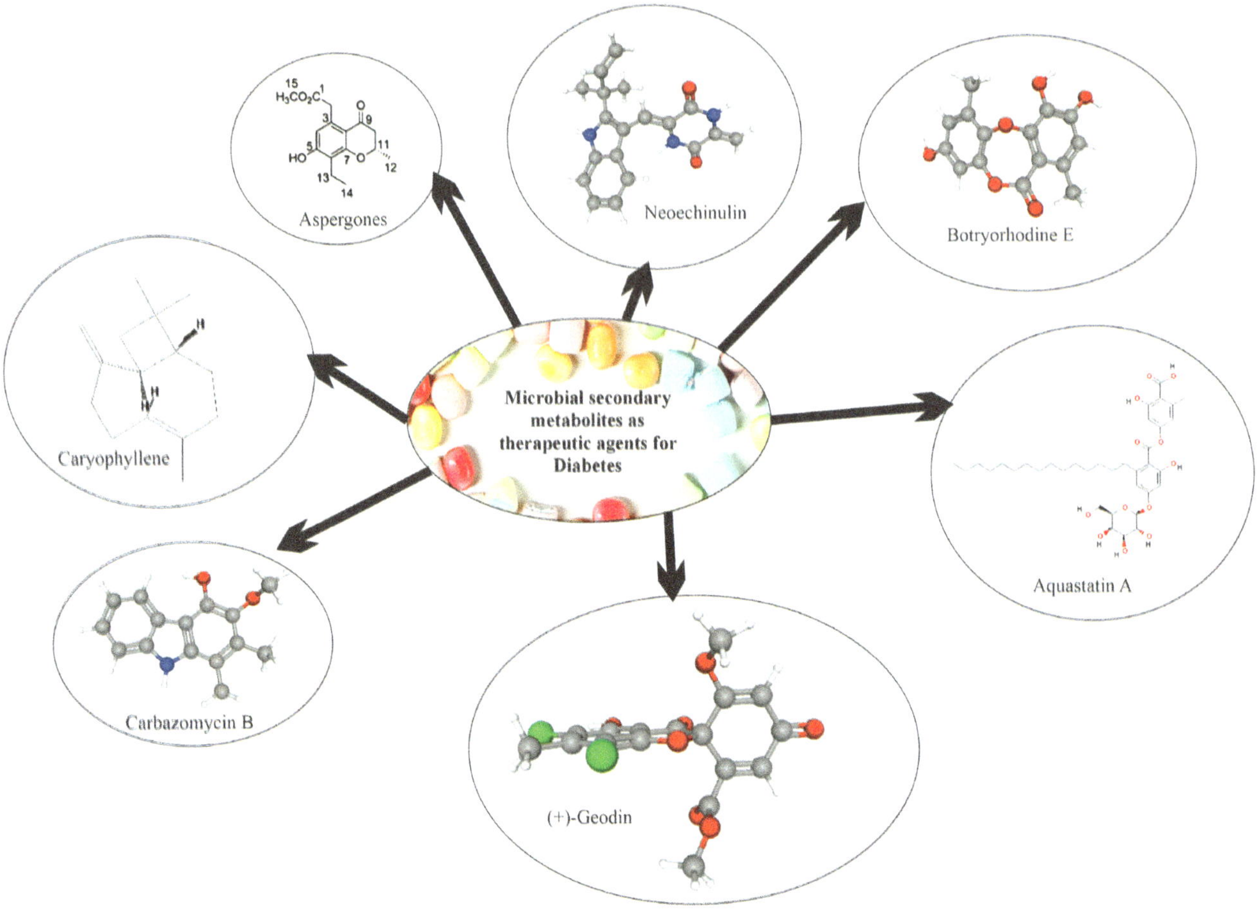

Fig. 1 Secondary metabolites from microbial origin and used as therapeutic agents for diabetes

Aspergillus saitoi aided in the hydroxylation of aglycones which were previously converted from the plant extracts of *Citrus grandis* (L.) Osbeck. This interaction allowed for a higher antidiabetic and antioxidant activity to be exhibited compared to their glycosides and aglycones (Kim et al., 2009). TNF-α-treated L6 cells were treated with *Bifidobacterium lactis* HY8101. It presented improvement in insulin-stimulated glucose uptake and translocation of GLUT4. It decreased fasting insulin and blood glucose (Kim et al., 2014).

Extracellular polysaccharides (EPS) from *Bacillus coagulans* possessed significant antioxidant activity when compared to standard antioxidants (Kodali & Sen, 2008). Bioactive peptides with antidiabetic and antioxidant properties were obtained via hydrolysis of casein by *Bacillus sp.* metalloendopeptidase (Megrous et al., 2020). Exopolysaccharide (EPSs) from *Bacillus amyloliquefaciens* contain glucose, galactose and mannose. They reduced blood glucose elevation and enhanced glucose uptake via Akts (Chen et al., 2020). Soybean extract is fermented by *Bacillus subtilis* MORI, and it has shown antidiabetic properties in non-insulin-dependent 2TD mouse models (Nam et al., 2012). *Bacillus licheniformis* has anti-α-glucosidase properties (AAG), it the protein TKU004P was obtained, which, in comparison with acarbose, a commercial drug antidiabetic drug, shows stronger AAG values (Doan et al., 2019). Eight strains of *Lactobacillus paracasei* were assessed where the strain *Lactobacillus paracasei* TD062 had shown high α-glucosidase inhibitory activity. It was administered to diabetic rats which further exhibited regulation of the levels of fasting blood glucose, glucose tolerance, postprandial blood glucose and lipid metabolism (Dang et al., 2018). *Clostridium butyricum* CGMCC0313.1 acts as a probiotic for hyperglycemia by increasing insulin sensitivity, reducing pancreatic damage and enhancing insulin signaling in HF diabetic mice (Jia et al., 2017). *Streptococcus salivarius* subsp. *thermophiles* fmb5 fermented γ-Aminobutyric acid (GABA)-rich yogurt. This exhibited improvement of impaired glucose tolerance, enhanced levels of serum insulin and restored pancreas and ameliorated hyperglycaemia of streptozotocin-induced diabetic mice (SIDM) (Chen et al., 2016a, 2016b).

Seven strains obtained from 55 marine actinomycetes were found to have antidiabetic activity. The strain with the highest potential was *Streptomyces coelicolor*. This microorganism, when exposed to UV radiation, underwent genetic improvement in the form of an inhibitory protein being highly active and remaining stable for 5 days at 37 °C. Inoculum size, incubation period, incubation temperature, agitation speed, and pH were the factors that enabled the high activity and stability of the protein (Abdulkhair et al., 2018).

2.2 Secondary Metabolites for the Treatment of Cancer

Various bioactive compounds isolated from fungi show significant anticancer properties targeting various processes involved in the onset of cancer (Banerjee & Paruthy, 2017) (Fig. 2). Like any other organism, fungi also produce various secondary metabolites which can be further explored for pharmaceutical applications. Studies done on some of these fungal metabolites indicate that they can be a useful source of antimicrobial agents, tumor-suppressing agents, anti-oxidants, anti-diabetic agents and various other important compounds (Elkhouly et al., 2021). Sorbicillin-derived compounds, dihydrotrichodimerol, bislongiquinolide, trisorbicillinoids, sorbicillamines, rezishanone and bisvertinol are often isolated from the fungal sources, are known to inhibit the apoptosis resistance and cause cytotoxicity to the cancerous cells (Balde et al., 2010; Guo et al., 2013; Li et al., 2010a, 2010b; Maskey et al., 2005).

Aspergillus tubinginses synthesizes anofinic acid, an endophyte isolated from *Hyoscyamus muticus*, a medicinal plant, which has strong antiproliferative activity against hepatocellular, colon and breast cancer (Elkhouly et al., 2021). Three compounds isolated from an ascomycete *Epicoccum nigrum*, endophytic fungi found in *Ferula sumbul* showed inhibitory activity on two of the human melanoma cell lines (Perveen et al., 2017). *Taxomyces andreanae*, an endophytic fungus isolated from the plant *Taxus brevifolia*, secretes taxol. This is extensively used in chemotherapy (Stierle et al., 1993). Sclerotiorin which is synthesized by *Penicillium sclerotiorum* and endophyte *Cephalotheca faveolata* induces cytotoxicity to the colon

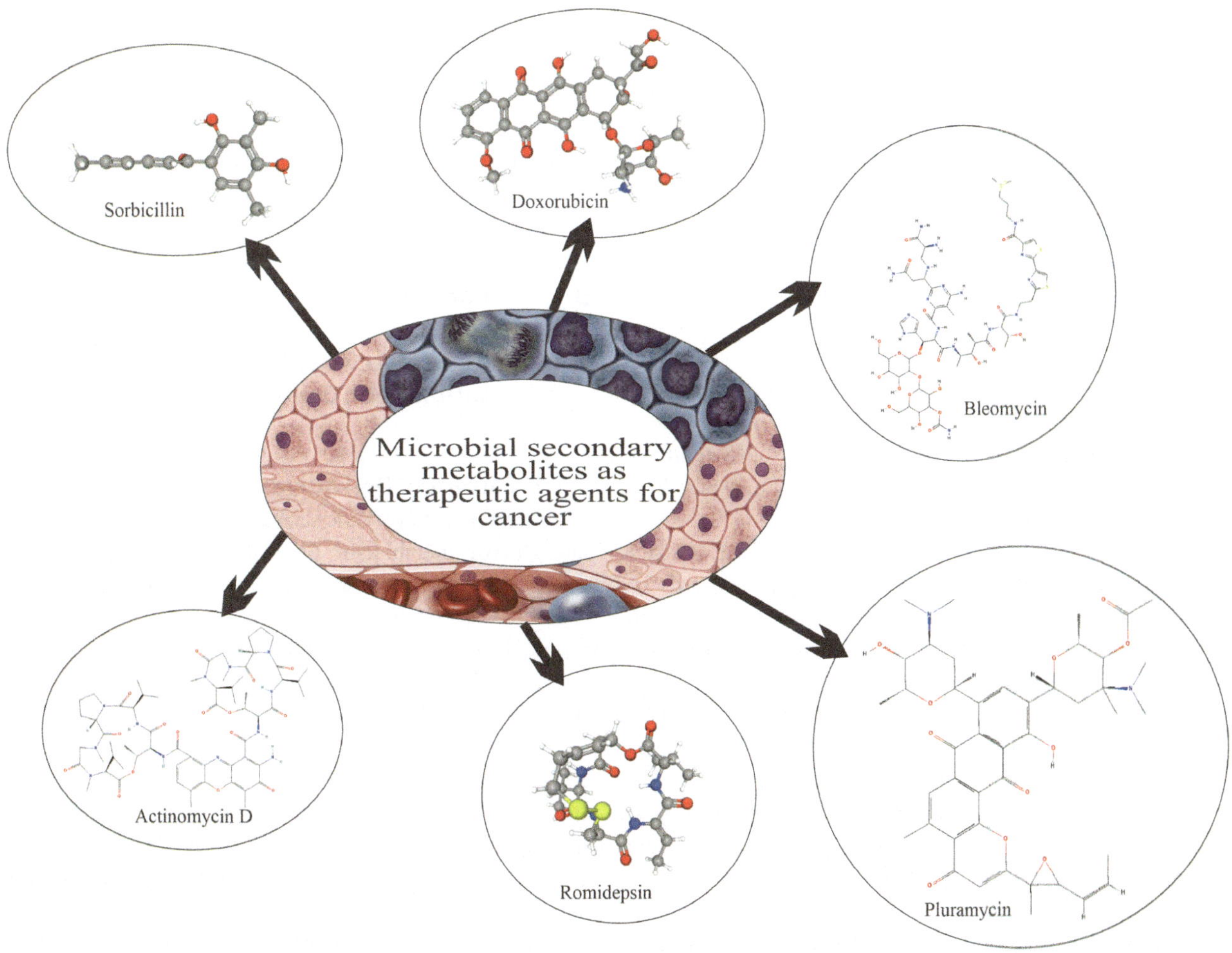

Fig. 2 Secondary metabolites from microbial origin and used as therapeutic agents for cancer

cancer cells indicated by an increase in lactate dehydrogenase levels (Giridharan et al., 2012). Camptothecin, one of the most effective biological chemotherapeutic agents, three endophytic fungi *Alternaria alternata, Fomitopsis sp.* and *Phomopsis* sp. isolated from *Miquelia dentata* synthesize its derivatives (Shweta et al., 2013). Derivatives of anthracenedione isolated from mangroves endophytes *Guignardia sp.* and *Halorosellinia sp.* showed cytotoxicity (Zhang et al., 2010). *Penicillium oxalicum* showed anticancer action by disrupting the integrity of the membrane which is an endophytic fungus isolated from *Chaetomorpha antenna*, a green alga (Parthasarathy et al., 2020). Endophytic *Aspergillus sp.* isolated from various plants like *Gloriosa superba, Hyoscyamus muticus, Taxus cuspidata, Rosa damascena* produce compounds like colchatetralene, anofinic acid, taxol, 2-Phenylethanol respectively which showed cytotoxicity against various cell lines (Budhiraja et al., 2013; Elkhouly et al., 2021). Some of the secondary metabolites and their respective endophytic fungi include depsipeptide PM181110 synthesized by *Phomopsis glabrae* (Verekar et al., 2014), griffithiiene from *Hypomontagnella monticulosa* (Lutfia et al., 2021).

The members of the genus *Streptomyces* are widely known for the production of many pharmaceutically important secondary metabolites. They also show activity against cancer cell lines, for instance in colon cancer cell lines, Tangerina et al. (2020) used extracts made from *Streptomyces sp.* demonstrated positive growth inhibition, and examination of the extract's chemical makeup revealed that surugamides and a few other chemicals may have contributed to the growth suppression. Due to their tendency to generate unique compounds that enable them to survive in severe environments, microorganisms from harsh habitats present enormous prospects for the study of pharmaceutically significant drugs. Compounds like streptocarbazoles, azalomycines, elaiophylins isolated from *Streptomyces sp.* collected from mangrove soil samples also show significant cytotoxicity effects on cancer cell lines (Fu et al., 2012; Han et al., 2016; Yuan et al., 2011).

Some of the marine actinomycetes from oceanic sediments like *Salinispora arenicola, Marinispora sp., Nocardiopsis lucentensis, Streptomyces sp.* produces antibiotics saliniquinones A-F, marinomycins A-D, lucentamycins A-D, piperazimycins respectively which also shows potential activity against human colon adenocarcinoma cell lines (Cho et al., 2007; Kwon et al., 2006; Murphy et al., 2010). Various other antitumor compounds synthesized by actinomycetes include rubiflavin, kidamycin, hedamycin, largomycin, hydramycin, altromycins, and pluraflavins (Heo et al., 2022; Lee et al., 2023).

Bacillus is recognized for producing several significant secondary metabolites, some of which have anticancer characteristics. For instance, cytotoxicity against HL-60 cell lines was observed by a lipopeptide, mojavensin which was isolated from *Bacillus mojavensis* (Ma et al., 2012) (Table 2). Exo-polysaccharides synthesized by marine Bacilli, *Bacillus flexus, Bacillus megaterium* and *Bacillus subtilis* revealed inhibitory activity against various cancer cell lines (Abdelnasser et al., 2017; Syed et al., 2019). Cytotoxicity against colon cancer cell lines has been demonstrated by cyclic acyl peptides, mixirins, isolated from *Bacillus sp.* (Agrawal et al., 2017; Zhang et al., 2004). Another anticancer compound, echercharmycin A with strong activity against lung cancer and leukemia cell lines, was isolated from *Thermoactinomyces sp.* (Agrawal et al., 2017; Kanoh et al., 2005). Inhibitory activity against lung cancer was observed by the activity of biosurfactants synthesized by *Acinetobacter indicus* (Karlapudi et al., 2020). *Bacillus licheniformis* obtained from the Red Sea, with significant production of L-asparaginase, a chemotherapeutic drug, also showed antiproliferative activity against liver, breast and colon cancer cell lines (Alrumman et al., 2019). The metabolites from *Bacillus subtilis*, from deer gut, showed potential cytotoxic activity against lung cancer cell lines (Shao et al., 2021) and sea origin isolate showed activity of ε-poly-L-lysine against cervical cancer cell lines (El-Sersy et al., 2012). A probiotic bacteria helpful in GI-related issues, *Bacillus polyfermenticus* also showed a reduction in tumor cells for three of the colon cancer cell lines (Ma et al., 2010). Matsuo et al., (2007), isolated a potential cytotoxic compound urukthapelstatin A from *Mechercharimyces asporophorigenens*.

3 Mode of Action

3.1 Mode of Action of Some Microbial Antidiabetic Compounds (Fig. 3)

3.1.1 Carbazole Derivatives

Carbazole is a polycyclic aromatic hydrocarbon that regulate glucose metabolism, prevention of damage to β cells of the pancreas, oxidative mediation and inhibition of α-glucosidase are a few instances of antidiabetic activities reported on carbazole derivatives. Carvedilol, a derivative of carbazole, can increase insulin receptor sensitivity (Grande et al., 2022).

3.1.2 Aspergones

It is a chromanone derivative that was isolated from the fermentation broth of *Aspergillus sp.* OUCMDZ-1583. It possessed α-glucosidase inhibitory effects and demonstrated stronger α-glucosidase inhibition than acarbose which acted as a control compound (Kong et al., 2015).

Table 2 List of anticancer microbial secondary metabolites and their source organism

Compound	Microorganism	Cell line type	Clinical use	References
Bleomycins	*Streptomyces verticillus, Streptomyces mobaraensis*	Germ cell tumors, squamous cell carcinomas and lymphomas	Blenoxane	Giddings and Newman (2013), Hindra et al. (2017)
Lucentamycins	*Nocardiopsis lucentensis*	Colon cancer		Cho et al. (2007)
Mechercharmycin, Echercharmycin A	*Thermoactinomyces sp.* YM3-251	Leukaemia, lung cancer		Kanoh et al. (2005)
L-asparaginase	*Bacillus licheniformis, Escherichia coli, Pseudomonas fluorescens, Streptomyces venezuelae*	Liver cancer, breast cancer, colorectal cancer, cervical cancer, leukaemia	Asparaginase	Alrumman et al. (2019)
Halichoblelide	*Streptomyces* sp. 219,807	Breast cancer, cervical cancer		Han et al. (2016)
Colchatetralene	*Aspergillus Tubenginses* ASH4	Breast cancer, hepatocellular cancer, colon cancer		Elkhouly et al. (2021)
Mojavensin	*Bacillus mojavensis, Gordonia sp.*	Leukaemia, nasopharyngeal carcinoma, hepatocellular carcinoma		Ma et al. (2012), Ma et al. (2021)
Piperazimycins	*Streptomyces* sp., *Streptomyces alboflavus, Streptomyces hygroscopicus*	Colon cancer		Cai et al. (2013)
Actinomycin-D	*Streptomyces griseoruber, Streptomyces smyrnaeus, Streptomyces hydrogenans, Gordonia sp.*	Hepatocellular carcinoma, nasopharyngeal carcinoma	Dactinomycin	Qureshi et al. (2021), Ma et al. (2021)
Griffithiiene	*Hypomontagnella monticulosa*	Pancreatic cancer, bladder cancer, colon cancer		Lutfia et al. (2021)
Bisorbicillinoids (dimeric sorbicillinoids)	*Trichoderma citrinoviride, Phialocephala sp.*	Lymphomas, leukaemia, hepatocelleular carcinoma		Balde et al. (2010)
Mitomycin- C	*Streptomyces griseoviridis, Streptomyces lavendulae*	Prostate cancer, lung carcinoma	Mutamycin, Mitosol	Saleem et al. (2023)
Daunorubicin	*Streptomyces peucetius*	Lymphomas	Cerubidine, Vyxeos	Giddings and Newman (2013)
Doxorubicin	*Streptomyces peucetius* var. *caesius*	Breast cancer, soft tissue sarcomas, lymphomas respectively	Doxorubicin	Giddings and Newman (2013)
Calicheamicin	*Micromonospora echinospora* subsp. *calichensis*	Leukemia		Giddings and Newman (2013)
Rapamycin	*Streptomyces hygroscopicus*	Brain, breast, kidney and pancreatic cancer	Everolimus, zotarolimus, sirolimus, temsirolimus, and	Giddings and Newman (2013)
Anthracenediones	*Guignardia sp., Halorosellinia sp., Streptomyces olivochromogenes, Fusarium solani*	Leukaemia, lung cancer, colon cancer		Zhang et al. (2010), Balachandran et al. (2016)
Geldanamycin	*Streptomyces hygroscopicus* var. *geldanus*	Thyroid cancer		Giddings and Newman (2013)
Staurosporines	*Streptomyces staurosporeus, Streptomyces sp.* QD518, *Actinomadura* sp. 007, *Streptomyces sanyensis*	Small cell lung cancer, soild tumors, bile duct tumors, leukemia, renal cancer, liver cancer etc.	Staurosporine	Giddings and Newman (2013), Cartuche et al. (2019)

(continued)

Table 2 (continued)

Compound	Microorganism	Cell line type	Clinical use	References
Romidepsins	*Chromobacterium violaceum, Burkholderia rinojensis, Burkholderia chassis*	Relapsed peripheral T-cell lymphoma, cutaneous cell lymphoma, Lung cancer	Romidepsin	Giddings and Newman (2013), Gong et al. (2023), Karthik et al. (2014)
Dolastatins	*Dolabella auricularia, Symploca sp.*	Pancreatic and kidney cancer		Giddings and Newman (2013)
Trabectedin	*Ecteinascidia turbinata*	Soft tissue sarcomas	Trabectedin	Giddings and Newman (2013)
Kahalalide F	*Vibrio mediterranei/shilonii*	Prostate cancer		Giddings and Newman (2013)
Salinosporamides	*Salinispora tropica*	Colon cancer, melanoma, breast cancer, lung cancer	Clinical trials	Giddings and Newman (2013)
Wortmannins	*Talaromyces wortmannin, Penicillium radicum*	Lung cancer, colon cancer	No clinical trials because of its toxicity and low stability	Boehle et al. (2002)
Pluramycins	*Streptomyces sp., Saccharothrix sp.* DSM 12,931, *Rhodococcus sp.*	Colon cancer		Lee et al. (2023)
Sclerotiorins	*Penicillium sclerotiorum*	Colon cancer		Giridharan et al. (2012)
Flavipin	*Chaetomium globosum*	Lung, human colorectal adenocarcinoma, breast cancer cell		Senthilkumar et al. (2019)
Pyrrole (1, 2, a) pyrazine. 1, 4, dione, hexahydro 3-(2-methylpropyl)	*Staphylococcus sp.* strain MB30	Lung and cervical (HeLa) cancer cells		Lalitha et al. (2016)
Fumigaclavine C	*Aspergillus fumigatus*	Breast cancer cells		Li et al. (2013)
Meroditerpenoids	*Stypopodium flabelliforme*	Human colon adenocarcinoma, human neuroblastoma, rat basophilic leukemia, murine macrophages and Chinese hamster fibroblasts		Pereira et al. (2011)

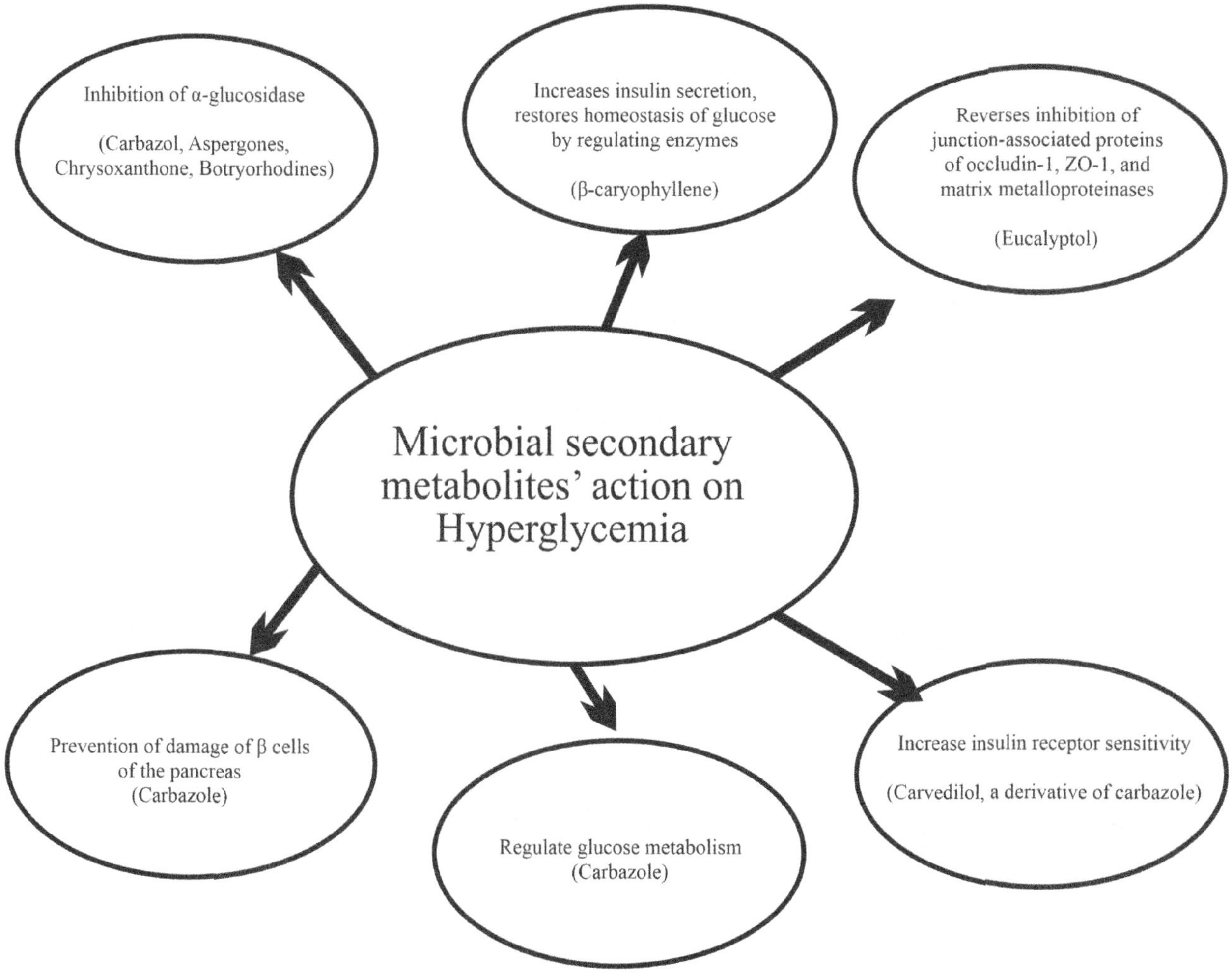

Fig. 3 Mode of action of different microbial secondary metabolites in treating hyperglycemia

3.1.3 β-caryophyllene

A bicyclic sesquiterpene increases insulin secretion and restores homeostasis of glucose by regulating enzymes that are found in glucose production. Experiments were performed on streptozotocin (STZ) induced diabetic rats. HbA1c is a biomarker to monitor glucose-lowering therapy. After treatment of β-caryophyllene (BCP), HbA1c levels decreased while hemoglobin levels increased which resulted in a decrease in fasting blood sugar levels. Gluconeogenic enzymes in diabetic conditions increase in activity due to reduction of insulin levels which leads to endogenous glucose production. Oral administration of BCP reverses gluconeogenic enzyme activity (Basha & Sankaranarayanan, 2014).

3.1.4 Chrysoxanthone

It is a heterodimer of the anthraquinone 2-hydroxychrysophanol and the tetrahydroxanthone blennolide A linked through a diaryl ether bridge (Schueffler & Anke, 2014). Obtained from *Penicillium chrysogenum*, it exhibits α glucosidase inhibition at IC50 = 0.04 mM. Inhibition of α-glucosidase leads to carbohydrates not being absorbed by the small intestine. Postprandial hyperglycemia and hyperinsulinemia are decreased and insulin sensitivity is increased (Scheen, 2003).

3.1.5 Eucalyptol

Eucalyptol (1,8-cineole), a terpenoid oxide, was isolated from *Nodulisporium sp.*, an endophyte present in *Eucalyptus sp.* Hyperglycemia causes tight junction disruption and blood-retinal barrier breakdown, leading to vision loss. In this regard, eucalyptol reverses the inhibition of junction-associated proteins of occludin-1, ZO-1, and matrix metalloproteinases in αβ-exposed RPE cells and in diabetic rat eyes. It weakened the generation of reactive oxygen species as well (Kim et al., 2020).

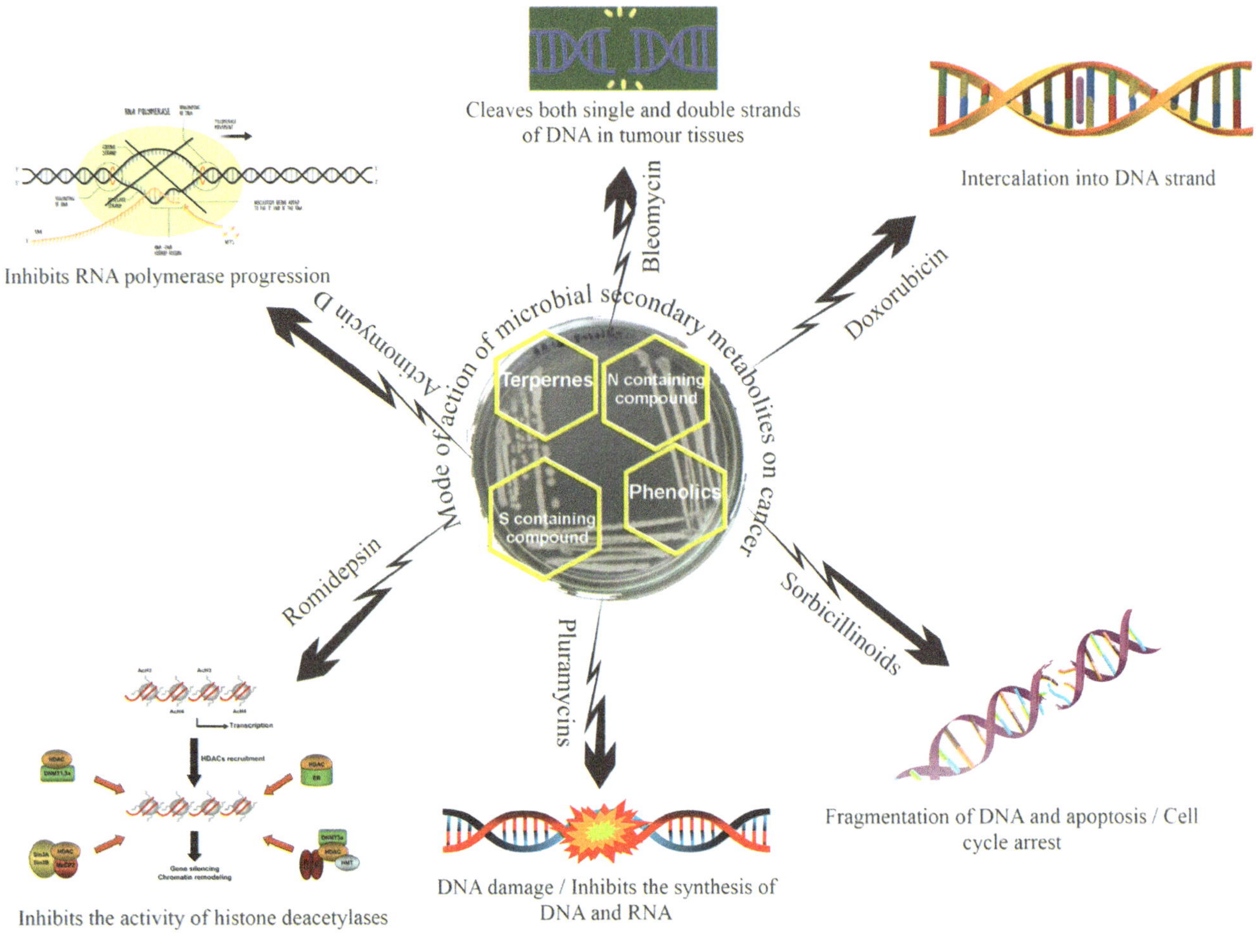

Fig. 4 Mode of action of different microbial secondary metabolites in treating cancer

3.1.6 Botryorhodines

A depsidone derivative which exhibits strong α-glucosidase inhibitory activity especially Botryorhodine D with IC50: 2.1 μM. It was obtained from mangrove endophytic fungi *Meyerozyma guilliermondii*. The kinetic study of Botryorhodine F revealed it to be a noncompetitive inhibitor of α-glucosidase (Hussain et al., 2021).

3.2 Mode of Action of Some Microbial Anticancer Compounds

Mode of action of different microbial secondary metabolites in treating cancer is presented in Fig. 4.

3.2.1 Sorbicillinoids

Sorbicillinoids include a vast range of bioactive compounds which are known to have antiproliferative activity against cancerous cells. Fungal sources of this compound include genera like *Penicillium, Verticillium, Trichoderma, Aspergillus, Phialocephala* and many more (Guo et al., 2013; Maskey et al., 2005; Li et al., 2010a, 2010b). Yao et al. (2015) studied the effect of sorbicillin in the leukemia cell lines for its apoptotic activity, results indicate that the compound increased the caspase 3 and caspase 7 activity which is responsible for the fragmentation of DNA and apoptosis, also the activity of sorbicillin showed some dependency on the presence of reactive oxygen species. Another study on sorbicillinoids isolated from marine *Penicillium allii-sativi* on colon cancer cell lines indicated the increase in cyclin B1 and phosphohistone H3 which induces cell cycle arrest at G2-M phase (Xie et al., 2021).

3.2.2 Bleomycins

Bleomycins is a glycopeptide antibiotic that is commonly used as FDA FDA-approved chemotherapeutic drug as a part of combination drug therapy, for treating various cancer conditions including lymphatic cancer, head and neck cancers and testicular cancers (Bennett & Reich, 1979). Bleomycins alone do not participate in the anti-tumor activity, the combined form of Bleomycins with metal ion Fe^{+2} and reactive oxygen species, cleaves both single and double strands of DNA in tumor tissues at preferably specific active gene sites which are associated with cell cycle arrest, ultimately inhibiting the growth of the tumor cells (Hecht, 2000; Murray et al., 2018).

3.2.3 Pluramycins and Its Derivatives

A broad class of substances synthesized by Actinomycetes, pleuramycins have anthrapyranones as their fundamental structural component (Lopez et al., 2022). Studies show that the cytotoxic activity of the compound is by virtue of the presence of epoxide at the second position of pluramycins. It is indicated that they get incorporated between the base pairs into the grooves and act as guanine electrophilic alkylators at the N7 position resulting in DNA damage in the tumor cells and interruption in the cell cycle (Sun et al., 1995). Another work done to study the activity of pluramycin, indicated that the compound inhibits the synthesis of nucleic acids like DNA and RNA (Nagai et al., 1967).

3.2.4 Doxorubicin

Doxorubicin is one of the clinically used anticancer drugs which belongs to the anthracycline family of antibiotic drugs. It is a tetracyclic compound synthesized by actinomycete *Streptomyces peucetius* (Giddings & Newman, 2013). The anticancer properties of the non-specific cell cycle agent, doxorubicin, due to its ability to block the activity of topoisomerase II, which is an important enzyme involved in various functions of the cell cycle (McClendon & Osheroff, 2007). Another way the compound shows antiproliferative activity is by disrupting the DNA replication by intercalating itself between the DNA bases.

3.2.5 Romidepsin

A bicyclic peptide, romidepsin is a FDA approved natural chemotherapeutic drug used against T-cell Lymphoma; it is isolated from microbes like *Chromobacterium violaceum* and *Burkholderia sp.* (Marchi et al., 2017). Romidepsin acts by inhibiting the activity of histone deacetylases, a group of enzymes that are involved in cell growth and cell proliferation; overexpression of HDACs due to mutation can result in tumor formation and its growth (Gong et al., 2023). Romidepsin not only inhibits HDACs but also downregulates the activity of matrix metalloproteinases (MMPs), which play an important role in metastasis and growth of cancerous tumors. It also promotes the synthesis of proteins like p21, p53 and caspase-3 (Karthik et al., 2014). The enzyme p21 is a cyclin-dependent kinase inhibitor which, on binding with kinases, arrests the cell cycle (Abbas & Dutta, 2009), whereas p53 is a tumor suppressor protein (Ryan et al., 2001) and caspase-3 (Zhou et al., 2018) promotes cell apoptosis.

3.2.6 Actinomycin D

Actinomycin D, or dactinomycin, is a clinically used anticancer and antimicrobial drug. Initial studies indicate that the compound shows inhibitory effects on transcription by interacting with DNA to form a stable complex; this inhibitory effect on transcription also results in insufficient protein synthesis due to the lack of ribosomal RNA, indirectly affecting cell proliferation. Other possible processes resulting in anti-tumor activity could be due to its ability to stabilize the topoisomerases-DNA complex and intercalation within the DNA bases (Koba & Konopa, 2005).

4 Discussion

Due to the importance of secondary metabolites, various pathways have been genetically altered using gene-encoding biosynthesis enzymes or regulatory proteins. This has been done to increase or decrease the quantity of certain compounds. In cases where the amount of a compound needs to be reduced, an enzymatic step in a pathway is halted using methods such as RNA interference technology, enzymes being suppressed due to the overexpression of antibodies. Antisense genes block competitive pathways, increasing the metabolic flux, which refers to the rate of turnover of molecules through a metabolic pathway (Verpoorte & Memelink, 2002).

Certain well-known microbial metabolites, such as bleomycin and doxorubicin, have proven to be highly effective in killing cancer cells and are frequently included in cancer therapy regimens. These well-known, microbiologically produced chemicals frequently have serious negative toxicities attached to them. Owing to these toxicities, the compound's treatment is frequently discontinued, abbreviated, or the dosage is decreased to the point where it no longer appreciably slows the patient's illness development (Giurini et al., 2024).

These secondary metabolites from microorganisms have already shown clinical promise, and it is expected that they will be improved by adding new targets for tumor antigens and more potent microbial metabolites. Only a small portion of the bioactive substances contained in microbes that have the potential to treat human illnesses, such as cancer, are represented by this fascinating generation of secondary metabolites, many more yet to be tried.

5 Conclusion

Various microbial organisms produce metabolites that enable them to operate in their respective environments. The metabolites exhibit a multitude of medicinal functions which have been discussed to an extent. The idea of using microbial metabolites functions on the basis of securing compounds that are not actively or chemically synthesized. This is done via culturing microbes in bioreactors to obtain large quantities of the organism, which is then refined to procure the said compound, transforming the genetic constitution of the microbe to either increase expression or production of the desired compound. All of these methods are done in order to obtain compounds that have medicinal qualities. Some of the secondary metabolites which have been clinically used are bleomycins, dactinomycin, mitomycin-C, daunorubicin, rapamycin, staurosporine, asparaginase, romidepsin, trabectedin, etc. Many microbes have compounds with anticancer and antidiabetic properties but a good number of these compounds have not been converted into usable drugs for the masses. Further studies are required to ascertain their effectiveness and safety when being incorporated into the human system.

Acknowledgements The authors wish to thank the management, especially Rev. Fr. Swebert D'Silva, SJ, Pro-Chancellor, Rev. Fr. Dr. Victor Lobo, SJ, Vice-Chancellor, Rev. Fr. Dr. Roshan Castelino, SJ, Dean, Research & Innovation Council of St. Joseph's University and Dr. V. J. Jacob Paul, Head, Dr. M. Vaishnavi PG coordinator of the Department of Botany, St. Joseph's University for providing all the necessary facilities, encouragement and congenial environment for research at SJU.

References

Abbas, T., & Dutta, A. (2009). P21 in cancer: Intricate networks and multiple activities. *Nature Reviews Cancer, 9*(6), 400–414. https://doi.org/10.1038/nrc2657

Abdelnasser, S. M., Yahya, S. M., Mohamed, W. F., Asker, M. M., Shady, H. M. A., Mahmoud, M. G., & Gadallah, M. A. (2017). Antitumor exopolysaccharides derived from Novel Marine *Bacillus*: Isolation, characterization aspect and biological activity. *Asian Pacific Journal of Cancer Prevention, 18*(7), 1847–1854. https://doi.org/10.22034/APJCP.2017.18.7.1847

Abdulkhair, W. M., Abdel-all, W. S., & Bahy, R. H. (2018). Genetic improvement of antidiabetic alpha-glucosidase inhibitor producing *Streptomyces sp. International Journal of Pharmacy and Pharmaceutical Science, 10*(5), 77–84. https://doi.org/10.22159/ijpps.2018v10i5.25338

Agrawal, S., Acharya, D., Adholeya, A., Barrow, C. J., & Deshmukh, S. K. (2017). Nonribosomal peptides from marine microbes and their antimicrobial and anticancer potential frontiers in pharmacology *8*, 828. https://doi.org/10.3389/fphar.2017.00828

Alberto, M. R., Gómez-Cordovés, C., & Manca de Nadra, M. C. (2004). Metabolism of gallic acid and catechin by *Lactobacillus hilgardii* from wine. *Journal of Agricultural and Food Chemistry, 52*(21), 6465–6469. https://doi.org/10.1021/jf049239f

Alrumman, S. A., Mostafa, Y. S., Al-Izran, K. A., Alfaifi, M. Y., Taha, T. H., & Elbehairi, S. E. (2019). Production and anticancer activity of an L-asparaginase from *Bacillus licheniformis* isolated from the Red Sea, Saudi Arabia. *Scientific Reports, 9*(1), 3756. https://doi.org/10.1038/s41598-019-40512-x

Balachandran, C., Duraipandiyan, V., Arun, Y., Sangeetha, B., Emi, N., Al-Dhabi, N. A., Ignacimuthu, S., Inaguma, Y., Okamoto, A., & Perumal, P. T. (2016). Isolation and characterization of 2-hydroxy-9, 10-anthraquinone from *Streptomyces olivochromogenes* (ERINLG-261) with antimicrobial and antiproliferative properties. *Revista Brasileira De Farmacognosia, 26*, 285–295. https://doi.org/10.1016/j.bjp.2015.12.003

Balde, E. S., Andolfi, A., Bruyere, C., Cimmino, A., Lamoral-Theys, D., Vurro, M., Damme, M. V., Altomare, C., Mathieu, V., Kiss, R., & Evidente, A. (2010). Investigations of fungal secondary metabolites with potential anticancer activity. *Journal of Natural Products, 73*(5), 969–971. https://doi.org/10.1021/np900731p

Banerjee, S. & Paruthy, S. B. (2017). Preclinical and clinical perspective on fungal metabolites and their analogs as anticancer agents–from bench to bedside. In J.-M. Mérillon & K. G. Ramawat (Eds.), *Fungal metabolites* (pp. 670–700). Springer.

Basha, R. H., & Sankaranarayanan, C. (2014). β-Caryophyllene, a natural sesquiterpene, modulates carbohydrate metabolism in streptozotocin-induced diabetic rats. *Acta Histochemica, 116*(8), 1469–1479. https://doi.org/10.1016/j.acthis.2014.10.001

Bennett, J. M., & Reich, S. D. (1979). Drugs five years later: Bleomycin. *Annals of Internal Medicine, 90*(6), 945–948. https://doi.org/10.7326/0003-4819-90-6-945

Bérdy, J. (2012). Thoughts and facts about antibiotics: Where we are now and where we are heading. *The Journal of Antibiotics, 65*(8), 385–395. https://doi.org/10.1038/ja.2012.27

Bibb, M. J. (2013). Understanding and manipulating antibiotic production in actinomycetes. *Biochemical Society Transactions., 41*(6), 1355–1364. https://doi.org/10.1042/BST20130214

Bidon-Chanal, A., Fuertes, A., Alonso, D., Pérez, D. I., Martínez, A., Luque, F. J., & Medina, M. (2013). Evidence for a new binding mode to GSK-3: Allosteric regulation by the marine compound palinurin. *European Journal of Medicinal Chemistry, 60*, 479–489. https://doi.org/10.1016/j.ejmech.2012.12.014

Boehle, A., Kurdow, R., Boenicke, L., Schniewind, B., Faendrich, F., Dohrmann, P., & Kalthoff, H. (2002). Wortmannin inhibits growth of human non-small-cell lung cancer in vitro and in vivo. *Langenbeck's Archives of Surgery, 387*, 234–239. https://doi.org/10.1007/s00423-002-0314-x

Budhiraja, A., Nepali, K., Sapra, S., Gupta, S., Kumar, S., & Dhar, K. L. (2013). Bioactive metabolites from an endophytic fungus of *Aspergillus* species isolated from seeds of *Gloriosa superba* Linn. *Medicinal Chemistry Research, 22*, 323–329. https://doi.org/10.1007/s00044-012-0032-z

Cai, G., Napolitano, J. G., McAlpine, J. B., Wang, Y., Jaki, B. U., Suh, J. W., Yang, H. Y., Lee, I., Franzblau, S. G., Pauli, G. F., & Cho, S. (2013). Hytramycins V and I, anti-Mycobacterium tuberculosis hexapeptides from a *Streptomyces hygroscopicus* strain. *Journal of Natural Products, 76*(11), 2009–2018. https://doi.org/10.1021/np400145u

Canturk, Z., Dikmen, M., Ozturk, N., Kocabiyik, E., & Ilhan, S. (2013). Antioxidant properties of halotolerant *Penicillium chrysogenum* var. *chrysogenum* secondary metabolites. In *2013 E-Health and bioengineering conference (EHB)*. https://doi.org/10.1109/ehb.2013.6707274

Cartuche, L., Sifaoui, I., Cruz, D., Reyes-Batlle, M., López-Arencibia, A., Javier Fernández, J., Díaz-Marrero, A. R., Piñero, J. E., & Lorenzo-Morales, J. (2019). Staurosporine from *Streptomyces sanyensis* activates programmed cell death in Acanthamoeba via the mitochondrial pathway and presents low in vitro cytotoxicity levels in a macrophage cell line. *Scientific Reports, 9*(1), 11651. https://doi.org/10.1038/s41598-019-48261-7

Chatterjee, S., Khunti, K., & Davies, M. J. (2017). Type 2 diabetes. *The Lancet, 389*(10085), 2239–2251. https://doi.org/10.1016/s0140-6736(17)30058-2

Chen, C. J., Hsu, W. L., Yang, H. I., Lee, M. H., Chen, H. C., Chien, Y. C., & You, S. L. (2014). In: M. Chang & K. T. Jeang (Eds.) *Viruses and human cancer. Recent results in cancer research* (Vol. 193, pp. 11–32). Springer. https://doi.org/10.1007/978-3-642-38965-8_2

Chen, F., Ren, C. G., Zhou, T., Wei, Y. J., & Dai, C. C. (2016a). A novel exopolysaccharide elicitor from endophytic fungus *Gilmaniella sp.* AL12 on volatile oils accumulation in *Atractylodes lancea*. *Scientific Reports, 6*, 34735. https://doi.org/10.1038/srep34735

Chen, L., Zhao, H., Zhang, C., Lu, Y., Zhu, X., & Lu, Z. (2016b). γ-Aminobutyric acid-rich yogurt fermented by *Streptococcus salivarius* subsp. *thermophiles* fmb5 apprars to have anti-diabetic effect on streptozotocin-induced diabetic mice. *Journal of Functional Foods, 20*, 267–275. https://doi.org/10.1016/j.jff.2015.10.030

Chen, Y. C., Huang, S. D., Tu, J. H., Yu, J. S., Nurlatifah, A. O., Chiu, W. C., Su, Y. H., Chang, H. L., Putri, D. A., & Cheng, H. L. (2020). Exopolysaccharides of *Bacillus amyloliquefaciens* modulate glycemic level in mice and promote glucose uptake of cells through the activation of Akt. *International Journal of Biological Macromolecules, 146*, 202–211. https://doi.org/10.1016/j.ijbiomac.2019.12.217

Cho, J. Y., Williams, P. G., Kwon, H. C., Jensen, P. R., & Fenical, W. (2007). Lucentamycins A–D, cytotoxic peptides from the marine-derived actinomycete *Nocardiopsis lucentensis*. *Journal of Natural Products, 70*(8), 1321–1328. https://doi.org/10.1021/np070101b

Dang, F., Jiang, Y., Pan, R., Zhou, Y., Wu, S., Wang, R., Zhuang, K., Zhang, W., Li, T., & Man, C. (2018). Administration of *Lactobacillus paracasei* ameliorates type 2 diabetes in mice. *Food and Function, 9*(7), 3630–3639. https://doi.org/10.1039/C8FO00081F

Deshmukh, S. K., Agrawal, S., & Gupta, M. K. (2023). Fungal metabolites: A potential source of antidiabetic agents with particular reference to PTP1B inhibitors. *Current Pharmaceutical Biotechnology, 24*(8), 927–945. https://doi.org/10.2174/1389201023666220506104219

Doan, C. T., Tran, T. N., Nguyen, M. T., Nguyen, V. B., Nguyen, A. D., & Wang, S. L. (2019). Anti-α-glucosidase activity by a protease from *Bacillus licheniformis*. *Molecules, 24*(4), 691. https://doi.org/10.3390/molecules24040691

Elkhouly, H. I., Hamed, A. A., El Hosainy, A. M., Ghareeb, M. A., & Sidkey, N. M. (2021). Bioactive secondary metabolite from endophytic *Aspergillus tubenginses* ASH4 isolated from *Hyoscyamus muticus*: antimicrobial, antibiofilm, antioxidant and anticancer activity. *Pharmacognosy Journal, 13*(2). https://doi.org/10.5530/pj.2021.13.55

El-Sersy, N. A., Abdelwahab, A. E., Abouelkhiir, S. S., Abou-Zeid, D. M., & Sabry, S. A. (2012). Antibacterial and Anticancer activity of ε-poly-L-lysine (ε-PL) produced by a marine *Bacillus subtilis sp. Journal of Basic Microbiology, 52*(5), 513–522. https://doi.org/10.1002/jobm.201100290

Fu, P., Yang, C., Wang, Y., Liu, P., Ma, Y., Xu, L., Su, M., Hong, K., & Zhu, W. (2012). Streptocarbazoles A and B, two novel indolocarbazoles from the marine-derived actinomycete strain *Streptomyces sp.* FMA. *Organic Letters, 14*(9), 2422–2425. https://doi.org/10.1021/ol3008638

Giddings, L. A., & Newman, D. J. (2013). Microbial natural products: Molecular blueprints for antitumor drugs. *Journal of Industrial Microbiology and Biotechnology, 40*(11), 1181–1210. https://doi.org/10.1007/s10295-013-1331-1

Giridharan, P., Verekar, S. A., Khanna, A., Mishra, P. D., & Deshmukh, S. K. (2012). Anticancer activity of sclerotiorin, isolated from an endophytic fungus *Cephalotheca faveolata* Yaguchi, Nishim. and Udagawa. *Indian Journal of Experimental Biology, 50*, 464–468. PMID: 22822525.

Giurini, E. F., Godla, A., & Gupta, K. H. (2024). Redefining bioactive small molecules from microbial metabolites as revolutionary anticancer agents. *Cancer Gene Therapy, 31*, 187–206. https://doi.org/10.1038/s41417-023-00715-x

Gong, K., Wang, M., Duan, Q., Li, G., Yong, D., Ren, C., Li, Y., Zhang, Q., Wang, Z., Sun, T., Zhang, H., Tu, Q., Wu, C., Fu, J., Li, A., Song, C., Zhang, Y., & Li, R. (2023). High-yield production of FK228 and new derivatives in a *Burkholderia chassis*. *Metabolic Engineering, 75*, 131–142. https://doi.org/10.1016/j.ymben.2022.12.002

Grande, F., Ioele, G., Caruso, A., Occhiuzzi, M. A., El-Kashef, H., Saturnino, C., & Sinicropi, M. S. (2022). Carbazoles: Role and functions in fighting diabetes. *Applied Sciences, 13*(1), 349. https://doi.org/10.3390/app13010349

Guo, W., Peng, J., Zhu, T., Gu, Q., Keyzers, R. A., & Li, D. (2013). Sorbicillamines A-E, nitrogen-containing sorbicillinoids from the deep-sea-derived fungus *Penicillium sp.* F23-2. *Journal of Natural Products, 76*(11), 2106–2112. https://doi.org/10.1021/np4006647

Han, Y., Tian, E., Xu, D., Ma, M., Deng, Z., & Hong, K. (2016). Halichoblelide D, a new elaiophylin derivative with potent cytotoxic activity from mangrove-derived *Streptomyces sp.* 219807. *Molecules, 21*(8), 970. https://doi.org/10.3390/molecules21080970

Hecht, S. M. (2000). Bleomycin: New perspectives on the mechanism of action. *Journal of Natural Products, 63*(1), 158–168. https://doi.org/10.1021/np990549f

Heo, K. T., Lee, B., Jang, J. H., & Hong, Y. S. (2022). Elucidation of the di-c-glycosylation steps during biosynthesis of the antitumor antibiotic, kidamycin. *Frontiers in Bioengineering and Biotechnology, 10*, https://doi.org/10.3389/fbioe.2022.985696

Hindra, Y. D., Teng, Q., Dong, L. B., Crnovcic, I., Huang, T., Ge, H., & Shen, B. (2017). Genome mining of *Streptomyces mobaraensis* DSM40847 as a bleomycin producer providing a biotechnology platform to engineer designer bleomycin analogues. *Organic Letters, 19*(6), 1386–1389. https://doi.org/10.1021/acs.orglett.7b00283

Huang, H., Feng, X., Xiao, Z. E., Liu, L., Li, H., Ma, L., Lu, Y., Ju, J., She, Z., & Lin, Y. (2011). Azaphilones and p-terphenyls from the mangrove endophytic fungus *Penicillium chermesinum* (ZH4-E2) isolated from the South China Sea. *Journal of Natural Products, 74*(5), 997–1002. https://doi.org/10.1021/np100889v

Hussain, H., Nazir, M., Saleem, M., Al-Harrasi, A., Elizbit, & Green, I. R. (2021). Fruitful decade of fungal metabolites as anti-diabetic agents from 2010 to 2019: emphasis on α-glucosidase inhibitors. *Phytochemistry Reviews, 20*(1), 145–179. https://doi.org/10.1007/s11101-020-09733-1

Jia, L., Li, D., Feng, N., Shamoon, M., Sun, Z., Ding, L., Zhang, H., Chen, W., Sun, J., & Chen, Y. Q. (2017). Anti-diabetic effects of *Clostridium butyricum* CGMCC0313. 1 through promoting the growth of gut butyrate-producing bacteria in type 2 diabetic mice. *Scientific reports, 7*(1), 7046. https://doi.org/10.1038/s41598-017-07335-0

Kanoh, K., Matsuo, Y., Adachi, K., Imagawa, H., Nishizawa, M., & Shizuri, Y. (2005). Mechercharmycins A and B, cytotoxic substances from marine-derived *Thermoactinomyces sp.* YM3–251. *The Journal of Antibiotics, 58*(4), 289–292. https://doi.org/10.1038/ja.2005.36

Karlapudi, A. P., Venkateswarulu, T. C., Srirama, K., Kota, R. K., Mikkili, I., & Kodali, V. P. (2020). Evaluation of anti-cancer, anti-microbial and anti-biofilm potential of biosurfactant extracted from an *Acinetobacter* M6 strain. *Journal of King Saud University-Science, 32*(1), 223–227. https://doi.org/10.1016/j.jksus.2018.04.007

Karthik, S., Sankar, R., Varunkumar, K., & Ravikumar, V. (2014). Romidepsin induces cell cycle arrest, apoptosis, histone hyperacetylation and reduces matrix metalloproteinases 2 and 9 expression in bortezomib sensitized non-small cell lung cancer cells. *Biomedicine and Pharmacotherapy, 68*(3), 327–334. https://doi.org/10.1016/j.biopha.2014.01.002

Kato, S., Kawasaki, T., Urata, T., & Mochizuki, J. (1993). In vitro and ex vivo free radical scavenging activities of carazostatin, carbazomycin B and their derivatives. *The Journal of Antibiotics, 46*(12), 1859–1865. https://doi.org/10.7164/antibiotics.46.1859

Kawakubo, J., Nishira, H., Aoki, K., & Shinke, R. (1993). Isolation of a gallic acid-producing microorganism with sake cake medium and production of gallic acid. *Bioscience, Biotechnology, and Biochemistry, 57*(8), 1360–1361. https://doi.org/10.1271/bbb.57.1360

Kim, D. Y., Kang, M. K., Lee, E. J., Kim, Y. H., Oh, H., Kim, S. I., Oh, S. Y., Na, W., & Kang, Y. H. (2020). Eucalyptol inhibits Amyloid-β-induced barrier dysfunction in glucose-exposed retinal pigment epithelial cells and diabetic eyes. *Antioxidants, 9*(10), 1000. https://doi.org/10.3390/antiox9101000

Kim, G. N., Shin, J. G., & Jang, H. D. (2009). Antioxidant and antidiabetic activity of Dangyuja (*Citrus grandis* Osbeck) extract treated with *Aspergillus saitoi*. *Food Chemistry, 117*(1), 35–41. https://doi.org/10.1016/j.foodchem.2009.03.072

Kim, S. H., Huh, C. S., Choi, I. D., Jeong, J. W., Ku, H. K., Ra, J. H., Kim, T. Y., Kim, G. B., Sim, J. H., & Ahn, Y. T. (2014). The anti-diabetic activity of *Bifidobacterium lactis* HY8101 in vitro and in vivo. *Journal of Applied Microbiology, 117*(3), 834–845. https://doi.org/10.1111/jam.12573

Koba, M., & Konopa, J. (2005). Aktynomycyna D i mechanizmy jej działania Actinomycin D and its mechanisms of action. *Postępy Higieny i Medycyny Doświadczalnej (Online), 59*, 290–298.

Kodali, V. P., & Sen, R. (2008). Antioxidant and free radical scavenging activities of an exopolysaccharide from a probiotic bacterium. *Biotechnology Journal, 3*(2), 245–251. https://doi.org/10.1002/biot.200700208

Komiyama, K., Funayama, S., Anraku, Y., Mita, A., Takahashi, Y., Omura, S., & Shimasaki, H. (1989). Isolation of isoflavonoids possessing antioxidant activity from the fermentation broth of *Streptomyces sp. The Journal of Antibiotics, 42*(9), 1344–1349. https://doi.org/10.7164/antibiotics.42.1344

Kong, F., Zhao, C., Hao, J., Wang, C., Wang, W., Huang, X., & Zhu, W. (2015). New α-glucosidase inhibitors from a marine sponge-derived fungus, *Aspergillus sp.* OUCMDZ-1583. *RSC Advances, 5*(84), 68852–68863. https://doi.org/10.1039/c5ra11185d

Kulkarni-Almeida, A. A., Brahma, M. K., Padmanabhan, P., Mishra, P. D., Parab, R. R., Gaikwad, N. V., Thakkar, C. S., Tokdar, P., Ranadive, P. V., Nair, A. S., Damre, A. A., Bahirat, U. A., Deshmukh, N. J., Doshi, L. S., Dixit, A. V., George, S. D., Vishwakarma, R. A., Nemmani, K. V., & Mahajan, G. B. (2011). Fermentation, isolation, structure, and antidiabetic activity of NFAT-133 produced by *Streptomyces* strain PM0324667. *AMB Express, 1*, 1–12. https://doi.org/10.1186/2191-0855-1-42

Kwon, H. C., Kauffman, C. A., Jensen, P. R., & Fenical, W. (2006). Marinomycins A–D, antitumor-antibiotics of a new structure class from a marine actinomycete of the recently discovered genus "*Marinispora*". *Journal of the American Chemical Society, 128*(5), 1622–1632. https://doi.org/10.1021/ja0558948

Lalitha, P., Veena, V., Vidhyapriya, P., Lakshmi, P., Krishna, R., & Sakthivel, N. (2016). Anticancer potential of pyrrole (1, 2, a) pyrazine. 1, 4, dione, hexahydro 3-(2-methylpropyl) (PPDHMP) extracted from a new marine bacterium, *Staphylococcus sp.* Strain MB30. *Apoptosis, 21*, 566–577. https://doi.org/10.1007/s10495-016-1221-x

Lee, B., Lee, G. E., Hwang, G. J., Heo, K. T., Lee, J. K., Jang, J. P., Hwang, B. Y., Jang, J. H., Cho, Y. Y., & Hong, Y. S. (2023). Rubiflavin G, photorubiflavin G, and photorubiflavin E: Novel pluramycin derivatives from *Streptomyces sp.* W2061 and their anticancer activity against breast cancer cells. *The Journal of Antibiotics, 76*(10), 585–591. https://doi.org/10.1038/s41429-023-00643-w

Li, D., Cai, S., Zhu, T., Wang, F., Xiao, X., & Gu, Q. (2010a). Three new sorbicillin trimers, trisorbicillinones B, C, and D, from a deep ocean sediment derived fungus, *Phialocephala sp.* FL30r. *Tetrahedron, 66*(27–28), 5101–5106. https://doi.org/10.1016/j.tet.2010.04.111

Li, G.-Y., Li, L.-M., Yang, T., Chen, X.-Z., Fang, D.-M., & Zhang, G.-L. (2010b). Four New Alkaloids, Brevianamides O-R, from the Fungus *Aspergillus versicolor*. *Helvetica Chimica Acta, 93*(10), 2075–2080. https://doi.org/10.1002/hlca.201000053

Liu, J., Wang, X., Zhu, Y., Deng, H., Huang, X., Jayavanth, P., Xiao, Y., Wu, J., & Jiao, R. (2023). Theabrownin from dark tea ameliorates insulin resistance via attenuating oxidative stress and modulating IRS-1/PI3K/Akt pathway in HepG2 cells. *Nutrients, 15*(18), 3862. https://doi.org/10.3390/nu15183862

Li, Y.-X., Himaya, S. W. A., Dewapriya, P., Zhang, C., & Kim, S.-K. (2013). Fumigaclavine C from a marine-derived fungus *Aspergillus fumigatus* induces apoptosis in MCF-7 breast cancer cells. *Marine Drugs, 11*, 5063–5086. https://doi.org/10.3390/md11125063

Liu, Z., Xia, G., Chen, S., Liu, Y., Li, H., & She, Z. (2014). Eurothiocin A and B, sulfur-containing benzofurans from a soft coral-derived fungus *Eurotium rubrum* SH-823. *Marine Drugs, 12*(6), 3669–3680. https://doi.org/10.3390/md12063669

Lopez, J., Lagunas-Rivera, S., Vargas-Diaz, M. E., Alcaraz-Contreras, Y., & Vazquez, M. A. (2022). Recent advances in pluramycins, biological activity, and synthetic strategies. *Studies in Natural Products Chemistry, 73*, 1–44. https://doi.org/10.1016/B978-0-323-91097-2.00005-4

Lutfia, A., Munir, E., Yurnaliza, Y., & Basyuni, M. (2021). Chemical analysis and anticancer activity of sesterterpenoid from an endophytic fungus *Hypomontagnella monticulosa* Zg15SU and its host *Zingiber griffithii* Baker. *Heliyon, 7*(2). https://doi.org/10.1016/j.heliyon.2021.e06292

Ma, E. L., Choi, Y. J., Choi, J., Pothoulakis, C., Rhee, S. H., & Im, E. (2010). The anticancer effect of probiotic *Bacillus polyfermenticus* on human colon cancer cells is mediated through ErbB2 and ErbB3 inhibition. *International Journal of Cancer, 127*(4), 780–790. https://doi.org/10.1002/ijc.25011

Ma, Y., Xu, M., Liu, H., Yu, T., Guo, P., Liu, W., & Jin, X. (2021). Antimicrobial compounds were isolated from the secondary metabolites of *Gordonia*, a resident of intestinal tract of *Periplaneta americana*. *AMB Express, 11*, 1–11. https://doi.org/10.1186/s13568-021-01272-y

Ma, Z., Wang, N., Hu, J., & Wang, S. (2012). Isolation and characterization of a new iturinic lipopeptide, mojavensin A produced by a marine-derived bacterium *Bacillus mojavensis* B0621A. *The Journal of Antibiotics, 65*(6), 317–322. https://doi.org/10.1038/ja.2012.19

Marchi, E., Raufi, A. G., & O'Connor, O. A. (2017). Novel agents in the treatment of relapsed or refractory peripheral T-cell lymphoma. *Hematology/oncology Clinics, 31*(2), 359–375. https://doi.org/10.1016/j.hoc.2016.11.002

Marx, J. (2002). Unraveling the causes of diabetes. *Science, 296*(5568), 686–689. https://doi.org/10.1126/science.296.5568.686

Maskey, R. P., Grün-Wollny, I., & Laatsch, H. (2005). Sorbicillin analogues and related dimeric compounds from *Penicillium notatum*. *Journal of Natural Products, 68*(6), 865–870. https://doi.org/10.1021/np040137t

Matsuo, Y., Kanoh, K., Imagawa, H., Adachi, K., Nishizawa, M., & Shizuri, Y. (2007). Urukthapelstatin A, a novel cytotoxic substance from marine-derived *Mechercharimyces asporophorigenens*

YM11-542. *The Journal of Antibiotics, 60*(4), 256–260. https://doi.org/10.1038/ja.2005.36

McClendon, A. K., & Osheroff, N. (2007). DNA topoisomerase II, genotoxicity, and cancer. *Mutation Research/fundamental and Molecular Mechanisms of Mutagenesis, 623*(1–2), 83–97. https://doi.org/10.1016/j.mrfmmm.2007.06.009

Megrous, S., Al-Dalali, S., Zhao, X., Chen, C., Cao, Y., Bourouis, I., Mekkaoui, A., Yang, Z., & Yang, Z. (2020). Evaluation of antidiabetic activities of casein hydrolysates by a *Bacillus* metalloendopeptidase. *International Journal of Peptide Research and Therapeutics, 26*, 2519–2527. https://doi.org/10.1007/s10989-020-10045-3

Mohan, C. D., Rangappa, S., Nayak, S. C., Jadimurthy, R., Wang, L., Sethi, G., Garg, M., & Rangappa, K. S. (2022). Bacteria as a treasure house of secondary metabolites with anticancer potential. *Seminars in Cancer Biology, 86*(pt 2), 998–1013. https://doi.org/10.1016/j.semcancer.2021.05.006

Murphy, B. T., Narender, T., Kauffman, C. A., Woolery, M., Jensen, P. R., & Fenical, W. (2010). Saliniquinones A-F, new members of the highly cytotoxic anthraquinone-γ-pyrones from the marine actinomycete *Salinispora arenicola. Australian Journal of Chemistry, 63*(6), 929–934. https://doi.org/10.1071/CH10068

Murray, V., Chen, J. K., & Chung, L. H. (2018). The interaction of the metallo-glycopeptide anti-tumour drug bleomycin with DNA. *International Journal of Molecular Sciences, 19*(5), 1372. https://doi.org/10.3390/ijms19051372

Nagai, K., Yamaki, H., Tanaka, N., & Umezawak, H. (1967). Inhibition by pluramycin A of nucleic acid biosynthesis. *The Journal of Biochemistry, 62*(3), 321–327. https://doi.org/10.1093/oxfordjournals.jbchem.a128669

Nam, H., Jung, H., Karuppasamy, S., Park, Y. S., Cho, Y. S., Lee, J. Y., Seong, S., & Suh, J. G. (2012). Anti-diabetic effect of the soybean extract fermented by *Bacillus subtilis* MORI in db/db mice. *Food Science and Biotechnology, 21*, 1669–1676. https://doi.org/10.1007/s10068-012-0222-y

Parthasarathy, R., Chandrika, M., Rao, H. Y., Kamalraj, S., Jayabaskaran, C., & Pugazhendhi, A. (2020). Molecular profiling of marine endophytic fungi from green algae: Assessment of antibacterial and anticancer activities. *Process Biochemistry, 96*, 11–20. https://doi.org/10.1016/j.procbio.2020.05.012

Pereira, D. M., Cheel, J., Areche, C., San-Martin, A., Rovirosa, J., Silva, L. R., Valentao, P., & Andrade, P. B. (2011). Antiproliferative activity of meroditerpenoids isolated from the brown alga *Stypopodium flabelliforme* against several cancer cell lines. *Marine Drugs, 9*, 852–862. https://doi.org/10.3390/md9050852

Perveen, I., Raza, M. A., Iqbal, T., Naz, I., Sehar, S., & Ahmed, S. (2017). Isolation of anticancer and antimicrobial metabolites from *Epicoccum nigrum*; endophyte of *Ferula sumbul*. *Microbial Pathogenesis, 110*, 214–224. https://doi.org/10.1016/j.micpath.2017.06.033

Qureshi, K. A., Bholay, A. D., Rai, P. K., Mohammed, H. A., Khan, R. A., Azam, F., Jaremko, M., Emwas, A. H., Stefanowicz, P., Waliczek, M., Kijewska, M., Ragab, E. A., Rehan, M., Elhassan, G. O., Anwar, M. J., & Prajapati, D. K. (2021). Isolation, characterization, anti-MRSA evaluation, and in-silico multi-target anti-microbial validations of actinomycin X2 and actinomycin D produced by novel *Streptomyces smyrnaeus* UKAQ_23. *Scientific Reports, 11*(1), 14539. https://doi.org/10.1038/s41598-021-93285-7

Ramírez-Rendon, D., Passari, A. K., Ruiz-Villafán, B., Rodríguez-Sanoja, R., Sánchez, S., & Demain, A. L. (2022). Impact of novel microbial secondary metabolites on the pharma industry. *Applied Microbiology and Biotechnology, 106*(5), 1855–1878. https://doi.org/10.1007/s00253-022-11821-5

Ravi, L., Ragunathan, A., & Krishnan, K. (2017). Antidiabetic and antioxidant potential of GancidinW from *Streptomyces paradoxus* VITALK03. *The Open Bioactive Compounds Journal, 5*(1). https://doi.org/10.2174/1874847301705010031

Ryan, K. M., Phillips, A. C., & Vousden, K. H. (2001). Regulation and function of the p53 tumor suppressor protein. *Current Opinion in Cell Biology, 13*(3), 332–337. https://doi.org/10.1016/S0955-0674(00)00216-7

Saleem, M., Hassan, A., Li, F., Lu, Q., Ponomareva, L. V., Parkin, S., Sun, C., Thorson, J. S., Shaaban, K. A., & Sajid, I. (2023). Bioprospecting of desert actinobacteria with special emphases on griseoviridin, mitomycin C and a new bacterial metabolite producing *Streptomyces sp.* PU-KB10–4. *BMC Microbiology, 23*(1), 69. https://doi.org/10.1186/s12866-023-02770-8

Sambandam, T., & Mahadevan, A. (1993). Degradation of catechin and purification and partial characterization of catechin oxygenase from *Chaetomium cupreum. World Journal of Microbiology and Biotechnology, 9*(1), 37–44. https://doi.org/10.1007/BF00656513

Scheen, A. J. (2003). Is there a role for α-glucosidase inhibitors in the prevention of type 2 diabetes mellitus? *Drugs, 63*, 933–951. https://doi.org/10.2165/00003495-200363100-00002

Schueffler, A., & Anke, T. (2014). Fungal natural products in research and development. *Natural Product Reports, 31*(10), 1425–1448. https://doi.org/10.1039/c4np00060a

SenthilKumar, V., Kumaresan, S., Tamizh, M. M., Islam, V. I. H., & Thirugnanasambantham, K. (2019). Anticancer potential of NFκB targeting apoptotic molecule "flavipin" isolated from endophytic *Chaetomium globosum*. *Phytomedicine*, 152830. https://doi.org/10.1016/j.phymed.2019.152830

Seo, C., Sohn, J. H., Oh, H., Kim, B. Y., & Ahn, J. S. (2009). Isolation of the protein tyrosine phosphatase 1B inhibitory metabolite from the marine-derived fungus *Cosmospora sp.* SF-5060. *Bioorganic and Medicinal Chemistry Letters, 19*(21), 6095–6097. https://doi.org/10.1016/j.bmcl.2009.09.025

Shao, Y., Wang, X. Y., Qiu, X., Niu, L. L., & Ma, Z. L. (2021). Isolation and purification of a new *Bacillus subtilis* strain from deer dung with anti-microbial and anti-cancer activities. *Current Medical Science, 41*, 832–840. https://doi.org/10.1007/s11596-021-2383-5

Shao, Z., Tian, Y., Liu, S., Chu, X., & Mao, W. (2023). Anti-diabetic activity of a novel exopolysaccharide produced by the mangrove endophytic fungus *Penicillium janthinellum* N29. *Marine Drugs, 21*(5), 270. https://doi.org/10.3390/md21050270

Shin-ya, K., Furihata, K., Teshima, Y., Hayakawa, Y., & Seto, H. (1993a). Benthocyanins B and C, new free radical scavengers from *Streptomyces prunicolor. The Journal of Organic Chemistry, 58*(15), 4170–4172. https://doi.org/10.1021/jo00067a069

Shin-ya, K., Hayakawa, Y., & Seto, H. (1993b). Structure of benthophoenin, a new free radical scavenger produced by *Streptomyces prunicolor*. *Journal of Natural Products, 56*(8), 1255–1258. https://doi.org/10.1021/np50098a008

Shweta, S., Gurumurthy, B. R., Ravikanth, G., Ramanan, U. S., & Shivanna, M. B. (2013). Endophytic fungi from *Miquelia dentata* Bedd., produce the anti-cancer alkaloid, camptothecine. *Phytomedicine, 20*(3–4), 337–342. https://doi.org/10.1016/j.phymed.2012.11.015

Singh, B., & Kaur, A. (2016). Antidiabetic potential of a peptide isolated from an endophytic *Aspergillus awamori. Journal of Applied Microbiology, 120*(2), 301–311. https://doi.org/10.1111/jam.12998

Soliman, M. (2018). Cancer causing chemicals. In: Atroshi, F. (Ed.) *Cancer causing substances*. IntechOpen. https://doi.org/10.5772/intechopen.71560

Stierle, A., Strobel, G., & Stierle, D. (1993). Taxol and taxane production by *Taxomyces andreanae*, an endophytic fungus of Pacific yew. *Science, 260*(5105), 214–216. https://doi.org/10.1126/science.8097061

Stephen, A., & Kumaresan, S. (2020). Antimicrobial activity of selected plants from southern India: A critical appraisal. In

Jayakumar, R. (Ed.) *Research trends in multidisciplinary research* (Vol. 14, pp. 159–176). AkiNik Publications.
Stephen, A., Ragupathi, V., & Kumaresan, S. (2017). Identification and utilization of pharmaceutically novel chemicals from plants for the treatment of cancer—A critical appraisal. *Voyager, 8*(2), 205–216.
Sun, D., Hansen, M., & Hurley, L. (1995). Molecular basis for the DNA sequence specificity of the pluramycins. A novel mechanism involving groove interactions transmitted through the helix via intercalation to achieve sequence selectivity at the covalent bonding step. *Journal of the American Chemical Society, 117*(9), 2430–2440. https://doi.org/10.1021/ja00114a007
Syed, C. S., Sairam, M., & Audipudi, A. V. (2019). Exploration of antibacterial and antiproliferative secondary metabolites from marine *Bacillus. The Journal of Microbiology, Biotechnology and Food Sciences, 9*(3), 628. https://doi.org/10.15414/jmbfs.2019/20.9.3.628-633
Tangerina, M. M., Furtado, L. C., Leite, V. M., Bauermeister, A., Velasco-Alzate, K., Jimenez, P. C., Garrido, L. M., Padilla, G., Lopes, N. P., Costa-Lotufo, L. V., & Pena Ferreira, M. J. (2020). Metabolomic study of marine *Streptomyces sp.* : Secondary metabolites and the production of potential anticancer compounds. *PLoS One, 15*(12). https://doi.org/10.1371/journal.pone.0244385
Verekar, S. A., Mishra, P. D., Sreekumar, E. S., Deshmukh, S. K., Fiebig, H. H., Kelter, G., & Maier, A. (2014). Anticancer activity of new depsipeptide compound isolated from an endophytic fungus. *The Journal of Antibiotics, 67*(10), 697–701. https://doi.org/10.1038/ja.2014.58
Verpoorte, R., & Memelink, J. (2002). Engineering secondary metabolite production in plants. *Current Opinion in Biotechnology, 13*(2), 181–187. https://doi.org/10.1016/S0958-1669(02)00308-7
Wang, Q., Gong, J., Chisti, Y., & Sirisansaneeyakul, S. (2014). Bioconversion of tea polyphenols to bioactive theabrownins by *Aspergillus fumigatus. Biotechnology Letters, 36*(12), 2515–2522. https://doi.org/10.1007/s10529-014-1632-0
Wang, M., Sperrin, M., Rutter, M. K., & Renehan, A. G. (2023). Cancer is becoming the leading cause of death in diabetes. *The Lancet, 401*(10391), P1849. https://doi.org/10.1016/S0140-6736(23)00445-2
Wang, J., & Wang, H. (2017). Oxidative stress in pancreatic beta cell regeneration. *Oxidative Medicine and Cellular Longevity, 2017*, 1930261. https://doi.org/10.1155/2017/1930261
Wang, Z., Xie, Z., Lu, Q., Chang, C., & Zhou, Z. (2016). Beyond Genetics: What causes type 1 diabetes. *Clinical Reviews in Allergy and Immunology, 52*(2), 273–286. https://doi.org/10.1007/s12016-016-8592-1
Xie, C. L., Zhang, D., Lin, T., He, Z. H., Yan, Q. X., Cai, Q., Zhang, X. K., Yang, X. W., & Chen, H. F. (2021). Antiproliferative sorbicillinoids from the deep-sea-derived *Penicillium allii-sativi. Frontiers in Microbiology, 11*, 636948. https://doi.org/10.3389/fmicb.2020.636948
Yadav, A. N. (2021). Microbial biotechnology for bio-prospecting of microbial bioactive compounds and secondary metabolites. *Journal of Applied Biology and Biotechnology, 9*(2), 1–6. https://doi.org/10.7324/JABB.2021.92ED
Yagi, R., & Doi, M. (1999). Isolation of an antioxidative substance produced by *Aspergillus repens. Bioscience, Biotechnology, and Biochemistry, 63*(5), 932–933. https://doi.org/10.1271/bbb.63.932
Yao, Y., Li, J., Jiang, C. S., Zhao, X. X., Miao, Z. H., Liu, H. T., Zheng, P., Ya, W. X., & Li, W. Q. (2015). Trichodimerol and sorbicillin induced apoptosis of HL-60 cells is mediated by reactive oxygen species. *Die Pharmazie-an International Journal of Pharmaceutical Sciences, 70*(6), 394–398. https://doi.org/10.1691/ph.2015.4868
Ye, G., Huang, C., Li, J., Chen, T., Tang, J., Liu, W., & Long, Y. (2021). Isolation, structural characterization and antidiabetic activity of new diketopiperazine alkaloids from mangrove endophytic fungus *Aspergillus sp.* 16-5c. *Marine Drugs, 19*(7), 402. https://doi.org/10.3390/md19070402
Yuan, G., Lin, H., Wang, C., Hong, K., Liu, Y., & Li, J. (2011). 1h and 13c assignments of two new macrocyclic lactones isolated from *Streptomyces sp.* 211726 and revised assignments of azalomycins f3a, f4aand f5a. *Magnetic Resonance in Chemistry, 49*(1), 30–37. https://doi.org/10.1002/mrc.2697
Zhang, H. L., Hua, H. M., Pei, Y. H., & Yao, X. S. (2004). Three new cytotoxic cyclic acylpeptides from marine *Bacillus sp. Chemical and Pharmaceutical Bulletin, 52*(8), 1029–1030. https://doi.org/10.1248/cpb.52.1029
Zhang, J. Y., Tao, L. Y., Liang, Y. J., Chen, L. M., Mi, Y. J., Zheng, L. S., Wang, F., She, Z. G., Lin, Y. C., To, K. K., & Fu, L. W. (2010). Anthracenedione derivatives as anticancer agents isolated from secondary metabolites of the mangrove endophytic fungi. *Marine Drugs, 8*(4), 1469–1481. https://doi.org/10.3390/md8041469
Zhou, M., Liu, X., Li, Z., Huang, Q., Li, F., & Li, C. Y. (2018). Caspase-3 regulates the migration, invasion and metastasis of colon cancer cells. *International Journal of Cancer, 143*(4), 921–930. https://doi.org/10.1002/ijc.31374

The *Cyanobacterial* Renaissance: Bioactive Discoveries and Future Prospects

Nital Nilesh Chankeshwara, Manu Pant, and Kumud Pant

Abstract

Blue-green algae, or cyanobacteria, are important to Earth's biological past and present ecosystem dynamics. These photosynthetic prokaryotes have been vital in forming the earth's biogeochemical cycles, with a fossil record spanning over 2.1 billion years. This abstract explores the many facets of cyanobacteria, including their crucial role in the cycling of nutrients, their capacity to fix nitrogen, and their methods for transforming carbon. One of the key mechanisms that keeps life on Earth alive is the capacity of cyanobacteria to fix nitrogen and transform atmospheric carbon dioxide into organic matter. Through photosynthesis, cyanobacteria utilize sunlight to change critical nutrients, contributing to the global cycle of nutrients. Their presence in maritime environments affects food webs and presents an opportunity as a sustainable food source for future generations. Moreover, the historical relevance of cyanobacteria is underscored by their role in the Great Oxidation Event, which resulted in the atmospheric oxygenation of Earth. The photosynthetic activity of cyanobacteria drove this critical point in Earth's history, opening the door for the emergence of aerobic life and the oxygen-rich environment we live in today. Concerns regarding the ecological effects and possible health dangers of cyanobacterial blooms have been raised by recent research that has revealed their rising frequency. To lessen the negative effects of these blooms on aquatic ecosystems, it is crucial to comprehend the biological characteristics of these blooms, the mechanisms underlying them, and potential control measures. In conclusion, from ancient times to the present, cyanobacteria have shaped Earth's ecosystems and now do. The fact that they can affect the dynamics of the food web, oxygen production, and nutrient cycling highlights how crucial they are to preserving ecological balance. With more research into the biological characteristics, dangers, and mitigation techniques related to cyanobacterial blooms, we can work toward a mutually beneficial coexistence between these venerable creatures.

Keywords

Cyanobacteria · Bioactive compounds · Pharmaceutical applications · Nutritional significance · Biotechnological applications

1 Introduction

Cyanobacteria, also called blue-green algae, have a special place in Earth's biological history and history that is multifaceted. With a fossil record spanning over 2.1 billion years, these photosynthetic prokaryotes are not only among the oldest living things on Earth, but they also play a significant role in determining the biogeochemical cycles of our planet. Even in the present, cyanobacteria remain essential, despite their historical significance. They fix nitrogen and transform atmospheric carbon dioxide into organic matter that can be used, which makes a major contribution to the cycle of nutrients. Because of their superior nutritional value, they are not only essential parts of marine food webs but also have the potential for a sustainable future. Their sensitivity to environmental changes and capacity to flourish in a variety of habitats makes them potential sources of food and feed.

These photosynthetic prokaryotes were essential to the development of the biosphere and atmosphere on Earth.

N. N. Chankeshwara · M. Pant · K. Pant (✉)
Department of Biotechnology, Graphic Era Deemed to Be University, Dehradun, Uttarakhand, India
e-mail: kumud.pant@geu.ac.in

K. Sharma (ed.), *Bio-prospecting of Novel Microbial Bioactive Compounds for Sustainable Development*, Sustainable Landscape Planning and Natural Resources Management, https://doi.org/10.1007/978-3-031-82178-3_13

The Great Oxidation Event, which occurred approximately 2.4 billion years ago and resulted in the oxygenation of the atmosphere, is thought to have been caused by cyanobacteria (Schirrmeister et al., 2011). The earliest clear-cut fossils of cyanobacteria are from between 1.89 and 1.84 billion years ago. However, because the fossil record is limited and occasionally unclear, there is continuous disagreement regarding the chronology and pattern of cyanobacterial evolution. Molecular phylogenetic studies have provided additional insights but have also revealed discrepancies with the geological record, leading to continued uncertainty about cyanobacteria's origin and early diversification (Schirrmeister et al., 2011).

Previously, cyanobacteria were categorized according to their morphological characteristics, including specialized cells, colony structure, and cell shape (Schirrmeister et al., 2011). The five main divisions of cyanobacteria in this traditional taxonomy are as follows: Oscillatoriales (filamentous without heterocysts), Nostocales (filamentous with heterocysts), Stigonematales (filamentous with branching), and Chroococcales (unicellular) (Schirrmeister et al., 2011). However, this morphology-based classification scheme has proven problematic since similar morphologies have evolved independently several times, and current phylogenetic analyses do not consider any of the traditional sections to be monophyletic (Schirrmeister et al., 2011). Consequently, efforts have been made to create new cyanobacteria classification systems that take into account a greater variety of traits, such as physiological, biochemical, and genetic information (Schirrmeister et al., 2011).

Cyanobacteria are of great significance due to their pivotal role in the evolution of the Earth's atmosphere and biosphere. They were the first living things to evolve oxygenic photosynthesis, which led to the atmosphere's gradual aeration and the subsequent emergence of aerobic life forms (Schirrmeister et al., 2011). Apart from their historical significance, cyanobacteria still have a vital role in contemporary ecosystems. In numerous aquatic habitats, they function as the main producers, forming the foundation of the food chain and being essential to the cycles of carbon and nitrogen (Willis et al., 2022). In order to transform atmospheric nitrogen into forms that other organisms can use, certain cyanobacteria can also fix nitrogen (Willis et al., 2022). Additionally, there has been a great deal of commercial and scientific interest in cyanobacteria. They are an abundant source of a wide variety of secondary metabolites, some of which may find use in industry, agriculture, and medicine. For example, because of its high protein content and other advantageous properties, the cyanobacterium Spirulina is widely grown as a nutritional supplement. Yet, certain species of cyanobacteria are known to produce toxins that are dangerous to people, animals, and other living things. As a result, cyanobacteria can also have unfavorable effects. These toxins can be produced by cyanobacterial blooms in eutrophic water bodies, endangering both the environment and public health.

This chapter explores the vast range of bioactive substances that cyanobacteria create in greater detail, offering a deeper look into this intriguing world. The production of these secondary metabolites varies greatly throughout strains, displaying an impressive chemical variety. Realizing the full potential of these chemicals requires an understanding of the biochemical mechanisms that produce them. The pharmaceutical and therapeutic uses of these metabolites—in particular, their anticancer, antibacterial, antioxidant, and anti-aging properties—are the subject of ongoing research. Their potential is not limited to the medical field; nutraceuticals, cosmetics, and even agriculture show promise in using them. The mechanisms of action underlying these various biological activities will be examined in this chapter, including how cyanobacterial metabolites elicit their anti-inflammatory, antioxidant, and anticancer actions. By offering a thorough summary of what is currently known about cyanobacteria, their bioactive substances, and their potential applications, this chapter aims to highlight the multifaceted importance of these ancient life forms, from their role in shaping our planet's past to their exciting potential in shaping a sustainable future for various industries and sectors.

2 Nutritional Significance of Cyanobacteria

Cyanobacteria, a group of photosynthetic bacteria, is a high-quality food resource for intertidal grazers due to their high nutritional value, particularly in protein, carbohydrate, and calorific content. This nutritional significance extends to human consumption, with potential biotechnological and therapeutic applications. In addition, cyanobacteria play a crucial role in biological nutrient cycling in various habitats, improving soil nutrient content and using efficiency in crop production (Kumar et al., 2015). Also, cyanobacteria are a key component of the aquatic ecosystem and can be used as a bioreporter to measure nutrient availability, particularly Phosphorus, Nitrogen, and Iron.

Cyanobacteria are rich sources of essential nutrients and bioactive compounds, making them valuable in various fields. Studies have highlighted the significance of cyanobacteria in nutrient cycling and availability in aquatic and terrestrial environments. The nutrient composition of cyanobacteria varies depending on factors such as species, environmental conditions, and nutrient availability. Cyanobacteria can accumulate high levels of essential nutrients, including nitrogen, phosphorus, iron, potassium, calcium, magnesium, manganese, zinc, copper, and others.

Nutrient availability strongly influences cyanobacteria's growth, productivity, and metabolic activities.

In aquatic systems, cyanobacteria play a crucial role in primary productivity and nutrient cycling, serving as indicators of ecosystem health and nutrient availability. Research has focused on developing cyanobacterial bioreporters to accurately measure nutrient bioavailability in environmental samples. Luminescent cyanobacterial bioreporters, such as *Synechococcus spp.* and *Synechocystis spp.*, have been utilized to assess the bioavailability of phosphorus, nitrogen, and iron in aquatic systems. Understanding cyanobacteria's nutrient composition and bioavailability is essential for exploiting their potential in various applications, including nutrition, biotechnology, and environmental management.

The nutrient composition of cyanobacteria can vary depending on factors such as environmental conditions, nutrient availability, and treatment processes. Cyanobacteria are known for their high protein content, with some strains containing up to 78% protein, making them valuable protein sources. For instance, cyanobacteria bioconversion by black soldier fly larvae (BSFL) can alter the substrate's nutrient composition and affect the larvae's growth performance and nutrient yield. The sun-drying treatment process benefits BSFL growth, maintaining the optimal moisture content and reducing the interference from cyanobacteria toxins (Gu et al., 2024).

The bioavailability of nutrients from cyanobacteria can also be influenced by factors such as dissolved organic matter (DOM) and the composition of bacterial communities. For example, heterotrophic bacteria can rapidly utilize labile autochthonous DOM, while recalcitrant DOM can persist for longer periods, affecting nutrient availability and microbial community composition (Hoikkala et al., 2016). Cyanobacteria are a valuable source of essential nutrients for plants and can contribute to sustainable agriculture practices by reducing the need for agrochemicals and promoting soil fertility.

Cyanobacteria are significant in human and animal nutrition due to their nutrient-rich composition and potential health benefits. Cyanobacteria holds promise as a sustainable source of food for humans and animals. Their rich protein content makes them a valuable plant-based protein source, offering essential vitamins, minerals, and amino acids. Historically consumed in certain regions, some cyanobacterial strains are explored for potential health benefits ranging from antiviral to immune-boosting effects. In animal nutrition, cyanobacteria can serve as a feed supplement for livestock and a natural food source in aquaculture, promoting growth and health. However, careful monitoring of cyanobacterial toxins like microcystins is necessary to avoid potential harm to both humans and animals consuming contaminated sources (Gantar & Svirčev, 2008).

In addition to their high nutritional content, cyanobacteria present a sustainable and promising alternative to traditional food sources. Some of them, like *Chlorella spp.* and *Arthrospira platensis,* are high in protein and have been used in animal feed and human nutrition. They can be used as sources of protein. They can also be used as sources of carbohydrates, like cyanobacteria and microalgae, which can grow on wastewater. In addition to being a source of vitamins, lipids, and other nutrients, cyanobacteria have many uses. Bioreactors and wastewater cultivation are sustainable farming practices that minimize land use and environmental impact.

Some species of Cyanobacteria with commercial and nutritional potential:

1. *Aphanizomenon Flos-Aquae* (AFA):

Aphanizomenon flos-aquae (AFA) is a filamentous cyanobacterium that has gained attention for its rich nutritional profile and potential applications in food and health supplements. Found primarily in freshwater ecosystems, particularly Klamath Lake in Oregon, USA, AFA is recognized for its high protein content, essential amino acids, vitamins, and bioactive compounds. AFA's remarkable nutritional makeup has earned it the title of "superfood". It is one of the most abundant plant-based protein sources available, with between 60 and 70% of its protein coming from protein. A range of important vitamins, minerals, and amino acids are added to this protein. Interestingly, alpha-linolenic acid (ALA), one of the omega-3 polyunsaturated fatty acids (PUFAs) abundant in AFA, is essential for lowering inflammation and preserving cardiovascular health (Scoglio et al., 2024). AFA contains a number of vitamins in addition to its macronutrients, such as the B vitamins (B1, B3, B5, B9), vitamin A, vitamin E, and vitamin K. In addition, it has 73 trace minerals, including iodine, calcium, and iron, all of which are essential for different physiological processes. The health benefits of AFA are further enhanced by the presence of bioactive compounds such as C-phycocyanin, which is an antioxidant and anti-inflammatory agent. Furthermore, AFA produces β-phenylethylamine (PEA), a compound associated with mood enhancement and cognitive function (Scoglio et al., 2024).

The consumption of AFA has been linked to numerous health benefits. Its high protein content supports muscle growth and repair, making it an excellent supplement for athletes and those engaged in physical activity. The omega-3 fatty acids found in AFA contribute to cardiovascular health by reducing triglyceride levels and improving overall heart function. Moreover, the antioxidant properties of C-phycocyanin and other pigments in AFA help combat oxidative stress, which is associated with chronic diseases such as cancer and heart disease. Additionally, AFA's potential as a natural source of vitamin B12 has garnered interest, particularly among vegetarians and vegans. While AFA primarily produces pseudocobalamin, studies have indicated that it may also contain bioactive forms of vitamin B12, thereby addressing the dietary needs of those who avoid animal products (Scoglio et al., 2024).

Currently, AFA is primarily harvested from natural blooms in Klamath Lake, where it is processed and marketed as a nutritional supplement. The global market for AFA supplements has been growing, positioning it as the second most consumed cyanobacterial supplement after Spirulina. However, the reliance on wild harvesting poses challenges, including seasonal availability and potential contamination with harmful toxins like microcystins, which are produced by other cyanobacteria in the same ecosystem. To overcome these limitations, researchers are exploring the cultivation of AFA in controlled environments, such as photobioreactors. This approach could provide a consistent supply of AFA biomass while minimizing the risks associated with wild harvesting. Additionally, advances in genetic engineering could lead to the development of bespoke AFA strains with enhanced nutritional profiles or increased resistance to environmental stressors, further expanding its commercial potential (Scoglio et al., 2024).

Nostoc

In addition to its ecological significance, nostoc has a high nutritional value. Many cultures consume food species like Nostoc commune and *Nostoc flagelliforme*, especially in Asian cuisines. They are an important food source because they are high in vitamins, minerals, and proteins. For example, research has indicated that *Nostoc sphaericum* contains roughly 28% protein, 62% carbohydrates, and important minerals like iron and calcium, which can help prevent anaemia and malnutrition in human populations (Méndez-Ancca et al., 2023).

3 Bioactive Compounds in Cyanobacteria

3.1 Biosynthesis and Diversity of Cyanobacterial Metabolites

Cyanobacteria are a prolific source of diverse bioactive compounds with a wide range of biological activities and applications. These photosynthetic bacteria are known to produce a variety of secondary metabolites, including toxins, antimicrobials, anticancer agents, and other bioactive molecules (Tiam et al., 2019). These include polyketides and non-ribosomal peptides with demonstrated cytotoxicity, antimicrobial, and antiviral activities. Additionally, cyanobacteria produce a range of fatty acids and lipids with nutraceutical and pharmaceutical potential. Notably, phycocyanin, a blue pigment produced by these organisms, holds promise for applications in fluorescent labelling, food colouring, and potentially even therapeutics. Beyond these well-characterized examples, cyanobacteria also generate a repertoire of other bioactive compounds like alkaloids, terpenoids, and saccharides, further highlighting their potential for drug discovery. Different cyanobacterial species have a vast variety of bioactive compounds mentioned in Table 1.

The diversity of bioactive compounds produced by cyanobacteria is closely linked to their geographic distribution and ecological niches. Tropical marine ecosystems, in particular, are characterized by a high biodiversity of cyanobacteria, with species adapted to various habitats such as intertidal and infralittoral mats, endolithic communities in carbonate substrates, and symbiotic associations with sponges and ascidians (Hoffmann, 1999). Studies have shown that the production of bioactive compounds by cyanobacteria is often strain-specific, with closely related strains exhibiting remarkable chemo diversity (Tiam et al., 2019). This observation has prompted the use of cellular oligopeptide compositions as biomarkers to identify distinct cyanobacterial subpopulations or chemotypes within a species (Agha & Quesada, 2014). Furthermore, the biosynthesis

Table 1 Cyanobacterial species and their bioactive compounds

Species name	Bioactive compound	References
Microcystis aeruginosa	Amorphadiene, farnesene, isoprene, beta-phellandrene, squalene, geranyllinalool, manoyl oxide, Microcystins, Aerucyclamide	Bouyahya et al. (2024)
Lyngbya majuscula	Aurilides B and C	Bouyahya et al. (2024)
Anabaena sp.	Anacyclamide	Nandagopal et al. (2021)
Nostoc sp.	Nostoc peptides	Bouyahya et al. (2024)
Marine cyanobacteria	Patellamide A and C	Bouyahya et al. (2024)

of these bioactive compounds is often mediated by non-ribosomal peptide synthetases (NRPS) and polyketide synthases (PKS), which are encoded by gene clusters in the cyanobacterial genomes (Bhaskar & Shahi, 2020). Bioinformatic analyses of these gene clusters have revealed the potential for the production of a wide range of natural products, including many yet to be discovered (Bhaskar & Shahi, 2020).

3.2 Pharmaceutical and Therapeutic Applications

Cyanobacteria are a rich source of natural products with anticancer properties, antimicrobial properties antioxidant and antiaging properties. Compounds such as Apratoxin, Symplostatin, Bartolosides, Caylobolide, Bisebromoamides, Carmaphycins, and Anaenamides have demonstrated potent cytotoxicity against various cancer cell lines (Verma et al., 2022). These compounds target different cellular processes, including microtubule dynamics, protein synthesis, and signaling pathways, leading to the inhibition of cancer cell proliferation and induction of apoptosis (Verma et al., 2022). The development of synthetic analogs of these natural products, such as the dolastatin 10 analogue, has further advanced their potential as anticancer agents (Verma et al., 2022). Ongoing research continues to uncover the vast anticancer potential of cyanobacterial metabolites, with promising results in both in vitro and in vivo studies. The exploration of cyanobacterial diversity and the elucidation of their biosynthetic pathways hold the key to the discovery of additional potent anticancer compounds.

Similarly, Cyanobacteria have also emerged as a promising source of novel antimicrobial compounds to address the growing threat of antimicrobial resistance. Numerous cyanobacterial secondary metabolites have exhibited potent antibacterial, antifungal, and antiviral activities, targeting many pathogenic microorganisms. Several species of cyanobacteria have been studied for their antimicrobial activities. *Arthrospira platensis*, also known as Spirulina, is a well-known cyanobacterium that has been studied for its bioactive compounds with potential antimicrobial activities (Liu et al., 2019). *Oscillatoria agardhii* is another cyanobacterial species studied for its bioactive compounds, including oscillapeptin G, with potential antimicrobial properties (Sano & Kaya, 1996). *Haematococcus pluvialis*, a microalgae species, has been researched for its bioactive compounds, including mono and diesters of astaxanthin, which have shown inhibitory effects on tyrosinase (Rao et al., 2013). Another microalgal species, *Nannochloropsis oculata*, has been studied for its compounds, such as zeaxanthin, which have demonstrated inhibitory effects on tyrosinase (Shen et al., 2011). These examples highlight the diverse antimicrobial potential of cyanobacteria and microalgae, which can be further explored for their bioactive compounds and applications in antimicrobial research and product development.

In addition to their therapeutic potential, cyanobacterial metabolites have also demonstrated beneficial effects in the context of skin health and aging. Compounds such as carotenoids, phenols, and chlorophyll derivatives have exhibited potent antioxidant activities, which can help mitigate the deleterious effects of UV radiation and oxidative stress on the skin (Ferreira et al., 2022). Furthermore, some cyanobacterial extracts have shown the ability to inhibit matrix metalloproteinases (MMPs) and tyrosinase, enzymes involved in the degradation of extracellular matrix components and melanin production, respectively (Ferreira et al., 2022).

These properties suggest the potential of cyanobacterial compounds in the development of anti-aging cosmetic and dermatological products. The antioxidant and skin-protective properties of cyanobacterial metabolites make them attractive candidates for use in various cosmetic and personal care applications, such as sunscreens, skin-lightening agents, and anti-aging formulations. The scalable production of cyanobacterial biomass and bioactive compounds using biotechnological approaches further enhances their potential as sustainable and eco-friendly sources of cosmetic ingredients. Several potential uses of bioactive compounds are mentioned in Table 2.

3.3 Biotechnological Applications of Cyanobacterial Bioactive Compounds

The various bioactive compounds produced by cyanobacteria have a wide range of potential biotechnological applications, including in the pharmaceutical, nutraceutical, cosmetic, and agricultural sectors. Cyanobacterial bioactive compounds, such as polyketides, non-ribosomal peptides, fatty acids, and pigments, have demonstrated potent pharmacological activities, including anticancer, antimicrobial, anti-inflammatory, and neuroprotective effects (Jeyaraj, 2024). Several cyanobacterial compounds have reached successful clinical trials for the development of new medicines, including potential treatments for viral infections like COVID-19 (Jeyaraj, 2024). In addition, cyanobacterial extracts have been found to have antioxidant, anti-inflammatory, and antimicrobial properties, making them potential candidates for use in functional foods and dietary supplements (Lau et al., 2015). Cyanobacterial metabolites, particularly pigments and

Table 2 Novel bioactive agents from cyanobacteria and their potential effects and applications

Cyanobacterial species	Bioactive compound	Potential effects	Applications	References
Arthrospira platensis	Phycocyanin	Antioxidant, anti- nflammatory, anticancer	Nutritional supplements, cosmetics	Zahra et al. (2020)
Spirulina (Arthrospira)	Beta-Carotene	Antioxidant, supports eye health, immune function	Food supplements, health products	Zahra et al. (2020)
Nostoc spp.	Cryptophycin	Anticancer properties, effective against multi- drug-resistant tumors	Cancer therapy research	Zahra et al. (2020)
Microcystis aeruginosa	Microcystins	Hepatotoxic, potential for cancer treatment under controlled conditions	Monitoring and bioremediation	Zahra et al. (2020)
Synechocystis sp. PCC 6803	Ethanol	Biofuel production, potential for metabolic engineering	Biofuel production	Zahra et al. (2020)
Anabaena spp.	Nitrogenase	Nitrogen fixation, enhances soil fertility	Biofertilizers in agriculture	Zahra et al. (2020)
Aphanothece sacrum	Sacran	Moisturizing agent, potential in cosmetics	Cosmetics and skincare products	Zahra et al. (2020)
Synechococcus elongatus	Fatty acids	Source for biofuels, high energy content	Biofuel production	Zahra et al. (2020)
Lyngbya spp.	Antimicrobial compounds	Antibacterial, antifungal	Pharmaceutical applications	Zahra et al. (2020), Mazard et al. (2016)
Chlorella spp.	Chlorophyll	Antioxidant, supports detoxification	Nutritional supplements	Zahra et al. (2020)
Pseudanabaena spp.	Exopolysaccharides	Enhances soil moisture retention	Soil conditioning and agriculture	Zahra et al. (2020)
Oscillatoria spp.	Antimicrobial compounds	Antibacterial, antifungal	Pharmaceutical applications	Mazard et al. (2016)
Anabaena cylindrica	Carotenoids	Antioxidant, anti- nflammatory	Nutraceuticals, cosmetics	Bhattacharya et al. (2016)
Calothrix spp.	Exopolysaccharides	Bioremediation of heavy metals	Environmental remediation	Parikh and Madamwar (2006)
Nostoc commune	Scytonemin	UV protection, antioxidant	Sunscreens, cosmetics	Soule et al. (2009)

other bioactive compounds like carotenoids, phenolics, and mycosporine-like amino acids, have shown promise for use in cosmetic and dermatological products, exhibiting potent antioxidant and UV-protective properties (Chiaiese et al., 2018). Furthermore, cyanobacterial extracts have demonstrated the ability to inhibit enzymes involved in skin aging, suggesting their potential in anti-aging and skin-lightening formulations (Chiaiese et al., 2018). Certain cyanobacterial metabolites have also shown biocidal activities against various pests and pathogens, indicating their potential as biopesticides and biofungicides in sustainable agriculture, while cyanobacterial exopolysaccharides and other bioactive compounds have demonstrated the ability to promote plant growth, enhance soil fertility, and improve crop yields, making them valuable for agricultural applications.

4 Mechanism of Action and Biological Activities

Cyanobacteria are a rich source of bioactive compounds that exhibit diverse biological activities, including anticancer, anti-inflammatory, antiviral, and antibacterial effects. Some of the key mechanisms of action of these cyanobacteria-derived bioactive compounds include:

Anti-inflammatory Effects: Polysaccharides extracted from Nostoc colonies have been shown to reduce inflammation in various biological models, exhibiting antioxidant properties that can help neutralize free radicals and reduce oxidative stress, contributing to their anti-inflammatory effects. Certain cyanobacteria-derived compounds have been found to inhibit the production of pro-inflammatory cytokines, such as interleukin-6 (IL-6) and tumor necrosis factor-alpha (TNF-α), thereby reducing inflammation (Tabarzad et al., 2020). Additionally, some bioactive compounds from cyanobacteria have been reported to modulate signalling pathways involved in inflammation, such as the NF-κB and MAPK pathways, which suppress inflammatory responses (Tabarzad et al., 2020). The antioxidant activity of cyanobacterial polysaccharides can also help mitigate oxidative stress and the associated inflammatory processes. Overall, cyanobacteria-derived bioactive compounds have demonstrated anti-inflammatory effects through multiple mechanisms, including the inhibition of pro-inflammatory mediators, modulation of signalling pathways, and antioxidant activity (Tabarzad et al., 2020).

Antioxidant Effect: Cyanobacteria are a rich source of bioactive compounds with potent antioxidant properties. Polysaccharides extracted from Nostoc colonies have demonstrated the ability to neutralize free radicals and reduce oxidative stress, contributing to their antioxidant activity (Perera et al., 2023). The methanolic extracts of *Leptolyngbya sp.* and *Desertifilum sp.* cyanobacteria also exhibited free radical scavenging activity against DPPH, indicating their antioxidant potential. GC-MS analysis of these extracts revealed the presence of phenolic compounds, saturated and unsaturated fatty acids, which likely contribute to their antioxidant properties. The antioxidant activity of cyanobacteria-derived compounds, particularly polysaccharides and phenolic compounds, can help mitigate oxidative stress and the associated inflammatory processes (Perera et al., 2023). These findings highlight the antioxidant mechanisms of cyanobacterial bioactive compounds and their potential to reduce oxidative stress-related diseases and promote overall health.

Anticancer Effects: Several bioactive compounds derived from cyanobacteria have been found to exhibit anticancer properties. For example, Apratoxin, Symplostatin-1, Bartolosides, Caylobolide, Bis-bromoamides, Carmaphycins, and Anaenamides have been reported to demonstrate anticancer effects against multiple cancer cell lines (Hussein et al., 2023). These compounds have been found to inhibit cell proliferation, induce apoptosis, and disrupt microtubules, leading to the suppression of cancer cell growth (Hussein et al., 2023).

Antimicrobial Effects: Cyanobacteria-derived bioactive compounds have also been found to exhibit antimicrobial properties. For instance, picocyanobacteria from the marine Cyanobium genus have been shown to inhibit the growth of marine microalgae, bacteria, and marine invertebrates (Costa et al., 2015). These compounds have been found to exhibit antimicrobial activity against a range of microorganisms, including bacteria and fungi (Costa et al., 2015).

The mode of action of cyanobacteria-derived bioactive compounds is diverse and depends on the specific compound and its target. For example, some compounds have been found to inhibit cell proliferation by disrupting microtubules, while others have been shown to induce apoptosis by activating specific signalling pathways (Hussein et al., 2023). Additionally, some compounds have been found to exhibit antioxidant and anti-inflammatory activities by neutralizing free radicals and reducing oxidative stress.

5 Genetic Modifications and Biotechnological Applications

Cyanobacteria gained significant interest in the field of genetic engineering due to their potential applications in biofuel production, carbon sequestration, and the synthesis of valuable compounds. Several techniques have been developed for their genetic modification, including transformation, the most commonly used method involving the uptake of DNA by cyanobacterial cells through various methods such as electroporation, conjugation, or natural competence; conjugation, the transfer of genetic material from a donor bacterium to a recipient cyanobacterium through cell-to-cell contact, often used to introduce large DNA fragments or entire plasmids; transduction, the use of bacteriophages as vectors to deliver genetic material into cyanobacterial cells, and genome editing, the development of tools like CRISPR-Cas9 allowing for precise, targeted modifications of the cyanobacterial genome, enabling specific gene insertions, deletions, or replacements.

5.1 Application of Genetic Engineering of Cyanobacteria

One of the key applications of genetic engineering in cyanobacteria is the production of biofuels. Cyanobacteria have garnered significant interest as potential biofuel feedstocks due to their ability to convert carbon dioxide and sunlight into energy-rich compounds directly. Genetic engineering approaches have enabled the manipulation of cyanobacterial metabolic pathways to enhance the production of various biofuel precursors, such as alcohols, alkanes, and lipids. By introducing heterologous genes or modifying endogenous pathways, researchers have been able to redirect the carbon flux in cyanobacteria towards the synthesis of these valuable compounds. While genetic engineering of cyanobacteria holds great promise, several challenges and considerations need to be addressed, such as genetic stability, optimization of metabolic pathways, scalability and bioreactor designs, and regulatory considerations.

5.2 Potential for Enhanced Productivity and Bioactive Compound Production

Cyanobacteria have demonstrated remarkable potential for enhanced productivity and the production of valuable bioactive compounds. Recent advancements in cyanobacterial research have unlocked new opportunities in this field. Under salt stress conditions, researchers have engineered the rapidly proliferating cyanobacterium Synechococcus UTEX 2973 to produce high titers of sucrose, with a productivity of up to 1.9 g/L/day (Lin et al., 2020). This is the maximum amount of sucrose that cyanobacteria have been known to produce when using CO_2 as their sole carbon source. By disrupting genes related to fatty acid recycling and surface layer proteins in *Synechocystis sp.* PCC 6803, researchers have significantly enhanced the secretion of free fatty acids, particularly under nitrogen-deficient conditions (Singh & Ducat, 2019). This demonstrates the potential for cyanobacteria to serve as efficient platforms for biofuel and biochemical production. Cyanobacteria have been engineered to produce the valuable terpene compound limonene, a key ingredient in many fragrances and flavours (Singh & Ducat, 2019). The breakthrough in limonene production allowed for systems biology analysis to reveal photosynthesis regulations under high-level limonene production.

In a notable study, researchers engineered *Synechocystis sp.* PCC 6803 to improve its ability to produce alkanes. The process involved the following steps (Singh & Ducat, 2019) (Fig. 1):

1. Identification of Biosynthetic Pathways: A two-step alkane biosynthetic pathway was identified, which includes:
   - Reduction of acyl-acyl carrier protein (ACP) to aldehyde by acyl-ACP reductase (AAR).
   - Oxidation of aldehyde to alkane by aldehyde-deformylating oxygenase (ADO).
2. Gene Overexpression: The genes encoding AAR and ADO were overexpressed in engineered strains of *Synechocystis* to enhance the flux of carbon toward alkane production.
3. Mutant Strain Development: Various mutant strains were constructed, and one particular strain (designated LX56) showed a significant increase in alkane production up to 1.3% of cell dry weight, which is an 8.3-fold increase compared to the wild-type strain.
4. Cultivation and Production: The engineered strains were cultivated in column photo-bioreactors, demonstrating enhanced alkane production rates, further validating the effectiveness of the genetic modifications.

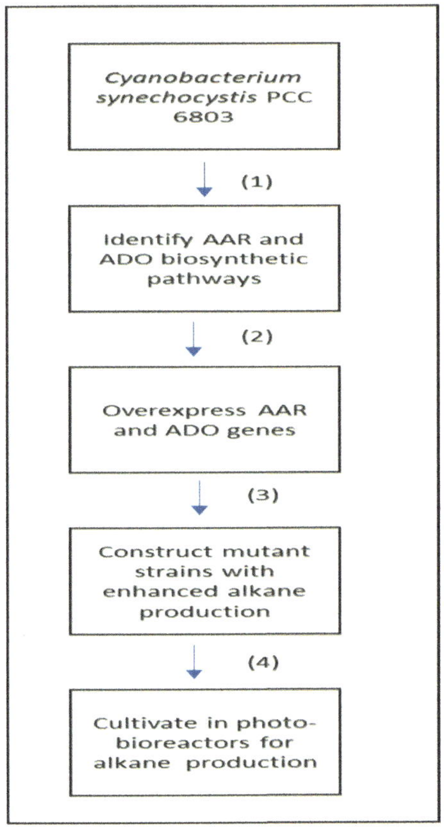

Fig. 1 Genetic modification process used for *Cyanobacterium synechocystis*—(1) Gene identification; (2) Genetic engineering; (3) Mutant development; (4) Cultivation

These advancements in cyanobacterial research highlight the immense potential for enhancing productivity and the production of diverse bioactive compounds, paving the way for sustainable biotechnological applications.

6 Ecological Role and Ecosystem Management

Cyanobacteria are primary producers in many ecosystems, contributing significantly to the overall productivity. They are found in a wide range of habitats, from marine environments to freshwater lakes, rivers, and even terrestrial environments.

6.1 Ecological Functions of Cyanobacteria in Aquatic and Terrestrial Environments

Cyanobacteria are important nitrogen fixers, converting atmospheric nitrogen into biologically available forms through nitrogenase enzymes (Bonab et al., 2019). This nitrogen fixation ability allows cyanobacteria to thrive in nitrogen-limited environments and provide essential nutrients to other organisms. Studies have shown that soil cyanobacteria can express the nif-H gene responsible for nitrogen fixation, with some species like *Nodularia sp.* exhibiting high nitrogen.

fixation activity. Cyanobacteria play a significant role in the global carbon cycle through photosynthesis, converting carbon dioxide into organic carbon compounds (Erratt & Freeman, 2024). Erratt and Freeman (2024) studied the carbon-concentrating mechanisms (CCMs) in cyanobacteria, facilitated by carbonic anhydrase enzymes, which allow them to efficiently utilize inorganic carbon sources and contribute to carbon sequestration. Cyanobacterial blooms can lead to the burial of organic carbon in sediments, potentially sequestering carbon in aquatic environments (Erratt & Freeman, 2024).

Cyanobacteria are involved in various nutrient cycling processes, including the cycling of nitrogen, phosphorus, and sulphur (Bonab et al., 2019). Their ability to fix nitrogen and produce biomass can contribute to the availability of nutrients in aquatic and terrestrial ecosystems (Bonab et al., 2019). Symbiotic relationships between cyanobacteria and other organisms, such as sponges and macroalgae, can further enhance nutrient cycling in marine environments. Cyanobacteria can form stable communities and contribute to the overall stability and resilience of aquatic and terrestrial ecosystems (Erratt & Freeman, 2024). Their adaptability to a wide range of environmental conditions, including extreme temperatures, salinity, and UV radiation, allows them to thrive in diverse habitats (Erratt & Freeman, 2024). The production of secondary metabolites by cyanobacteria can act as feeding deterrents, protecting them from grazers and contributing to the overall ecosystem balance. Cyanobacteria and their roles in ecosystem management are mentioned below in Table 3.

6.2 Impact of Cyanobacterial Blooms and Management Strategies

Cyanobacterial blooms, often referred to as harmful algal blooms (HABs), present significant environmental challenges in freshwater ecosystems globally. These blooms can severely impact aquatic biodiversity, water quality, and human health. Cyanobacteria can outcompete other phytoplankton and aquatic plants, leading to reduced biodiversity and disrupted food webs. The decomposition of cyanobacterial biomass can deplete dissolved oxygen levels in water, resulting in hypoxic or anoxic conditions that can cause fish kills and mortality in other aquatic organisms (Smucker et al., 2021). Additionally, cyanobacterial toxins can bioaccumulate in the food chain, affecting higher trophic levels and posing health risks to aquatic organisms and humans.

Cyanobacterial blooms can produce taste and odour compounds, rendering water unpalatable for drinking and recreational use. Some species of cyanobacteria produce potent toxins, such as microcystins and cylindrospermopsins, which can contaminate drinking water supplies and threaten human and animal health (Wasswa et al., 2024). Exposure to these toxins can occur through drinking water, recreational activities, or consumption of contaminated seafood, leading to various health issues, including gastrointestinal illness, liver damage, neurological effects, and potentially cancer. Moreover, aerosolized toxins can be inhaled, causing respiratory problems and skin irritation (Wasswa et al., 2024).

The frequency and severity of cyanobacterial blooms have been increasing, driven by factors such as nutrient pollution, climate change, and other anthropogenic environmental changes. Addressing this issue necessitates a multifaceted management approach. Key strategies include reducing nutrient inputs, particularly nitrogen and phosphorus, through improved wastewater treatment and

Table 3 Cyanobacteria and their roles in ecosystem management

Cyanobacteria species	Roles in ecosystem management
Anabaena	– Nitrogen fixation – Symbiotic relationships with plants, particularly in rice paddies – Enhancing soil fertility
Nostoc	– Nitrogen fixation – Symbiotic relationships with plants and fungi – Contributing to soil health and fertility – Stabilizing soil structure
Spirulina (*Arthrospira*)	– Carbon fixation – Oxygen production – Nutrient cycling in various ecosystems
Microcystis	– Contributing to nutrient cycling in freshwater ecosystems – Managing harmful algal blooms to maintain water quality and ecosystem health

agricultural best management practices (Chen et al., 2023). Biomanipulation, which involves using biological agents like filter-feeding fish to control cyanobacterial populations, has shown promise in some cases. However, the effectiveness of such strategies can vary based on the trophic status of the waterbody and the specific cyanobacterial species present (Chen et al., 2023).

6.3 Contributions to Nutrient Cycling and Ecosystem Stability

Cyanobacteria play a crucial role in nutrient cycling and maintaining ecosystem stability in both aquatic and terrestrial environments. Their ability to fix nitrogen and produce organic compounds through photosynthesis makes them essential components of nutrient cycles. One of the key contributions of cyanobacteria to nutrient cycling is their ability to fix atmospheric nitrogen (Lobus & Kulikovskiy, 2022). Many cyanobacterial species possess specialized cells called heterocysts that are capable of converting atmospheric nitrogen (N_2) into ammonia (NH_3), which can then be utilized by other organisms. This process of nitrogen fixation is particularly important in aquatic ecosystems where nitrogen may be a limiting nutrient, as it provides a source of bioavailable nitrogen that can be assimilated by other phytoplankton, aquatic plants, and higher trophic levels (Lobus & Kulikovskiy, 2022). Additionally, the decomposition of cyanobacterial biomass can release nitrogen and other nutrients back into the water, further contributing to the cycling of essential elements. Cyanobacteria also play a significant role in the cycling of carbon and other nutrients, such as phosphorus and sulphur (Joshi & Jackson, 2021; Lobus & Kulikovskiy, 2022). Through photosynthesis, cyanobacteria convert carbon dioxide (CO_2) into organic carbon compounds, which can be incorporated into the aquatic food web or sequestered in sediments. This carbon fixation process not only supports the growth of other organisms but also helps to regulate the overall carbon balance in the ecosystem.

Furthermore, cyanobacteria can contribute to the stability of aquatic ecosystems by outcompeting other phytoplankton species under certain environmental conditions (Zhao et al., 2023). Cyanobacteria are often well-adapted to thrive in nutrient-rich, eutrophic environments, where they can form dense blooms that can dominate the phytoplankton community. This can have both positive and negative implications for ecosystem stability. On the positive side, the dominance of cyanobacteria can help maintain the overall productivity and biomass of the ecosystem, ensuring a steady supply of energy and nutrients for higher trophic levels. However, the formation of harmful cyanobacterial blooms can also lead to the depletion of dissolved oxygen, the production of toxic secondary metabolites, and the disruption of the natural balance of the aquatic community, which can have detrimental effects on the overall ecosystem stability (Zhao et al., 2023).

7 Cultivation Strategies and Bioreactor Systems

Cyanobacteria are a group of photosynthetic microorganisms with significant potential for various applications, including biofuel production, wastewater treatment, and the generation of high-value bioproducts. The cultivation of cyanobacteria requires careful optimization of several factors to ensure efficient growth and biomass production. The cultivation of cyanobacteria can be achieved through various techniques, each with its advantages and applications. Photobioreactors (PBRs) are closed systems that allow for precise control over environmental factors such as light, temperature, and pH, enabling optimal growth conditions (Novoveská et al., 2023). Open ponds, on the other hand, are large, shallow, outdoor systems suitable for large-scale cultivation, although they are relatively inexpensive to set up and prone to contamination (Novoveská et al., 2023). Biofilms, which are complex communities of microorganisms adhering to surfaces, offer a stable environment for cyanobacteria, promoting efficient nutrient uptake and waste management, particularly in wastewater treatment and bioremediation applications. Additionally, cyanobacteria can be cultivated on various substrates, including industrial wastewater and agricultural byproducts, providing a cost-effective and sustainable method while reducing the environmental impact of these waste streams (Novoveská et al., 2023). Optimizing the growth of cyanobacteria requires a multifaceted approach. Nutrient availability, particularly the balance of nitrogen and phosphorus, is crucial for enhancing biomass production and improving the quality of the biomass, such as increasing the yield of polyhydroxybutyrate (PHB) (Novoveská et al., 2023). Optimizing light intensity and spectral distribution, using tailored light-emitting diodes (LEDs), can improve photosynthetic efficiency and reduce energy costs. Temperature control is also critical, as different cyanobacteria species thrive at varying optimal temperatures. Proper mixing and aeration are essential for maintaining uniform growth conditions and preventing the formation of hotspots or dead zones. Finally, regular monitoring of environmental parameters and early detection of contamination, using techniques like real-time PCR and flow cytometry, are crucial for maintaining optimal growth conditions (Novoveská et al., 2023).

Cyanobacterial growth and productivity are influenced by a complex interplay of factors, including nutrient availability, temperature, light, water chemistry, hydrological conditions, and human activities. Nutrient availability is a crucial factor, as evidenced by a massive bloom of *Dolichospermum lemmermannii* in Lake Baikal in 2019, which was attributed to elevated nutrient concentrations following heavy rains. The nutrient levels were significantly higher than the long-term averages, with phosphate concentrations reaching up to 0.200 mg L^{-1},

ammonium ions at 0.29 mg L^{-1}, and nitrates at 0.31 mg L^{-1}, highlighting the importance of nutrient availability in promoting cyanobacterial growth. Temperature and light are also essential environmental factors that influence cyanobacterial growth, as seen in Lake Baikal, where the optimal temperature range for the growth of *D. lemmermannii* was between 15 °C and 19.2 °C, with high air temperatures (from 13 to 15 °C at night and 29 °C during the day) also contributing to the bloom (Bondarenko et al., 2021). These conditions provide the necessary thermal energy for photosynthesis and metabolic processes. Water chemistry, including pH, salinity, and dissolved oxygen levels, can also significantly impact cyanobacterial growth, as observed in Lake Baikal, where the bloom occurred in the coastal zone, where the water chemistry is influenced by the proximity to the shoreline and the presence of nutrients. The absence of isolated septic tanks in the area also contributed to the high nutrient levels, which further supported the growth of *D. lemmermannii* (Bondarenko et al., 2021). Hydrological conditions, such as water flow and mixing, can affect the distribution and growth of cyanobacteria, as seen in Lake Baikal, where the absence of wind and the presence of long-lasting heavy rains create conditions favourable for the bloom. The calm water allowed the cyanobacteria to accumulate and proliferate without being dispersed by currents (Bondarenko et al., 2021). Human activities, including agricultural runoff and wastewater discharge, can introduce nutrients and pollutants into aquatic ecosystems, promoting the growth of cyanobacteria. In Lake Baikal, the presence of a biological station and the hosting of conferences at the time of the bloom may have contributed to the increased nutrient levels through human activities (Bondarenko et al., 2021). Lake Baikal, the world's deepest lake, is known for its unique ecosystem and biodiversity. However, the lake has experienced several cyanobacterial blooms in recent years, including a significant one in 2019. The bloom of *D. lemmermannii* in Lake Baikal was characterized by wide bands of cyanobacteria along the shoreline, stretching over 2 km. The abundance of cyanobacteria varied from 7.2 to 71.9 thousand cells per millilitre, with a biomass of 0.73 to 7.20 g per cubic meter (Bondarenko et al., 2021).

Sustainable practices for large-scale production of cyanobacteria should focus on optimizing growth conditions while minimizing environmental impact. Carefully control nutrient inputs, temperature, and light particularly nitrogen and phosphorus, to meet growth requirements without causing excessive nutrient loading in the environment. Implementation of water reuse and recycling strategies to minimize freshwater usage and wastewater discharge. By adopting these sustainable practices, large-scale cyanobacterial production can be conducted in an environmentally responsible manner, minimizing resource consumption and waste generation while maintaining high productivity.

8 Industrial and Commercial Uses

Cyanobacteria have a wide range of applications in various industries. Cyanobacteria are a promising source for biofuel production due to their ability to convert carbon dioxide into lipids and other biofuel precursors (Amadu et al., 2021). Certain cyanobacteria like Spirulina (*Arthrospira*) are considered a rich source of vitamins, minerals, proteins, and antioxidants, making them valuable as food supplements and nutraceuticals (Zahra et al., 2020). Furthermore, Cyanobacterial hydrogen has been considered a promising source of alternative energy and is now commercially available. Certain mutant forms can produce hydrogen under specific growth conditions (Amadu et al., 2021).

Cyanobacteria have a diverse variety of commercial uses; Cyanobacteria are used in skin treatments, anti-aging creams, detoxifying masks, and cleansers due to their thickening properties and claimed antioxidant, mineral, and moisturizing benefits. Diatom algae with glass-like shells are used as natural abrasives in toothpaste and other personal care products. Cyanobacterial pigments like carotenoids (e.g. canthaxanthin, beta-carotene) and phycobiliproteins have high commercial value as food colorants, animal feed supplements, and for human health applications. Cyanobacteria can be used to produce a wide range of pigments and dyes with industrial applications (Amadu et al., 2021). Although cyanobacteria have different applications mentioned in Fig. 2. Further research needed to be done for the successful use of cyanobacteria.

9 Future Perspectives and Emerging Trends

Cyanobacteria have gained significant attention in recent years due to their remarkable potential in various applications, from bioremediation to bioenergy production. As our understanding of these photosynthetic microorganisms continues to evolve, several key advancements have emerged in cyanobacterial research.

Researchers are now more successful in exploring the unexplored biosynthetic potential of cyanobacteria via developments in omics technology, bioinformatics, and

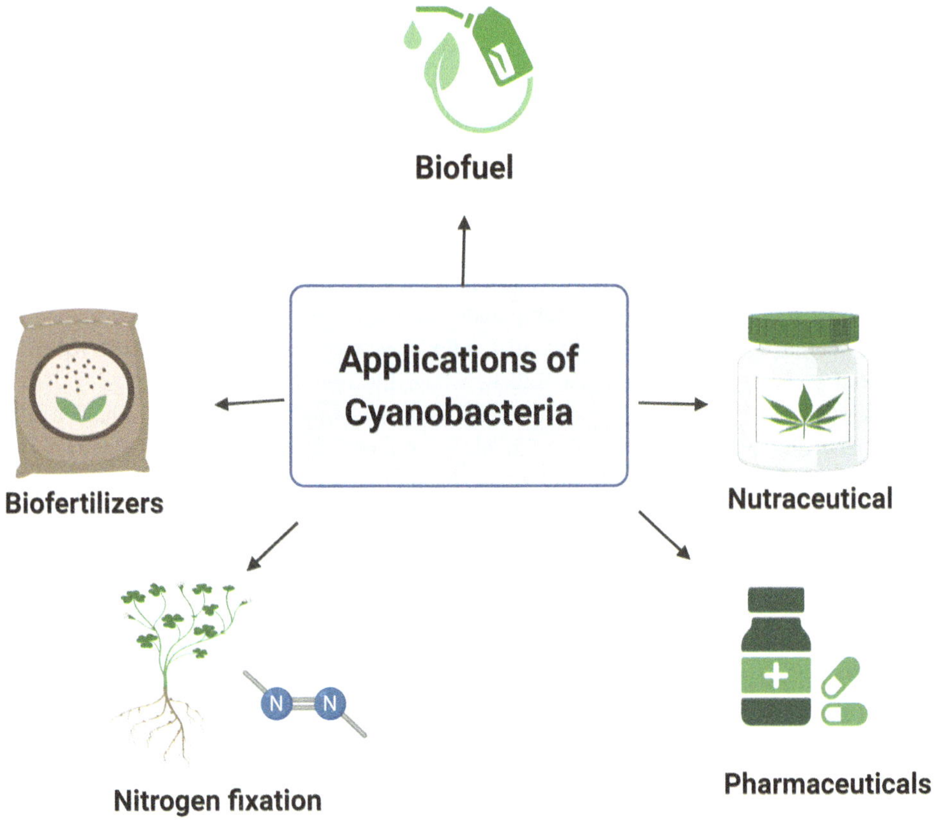

Fig. 2 Applications of cyanobacteria

machine learning. Numerous mysterious metabolic pathways have been revealed by genome analysis methods, which have resulted in the discovery of novel bioactive natural compounds with biotechnological and medicinal uses (Baunach et al., 2024). Furthermore, scientists are now able to engineer cyanobacteria to produce a variety of important substances, from specialized chemicals to biofuels, thanks to the development of strong genetic tools and synthetic biology approaches. According to Burford et al. (2020), cyanobacteria have shown an extraordinary capacity for resilience and adaptation in the face of a wide range of environmental stresses, including UV radiation, temperature changes, and nutrition availability. The ways in which cyanobacteria adapt to these difficulties have been the subject of recent research, which has shed light on their potential for bioremediation and survival in the face of global climate change (Burford et al., 2020; Zahra et al., 2020). Because these blooms produce strong toxins and upset aquatic food webs, the spread of hazardous cyanobacterial blooms (cyanoHABs) in aquatic ecosystems has drawn increased attention from around the world. Advances in our knowledge of the factors that contribute to the production of cyanoHABs, such as eutrophication and climate change, have been made (Burford et al., 2020; Zahra et al., 2020).

9.1 Potential Applications in Bioremediation, Bioenergy, and Beyond

According to studies by Zahra et al. (2020) and Baunach et al. (2024) cyanobacteria have shown the capacity to eliminate a variety of contaminants from contaminated environments, including heavy metals, organic compounds, and nutrients. Their ability to photosynthesize and their resilience to several kinds of stresses make them desirable options for bioremediation projects, especially when it comes to cleaning up soil and water contamination. The potential of cyanobacteria for biofuel and biochemical production has received attention due to their autotrophic nature, which enables them to convert carbon dioxide and sunlight into organic compounds. The goal of ongoing research is to design cyanobacteria to produce more value-added chemicals and biofuels, like hydrogen, ethanol, and biodiesel. Numerous chemicals, such as carotenoids, phycobiliproteins, and polyunsaturated fatty acids (PUFAs), are abundant in cyanobacteria (Zahra et al., 2020). These substances are driving the creation of effective cyanobacterial production platforms because of their many uses in the food, pharmaceutical, and cosmetic industries. Because cyanobacteria are sensitive to a wide range of environmental stresses, they are used as bioindicators to track the health of ecosystems, water quality, and the effects of climate change (Zahra et al., 2020). Developments in early warning systems and cyanobacterial biosensors can help improve environmental management and decision-making.

9.2 Challenges and Opportunities for Future Development

Although cyanobacteria exhibit great potential for a range of uses, several obstacles must be overcome before they can be widely used. These include raising their stress tolerance and productivity, expanding the genetic toolkit to accommodate a variety of strains, and refining the biomass collection and processing techniques. Commercialization also requires overcoming regulatory obstacles, guaranteeing cost-effectiveness, and winning over customers. Furthermore, new avenues for study on the integration of cyanobacteria into sustainable systems, such as agriculture and wastewater treatment, may become available. Finally, the development of computational and analytical tools will hasten this field's research and progress.

In conclusion, there is a great deal of promise for cyanobacterial study and uses in the future. Through tackling the obstacles and leveraging the burgeoning patterns, scientists and businesses can fully utilize these exceptional microbes to propel sustainable solutions for energy, environmental restoration, and other domains. Future research will be heavily focused on integrating cyanobacteria into various biotechnological and environmental applications, which will pave the way for a future that is more resilient and sustainable.

10 Conclusion

Cyanobacteria, often referred to as blue-green algae, have played a pivotal role in shaping Earth's ecosystems and continue to influence contemporary ecological dynamics. Their extensive fossil record, spanning over 2.1 billion years, underscores their significance in the evolution of life on our planet, particularly through their contributions to atmospheric oxygenation during the Great Oxidation Event. As primary producers in various aquatic environments, cyanobacteria are crucial for nutrient cycling, particularly nitrogen and carbon, which support food webs and sustain biodiversity.

The nutritional significance of cyanobacteria extends beyond their ecological roles; they are rich in proteins, vitamins, and essential nutrients, making them valuable for human and animal consumption. Their potential as sustainable food sources and feed supplements highlights their importance in addressing global food security challenges. Furthermore, ongoing research into their bioactive compounds reveals promising applications in pharmaceuticals, nutraceuticals, and biotechnology. These metabolites exhibit a range of biological activities, including anticancer, antimicrobial, and antioxidant properties, positioning cyanobacteria as a vital resource for drug discovery and development.

However, the rise of harmful cyanobacterial blooms poses significant ecological and health risks, necessitating a deeper understanding of their biology and the factors that trigger such events. Future research should focus on elucidating the mechanisms underlying these blooms and developing effective management strategies to mitigate their impacts.

In summary, cyanobacteria are not only integral to Earth's past but also hold immense potential for a sustainable future. Their dual role as both essential ecosystem players and sources of valuable bioactive compounds emphasizes the need for continued exploration and responsible management of these ancient organisms. By fostering a more profound understanding of cyanobacteria, we can harness their capabilities to benefit both human health and environmental sustainability.

Acknowledgements Thankful to the Department of Biotechnology of Graphic Era deemed to be University, Dehradun for providing this opportunity to compile this chapter.

References

Agha, R., & Quesada, A. (2014). Oligopeptides as biomarkers of cyanobacterial subpopulations. Toward an understanding of their biological role. *Toxins*, *6*(6), 1929–1950.

Amadu, A., & de Graft-Johnson, K., & Ameka, G. (2021). *Industrial Applications of Cyanobacteria.* https://doi.org/10.5772/intechopen.98859.

Baunach, M., Guljamow, A., Miguel-Gordo, M., & Dittmann, E. (2024). Harnessing the potential: Advances in cyanobacterial natural product research and biotechnology. *Natural Product Reports, 41*(3), 347–369.

Bhaskar, K. D., & Shahi, S. K. (2020). In-vitro evaluation of antimicrobial and anticancer activity of two diazotropic Cyanobacteria from Bilaspur Chhattisgarh, India. *Plant Archives (09725210), 20*(2).

Bhattacharya, S., Saha, S., & Saha, S. (2016). Cyanobacteria: A potential source of bioactive compounds. *Journal of Applied Phycology, 28*(5), 2991–3000. https://doi.org/10.1007/s10811-016-0873-2

Bonab, Z. H., Mohammadi, P., Asgarani, E., & Ghorbanmehr, N. (2019). The evaluation of nitrogen fixation activity of soil cyanobacteria via reduction assay and real-time PCR.

Bondarenko, N. A., Tomberg, I. V., Shirokaya, A. A., Belykh, O. I., Tikhonova, I. V., Fedorova, G. A., Netsvetaeva, O. G., Eletskaya, E. V., & Timoshkin, O. A. (2021). Dolichospermum lemmermannii (Nostocales) bloom in world s deepest Lake Baikal (East Siberia): abundance, toxicity and factors influencing growth. *Limnology and Freshwater Biology*.

Bouyahya, A., Bakrim, S., Chamkhi, I., Taha, D., El Omari, N., El Mneyiy, N., et al. (2024). Bioactive substances of cyanobacteria and microalgae: Sources, metabolism, and anticancer mechanism insights. *Biomedicine and Pharmacotherapy, 170*, 115989.

Burford, M. A., Carey, C. C., Hamilton, D. P., Huisman, J., Paerl, H. W., Wood, S. A., & Wulff, (2020). Perspective: Advancing the research agenda for improving understanding of cyanobacteria in a future of global change. *Harmful Algae, 91*, 101601.

Chen, J., Liu, J., Han, S., Su, H., Xia, W., Wang, H., Liu, Y., Zhang, L., Ke, Z., Zhang, X., Tang, H., Shen, H., Tao, M., Shi, P., Zhang, W., Wang, H., Zhang, J., Chen, Y., Rao, Q., Rao, X., Xie, L., Chen, F., Deng, X., Fan, X., He, J., Chen, L., Xu, J., & Xie, P. (2023). Nontraditional biomanipulation: A powerful ecotechnology to combat cyanobacterial blooms in eutrophic freshwaters. *The Innovation Life*.

Chiaiese, P., Corrado, G., Colla, G., Kyriacou, M. C., & Rouphael, Y. (2018). Renewable sources of plant biostimulation: Microalgae as a sustainable means to improve crop performance. *Frontiers in Plant Science, 9*, 1782.

Costa, M. S., Costa, M., Ramos, V., Leão, P. N., Barreiro, A., Vasconcelos, V., & Martins, R. (2015). Picocyanobacteria from a clade of marine Cyanobium revealed bioactive potential against microalgae, bacteria, and marine invertebrates. *Journal of Toxicology and Environmental Health. Part A, 78*(7), 432–442.

Erratt, K. J., & Freeman, E. C. (2024). Cyanobacteria in the Anthropocene: Synanthropism forged in an era of global change. *Water Environment Research, 96*.

Ferreira, N., Menezes, C., Dias, I. E., Azevedo, J., Rosado, T., Coelho, C. A., Gonçalves, P., da Silva, J. A., Martins, R., & Dias, E. (2022). Antibacterial Activity of Cyanobacterial Extracts against Legionella spp. In *The 7th Iberian congress on cyanotoxins/3rd iberoamerican congress on cyanotoxins*.

Gantar, M., & Svirčev, Z. (2008). Microalgae and cyanobacteria: Food for thought 1. *Journal of Phycology, 44*(2), 260–268.

Gu, P., Chen, L., Zhang, K., Hu, H., Yang, K., Zhang, Z., Ren, X., Liu, C., Zhang, Z., Zhou, J., & Miao, H. (2024). Effects of treatment process on larvae growth performance and nutrient yield during cyanobacteria bioconversion by black soldier fly. *Journal of Insects as Food and Feed*.

Hoffmann, L. (1999). Marine cyanobacteria in tropical regions: Diversity and ecology. *European Journal of Phycology, 34*, 371–379.

Hoikkala, L., Tammert, H., Lignell, R., Eronen-Rasimus, E., Spilling, K., & Kisand, V. (2016). Autochthonous dissolved organic matter drives bacterial community composition during a bloom of filamentous cyanobacteria. *Frontiers in Marine Science, 3*.

Hussein, H. A., Khaphi, F. L., & Saeed, Z. K. (2023). Cytotoxicity of bioactive compounds derived from cyanobacteria. *INNOSC Theranostics and Pharmacological Sciences*.

Jeyaraj, G. (2024). Advancing cyanobacterial biotechnology: new frontiers in natural product discovery and production. In *Natural product research* (pp. 1–3). Advance Online Publication.

Joshi, S. M., & Jackson, L. J. (2021). How might changing climate limit cyanobacteria growth in Shallow Prairie Lakes? An empirical space-for-time evaluation of the potential role of increasing sulfate. *Advances in Environmental and Engineering Research*.

Kumar, M., Singh, D. P., Prabha, R., & Sharma, A. K. (2015). Role of cyanobacteria in nutrient cycle and use efficiency in the soil. *Nutrient Use Efficiency: From Basics to Advances*, 163–171.

Lau, N. S., Matsui, M., & Abdullah, A. A. A. (2015). Cyanobacteria: Photoautotrophic microbial factories for the sustainable synthesis of industrial products. *BioMed Research International, 2015*(1), 754934.

Lin, P. C., Zhang, F., & Pakrasi, H. B. (2020). Enhanced production of sucrose in the fast-growing cyanobacterium Synechococcus elongatus UTEX 2973. *Scientific Reports, 10*(1), 390.

Liu, P., Lee, M. K., Choi, J. W., Choi, Y. H., & Nam, T. J. (2019). Crude protein from spirulina increases the viability of CCD-986sk cells via the EGFR/MAPK signaling pathway. *International Journal of Molecular Medicine, 43*(2), 771–778.

Lobus, N. V., & Kulikovskiy, M. S. (2022). The co-evolution aspects of the biogeochemical role of phytoplankton in aquatic ecosystems: A review. *Biology, 12*(1).

Mazard, S., Penesyan, A., Ostrowski, M., Paulsen, I., & Egan, S. (2016). Tiny microbes with a big impact: The role of cyanobacteria and their metabolites in shaping our future. *Marine Drugs, 14*(6), 97. https://doi.org/10.3390/md14060097

Méndez-Ancca, S., Pepe-Victoriano, R., Gonzales, H. H. S., Zambrano-Cabanillas, A. W., Marín-Machuca, O., Rojas, J. C. Z., Maquera, M. M., Huanca, R. F., Aguilera, J. G., Zuffo, A. M., & Ratke, R. F. (2023). Physicochemical Evaluation of Cushuro (*Nostoc sphaericum* Vaucher ex Bornet & Flahault) in the Region of Moquegua for Food Purposes. *Foods (Basel, Switzerland), 12*(10), 1939. https://doi.org/10.3390/foods12101939

Nandagopal, P., Steven, A. N., Chan, L. W., Rahmat, Z., Jamaluddin, H., & Mohd Noh, N. I. (2021). Bioactive metabolites produced by cyanobacteria for growth adaptation and their pharmacological properties. *Biology, 10*(10), 1061.

Novoveská, L., Nielsen, S. L., Eroldoğan, O. T., Haznedaroglu, B. Z., Rinkevich, B., Fazi, S., et al. (2023). Overview and challenges of large-scale cultivation of photosynthetic microalgae and cyanobacteria. *Marine Drugs, 21*(8), 445.

Parikh, S. J., & Madamwar, D. (2006). Cyanobacterial biofertilizers: A sustainable alternative to chemical fertilizers. *Current Science, 90*(12), 1667–1672.

Perera, R. M. T. D., Herath, K. H. I. N. M., Sanjeewa, K. A., & Jayawardena, T. U. (2023). Recent reports on bioactive compounds from marine Cyanobacteria in relation to human health applications. *Life, 13*(6), 1411.

Rao, A. R., Sindhuja, H. N., Dharmesh, S. M., Sankar, K. U., Sarada, R., & Ravishankar, G. A. (2013). Effective inhibition of skin cancer, tyrosinase, and antioxidative properties by astaxanthin and astaxanthin esters from the green alga Haematococcus pluvialis. *Journal of Agricultural and Food Chemistry, 61*(16), 3842–3851.

Sano, T., & Kaya, K. (1996). Oscillapeptin G, a tyrosinase inhibitor from toxic Oscillatoria agardhii. *Journal of Natural Products, 59*(1), 90–92.

Schirrmeister, B. E., Antonelli, A., & Bagheri, H. C. (2011). The origin of multicellularity in cyanobacteria. *BMC Evolutionary Biology, 11*, 1–21.

Scoglio, G. D., Jackson, H. O., & Purton, S. (2024). The commercial potential of Aphanizomenon flos-aquae, a nitrogen-fixing edible cyanobacterium. *Journal of Applied Phycology*, 1–25.

Shen, C. T., Chen, P. Y., Wu, J. J., et al. (2011). Purification of algal anti-tyrosinase zeaxanthin from nannochloropsis oculata using supercritical anti-solvent precipitation. *The Journal of Supercritical Fluids, 55*, 955–962.

Singh, A. K., & Ducat, D. C. (2019). New applications of synthetic biology tools for cyanobacterial metabolic engineering. *Frontiers in Bioengineering and Biotechnology, 7*, 440298. https://doi.org/10.3389/fbioe.2019.00033

Smucker, N. J., Beaulieu, J. J., Nietch, C. T., & Young, J. L. (2021). Increasingly severe cyanobacterial blooms and deep water hypoxia coincide with warming water temperatures in reservoirs. *Global Change Biology, 27*(11), 2507–2519.

Soule, T., Heller, M., & Gibbons, H. (2009). The role of cyanobacteria in the production of bioactive compounds. *Marine Drugs, 7*(4), 435–454. https://doi.org/10.3390/md7040435

Tabarzad, M., Atabaki, V., & Hosseinabadi, T. (2020). Anti-inflammatory activity of bioactive compounds from microalgae and cyanobacteria by focusing on the mechanisms of action. *Molecular Biology Reports, 47*(8), 6193–6205.

Tiam, S. K., Gugger, M., Demay, J., Manach, S. L., Duval, C., Bernard, C., & Marie, B. (2019). Insights into the diversity of secondary metabolites of planktothrix using a biphasic approach combining global genomics and metabolomics. *Toxins, 11*(9).

Verma, S., Thapa, S., Siddiqui, N., & Chakdar, H. (2022). Cyanobacterial secondary metabolites towards improved commercial significance through multiomics approaches. *World Journal of Microbiology and Biotechnology, 38*(6), 100.

Wasswa, J., Perkins, M., Matthews, D. A., & Zeng, T. (2024). Characterizing the impact of cyanobacterial blooms on the photoreactivity of surface waters from New York Lakes: A combined statewide survey and laboratory investigation. *Environmental Science and Technology, 58*(18), 8020–8031.

Willis, A., Woodhouse, J. N., Neilan, B. A., & Burford, M. A. (2022). Comparative genomics for understanding intraspecific diversity: a case study of the cyanobacterium raphidiopsis raciborskii. In *Advances in phytoplankton ecology* (pp. 415–434). Elsevier.

Zahra, Z., Choo, D. H., Lee, H., & Parveen, A. (2020). Review of current potentials and applications of cyanobacteria. *Environments, 7*(13), 1–17. https://doi.org/10.3390/environments7020013

Zhao, A., Lu, Y., Li, Q., Li, T., & Zhao, J. (2023). Metagenomics reveals the diversity and role of surface-water microbes in biogeochemical cycles in lakes at different terrain ladders. *Frontiers in Environmental Science.*

Microbial Production of Biofuels: Technological Advancement and Perspectives

Kalpana Sagar and Km. Priti

Abstract

The excessive population growth and growing energy demand in recent years have led to increased usage of fossil fuels, which contribute to climate change and global warming. This is primarily due to the greenhouse gas emissions from these types of sources. Sustainable biofuels, particularly biodiesel, are preferred as a substitute for fossil fuels to fulfil energy demands. Microorganisms are a vital subject of study in several scientific disciplines. They are everywhere in most environments on Earth, and their use contributes to a minor but notable technological advancement. Microorganisms are being used to produce lignocellulose-derived ethanol for biofuels, a mixture of cellulose, hemicellulose, and lignin found in plant cell walls. Biodiesel from microbes is widely considered due to its advantages, including a short life span, easy access to various environments, and the capacity to remediate several polluted environmental conditions. Presently, the purpose of this chapter is to provide an overview of the different microbial species that are used to produce biofuel.

Keywords

Biofuel · Microbes · Energy · Climate · Gases · Biodiesel

K. Sagar (✉) · K. Priti
Department of Botany and Microbiology, Gurukul Kangri (Deemed to Be University), Haridwar, Uttarakhand, India
e-mail: kalpana.sagar@gkv.ac.in

K. Priti
e-mail: preetisaini7668@gmail.com

1 Introduction

The need for energy has grown tremendously recently due to rapid industrialization and population growth. The global energy consumption is predicted to increase by 49% between 2007 and 2035. Fossil fuels are the leading energy supplier and can be more efficient. Moreover, it is believed that emissions from fossil fuels play a vital role in global warming. As a result, there is a growing concern worldwide regarding the need for green fuels that can reach future energy demands, such as sustainable, alternative, and renewable fuels. In recent years, we have seen a significant focus on storing and capturing carbon dioxide emissions and shifting towards renewable energy sources. Therefore, a global focus has been on chemicals and alternative energy products derived from biomass. In numerous fields, biodiesel research is vast and quickly expanding to meet the ever-increasing demands of the automotive sector. Several advanced biofuels have recently been implemented, including aviation fuels, syngas, and biohydrogen. New approaches are being explored to develop new biological processes or for the in-depth investigation and engineering of metabolic processes. Fuel research focuses on creating sustainable and economically viable biological refineries that produce value-added biochemicals such as coloured compounds and dietary supplements. Massive agricultural and industrial waste can be used to develop oleaginous microbes with hyperproduction that consume a variety of sugars derived from hydrolysed waste, therefore alleviating environmental concerns about waste management. The use of versatile microbes in the synthesis of fuels for renewable energy sources has generated a lot of interest. Because of to their versatility in developing biofuels and other useful products, as well as their capacity to obtain nutrition and energy from waste. The use of stable bio/nano-catalyst systems, robust microorganisms, and economically viable novel feedstock has led to recent improvements in microbial biofuel production (Arora

K. Sharma (ed.), *Bio-prospecting of Novel Microbial Bioactive Compounds for Sustainable Development*,
Sustainable Landscape Planning and Natural Resources Management, https://doi.org/10.1007/978-3-031-82178-3_14

et al., 2020) (Fig. 1). The bioenergy industry currently involves biorefineries that produce additional value-added products from waste or vacant biomass and fuel production to increase the overall effectiveness of processes (Fig. 2). To address fundamental challenges like rising costs of fuel and run-out of fossil fuels, researchers are currently putting affordable bioenergy strategies into practice. A practical approach must be developed to rapidly assess the viability of any novel or improved microbial strain for bioenergy production improved screening techniques and strains of oleaginous yeast are currently being studied for biodiesel production. Researchers worldwide focus on biofuel production from biomass resources via eco-friendly methods. Biomass can produce several liquid and gaseous biofuels, including Fischer–Tropsch H_2, biodiesel, ethanol, methanol, and methane. According to Choi et al. (2010), biofuels hold enormous potential for providing secure and sustainable energy sources and a

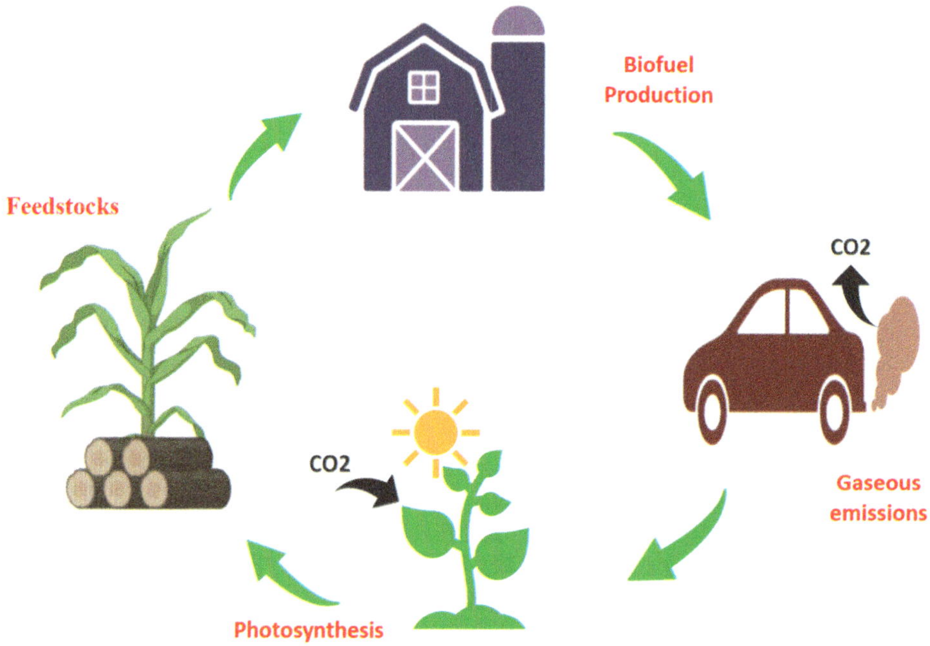

Fig. 1 Microbial sources for biofuels production

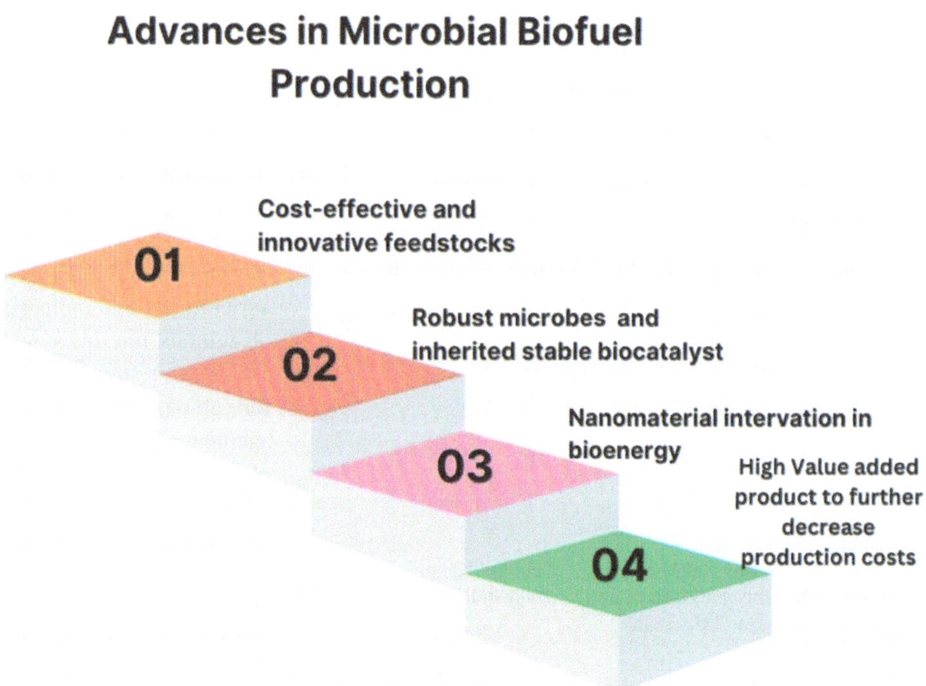

Fig. 2 Overview of advances made in microbial biofuel production

future supply. Because biofuels release less CO_2 during combustion and produce a smaller amount of other air pollutants overall, they are a renewable energy source that has a positive environmental impact. Since biomass absorbs the majority of CO_2 released into the atmosphere, producing biofuels and using them are also considered carbon–neutral activities. It is believed that biomass will supply 10–50% of global energy needs by 2050. According to the scenario, biomass will be among the world's primary sources of renewable energy (Moreira, 2006). There are currently four distinct types of biofuel generation technologies employed worldwide: first, second, third, and fourth generation (Figs. 3 and 4). According to Surriya et al. (2015), there are several challenges to generating biofuel with first-generation technology, which include air pollution, GHG emissions, biodiversity, cost of production and implementation, subsidies, etc. The potential for producing biofuels from second and third-generation processes is more significant than usual. As a result, the primary goal of this chapter is to investigate and assess various large-scale biofuel production technologies that utilize several kinds of biological resources. We tried to connect biofuel production difficulties with environmental factors, particularly the outcomes and effects of using them as fuel in a sustainable setting (Ambaye et al., 2021).

2 The Accessibility of Biomass for the Manufacturing of Biofuels

Biomass can be used as raw material, directly produced as biofuel, or processed into other bioenergy forms, agro-industry waste, municipal solid waste, biomass residues and waste agriculture waste products, and forest by-products all constitute viable and sustainable sources of biomass used in the manufacturing of biofuel. Popp et al. (2014) state that farm and forestry residues, along with municipal sewage, can generate 50–150 Exajoule (EJ) per year. It is expected that the global core energy market will expand to 600–1000 EJ yearly by 2050. 200–500 EJ of biomass could be produced annually globally, mainly from intensive residues (140 EJ), agricultural waste (100 EJ), and forest waste (80 EJ). According to Lal (2005), the amount of crop residue produced globally in the year 1991 from cereals sugarcane crops, soybeans, and lentils were estimated to be 6411 million tonnes (MT), and in 2001 it increased to 6973 MT (8.7% more than in 1991). The FAO predicts that in 2018, agriculture comprised around a third of all land area on Earth. According to OECD estimates from 2017, Brazil, China, the US, and India burned nearly 182 metric tonnes of crop residues (made from rice, wheat, sugarcane, and maize) in 2016, which is equivalent to 15.77 metric tonnes of CO_2. According to Nakada et al. (2014), the

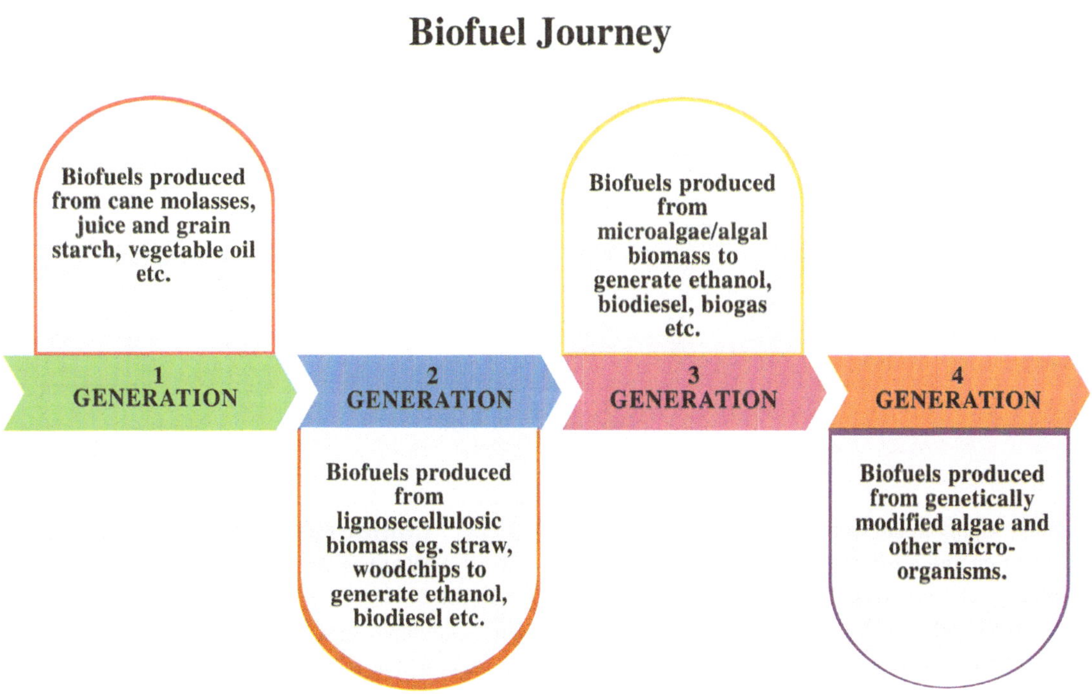

Fig. 3 Graphical demonstration of the advancement of biofuel production

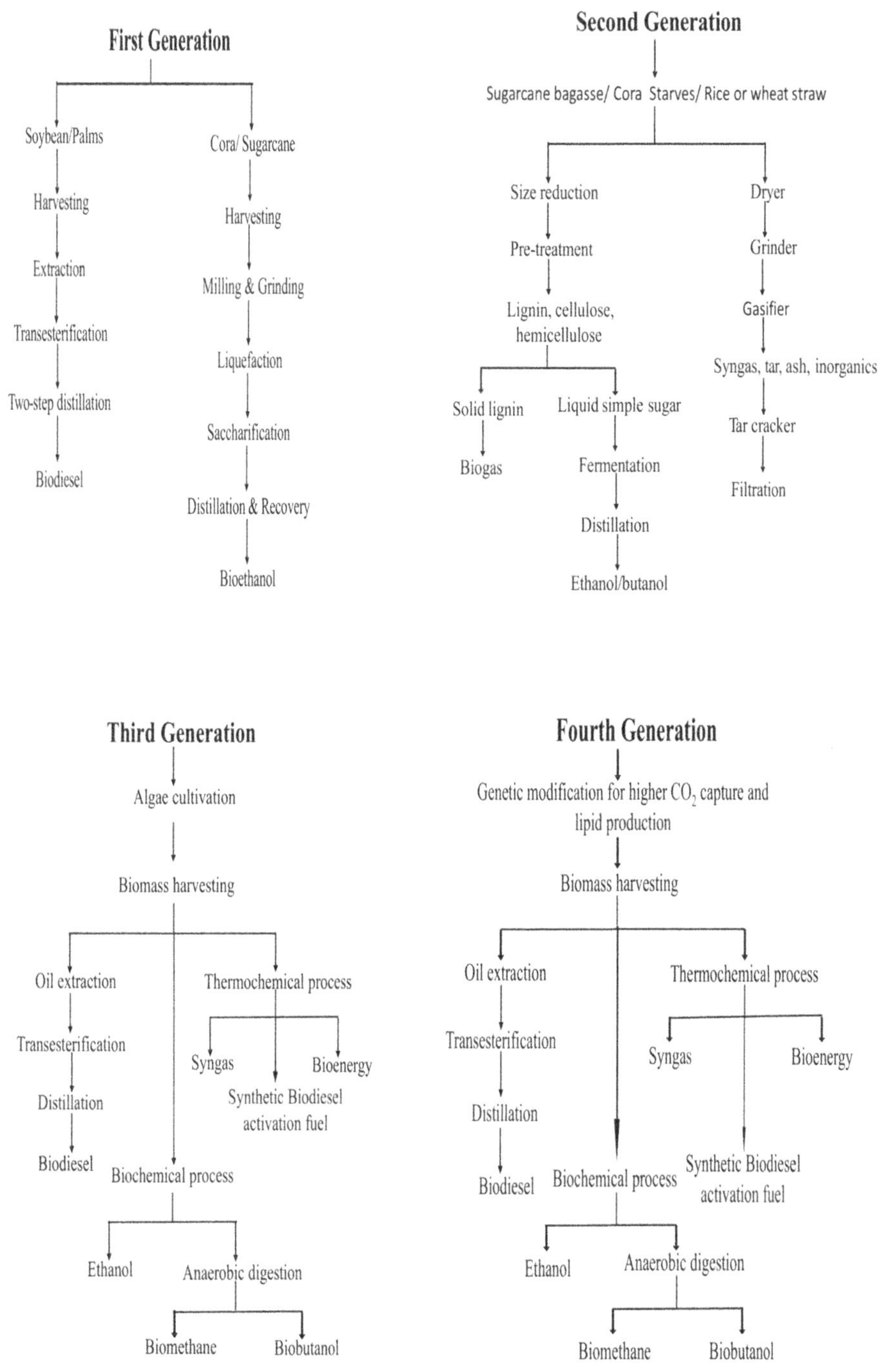

Fig. 4 Flowchart representation of different biofuel production technologies

overall biomass supply in the world is projected to vary about 97–147 EJ/y (in energy terms) by 2030. The majority of the biomass supply will come from agricultural residues, which can produce up to 550 EJ every year. However, the accessibility of land and the difficulties like water, natural stability, biomass production costs, and current supply chain infrastructure determines the strength of energy crop production. It is crucial to find low-cost residues because the palletization of post-harvest residues, wood flakes, and straw adds value to the final biomass-derived product.

3 Composition of Other Biomass-Derived Wastes and Crop Residues

The chemical structure of biomass must be recognized because it is a flexible energy source for biofuel. The productivity of biofuel is significantly affected by the biological structure and arrangement of biomass. Plant cell walls consist of lignin, cellulose, hemicellulose, and trace amounts of inorganic matter, which together make up lignocellulosic biomass. The biochemical arrangement and mass distribution of lignin biomass vary depending on the type of plant. Zhao et al. (2012) state that the rate at which biomass decomposes during pyrolysis and hydrolysis is influenced by the biochemical processes, equilibrium and structural durability of the biomass component. For example, According to Ahorsu et al. (2018), cellulose pyrolysis has the potential to generate greater amounts of oil compared with lignin. Yang et al. (2007) investigated the thermal degradation of cellulose, hemicellulose, and lignin. They observed that the cellulose lost 80% of its weight at 268 °C and about 94.5% of its weight at 400 °C, lignin only lost 54% of its weight at 900 °C. The components of biomass caused the variation in oil production yield and pyrolysis temperature. Moreover, lignocellulosic biomass contains lignin, therefore, pyrolysis of the biomass yields a significant quantity of biochar as a byproduct.

4 Microalgal Biomass

Biomass of algae has been extensively explored for feasible biofuel and value-added material sources. The enormous discharges of CO_2 into the atmosphere, sludge, and wastewater severely endangered the natural environment. In this scenario, algal biomass production from wastewater seems to be a viable way to generate renewable energy. Algal species have high protein, lipid, and carbohydrate percentages based on a dry weight basis and various other constituents (such as ash). These ingredients are used to create biofuels, including biodiesel and biofuel oil. It is feasible to valorise and incorporate additional molecules produced during adsorption into the main biofuel production chain. Researchers are looking through biobutanol and biohydrogen. Compared with crops, algae have several benefits, such as high oil/lipid content, yield, fast growth, and high photosynthetic efficiency for fixing CO_2. Algal biomass comprises 11–56% carbohydrates, 8–70% lipids, 40–70% proteins, and 3–5% coloured substances. Harvesting and cultivation possess many of drawbacks that contribute 20–30% to the total cost of production, including less cell size (2–20 μm), proliferation in diluted culture media, high cell density near water, dehydration, and water removal.

5 Development of New Technologies to Increase the Production of Biofuels

Globally, the production of biofuel from various biological resources, combined with the applications of novel innovations and biological processes, is currently growing. It is expected that the manufacturing of biofuels from crop residues and biomass materials will address a number of issues, including managing waste, and decreasing the adverse environmental impacts. Therefore, more studies are being done on innovative techniques for producing biofuel as an alternative form of energy. Based on the biomass-based resources which are utilised in their production the first-, second-, third-, and fourth-generation biofuels are shown in Fig. 3. First-generation biofuels, such as biodiesel and bioethanol, are produced from edible food crops such as sugarcane, potatoes, oil seeds, maize, wheat, barley, sunflower, and soybeans. According to Qin et al. (2018), The first chemical energy generated by fermentation with raw maize and sugarcane was ethanol. This was achieved by employing fungal mycelia. Wang et al. (2015) reported that starch-digesting microorganisms like *Saccharomyces cerevisiae* and *Rhizopus sp.* can be employed to ferment raw maize flour and produce ethanol. As a result, in the first generation, enzymatic hydrolysis techniques were employed to produce a substantial volume of bioethanol on a broad scale from cellulose. According to Lee et al. (2019), the term "second-generation" describes the process of producing biofuels from readily available lignocellulosic and various organic waste products, including oilseed-bearing plants like jatropha, timber, straw, and switchgrass. Algae are used as feedstock in the third generation of biofuels because they are a significant source of the lipids required for the production of biodiesel and other biofuels. Moreover, genetically modified organisms, modified metabolic pathways, enhanced CO_2 fixation capacity of microalgae, and post-genome technology have all been necessary for the production of fourth-generation biofuels.

5.1 Recent Advancements in the Field of Biofuels Using Genetic Engineering

The National Renewable Energy Laboratory (NREL) of the US Department of Energy has identified lignin bioengineering-based biomass sources with lower lignin content as the highest priority (Larson, 2008). A high lignin content generally prevents the fermentation of ethanol and reduces the amount of sugar produced during hydrolysis and saccharification. The production rate of ethanol and other biofuels after processing is enhanced by the amount and percentage of polysaccharides in lignin-based biomass, such as cellulose, galactan, and mixed-linkage glucans, and also through the composition of these substances in cell plant walls. Facilitating hydrolysis is also based on genetically engineering cellulolytic enzymes to become more efficient. Historically, yeast that ferments ethanol has certain limitations. They are restricted to breaking down sugars with six or hexose carbons. Novel varieties that can ferment both 5-carbon and 6-carbon sugars and withstand high temperatures and ethanol concentrations have consequently been developed. The expression of several biosynthesis genes can be reduced or inhibited through the use of genetic engineering, which could lead to a major advancement in achieving them. Enzymes which alter the polysaccharides in the cell wall before they fully mature can be transgenically expressed, leading to biomass with higher quality. In crop plants, glycosyl hydrolases are highly expressed and play a role in the development, remodelling, and construction of tissue cell walls. It has been reported that the native GH9B genes (OsGH9B1 and OsGH9B3) in rice crops can enhance biomass quality without influencing growth or development (Huang et al., 2019). Moreover, transgenic lines displayed a 23% decrease in the cellulose's degree of polymerisation (DP) and an 11–23% decrease in its crystalline index. After biomass pre-treatment procedures, transgenic strains GH9B1 and GH9B3 released 63% more glucose than the untreated control line.

Another study found that when compared to rice of the wild type, rice with OsAT10 overexpression possessed cell wall glucose levels 8–19% higher. In genetically modified rice and switchgrass lines, this gene was utilised to enhance total sugar production by approximately 40% following enzymatic saccharification. In *Populus alba*, overexpression of the *Aspergillus aculeatus*, xyloglucanase gene (AaXEG2) led to increased growth and cellulose deposition, as well as up to 81% more glucose released after enzymatic hydrolysis. Fonseca et al., (2020) generated cellulase enzymes at a rate of 80.6 g L^{-1} (0.24 g/L/h) using a modified strain of *T. reesei* RUT-C30. It was found that the produced enzyme had a very high saccharification productivity. It is capable of producing several kinds of liquid and gaseous biofuels. Moreover, biodiesel production can be raised through genetic engineering. Janben and Steinbuchel (2014) investigated the synthesis of triglycerides (TAG) 530 mg/L by genetically engineered *E. coli* in fed-batch fermentation.

CRISPR-Cas technology is a potential strategy to enhance biofuel production capability. This technology has enhanced intracellular enzyme efficiency in bacteria, fungi, and microalgae by modifying genes related to enzymatic activity, such as cellulase and hemicellulose. Figure 5 shows the overview of CRISPR-Cas technology in improving the production of biofuel. Extremophilic microorganisms such as *Sulfolobus sp.*, *Pyrococcus furiosus*, *Thermus thermophilus*, *Thermococcus kodakarensis*, and *Thermotoga maritima* extensively investigated for their potential in biofuel synthesis. The recent discovery of the CRISPR-Cas system could create new microbial biorefineries by allowing targeted gene editing. This could make it easier to produce biofuels from extremophiles. This method could facilitate the adjustment of several components in biofuel generation, including improved hydrolysis, increased suppression, and the selective targeting of competing metabolic pathways (Shanmugam et al., 2020).

Nevertheless, integrating heterologous CRISPR-Cas systems into prokaryote systems is often challenging due to native CRISPR-Cas in most microorganisms, which may interfere with the external CRISPR-Cas systems (Marino et al., 2020).

The design of a CRISPR-Cas system for multiplex genome editing can be approached in three ways:

1. Expressing multiple guide RNAs in a single guide RNA expression cassette.
2. Expressing multiple guide RNAs in multiple guide RNA expression cassettes.
3. Expressing a single specific guide RNA to target and modify various sequences in the genome.

Zhang et al. (2018) developed a CRISPR-Cas approach to alter *Clostridium tyrobutyricum* genes, including spo0A, pyrF, dehydrogenase (adhE1 or adhE2), and cat1 under a lactose-inducible promoter to reduce toxicity. Two mutants with the maximum butanol production (26.2 g/L) in batch fermentation were generated by replacing the adhE1 and adhE2 genes with cat1.

Wen et al. (2017) used the CRISPR-Cas system to target the genes encoding acetate kinase (Clocel_1892 and Clocel_3674) in a consortium of *Clostridium cellulovorans* DSM 743B and *Clostridium beijerinckii* NCIMB 8052. This targeted genome engineering resulted in the disruption of the carbon flux trigger towards butyrate production. Furthermore, Wen's team used CRISPR interference, a programmable CRISPR technology designed for gene suppression, to enhance ethanol production by inhibiting the suspected hydrogenase gene (Clocel_2243).

Escherichia coli is a crucial alternative for biofuel production due to its wide availability of genomic and metabolic data. Through metabolic engineering of *E. coli* using CRISPR and CRISPRi, six genes (cat1, cat2, bdh, bld, sucD, and 4hbd) drove an artificial pathway that produced 1, 4-butanediol (1, 4-BDO) (Wu et al., 2017).

By changing the metabolic pathways of many bacteria, like *Clostridium*, *E. coli*, *B. subtilis*, and *C. reinhardtii*, CRISPR-Cas-based genome editing systems have been successfully made to improve biofuel production. Modifying the genetic material of microorganisms might enhance the production of biofuels, but it also raises environmental and safety issues.

1. **Environmental impact**: Biofuel production using genome-edited microbes is dependent on the modifications made to the genomes (Hemalatha et al., 2023). Interactions between released genetically altered bacteria and native species must be rigorously monitored. Implementing containment and risk assessments before widespread deployment might mitigate ecological harm.

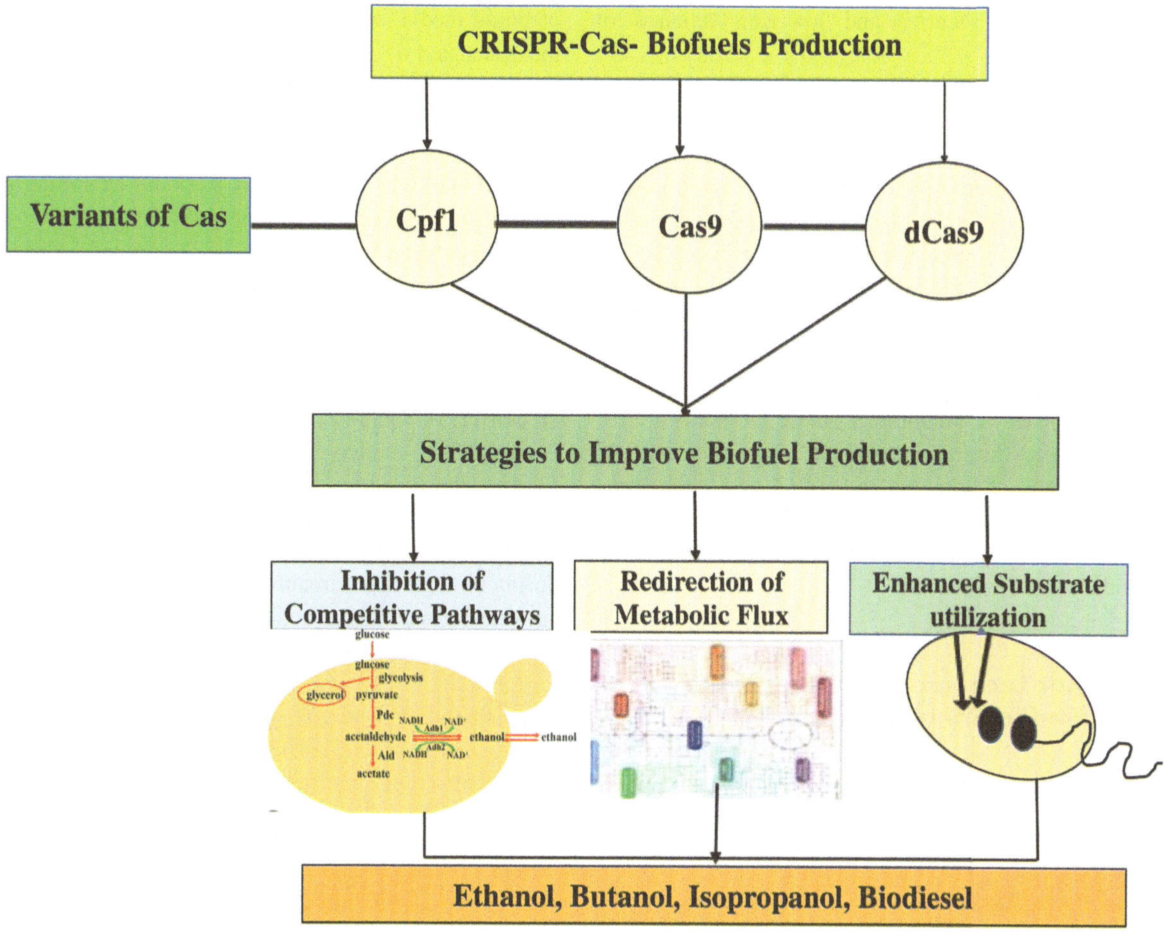

Fig. 5 Overview of the CRISPR-Cas system in improving the production of various biofuels

2. **Enhanced Efficiency**: CRISPR-Cas9 can modify the genetic material of bacteria that produce biofuel molecules. The enhancement of metabolic pathways results in increased efficiency of biomass-to-biofuel conversion. Advancements in genome editing enhance the efficiency and profitability of biofuel production (Lu et al., 2022).
3. **Biosafety criteria**: Organisms modified by genome editing to produce biofuels must adhere to biosafety standards (Xu et al., 2022). Before commercial distribution, modified microorganisms must systematically evaluate their safety risks. Country-specific regulatory regimes vary.
4. **Genetic stability**: Microbial genome editing, although capable of delivering precise alterations, raises questions about its long-term sustainability. Unintentional genetic changes or the loss of certain features might result from genetic recombination or other genetic processes. Retaining stable genetic alterations is necessary to guarantee the safety and effectiveness of biofuel-producing transgenic bacteria (Javed et al., 2019).
5. **Ethical Considerations**: In addition to environmental and safety concerns, ethical measures should be considered. These may include evaluating the ecological consequences of introducing modified bacteria and addressing ethical considerations related to the biomodification of living organisms (Zhang et al., 2020).

5.2 Prospects and Challenges for the Production of Biofuel

Biomass-derived fuels are increasingly being recognized as a highly viable alternative to fossil fuels, due to their renewable and ecologically sustainable characteristics.

The potential of advanced biofuel is dependent on socioeconomically and eco-friendly technologies. The demand for biofuel is proliferating because of its attractive and ecologically sustainable benefits, which include:

1. Being sustainable;
2. Enhancing carbon fixation and minimizing carbon dioxide indirectly;
3. Stimulating local economic growth;
4. Lowering the amount of air pollution caused by burning and decomposing biomass in fields;

5. Bringing energy security and lowering reliance on imported oil;
6. Developing high-tech jobs and jobs for farmers.

However, several challenges in biofuel production include biomass yield, transportation to the energy generation location, conversion process, storage, and the important parameters for biofuel commercialization. However, several significant challenges still need to be addressed, including economic conditions, limited resources, availability of food, disagreements over safety and health, water usage, challenges with land use, and several environmental effects that might impact nearby communities. These problems are frequently evaluated through a life-cycle assessment (LCA). The initial stages of applying the Life Cycle Assessment (LCA) method to advanced biofuels involve understanding the demands of farmers, the varieties and options of feedstock, and changes in land use.

6 Major Technological Issues in the Manufacturing of Biofuel

As previously mentioned, there are currently many options for producing biofuels. According to research, energy demand will account for about 37% of global demand by 2040 due to its critical role in the economic and social growth of the whole world. The petroleum industry supplies over 80% of the world's energy demands. To meet the world's energy needs, researchers are now scrambling to find more effective and sustainable sources of energy, like biofuels, as a result of the natural resources associated with petroleum fuels becoming scarce and depleted (Rodionova et al., 2017). The biofuel production sector faces a number of technological challenges, such as:

1. The utilization of inedible feedstock, including lignin-based biomass production and other organic compounds;
2. Manufacturing logistics;
3. Affordable pre-treatment, hydrolysis of enzymes, and fermentation processes;
4. Efficient use of co-products;
5. The development of biofuel standards;
6. Distribution logistics;
7. Economic and social benefits and acceptance.

The physical characteristics of biomass, including mass density, flexible properties, water content, size of particles, microscopic structure, and bulk density, are important determinants of biofuel conversion. This indicates that the biochemical mechanisms require a great deal of size reduction and the total amount which can be utilised is based on the manufacturing technique used. The high heating rate of liquid media in the thermochemical method implies that the dimension of the biomass particles has no impact on the total amount of biomass transformed. Therefore, the size of the particles becomes less significant. However, the biomass sludge must be significantly reduced in size before being pumped through a continuous system. Demirbas (2004) showed in a different study that particles with sizes of less than 0.5 mm can have a lower char yield and a faster heating rate, which is helpful for biomass conversion based on pyrolysis. According to Spliethoff and Hein (1998), different feedstock types have different ideal sizes for consuming biomass. Particle size reduction is usually needed for all conversion methods. Another physical characteristic that is particularly challenging and affects the process of transforming the moisture content of biomass waste converted into biofuel. This property also affects feedstock supply and bio-refining operations. The addition of more moisture during biological pre-treatment can lower the thermochemical transformation, the quality of the bio-oil, and the thermal effectiveness of the burning process. Other physical characteristics that affect biomass conversion to biofuels include mass density (BD), flexible characteristics, microstructure, moisture content, and size of particles. Weiss et al. (2010) state that the microscopic structure and elastic properties of biomass have a significant impact on the rate at which it transforms to fuels and biochemicals. These characteristics can improve the flexibility of biomass and the interactions between particles. They also declared that bulk density impacts feedstock handling and transportation. Transporting feedstock is more expensive, regardless of the bulk density. The quality of the biofuel and the composition of high-value chemicals produced during the conversion process are influenced by the chemical properties of biomass. Li et al. (2019) reported that cellulose, volatiles, and ash content are the primary chemical properties of biomass that influence its conversion into biofuel. Tumuluru et al. (2012) studied the impact of increasing ash content during biological pre-treatment on biomass transformation processes like pyrolysis, gasification, condensation, and combustion, as well as char yield. However, leaching and air sorting may reduce the ash content. The volatile organic components of furan and acetic acid are two additional chemical elements that influence the way biomass is converted into biofuel. Palmqvist and Hahn-Hagerdal (2000) discovered that during the biological process of converting biomass into biofuel, the presence of furan organic material can reduce the efficiency of the fermentation process. Carpenter et al. (2014) found the energy content and stability of bio-oils produced through thermochemical synthesis may be impacted by high concentrations of volatile organic matter. The conversion of biomass to biofuel is also influenced by the lignin content. On the other hand, the outcome varies based on the kind of procedure applied. According to

Sun and Cheng (2002), lignin content may increase the production of biomass oil and also the values for cooling and heating during the conversion process, which involves combustion and pyrolysis. However, it decreases the amount of ethanol production during fermentation, because it remains in enzymes that cannot interact with cellulose. The elimination of lignin and hazardous compounds formed during the pre-treatment of biomass is another major issue. These include furfural, acetate, levulinic acid, sulphuric acid, and 5-hydroxymethylfurfural (HMF). These compounds prevent the growth of microorganisms and have an adverse impact on the fermentation process and saccharification. Many researchers have employed methods to extract inhibitors from hydrolysates that have been pre-treated before fermentation to address the issues of inhibitors produced during biofuel production. Identifying effective microbes that resist the adverse effects of inhibitors could be a successful strategy for enhancing biofuels production. Genetically engineered microbial strains can also be developed to clean up hydrolysates that are currently undergoing onsite treatment and transform them into gaseous and liquid biofuels through fermentation and enzyme-based saccharification. Large-scale biofuel production can be enhanced by implementing these strategies More research is necessary to maximise biofuel production through the development of new strains which combine diverse genetic engineering techniques.

6.1 Environmental Issues in Biofuels Production

There are many benefits to using biofuels in place of fossil fuels. However, environmental concerns along with issues such as air quality, emissions of greenhouse gases, biodiversity, soil and water resources, land use, and sustainability are important aspects to consider when developing guidelines to evaluate the impacts of the biofuel industry. To improve socioeconomic development, it is critical to capitalize on opportunities and find long-term solutions to environmental challenges. According to LCA is a popular tool for assessing these ecological effects. Most biofuel manufacturing organizations take life cycle assessment (LCA) very seriously, considering the environmental impact of their decisions before making process changes that should be directed towards reducing emissions during the biofuel production process.

6.2 Biofuel Production and Land Use Issues

The primary challenge in the production of biofuels is managing efficient land use while reducing greenhouse gas emissions in order to prevent disputes over food and energy, global warming, and other adverse environmental effects. However, growing crops on different types of land addresses similar environmental issues, regardless of whether they are produced for food, industry, or fuel. Monocultures can harm pathogens, pests, and biodiversity in the field cultivating crops or forests. It may be more effective to use land for growing woody species like plants or perennial lawn grasses along with mixed crops in order to increase biodiversity without compromising yield (Tilman et al., 2006). Several types of land serve as crucial carbon sources, including tropical forests and peatlands. Carbon emissions from the soil are expected to increase when these lands are converted to crops. The application of synthetic fertilizers must be strictly limited to reduce greenhouse gas emissions from agricultural land, including N_2O (Lewandowski & Schmidt, 2006). Reducing adverse environmental effects is mainly dependent on effective agronomic practices. Managing soil and rhizosphere microbial diversity aids crop production over the long term. The combustion of fossil fuels in automobiles produces a greater amount of greenhouse gases, which are a major contributor to air pollutants and global warming. It is essential to transform other residues and lignocellulosic biomass waste into alternative energy sources, particularly oxygenated biofuels. Renewable energy products are used to treat high nitrate polluted water from sewage plants, animal waste, and drainage ditches. Examples of these products are Miscanthus (*Miscanthus giganteus*) and willows (*Salix* spp.) Thus, it can lead to increased biomass production and decreased emissions of greenhouse gases. On the other hand, some experts believe that biofuel-growing crops could replace current products on land used for fibre, forage, and food. This could result in increased food costs in the marketplace. GHG emissions need to be carefully evaluated. Less than 1% of all land is used to produce biofuel crops, whereas changes in land use patterns are responsible for nearly 15% of global carbon emissions. Changes in land use related to biofuels make up a tiny percentage of land use overall.

6.3 Impact of Biofuel Production on Greenhouse Gas Emissions

According to recent findings from the California Air Resources Board (CARB) and the US Environmental Protection Agency (EPA) (Sacramento Rivero et al., 2022) indicate that utilising biofuels instead of fossil fuels could decrease tailpipe greenhouse gas emissions by 50 percent. Utilizing 80% low-sulfur diesel fuel (ULSD) and 20% biodiesel may decrease greenhouse gas emissions by 50–85% without the need for extra funding. The most recent information from the California Air Resources Board shows that ULSD blends in biodiesel possess an opportunity to significantly reduce greenhouse gas emissions related to transportation (Sacramento Rivero et al., 2022). For example, the Low Carbon Fuel Standard (LCFS) programs in California have prevented over 18 metric tonnes of CO_2 by using biodiesel. Customers can operate vehicles without having to replace their current vehicle models due to the inexpensive fuel that is produced domestically. As a result, greenhouse gas emissions can be substantially decreased. According to recent reports that by utilising ULSD in biodiesel, California's annual CO_2 emissions are reduced by about 20 million tonnes; as a result, society may become less reliant on fossil fuels. At the same time, it still protects the environment. These biofuels have enormous potential to reduce greenhouse gas emissions and other environmental problems. Plants require CO_2 from the atmosphere to start the process of photosynthesis, which generates biomass and energy. A particular amount of carbon dioxide is released during biomass burning to generate biofuel, and crop plants use this carbon dioxide during their subsequent growth and development phases. Thus, biofuel from other organic waste and lignocellulosic biomass resources can be considered ecologically friendly. The carbon in biofuel is an element of the continually occurring carbon cycle (C-cycle), which is the transfer, recycle, and exchange of carbon from the atmosphere to the biosphere. Furthermore, employing other waste organic substances and lignocellulosic biomass resources may reduce greenhouse gas emissions by 39–52%. The use of bioethanol instead of gasoline, helps to reduce atmospheric CO_2 in three ways: (1) avoid the emissions associated with gasoline; (2) allow the CO_2 content of the fossil fuels to remain in storage; and (3) provide a mechanism for CO_2 absorption by growing new biomass for fuels. Biofuels have superior combustion characteristics compared to gasoline, leading to reduced greenhouse gas emissions, and possess complete biodegradability, in contrast to some fuel additives. The potential of ethanol and biofuels to substantially mitigate CO_2 emissions in comparison to fossil fuels is considerable. Employing these sustainable fuels in transportation, power production, and industrial operations aids in reducing the emission of greenhouse gases into the environment. In contrast to fossil fuels, biofuels emit carbon that has been recently assimilated from the atmosphere, instead of releasing long-stored carbon dating back millions of years. Carbon neutrality makes biofuels a crucial tool in combating climate change.

Biodiesel is derived from plants and other microbial sources and is often used in diesel mixtures.

Biodiesel is a fuel with low carbon intensity, capable of decreasing atmospheric greenhouse gas (GHG) emissions by as much as 20% in comparison to pure diesel. Corn-based ethanol provides a 40% reduction in life cycle greenhouse gas emissions compared to gasoline. Cellulosic ethanol may provide an even more substantial advantage. Cellulosic ethanol has the capacity to reduce greenhouse gas emissions by as much as 86%. Ethanol is easily biodegradable and resulting no damage to the environment, making it a secure and highly efficient substitute for gasoline additives like MTBE. Because of their compatibility with the natural carbon cycle, biofuels offer the most beneficial alternative for reducing greenhouse gases from the transportation sector. Using more biomass would expedite the substitution of fossil fuels and the mitigation of greenhouse gas emissions from transportation. There are currently few specialized decarbonization markets in which the environmental benefits of biofuel can be recognized; their contribution to the reduction of GHG emissions has been overlooked (GAIN Report, 2019).

It could be evident that market formalization is required, given that bioenergy remains a significant renewable energy source globally. In order to achieve the long-term carbon reduction of end-use sectors, renewable energy sources must be fully utilised. Policy support and innovation are crucial to reducing costs. The International Civil Aviation Organization (ICAO) has proposed that carbon offsets would become mandatory starting in 2027, after a period of optional participation from 2020 to 2027. In 2019, the aviation industry worldwide emitted 915 million tonnes of carbon dioxide (CO_2). Emissions from this sector account for about 2% of global CO_2 emissions and for 12% of all transportation emissions. Aviation-related non-CO_2 emissions also have a substantial influence on the climate, providing around two-thirds of the total radiative forcing. This is expected to increase twofold by 2050 on a business-as-usual projection. While the primary trade group of the aviation industry (IATA) has made a commitment to carbon mitigation objectives, the sector has few technological alternatives to accomplish the significant reductions in emissions required. Biomass-derived alternatives to kerosene jet fuel (biojet) are a limited resource for mitigating airlines' direct carbon emissions and might be a very significant use of bioenergy in the future. The offset program aims to cover the estimated 80% and optional ranges of increasing emissions from 2027 to 2035, or 65% above 2020 levels. Since there is not currently another method to reach nearly zero emissions, they suggested giving priority to providing access to biofuels over other aviation industry sectors.

6.3.1 Current Challenges in Scaling Up Aviation Biofuels

There are many challenges to scaling up aviation biofuels, including:

(i) **Cost**: The investment required to produce sustainable aviation fuel (SAF) is substantial.
(ii) **Competition for resources**: The availability of sustainable feedstock is a formidable challenge in the context of resource competition.
(iii) **Infrastructure**: Substantial infrastructure is required.
(iv) **Technological innovation**: Sustained technical advancement is crucial.
(v) **Feedstock**: The current availability of feedstock is insufficient to generate the required amounts of SAF to achieve the goals.
(vi) **Market competitiveness**: Cost competitiveness, infrastructural compatibility, and policy support are crucial characteristics for the expansion of biofuel production.
(vii) **International collaboration**: International collaboration is required.
(viii) **Alternative technologies**: Hydrogen and electric aircraft, examples of alternative technology, have restricted capacities.

6.4 Socio-Economic Concern

Demirbas (2017) states that biomass and other bioenergy resources are essential for social and economic development. They enhance the value of already-existing economic products. Future requirement for liquid fuels, which is projected to improve wealth and income as well as socioeconomic advantages, is linked to biofuel's social and economic aspects. Many nations expand their source of energy to ensure local and rural service from the production and consumption of domestic biofuels. Biofuel offers several career opportunities in biomass harvesting and cultivation, handling and transportation, plant execution, machinery, maintenance and fabrication. Ambaye et al. (2021) reported that 2,063,000 new biofuel jobs were created globally in 2018. Brazil has created the greatest number of jobs (832,000), more than China (51,000), India (35,000), the US (311,000), and the EU (208,000). Energy supply diversification through industry establishment, import substitution influencing the trade balance, proliferation of rural socioeconomic diversity, economic growth through business earnings and employment, and global and national energy safety are all facilitated by the manufacturing and consumption of biofuels.

7 Conclusion and Future Perspectives

This chapter addresses possibilities for producing biofuels, challenges utilizing various feedstocks, and advancements in process technology. However, the feedstock, operational conditions, process type, and cost–benefit analysis must restrict the biomass conversion process. The productivity of fungal enzymes that break down cellulosic materials and bacteria strains that ferment various sugars, including sugars with 6 and 5 carbons, determines biofuel production's cost-effectiveness using biomass. Therefore, novel strains developed via genetic engineering and advanced technologies must be further developed and refined to facilitate and address these issues. It is accurate to say that biofuels produced from microalgae and energy crops are a desirable and practical solution. Genetic engineering needs more research or validation before it can be employed to enhance biofuel manufacturing on a large commercial scale. Biofuels offer several advantages for the environment over fossil fuels. CRISPR-Cas technology is a potential strategy to improve biofuel production capability. This technology has enhanced intracellular enzyme efficiency in bacteria, fungi, and microalgae by modifying genes related to enzymatic activity, such as cellulase and hemicellulose. The recent discovery of the CRISPR-Cas system could create new microbial biorefineries by allowing targeted gene editing. This method could make it easier to change several factors involved in producing biofuels, such as improving hydrolysis, increasing suppression, and targeting specific metabolic pathways competing with each other. Airborne pollutants, transportation-related airborne pollutants, greenhouse gas (GHG) emissions and carbon dioxide emission levels may all be substantially decreased using biodiesel blends in low-sulfur diesel (ULSD) fuel. However, when developing guidelines to assess the impacts of the biofuel sector, biodiversity, water use, and land use must be considered. This chapter emphasizes the critical social and economic contributions provided by biofuels. Apart from providing a sustainable solution, biofuel production helps diversify and develop jobs in rural areas, promotes revenue growth, addresses national and local energy security issues, and may replace imports with less harmful effects on the energy supply and trade imbalance.

Acknowledgement and Contributions The authors gratefully acknowledge Gurukul Kangri (Deemed to be University), Haridwar, for giving necessary facilities for paper writing. Both the authors wrote, checked and edited the present manuscript.

References

Ahorsu, R., Medina, F., & Constantí, M. (2018). Significance and challenges of biomass as a suitable feedstock for bioenergy and biochemical production: A review. *Energies, 11*(12), 3366.

Ambaye, T. G., Vaccari, M., van Hullebusch, E. D., Amrane, A., & Rtimi, S. J. I. J. O. E. S. (2021). Mechanisms and adsorption capacities of biochar for the removal of organic and inorganic pollutants from industrial wastewater. *International Journal of Environmental Science and Technology, 18*(10), 3273–3294.

Arora, V. K., Katavouta, A., Williams, R. G., Jones, C. D., Brovkin, V., Friedlingstein, P., Schwinger, J., Bopp, L., Boucher, O., Cadule, P., & Chamberlain, M. A. (2020). Carbon–concentration and carbon–climate feedbacks in CMIP6 models and their comparison to CMIP5 models. *Biogeosciences, 17*(16), 4173–4222.

Carpenter, D., Westover, T. L., Czernik, S., & Jablonski, W. (2014). Biomass feedstocks for renewable fuel production: A review of the impacts of feedstock and pretreatment on the yield and product distribution of fast pyrolysis bio-oils and vapors. *Green Chemistry, 16*(2), 384–406.

Choi, G. W., Um, H. J., Kim, M. N., Kim, Y., Kang, H. W., Chung, B. W., & Kim, Y. H. (2010). Isolation and characterization of ethanol-producing *Schizosaccharomyces pombe* CHFY0201. *Journal of Microbiology and Biotechnology, 20*(4), 828–834.

Demirbas, A. (2004). Effects of temperature and particle size on bio-char yield from pyrolysis of agricultural residues. *Journal of Analytical and Applied Pyrolysis, 72*(2), 243–248.

Demırbas, A. (2017). The social, economic, and environmental importance of biofuels in the future. *Energy Sources, Part B: Economics, Planning, and Policy, 12*(1), 47–55.

Fao, F.A.O.S.T.A.T., 2018. Food and agriculture organization of the United Nations. *Rome, fao. org*, 403–403.

Fonseca, L. M., Parreiras, L. S., & Murakami, M. T. (2020). Rational engineering of the *Trichoderma reesei* RUT-C30 strain into an industrially relevant platform for cellulase production. *Biotechnology for Biofuels, 13*, 1–15.

GAIN Report. (2019). Global Agricultural Information Network Report.

Hemalatha, P., Abda, E. M., Shah, S., Venkatesa Prabhu, S., Jayakumar, M., Karmegam, N., Kim, W., & Govarthanan, M. (2023). Multi-faceted CRISPR-Cas9 strategy to reduce plant-based food loss and waste for sustainable bio-economy—a review. *Journal of Environmental Management, 332*, 117382.

Huang, J., Xia, T., Li, G., Li, X., Li, Y., Wang, Y., Wang, Y., Chen, Y., Xie, G., Bai, F. W., & Peng, L. (2019). Overproduction of native endo-β-1, 4-glucanases leads to largely enhanced biomass saccharification and bioethanol production by specific modification of cellulose features in transgenic rice. *Biotechnology for Biofuels, 12*, 1–15.

Janben, H. J., & Steinbuchel, A. (2014). Fatty acid synthesis in Escherichia coli and its applications towards the production of fatty-acid based biofuels. *Biotechnology for Biofuels, 7*, 1–26.

Javed, M. R., Noman, M., Shahid, M., Ahmed, T., Khurshid, M., Rashid, M. H., Ismail, M., Sadaf, M., & Khan, F. (2019). Current situation of biofuel production and its enhancement by CRISPR/Cas9-mediated genome engineering of microbial cells. *Microbiological Research, 219*, 1–11.

Lal, R. (2005). World crop residues production and implications of its use as a biofuel. *Environment International, 31*(4), 575–584.

Larson, E. D. (2008). *United Nations Conference on Trade and Development Biofuel Production Technologies: Status, Prospects and Implications for Trade and Development* (p. 1). UNCTAD secretariat at Palais des Nations, Switzerland.

Lee, Y., Kwon, E. E., & Lee, J. (2019). Polymers derived from hemicellulosic parts of lignocellulosic biomass. *Reviews in Environmental Science and Bio/Technology, 18*, 317–334.

Lewandowski, I., & Schmidt, U. (2006). Nitrogen, energy and land use efficiencies of miscanthus, reed canary grass and triticale as determined by the boundary line approach. *Agriculture, Ecosystems and Environment, 112*(4), 335–346.

Li, P., Sakuragi, K., & Makino, H. (2019). Extraction techniques in sustainable biofuel production: A concise review. *Fuel Processing Technology, 193*, 295–303.

Lu, L., Shen, X., Sun, X., Yan, Y., Wang, J., & Yuan, Q. (2022). CRISPR-based metabolic engineering in non-model microorganisms. *Current Opinion in Biotechnology, 75*, 102698.

Marino, N. D., Pinilla-Redondo, R., Csorgo, B., & Bondy-Denomy, J. (2020). Anti-CRISPR protein applications: Natural brakes for CRISPR-Cas technologies. *Nature Methods, 17*(5), 471–479.

Moreira, J. R. (2006). Global biomass energy potential. *Mitigation and Adaptation Strategies for Global Change, 11*, 313–342.

Nakada, S. H. U. N. I. C. H. I., Saygin, D., & Gielen, D. (2014). Global bioenergy: Supply and demand projections. In *International renewable energy agency (IRENA)*. Abu Dhabi.

OECD 2017, "Biofuels", in *OECD-FAO Agricultural Outlook* (pp. 2017–2026). OECD Publishing, Paris.

Palmqvist, E., & Hahn-Hagerdal, B. (2000). Fermentation of lignocellulosic hydrolysates. II: Inhibitors and mechanisms of inhibition. *Bioresource Technology, 74*(1), 25–33.

Popp, J., Lakner, Z., Harangi-Rakos, M., & Fari, M. (2014). The effect of bioenergy expansion: Food, energy, and environment. *Renewable and Sustainable Energy Reviews, 32*, 559–578.

Qin, L., Zhao, X., Li, W. C., Zhu, J. Q., Liu, L., Li, B. Z., & Yuan, Y. J. (2018). Process analysis and optimization of simultaneous saccharification and co-fermentation of ethylenediamine-pretreated corn stover for ethanol production. *Biotechnology for Biofuels, 11*, 1–10.

Rodionova, M. V., Poudyal, R. S., Tiwari, I., Voloshin, R. A., Zharmukhamedov, S. K., Nam, H. G., ... & Allakhverdiev, S. I. (2017). Biofuel production: challenges and opportunities. *International Journal of Hydrogen Energy, 42*(12), 8450–8461.

Sacramento Rivero, J. C., Mwampamba, T. H., Navarro-Pineda, F. S., Musule, R., García, C. A., Martínez-Bravo, R. D., Morales-García, A. L., Equihua-Sánchez, M., Fuentes-Gutiérrez, A. F., Gallardo-Alvarez, R. M., & Ruiz Camou, C. R. (2022). A methodological framework for assessing the sustainability of solid biofuels systems. *Bioenergy Research, 15*(4), 1797–1819.

Shanmugam, S., Ngo, H. H., & Wu, Y. R. (2020). Advanced CRISPR/Cas-based genome editing tools for microbial biofuels production: A review. *Renewable Energy, 149*, 1107–1119.

Spliethoff, H., & Hein, K. R. G. (1998). Effect of co-combustion of biomass on emissions in pulverized fuel furnaces. *Fuel Processing Technology, 54*(1–3), 189–205.

Sun, Y., & Cheng, J. (2002). Hydrolysis of lignocellulosic materials for ethanol production: A review. *Bioresource Technology, 83*(1), 1–11.

Surriya, O., Saleem, S. S., Waqar, K., Gul Kazi, A., & Ozturk, M. (2015). Bio-fuels: a blessing in disguise. *Phytoremediation for Green Energy*, 11–54.

Tilman, D., Hill, J., & Lehman, C. (2006). Carbon-negative biofuels from low-input high-diversity grassland biomass. *Science, 314*(5805), 1598–1600.

Tumuluru, J. S., Hess, J. R., Boardman, R. D., Wright, C. T., & Westover, T. L. (2012). Formulation, pretreatment, and densification options to improve biomass specifications for co-firing high percentages with coal. *Industrial Biotechnology, 8*(3), 113–132.

Wang, Y. Z., Zou, S. M., He, M. L., & Wang, C. H. (2015). Bioethanol production from the dry powder of Jerusalem artichoke tubers by recombinant Saccharomyces cerevisiae in simultaneous saccharification and fermentation. *Journal of Industrial Microbiology and Biotechnology, 42*(4), 543–551.

Weiss, N. D., Farmer, J. D., & Schell, D. J. (2010). Impact of corn stover composition on hemicellulose conversion during dilute acid pretreatment and enzymatic cellulose digestibility of the pretreated solids. *Bioresource Technology, 101*(2), 674–678.

Wen, Z., Minton, N. P., Zhang, Y., Li, Q., Liu, J., Jiang, Y., & Yang, S. (2017). Enhanced solvent production by metabolic engineering of a twin-clostridial consortium. *Metabolic Engineering, 39*, 38–48.

Wu, M. Y., Sung, L. Y., Li, H., Huang, C. H., & Hu, Y. C. (2017). Combining CRISPR and CRISPRi systems for metabolic engineering of E. coli and 1, 4-BDO biosynthesis. *ACS Synthetic Biology, 6*(12), 2350–2361.

Xu, W., Brynjolfsson, S., & Fu, W. (2022). Chapter 9. Use of genetically modified algae: Consumption safety challenges, current legislations, and socio-economic and environmental concerns. In A. Shekh & S. Dasgupta (Eds.), *Microalgae for sustainable products* (pp. 178–190). Royal Society of Chemistry.

Yang, H. H., Chien, S. M., Lo, M. Y., Lan, J. C. W., Lu, W. C., & Ku, Y. Y. (2007). Effects of biodiesel on emissions of regulated air pollutants and polycyclic aromatic hydrocarbons under engine durability testing. *Atmospheric Environment, 41*(34), 7232–7240.

Zhang, J., Petersen, S. D., Radivojevic, T., Ramirez, A., Perez-Manríquez, A., Abeliuk, E., S´anchez, B. J., Costello, Z., Chen, Y., Fero, M. J., Martin, H. G., Nielsen, J., Keasling, J. D., Jensen, M. K. (2020). Combining mechanistic and machine learning models for predictive engineering and optimization of tryptophan metabolism. *Nature Communications, 11*, 4880.

Zhang, J., Zong, W., Hong, W., Zhang, Z. T., & Wang, Y. (2018). Exploiting endogenous CRISPR-Cas system for multiplex genome editing in Clostridium tyrobutyricum and engineer the strain for high-level butanol production. *Metabolic Engineering, 47*, 49–59.

Zhao, X., Zhang, L., & Liu, D. (2012). Biomass recalcitrance. Part I: the chemical compositions and physical structures affecting the enzymatic hydrolysis of lignocellulose. *Biofuels, Bioproducts and Biorefining, 6*(4), 465–482.

Microbial Bioactive Metabolites: Biochemical Interface for Diverse Interactions

Satyanshu Kumar and Prafulla Parikh

Abstract

Microorganisms are emerging as an enormous renewable reservoir of novel bioactive compounds. Microorganisms find applications as food and feed supplements, flavouring agents, enzymes, antibiotics and antimicrobial agents. There exist opportunities to explore the enormous wealth of microorganisms with a holistic approach to the benefit of humans as well as environmental health. In this chapter, an attempt was made to summarize their important contributions to human, animal health care, agricultural and industrial processes.

Keywords

Microorganisms · Secondary metabolites · Bioprocess · Cosmetology

1 Introduction

Natural products in different forms were therapeutic reliefs for man and animals before the advent of alchemy (Kieslich, 1986). A huge number of natural products are produced by microorganisms. These products have been beneficial to humans, animals and plants (Singh et al., 2017). Secondary metabolites or specialized metabolites are produced by different classes of microorganisms and plants. Fungi and bacteria are the most inexhaustible producers of SM among microorganisms. Despite the fact that plants produce diverse classes of biologically active compounds, longer cultivation periods and high environmental susceptibility of plants are the limiting factors when the production of SM is taken into account (Park et al., 2018). Moreover, working and genetic manipulation in microorganisms for the production of SM is easier in comparison to plants and animals which makes it possible to intensify research on regulation as well as synthesis of SM by microorganisms (Siddiqui et al., 2012).

Researchers from natural product discovery have been fascinated by the chemical and functional diversity of SM. SM produced by microorganisms in the form of pharmaceuticals, flavouring agents, food additives have imparted significant contributions in the areas of nutritional health and the economy of humans. Furthermore, the increasing resistance towards the existing drugs against *Staphylococcus, Streptococus, Klebsiella, Mycobacterium*, etc. which are highly infectious microorganisms and also, high incidences of different types of cancer, warrants innovative approaches towards search and sustainable supply of pharmacologically active agents (Willems et al., 2020). Microorganisms have emerged as alternative renewable sources for the production of SM.

Microorganisms possess an amazing inventiveness of synthesising chemical moiety and arraying the intricate steps for their production. These capacities of microorganisms result in the synthesis of large number of original and bioactive molecules. It is nearly impossible or rare to find this diversity in chemical and combinatorial library. Microbial sources are still being reported as a huge repository of bioactive compounds (Fig. 1). These compounds exhibit wide structural and medicinal properties. Role played by microorganisms in human and animal health care, agriculture and industrial applications are briefly described in the present chapter.

S. Kumar (✉)
ICAR-Directorate of Medicinal and Aromatic Plants Research, Boriavi, Anand, Gujarat, India
e-mail: satyanshu66@gmail.com

P. Parikh
ACTREC, Tata Memorial Centre, Kharghar, Navi Mumbai, Maharashtra, India

K. Sharma (ed.), *Bio-prospecting of Novel Microbial Bioactive Compounds for Sustainable Development*, Sustainable Landscape Planning and Natural Resources Management, https://doi.org/10.1007/978-3-031-82178-3_15

Fig. 1 Microbial bioactive metabolites interface in human, animal, plant and industrial process

2 Role of SM and Their Production

About 23,000 SM are produced by microbes. Out of the total number of SM produced by microbes, *Actinomycetes* contribute to about 45% (10,000) and *Streptomyces* contribute to about 33% (7600) (Abd-Elnaby et al., 2016; Berddy, 2005). Chemically, SM are immensely diverse and belong to the major chemical groups including terpenoids, alkaloids, phenylpropanoids, polyketides and non-ribosomal polypeptides (He et al., 2018). SM produced by microorganisms mainly belong to polyketides and non-ribosomal polypeptides classes. Usually, SM is produced in low amounts. This restriction is due to an inbuilt mechanism. Transcriptional suppression of biosynthesis protects from excessive production of SM by these organisms. Excess of SM may otherwise cause autotoxicity in the microorganisms (Maithani et al., 2022). Elicitor could trigger the synthesis of SM (Chaing et al., 2011; Fisch et al., 2009). Researchers have tried classical techniques namely media composition optimization, aeration, etc., to overcome restrictions related to SM synthesis (Bode et al., 2002; Frisvad, 2012). Microorganism's growth and development do not have any significant influence of SM produced by them, however, processes like sporulation, formation of mycelium and defence against stress stimuli have the influence of SM (Kondo et al., 1998; Siddiqui et al., 2012). As a defence agent against other organisms, SM plays a role as competitive weapons. Several microalgae produce toxins and allelopathic compounds which help them to survive against other competitors.

Generally, different microbial strains produce different compounds (Tables 1 and 2). Chemically diverse SM is made by filamentous actinomycetes, myxobacteria, pseudomonads, cyanobacteria and filamentous fungi (Donadio et al., 2002).

Strain dereplication is a rapid analysis to discard identical isolates from the same or similar sources, is used for building strain collection. SM are produced in late growth phase.

3 Microbial Bioactive Metabolites in Human Health and Clinical Issues

Chemical syntheses have been used as a means of obtaining drugs for little more than 100 years (Kieslich, 1986). For microbes and microbial products, healthcare is the largest end user market. Penicillin was discovered from

Table 1 Production of selected polyphenols, amino acids and vitamins using genetically engineered microorganisms

Compound	Strain	References
Polyphenols		
Curcuminoid	*E. coli* coculture	Fang et al. (2018)
Cyanidin 3-O-glucoside	*E. coli*	Shrestha et al. (2019)
Calistephin	*E. coli* polyculture	Jones et al. (2017)
Eriodicytyol	*E. coli*	Zhu et al. (2014)
Naringenin	*S. cerevisiae*	Koopman et al. (2012)
Cyanidin 3-O-glucoside	*S. cerevisiae*	Takagi et al. (2009)
Resveratrol	*E. coli*	Lim et al. (2011)
Pterostilbene	*C. glutamicum*	Kallscheuer et al. (2017)
Quercetin	*C. glutamicum*	Kallscheuer et al. (2017)
Amino acids		
L-glutamate	*C. glutamicum*	Wada et al. (2016)
L-lysine	*C. glutamicum*	Becker et al. (2011)
L-threonine	*E. coli*	Ma et al. (2017)
L-phenylalanine	*E. coli*	Wu et al. (2019)
Vitamins		
Vitamin B_{12}	*Psedomonas denitrificans*	Li et al. (2008)
Vitamin B_2	*B. subtillis*	Schwechheimer et al. (2016)
Vitamin C	*Ketogulonicigenium vulgare* *Bacillus endophyticus* *Gluconobacter oxydans*	Abbas and Sibirny (2011), Takagi et al. (2009)
Pro-vitamin A	*Blakesiea trispora*	Papaioannou and Liakopoulou-Kyriakides (2010)
Pro-vitamin A	*E. coli*	Yang and Guo (2014)
Pro-vitamin A	*S. cerevisiae*	Verwaal et al. (2007)

fungal *sp. Penicillin notatum* in 1829. The number and volume of drugs prepared by microbial biosynthesis and biotransformation is increasing (Kieslich, 1986). In recent times, significant progress has been made in microbial production of polyphenols, polysaccharides, amino acids and vitamins (Wu et al., 2020, Table 2, Fig. 2). Antibiotics, antitumor agents (adriamycin, bleomycin, daunomycin, and mithramycin, Table 3), immunosuppressants, alkaloids and enzyme inhibitors are produced by microorganisms. Worldwide these are used in the treatment of infectious and fatal diseases thereby reducing mortality rate and increasing human life expectancy (Singh et al., 2017). Many antiparasitic drugs are also produced by microorganisms (Table 4, Shiomi & Omura, 2004).

4 Microbial Bioactive Metabolites in Animal Health

Many useful antiparasitic agents are produced by microorganisms. Spiramycin used as an antibacterial agent in the veterinary medicine, is produced by *Streptomyces ambofaciens* (Pinnert-Sindico et al., 1955). Ivermectin is used for antiparasitic control of the economically important gastrointestinal, lung and kidney nematodes as well as other nematodes namely *Thelazia* and *Parafilaria* in livestock (Shoop & Soli, 2002). Destomycin A and hygromycin B are produced by *Streptomyces rimofaciens* and *Streptomyces hygroscopicus,* respectively are used as feed additives for the treatment of nematode infection by *Ascaris, Oesophagostomum, Heterakis, Trichuris* and *Capillaria* (Ishii, 1998; Kondo et al., 1965; Mann & Bromer, 1958).

5 Microbial Bioactive Metabolites in Agriculture

Microbial byproducts have been identified as novel and readily accessible sources of novel biologically active compounds for plant health (Saxena & Pandey, 2001). Both insecticides as well as miticides are produced by some *Streptomyces sp.* Insecticidal as well as acaricidal properties of Avermectins have been reported. Milbemectins is derived from *Streptomyces hygroscopicus* (Dayan et al., 2009). *Streptomyces* group produces many toxins including

Table 2 Source organisms, secondary metabolites and biological activity

Source organisms	Secondary metabolites	Biological activity	References
Streptomyces strain Acta 2930	Warkmycin	Antimicrobial	Helalay et al., (2013)
Streptomyces nodosus	Deoxyamphoteronolides	Antimicrobial	Byrne et al. (2003)
Streptomyces atratus	Atratumycin	Antimicrobial	Sun et al., (2019)
Streptomyces griseochromogens	Blasticidin-S	Antimicrobial	Dayan et al., (2009)
Streptomyces chattanoogensis	Natamycin	Antimicrobial	TeWelcher et al., (2008)
Streptomyces peucetius	Daunorubicin	Anticancer/Antitumor	Katz and Baltz (2016)
Streptomyces viridochromogens	Bialaphos	Herbicidal	Li et al., (2003)
Streptomyces sp. A6308	Coaristeromycin	Herbicidal	Saxena and Pandey (2001)
Streptomyces saganonensis	Herbimycins	Herbicidal	Duke and Lydon (1987)
Streptomyces sp.	Gougerotin	Herbicidal	Saxena and Pandey, (2001)
Streptomyces sp.	Tacrolimus	Immunosuppressive	Barrios-Gonzalez et al., (2003)
Streptomyces avermitillis	Abamectin	Insecticidal	Dayan et al., (2009)
Streptomyces clavuligerus	Clavulanic acid	Enzyme inhibitor	Reading and Cole (1977)
Aspergillus terreus	Lovastatin	Antichlolesterolemics	Hendrickson et al., (1999)
Aspergillus sclerotiorum	Sclerotide A and B	Antimicrobial	Zheng et al., (2009)
Aspergillus flavus	Aflatoxin	Toxin/mycotoxin	Klich (2007), Pitt (2000)
P. syringae	Coronatine	Toxin	Gross and Loper (2009)
P. syringae	Phaseolotoxin	Toxin	Saxena and Pandey (2001)
P. syringae	Syringomycin	Toxin	Gross and Loper (2009)
Penicillium chrysogeum	Pravastatin	Antichlolesterolemics	McLean et al., 2015
Penicillium simplicissimum AK-40	Okaramine	Insecticidal	Matsuda (2018)
Penicillium brevicompactum	Mycophenolic acid	Immunosuppressive	Katz and Baltz (2016)
Penicillium roqueforti	Roquefortine C	Mycotoxin	Garcia-Estrada et al. (2011)
Cyanobacteria	Cryptophycins	Anticancer/Antitumor	Bhatnagar and Kim (2010)
Vibrio sp. EA348	Amifloxacin	Antimicrobial	Bibi et al., 2020
Vibrio neocaledonicus	1H-indole-3-carboxaldehyde	Enzyme inhibitor	Gomez-Betancur et al., (2019)
Entrophospora infrequens	Camptothecin	Anticancer/Antitumor	Puri et al., (2006)
Taxomyces andrenae	Paclitaxel	Anticancer/Antitumor	Strobel and Daisy (2003)
Sorangium cellulosum	Epothilone	Anticancer/Antitumor	Katz and Baltz (2016)
Chromobacterium violaceum	Depsipeptide	Anticancer/Antitumor	Suarez-Jimenez et al., (2012)
Cytonaema sp.	Cytonic acid A and B	Antiviral	Strobel and Daisy (2003)
Geobacillus thermodenitrificans	EPS-2	Antiviral	Arena et al., (2009)
Fusarium heterosporum	Equisetin	Antiviral	Singh et al., (1998)
Phom sp.	Phomanolide	Antiviral	Arena et al., (2009)
Isarias inclairti	Myriocin	Immunosuppressive	Gullo et al., (2006)

Bialaphos, phosphinothricin which are used as natural and ecofriendly herbicides (Table 5).

In comparison to synthetic, halogenated chemical structure, herbicidal compounds produced by microbes have relatively short lives. Because of their biodegradable nature, they do not have residual toxicity to the environment. Additionally, their working concentration is also low. Furthermore, the shelf-life of microbial phytotoxins is limited as compared to synthetic chemicals.

For sustainable agriculture and a green environment, microbial-origin pesticides and growth regulators have played a significant role. Microorganisms also produce plant hormones (Karadeniz et al., 2006; MacMillan, 2001; Tsavkelova et al., 2006; Yurekli et al., 1999). Microbial

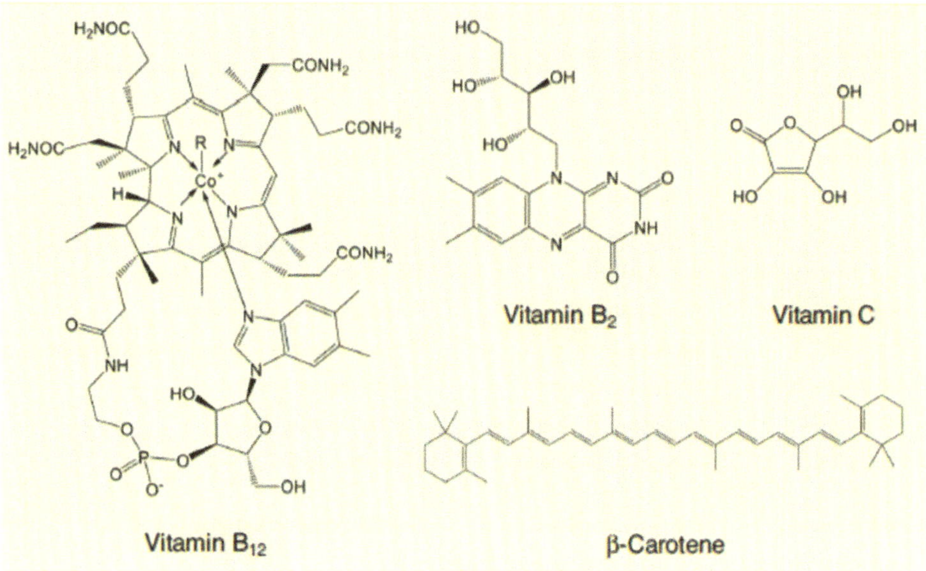

Fig. 2 Structures of some typical vitamins produced by engineered microbes, including vitamin B_{12}, B_2, vitamin C, and pro-vitamin A (β-carotene). R = 5'-deoxyadenosyl, Me, OH, CN. Adapted from Current Opinion in Food Science 2020, 32:9–16 with permission

Table 3 Microorganisms derived anticancer agents

Type of cancer	Anticancer agent	Function	Source
Lung	Aclacinomycin	Inhibition of nucleic acids and protein synthesis	*Streptomyces galilaeus*
Acute myelogenous leukemia	Calicheamicin	Cleavage of DNA	*Micromonospora echinospora*
Uterus and ovarian	Adriamycin (doxorubicin)	Inhibition of DNA replication	*Streptomyces peucetius*
Breast and urinary bladder	Chromomycin A_3 (toyamycin and aburamycin)	Inhibition of RNA synthesis	*Streptomyces griseus*
Wilm's tumour in children	Actinomycin D	Inhibition of RNA synthesis	*Streptomyces antibioticus*
Breast cancer, Kaposi sarcoma	Taxol (paclitaxel)	Microtubule depolymerization	*Taxomyces andreanae, Nodulisporium sylviforme*
Breast, stomach, oesophagus, bladder	Mitomycin C	Inhibition of DNA replication	*Streptomyces caespitosus*
Bone and testicular tumour	Mithramycin	Inhibition of RNA synthesis, mRNA expression and protein synthesis	*Streptomyces plicatus, S. argillaceus, S. atroolivaceus*
Acute myeloid leukemia, acute lymphocytic leukemia, neuroblastoma	Daunomycin	Inhibition of DNA replication	*Streptomyces peucetius*
Hodgkin's disease, squamous cell carcinoma, testicular, pleurodesis	Bleomycin	Cleavage of DNA	*Streptomyces verticillus*
Liver	Neocarzinostatin	Proteolysis of histone, degradation of DNA	*Streptomyces carzinostaticus*
Colon	Thiocoraline	Inhibition of DNA replication	*Micromonospora marina*
Leukemia, lymphoma, epidermal carcinoma	Alteramide	DNA fragmentation	Alteromonas sp.
Breast	Epothilone	Microtubule depolymerization	*Sorangium cellulosum*
Hairy cell leukemia, acute lymphocytic leukemia, prolymphocyclic leukemia	Pentostatin (deoxycoformycin)	Inhibition of DNA	*Streptomyces antibioticus*
Pancreatic islet cell	Streptozotocin	Inhibition of DNA	*Streptomyces achromogenes*

Source Singh et al. (2017)

Table 4 Antiparasitic drugs produced by microorganisms

Antibiotics	Producing strain	Disease	Mode of action
Antiprotozoals			
Tetracycline	*Streptomyces sp.*	Malaria	Protein synthesis inhibitor
Polyethers	*Actinomycetes*	Coccidioisis	Increase the influx of cations
Marinopyrrole	*Actinomycetes*	Human foreskin fibroblast/human hepato-carcinoma	-
Spiramycin	*Streptomyces ambofaciens*	Cryptosporidiosis	Protein synthesis inhibitor
Acetylspiramycin	Semi synthetic	Toxoplasmosis	Protein synthesis inhibitor
Clindamycin	Semi synthetic	Toxoplasmosis and babesiosis	Protein synthesis inhibitor
Paromomycin	*Streptomyces rimosus*	Cryptosporidiosis, amebissis and giardiasis	Protein synthesis inhibitor
Amphotericin B	*Streptomyces nordus*	Leishmaniasis	Bind to membrane
Trichomycin	*Streptomyces hachijoensis*	Trichomoniasis	Bind to membrane
Fumagillin	*Aspergillus fumigatus*	Microsporidiasis	Met aminopeptidase 2 inhibitor
Hoshinoamide	*Caldora penicillata*	Human cancer cells, HeLa and HL 60	

Source Shiomi and Omura (2004), Zhang et al., (2023)

herbicides include microbial preparations and microbial-derived herbicides. For weed control in agriculture, microorganisms (*Colletotrichum* spp. and *Phytophthora*) based products are also used (Chandler et al., 2011; Demain & Sanchez, 2009; Table 4). White fly strains that are resistant to pesticides are managed effectively using *B. bassiana*. Spittlebugs on sugarcane and grassland are managed using *M. anisopliae*.

6 Microbial Bioprocess Technology

Microorganisms have immensely contributed to the development of bioprocess technologies. Many industries utilize microbial enzymes up to a significant extent. *Aspergillus*, *Bacillus* are being utilized for enzyme preparation (Gurung et al., 2013; Mahmood, 2015; Shimizu, 2001; Sun et al., 2015). Selective use of individual enzymatic transformation stages with microorganisms in chemical production pathways has expanded the process of biotechnology-based production of pharmaceuticals (Kieslich, 1986). For the production ranging from hydrolytic enzymes such as proteases and alpha-enzymes such as proteases and alpha-amylases to speciality chemicals such as amino acids and vitamins *Bacillus* species have long been established as industrial resources (Debabov, 1982; Harwood, 1992; Itano & Kimata, 1996; Sauer et al., 1998).

Microorganisms find use in the preparation of food, feed and also for their supplement preparation. For the preparation of vitamins on a commercial scale, appropriate microorganisms are used. Microbial fermentation is used for commercial preparation of Vitamin B_{12} (Cyanocobalamin). Gram-positive bacilli are producers of purine nucleosides namely inosine and guanosine (Kunika, 1996). These nucleosides find use as raw material for the synthesis of inosine monophosphate and guanosine monophosphate which are flavour enhancers. *B. subtilis* also produces folic acid.

7 Microbial Products in Cosmetology

Many cosmetic products (Fig. 3) have biological origins and these have been acquired from bacteria, fungi and algae (Gupta et al., 2019). Underutilized biodiversity of microbes represents a potential opportunity for their application in cosmetics in the future (Satyanarayana et al., 2012).

Cyclodextrins have a significant contribution in cosmetic formulations (Del Valle, 2004). It finds use in talcum, diapers, menstrual discs, pads, napkins, etc. Bacterial strains namely *Bacillus subtillis* strain 313, *Brevibacterium sp.* Strain 9605, *Brevibacillus brevis* strain CD 162, *Microbacterium terrae* KNR 9 are used for Cyclodextrin production (Kato & Horikoshi, 1986; Kim et al., 1998; Mori et al., 1994; Rajput et al., 2016).

Biosurfactants are used in diverse cosmetic products. For the preparation of biosurfactants, bacteria, fungi and microbes are used. *Pseudozyma antarctica*, *P. aphidis*, *P. regulosa* and *P. parantarctica* produce mannosylerythritol lipid (MEL) which is one of the most prominent biosurfactants. MEL is used for cosmetic products such as lipsticks, lip makers, eye shades, soap, sprays, powders, nail care, and body massage oil (Ueno et al., 2007; Villeneuve, 2007).

Polysaccharides obtained from natural sources are both biodegradable and biocompatible. *Leuconostocaceae* family of microbes namely *Leuconostoe mesenteriodes* and *Streptococcus mutans* are sources of Dextran, an

Table 5 Promising phytotoxins for use as natural herbicides

Microbial source	Phytotoxin	Target weed	Site of action
Fungi			
Alternaria alternata	Tenuazonic acid	*Datura innoxia*	
A. alternata	Tentoxin	Grass, broad-leaved weeds	CF1-ATPase
A. alternata f. sp. lycopersici	AAL-toxin	Garden cress	
A. alternata f. sp maculosa	Maculosins	Spotted knapweed	
A. zinniae	Zinniol	Lettuce seedlings	Disruption of calcium-regulated cell process
Ascochyta hyalospora	Ascochytine, hyalopyrone	Lambsquaters, prickly sida	Electrolyte leakage and inhibition of root growth
Ascochyta caulina	Trans-4-aminoproline	Chenopodium rubrum	Unknown
Bipolaris cynodontis	Bipolaroxin	Velvet leaf and pigweed	
Cepahlosporium spp.	1233A		HMG CoA synthase
Cercospora kikuchii	*Cercosporin*	Higher plants	Lipid peroxidation and photosensitizing action
Fusarium sp.	Fusaric acid	Broad spectrum	Jimsonweed and Duckweed
Gliocladium virens	Virdol		
Helminthosporium sativum	Prehelminthosporal	Johnsongrass	Unknown
Irpex polyhedon	Irpexil		Enzyme inhibitor
Paecilomyces variotiiSANK 21,086	Carnexistin	Dicotyledonous weeds and some mono cotyledonous weeds	Aspartate amino transferase inhibition
Ascochyta agropyrina var. nana	Papyracillic acid	*Orobanche crenata, O. cumana, O. minor, ramosa*	Active against bacteria *Xanthomonas campestris* and *Bacillus subtills* and fungus *Candida tropicalis*
Alternaria alternata	Tenuazonic acid	*Ageratina adenophora*	
Actinomycetes			
Nocardia sp. no.2–200	Thiolactomycin	Higher plants	Type II fatty acid synthetase
Streptomyces hygroscopicus	Hydantocidin	Broad spectrum	
S. hygroscopicus	Polyethrin A	Garden cress	Inhibits ceramide synthase
S. hygroscopicus var. geldanus	Geldanamycin	Garden cress	
Streptomyces sp.	*Anisomycin*	Barnyard grass and crabgrass	Inhibits photo synthesis
Streptomyces suganonensis	Herbicidins A/B	Dicotyledonous and mono cotyledonous weeds	
Streptomyces toyacaensis	Toyocamycin		
Streptomyces hygroscopicus	Nigericin		Potassium ionphore and photophosphorylation inhibitor
Streptomyces sp. A7847	Herboxidiene	Broad spectrum	Unknown
Streptomyces sp.	SF-701	Barnyard grass	Inhibition via starch synthesis
Streptomyces sp. AM-3672	Herbimycin	Digitaria spp., Giant foxtail (*Echinocloa crusgalli*), *Chenopodium* and *Portulacea*	Unknown
Streptomyces sp. 620,061	Pyrizadocidin	Giant foxtail (*Echinocloa crusgalli*)	Electron transport inhibition
Bacteria			
Pseudomonas syringae var. tabaci	Tabtoxin	Broad spectrum	GS-GOGAT pathway
Pseudomonas syringae var. phaseolicola	Phaseolotoxin	*Glycine wightii* and *Macrophillium atropurpureum*	Ornithine carbamoyl transferase inhibitor
Cyanobacteria			
Scytonema hofmanni	Cynobacterin	*Lemna gibba*	Inhibits PSII site

Source Cimmino et al., (2015), Saxena and Pandey (2001)

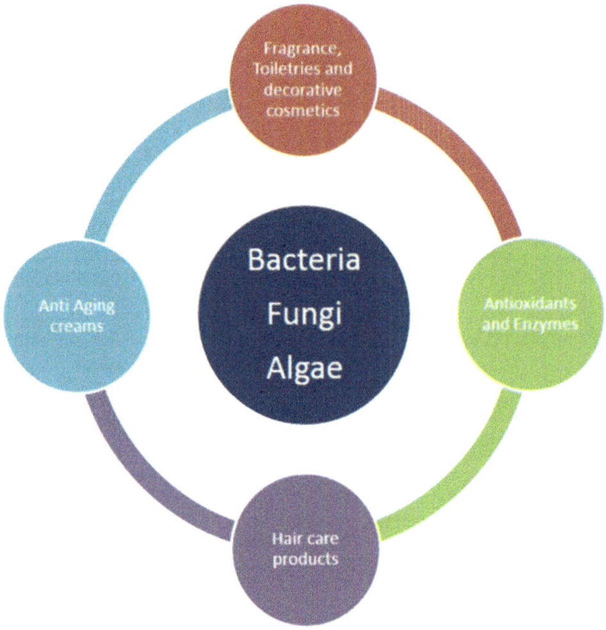

Fig. 3 Main categories of cosmetics. Reproduced from Nat. Prod. Bioprosp., (2019) 9: 267–278 with permission

exopolysaccharide (EPS) from glucose polymer (Yildiz & Karatas, 2018). Bacteria from the genera Pseudomonas are well-known producer of abundant amounts of alginate as EPS. Xanthan is a complex hetero-polymer EPS. It is produced by the bacterial genus Xanthomonas. Xanthan has thickening properties and assists in gelling. It is used in skin formulations for skin smoothing and moisturizing. In skin formulations, Xanthan finds application as emulsifiers and foaming agent (Bilanovic et al., 2016; Kumar et al., 2018).

Hyaluronic acid (HA) is a glucosaminoglycan. It is extensively used as a dermal filler and skin lotion. Naturally, it exits as a hydrogel and it is ubiquitous in human and animal tissues (Swan & Kuo, 1991). Skin serum containing HA increases moisture retention, reduces skin wrinkles and improves skin firmness and elasticity. HA is synthesized by *Streptococcus* genus.

Ceramides are used as skin-hydrating agents in cosmetics. Fungal species produce ceramides (Gao et al., 2004) and *Candida albicans*, *Agaricus bisporus*, *Armillaria tabescens*, etc. are sources of glycosyl ceramides (Prasad & Ghannoum, 2017).

Chitin-glucans have good moisturizing properties and are obtained from fungi (Gautier et al., 2008). Chitosan, a chitin-glucan with HA and collagen is used in hair setting lotions and gels. Minoxidil is a pyridine derivative and it finds application as a standard treatment for male and female androgenic alopecia. Clinical trials reported that the success rate of androgenic alopecia treatment with minoxidil (5%) barely exceeds 40%. Chitosan nanoparticles loaded with minoxidil are used for sustained release of minoxidil for effective transdermal transport and hair growth (Matos et al., 2017). Skin ageing gets slowed down by collagen and it enhances the suppleness of skin. *Chlorella*, a green microalga restores the firmness of the skin by protecting collagen and elastin fibres against the enzymes collagenase and elastase which degrades the skin (Wang et al., 2015).

Astaxanthin, a carotenoid, is obtained from *Haematococcus pluvialis*, restores skin tone, elasticity and retains the moisture content of corneocyte layer. These beneficial properties result in the lowering of skin wrinkles (Tominaga et al., 2012). Astaxanthin also decreases melanin in trans-epidermal cell. Zeaxanthin from *Nannochloropsis oculata*, a microalgae, exhibited anti-tyrosinase activity, which could be used for the management of hyperpigmentation (Shen et al., 2011).

8 Conclusion

Microorganisms as a source of natural compounds have significantly contributed to nutrition, health care, cosmetics, agricultural practices, environment, and industries. Microorganisms have a significant role in the preparation of food and feed. Microorganisms are used as sources for the production of antibiotics, antitumor drugs, immunosuppressants, enzyme inhibitors and antimicrobial agents. Worldwide these products find application in the treatment of both infectious and non-infectious diseases which leads to reduced mortality and enhanced life expectancy. Microbial enzymes have many industrial applications for the production of food and nutrition supplements, high-value chemicals, pharmaceuticals. The role of microbial pesticides and growth promoters in the promotion of sustainable agriculture and development of green environment is also very significant. The undesirable consequences of the rapid changing global scenario could be minimized with a holistic approach toward further exploration microorganisms.

References

Abbas, C. A., & Sibirny, A. A. (2011). Genetic control of biosynthesis and transport of biboflavin and flavin nucleotides and construction of robust biotechnological producers. *Microbiology and Molecular Biology Reviews, 75*, 321–360.

Abd-Elnaby, H., Abo-Elala, G., Abdel-Raouf, U., Abd-elwahab, A., & Hamed, M. (2016). Antibacterial and anticancer activity of marine *Streptomyces parvus*: Optimization and application. *Biotechnology and Biotechnological Equipment, 30*, 180–191.

Arena, A., Gugliandolo, C., Stassi, G., Pavone, B., Iannello, D., Bisignano, G., & Maugeri, T. L. (2009). An exopolysaccharides produced by *Geobacillusthermodenitrificans* strain B-3-72:

Antiviral activity on immunocompetent cells. *Immunology Letters, 123*, 132–137.

Barrios-Gonzalez, J., Fernandez, F. J., & Tomasini, A. (2003). Microbial secondary metabolites production and strain improvement. *Indian Journal of Biotechnology, 2*, 322–333.

Becker, J., Zelder, O., Hafner, S., Schroder, H., & Wittmann, C. (2011). From zero to hero-design based systems metabolic engineering of *Corynebacterium glutamicum* for L-lysine production. *Metabolic Engineering, 13*, 159–168.

Berddy, J. (2005). Bioactive microbial metabolites. *The Journal of Antibiotics, 58*, 1–26.

Bhatnagar, L., & Kim, S. K. (2010). Marine antitumor drugs status, shortfalls and strategies. *Marine Drugs, 8*, 2702–2720.

Bibi, F., Yasir, M., Al-Sofyani, A., Naseer, M. L., & Azhar, E. L. (2020). Antimicrobial activity of bacteria from marine sponge subereamollis and bioactive metabolites of *Vibrio sp.* EA348. *Saudi Journal of Biological Sciences, 27*, 1139–1147.

Bilanovic, D., Starosvetsky, J., & Armon, R. H. (2016). Preparation of biodegradable xanthan-glycerol hydrogel, foam, film, aerogel and xerogel at room temperature. *Carbohydrate Polymers, 148*, 243–250.

Bode, H. B., Bethe, B., Hofs, R., & Zeeck, A. (2002). Big effects from small changes: Possible ways to explore nature chemical diversity. *Chem BioChem, 3*, 619–662.

Byrne, B., Carmody, M., Gibson, E., Rawlings, B., & Caffrey, P. (2023). Biosynthesis of deoxyamphotericins and deoxyamphoteronolides by engineered strains of *Streptomyces nodosus*. *Chemistry and Biology, 10*, 1215–1224.

Chandler, D., Bailey, A. S., Tatchell, G. M., Davidson, G., Greaves, J., & Grant, W. P. (2011). The development regulation and use of biopesticide for integrated pest management. *Philosophical Transactions of the Royal Society b: Biological Sciences, 366*, 1987–1998.

Chiang, Y. M., Chang, S. L., Okaley, B. R., & Wang, C. C. (2010). Recent advances in awakening silent biosynthetic gene clusters and linking orphan clusters to natural products in microorganisms. *Current Opinion in Chemical Biology, 15*, 137–143.

Cimmino, A., Masi, M., Evidente, M., Superchi, S., & Evidentle, A. (2015). Fungal phytotoxins with potential herbicidal activity: Chemical and biological characterization. *Natural Product Research, 32*, 1629–1653.

Dayan, F. E., Cantrell, C. L., & Duke, S. O. (2009). Natural products in crop protection. *Bioorganic and Medicinal Chemistry, 17*, 4022–4034.

Debabov, V. G. (1982). The industrial use of *Bacilli*. In D. A. Dubnau (Ed.), Biology of the *Bacilli* (pp. 331–370). Wiley

Del Valle, E. M. M. (2004). Cyclodextrins and their uses: A review. *Process Biochemistry, 39*, 1033–1046.

Demain, A. L., & Sanchez, S. (2009). Microbial drug discovery: 80 years of progress. *Journal of Antibiotics, 62*, 5–16.

Donadio, S., Monciardini, P., Alduina, R., Mazza, P., Chiocchini, C., Cavaletti, L., Sosio, M., & Puglia, A. M. (2002). Microbial technologies for the discovery of novel bioactive metabolites. *Journal of Biotechnology, 99*, 187–198.

Duke, S. O., & Lydon, J. (1987). Herbicides from natural compounds. *Weed Technology, 1*, 122–128.

Fang, Z., Jones, J. A., Zhou, J., & Koffas, M. A. G. (2018). Engineering *Escherichia coli* co-cultures for oroduction of curcuminoids from glucose. *Biotechnology Journal, 13*, 1700576.

Fisch, K. M., Gillaspy, A. F., Gipson, M., Henrikson, J. C., Hoover, A. R., Jackson, L., & Cichewicz, R. H. (2009). Chemical induction of silent biosynthetic pathway transcription in *Aspergillus niger*. *Journal of Industrial Microbiology, 36*, 1199–1213.

Frisvad, J. C. (2012). Media and growth conditions for induction of secondary metabolite production. In N. Keller & G. Turner (Eds.), *Fungal secondary metabolism* (pp. 47–58). Humana Press.

Gao, J. M., Zhang, A. L., Chen, H., Liu, J.-K. (2004). Molecular species of ceramides from the ascomycete truffle *Tuber indicum*. *Chemistry and Physics of Lipids, 131*, 205–213.

Garcia-Estrada, C., Ullan, R. V., Albillos, S. M., Fernandez-Bodega, M. A., Durek, P., von Dohren, H., & Martin, J. F. (2011). A single cluster of coregulated genes encodes the biosynthesis of the mycotoxinsroquefortine C and meleagrin in *Penicilliumchrysogenum*. *Chemistry and Biology, 18*, 1499–1512.

Gautier, S., Xhaufflaire-Uhoda, E., Gonry, P., & Pierard, G. E. (2008). Chitin-glucan, a natural cell scaffold for moisturization and rejuvenation. *International Journal of Cosmetic Science, 30*, 459–469.

Gomez-Betancur, L., Zhao, J., Tan, L., Chen, C., Yu, G., Rey-Suarez, P., & Preciado, L. (2019). Bioactive compounds isolated from marine bacterium *Vibro neocaledonicus* and their enzyme inhibitory activities. *Marine Drugs, 17*, 401.

Gross, H., & Loper, J. E. (2009). Genomics of secondary metabolites production by *Pseudomonas* spp. *Natural Products Reports, 26*, 1408–1446.

Gullo, V. P., McAlpine, J., Lam, K. S., Baker, D., & Petersen, F. (2006). Drug discovery from natural products. *Journal of Industrial Microbiology and Biotechnology, 33*, 523–531.

Gupta, P. L., Rajput, M., Oza, T., Trivedi, U., & Sanghvi, G. (2019). Eminence of microbial products in cosmetic industry. *Natural Products and Bioprospecting, 9*, 267–278.

Gurung, N., Ray, S., Bose, S., Rai, V. (2013). A broader view: microbial enzymes and their relevance in industries, medicine and beyond, *BioMed Research International*, Article ID 329121.

Harwood, C. R. (1992). *Bacillus subtilis* and its relatives: Molecular biological and industrial workhorses. *Trends in Biotechnology, 10*, 247–256.

He, Y., Wang, B., Chen, W., Cox, R. J., He, J., & Chen, F. (2018). Recent advances in reconstructing microbial secondary metabolites biosynthesis in Aspergillus spp. *Biotechnology Advances, 36*, 739–783.

Helalay, S. E., Goodfellow, M., Zinecker, H., Imhoff, J. F., Siissmuth, R. D., & Fiedler, H. P. (2013). Warkmycin, a novel angucycline antibiotic produced by *Streptomyces sp.* Acta 2930. *The Journal of Antibiotic, 66*, 669–674.

Hendrickson, L., Davis, C. R., Roach, C., Aldrich, T., Mcada, P. C., & Reeves, C. D. (1999). Lovastin biosynthesis in *Aspergillusterreus* : Characterization of blocked mutants, enzyme activities and a multifunctional polyketide synthase gene. *Chemistry and Biology, 6*, 429–439.

Ishii, T. (1998). *Veterinary parasitology* (Vol. 1). Kodansha Scientific.

Itano, N., & Kimata, K. (1996). Expression cloning and molecular characterization of HAS protein, a eukaryotic hyaluronan synthase. *Journal of Biological Chemistry, 271*, 9875–9878.

Jones, J. A., Vernacchio, V. R., Collins, S. M., Shirke, A. N., Xiu, Y., Englaender, J. A., Cress, B. F., McCutcheon, C. C., Linhardt, R. J., Gross, R. A., Koffas, M. A. G. (2017). Complete biosynthesis of anthocyanins using *E. coli* polycultures. *mBio, 8*, e00621–00617.

Kallscheuer, N., Vogt, M., Bott, M., & Marienhagen, J. (2017). Functional expression of plant derived O-methyltransferase, flavone 3-hydroxylase and flavonol synthase in *Corynebacterium glutamicum* for production of pterostilbene, kaempferol and quercetin. *Journal of Biotechnology, 258*, 190–196.

Karadeniz, A., Topcuoglu, S. F., & Inan, S. (2006). Auxin, gibberellin, cytokinin and abscisic acid production in some bacteria. *World Journal of Microbiology and Biotechnology, 22*, 1061–1064.

Kato, T., & Horikoshi, J. (1986). A new γ-cyclodextrin forming enzyme produced by *Bacillus subtillis*. *Journal of Japanese Society of Starch Science, 33*, 137–143.

Katz, L., & Baltz, R. H. (2016). Natural product discovery: Past, present and future. *Journal of Industrial Microbiology and Biotechnology, 43*, 155–176.

Kieslich, K. (1986). Production of drugs by microbial biosynthesis and biotransformation: Possibilities, limits and future developments. *Ist Communication Arzneimittelforschung, 36*, 774–778.

Kim, M. H., Sohn, C. B., & Oh, T. K. (1998). Cloning and sequencing of a cyclodextrin glycosyl transferase gene from *Brevibacillus brevis* CD162 and its expression in *Escherichia coli*. *FEMS Microbiology Letters, 164*, 411–418.

Klich, M. A. (2007). *Aspergillus flavus* : The major producer of aflatoxin. *Molecular Plant Pathology, 8*, 713–722.

Kondo, S., Sezaki, M., Kolke, M., Shimura, M., Akita, E., Satoh, K., & Hara, T. (1965). Destomycins A and B, two new antibiotics produced by *Streptomyces*. *The Journal of Antibiotics, Series A, 18*, 38–42.

Kondo, S., Yasui, K., Natsume, M., Katayama, M., & Marumo, S. (1998). Isolation, physicochemical properties and biological activity of paramycin-607, an aerial mycelium-inducing substance from *Streptomyces alboniger*. *Journal of Antibiotics, 41*, 1196–1204.

Koopman, F., Beekwilder, J., Crimi, B., van Houwelingen, A., Hall, R. D., Bosch, D., van Maris, A. J., Pronk, J. T., & Daran, J.-M. (2012). De novo production of the flavonoid naringenin in engineered *Saccharomyces cerevisiae*. *Microbial Cell Factories, 11*, 155.

Kumar, A., Rao, K. M., & Han, S. S. (2018). Application of xanthan gum as polysaccharide in tissue engineering: A review. *Carbohydrate Polymers, 180*, 128–144.

Kunika, A. (1996). Nucleotides and related compounds. In M. Roehr (Ed.), *Products of primary metabolism* (Vol. 6, pp. 561–612). VCH.

Li, K.-T., Liu, D.-H., Chu, J., Wang, Y.-H., & Zhang, S.-L. (2008). An effective and simplified pH-stat control strategy for the industrial fermentation of vitamin B12 by *Psedomonas denitrificans*. *Bioprocess and Biosystems Engineering, 31*, 605–610.

Li, Y., Sun, Z., Zhuang, X., Xu, L., Chen, S., & Li, M. (2003). Research progress on microbial herbicides. *Crop Protection, 22*, 247–252.

Lim, C. G., Fowler, Z. L., Hueller, T., Schaffer, S., & Koffas, M. A. G. (2011). High yield resveratrol production in engineered *Escherichia coli*. *Applied and Environment Microbiology, 77*, 3451–3460.

Ma, Q., Zhang, Q., Xu, Q., Zhang, C., Li, Y., Fan, X., Xie, X., & Chen, N. (2017). Systems metabolic engineering strategies for the production of amino acids. *Synthetic and Systems Biotechnology, 2*, 87–96.

MacMillan, J. (2001). Occurrence of gibberellins in vascular plants, fungi and bacteria. *Journal of Plant Growth Regulation, 20*, 387–442.

Mahmood, Z. A. (2015). *Microbial amino acids production*. CRC Press, Taylor & Francis Group.

Maithani, D., Sharma, A., Gangola, S., Chaudhary, P., & Bhatt, P. (2022). Insights into applications and strategy for discovery of microbial metabolites. *Microbiological Research, 261*, 127053.

Mann, R. L., & Bromer, W. W. (1958). Isolation of a second antibiotic from *Streptomyces hygroscopicus*. *Journal of the American Chemical Society, 80*, 2714–2716.

Matos, B. N., Melo, L. M. A., Pereira, M. N., Reis, T. A., Cunha-Filho, M., Gratieri, T., & Gelfuso, G. M. (2017). Iontophoresis on minoxidil sulphate loaded chitosan nanoparticles accelerates drug release, decreasing their target effect to hair follicles. *Quimica Nova, 40*, 154–157.

Matsuda, K. (2018). Okramines and other plant fungal products as new insecticide leads. *Current Opinion in Insect Science, 30*, 67–72.

McLean, K. J., Hans, M., Meijrink, B., van Scheppingen, W. B., Vollebgret, A., Tee, K. L., & van den Berg, M. A. (2015). Single step fermentative production of thw cholesterol lowering drug pravastatin via reprogramming of *Penicilliumchrysogenum*. *Proceedings of the National Academy of Sciences, 112*, 2847–2852.

Mori, S., Hirose, S., Oya, T., & Kitahata, S. (1994). Purification and properties of cyclodextrin glucanotransferase from *Brevibacterium sp.* No. 9605. *Bioscience, Biotechnology, and Biochemistry, 58*, 1968–1972.

Papaioannou, E. H., & Liakopoulou-Kyriakides, M. (2010). Substrate contribution on caretonoids production in *Blakeslea trispora* cultivations. *Food and Bioproducts Processing, 88*, 305–311.

Park, S. Y., Yang, D., Ha, S. H., & Lee, S. Y. (2018). Metabolic engineering of microorganisms for the production of natural compounds. *Advanced Biosystems, 2*, 1–41.

Pinnert-Sindico, S., Ninet, L., Pred'homme, J., Cosar, C. (1955). A new antibiotic, spiramycin. *Antibiotics ANN, 1954–1955*, 724–727.

Pitt, J. I. (2000). Toxigenic fungi and mycotoxins. *British Medical Bulletin, 56*, 184–192.

Prasad, R., & Ghannoum, M. A. (2017). *Lipids of pathogenic fungi*. CRC Press.

Puri, S. C., Nazir, A., Chawla, R., Arora, R., Riyaz-ul-Hasan, S., Amna, T., Ahmad, B., Verma, V., Singh, S., Sagar, R., Sharma, A., Kumar, R., Sharma, R. K., & Qazi, G. N. (2006). The endophytic fungus Trameteshirsuta as a novel alternative source of podophyllotoxin and related aryl tetralinlignans. *Journal of Biotechnology, 122*, 494–510.

Rajput, K. N., Patel, K. C., Trivedi, U. B. (2016). A novel cyclodextrin glucanotransferese from alkaliphile *Microbacterium terrae* KNR 9: Purification and properties. *3 Biotech, 6*, 168.

Reading, C., & Cole, M. (1977). Clavulanic acid: A beta lactamase-inhibiting beta-lactum from *Strptomyces clavuligerus*. *Antimicrobial Agents and Chemotherapy, 11*, 852–857.

Satyanarayana, T., Johri, B. N., & Prakash, A. (2012). *Microorganisms in environment management: Microbes and environment*. Springer.

Sauer, U., Cameron, D. C., & Balley, J. E. (1998). Metabolic capacity of *Bacillus subtilis* for the production of purine nucleosides, riboflavin and folic acid. *Biotechnology and Bioengineering, 59*, 227–238.

Saxena, S., & Pandey, A. K. (2001). Microbial metabolites as ecofriendly agrochemicals for the next millennium. *Applied Microbiology and Biotechnology, 55*, 395–403.

Schwechheimer, S. K., Park, E. Y., Revuella, J. L., Becker, J., Wittmann, C. (2016). Biotechnology of riboflavin. *Applied Microbiology and Biotechnology*, 2107–2119.

Shen, C. T., Chen, P. Y., Wu, J. J., Lee, T. M., Hsu, S. L., Chang, C. M. J., Young, C. C., & Shiejh, C. J. (2011). Purification of algal anti-tyrosinase zeaxanthin from *Nannochloropsis oculate* using supercritical anti-solvent precipitation. *Journal of Supercritical Fluids, 55*, 355–362.

Shimizu, S. (2001). Vitamins and related compounds: microbial production. In H.-J. Rehm, G. Reed (Eds.), Biotechnology: Special processes (2nd ed., Vol. 10). Wiley-VCH Verlag GmbH, Weinheim.

Shiomi, K., & Omura, S. (2004). Antiparasitic agents produced by microorganisms. *Proceedings of the Japan Academy, Series B, 80*, 245–258.

Shoop, W., & Soli, M. (2002). Ivermectin, abamectin, and eprinomectin. In J. Vercruysse & R. S. Rew (Eds.), *Macrocyclic lactones in antiparasitic therapy* (pp. 1–29). CABI Publishing.

Shrestha, B., Pandey, R. P., Darsadhari, S., Parajuli, P., & Sohng, J. K. (2019). Combinatorial approach for improved cyanidin 3-O-glucoside production in *Escherichia coli*. *Microbial Cell Factories, 18*, 7.

Siddiqui, M. S., Thodey, K., Trenchardl, S. C. D. (2012). Advancing secondary metabolite biosynthesis in yeast with synthetic biology tools. *FEMS Yeast Research, 12*, 144–170.

Singh, R., Kumar, M., Mittal, A., Mehta, P. K. (2017). Microbial metabolites in nutrition, healthcare and agriculture. *3 Biotech, 7*, 15.

Singh, S. B., Zink, D. L., Goetz, M. A., Dombrowski, A. W., Polishook, J. D., & Hazuda, D. J. (1998). Equistein and a novel opposite sterochemical homlog phomastein, two fungal metabolites as inhibitor of HIV-1 integrase. *Tetrahedron Letters, 39*, 2243–2246.

Strobel, G., & Daisy, B. (2003). Bioprospecting for microbial endophytes and their natural products. *Microbiological Reviews, 67*, 491–502.

Suarez-Jimenez, G. M., Burgos-Hernandez, A., & Ezquerra-Brauer, J. M. (2012). Bioactive peptides and desipeptides with anticancer potential: Sources from marine animals. *Marine Drugs, 10*, 964–986.

Sun, C., Yang, Z., Zhang, C., Liu, Z., He, J., Liu, Q., & Ma, J. (2019). Genome miming of *Streptomyces atratus* SCIO ZH 16: Discovery of atratumycin and identification of its biosynthetic gene cluster. *Organic Letters, 21*, 1453–1457.

Sun, X., Shen, X., & Jain, R. (2015). Synthesis of chemicals by metabolic engineering of microbes. *Chemical Society Reviews, 44*, 3760–3785.

Swan, D. A., & Kuo, J.-W. (1991). Hyaluronic acid. In D. Byrom (Ed.), *Biomaterials-novel materials from biological sources* (pp. 286–305). Stockton Press.

Takagi, Y., Sugisawa, T., & Hoshino, T. (2009). Continuous 2-keto-L-gulonic acid fermentation from L-sorbose by *Ketogulonigenium vulgare* DSM 4025. *Applied Microbiology and Biotechnology, 82*, 1049–1056.

TeWelcher, Y. M., Hendrik, H., Balague, M. M., Souza, C. M., Riezman, H., De Kruijff, B., & Breukink, E. (2008). Natamycin blocks fungal growth by binding specifically to ergosterol without permeabilizing the membrane. *Journal of Biological Chemistry, 283*, 6393–6401.

Tominaga, K., Hongo, N., Karato, M., & Yamashita, E. (2012). Cosmetic benefits of astaxanthin on human subjects. *Acta Biochimica Polonica, 59*, 43–47.

Tsavkelova, E. A., Klimova, S.-Y., Cherdyntseva, T. A., & Netrusov, A. I. (2006). Hormones and hormone like substances of microorganisms: A review. *Applied Biochemistry and Microbiology, 42*, 229–235.

Ueno, Y., Hirashima, N., Inoh, Y., Furuno, T., & Nakanishi, M. (2007). Characterization of biosurfactant-containing liposomes and their efficiency for gene transfection. *Biological and Pharmaceutical Bulletin, 30*, 169–172.

Verwaal, R., Wang, J., Meijnen, J.-P., Visser, H., Sandmann, G., van den Berg, J. A., & van Ooyen, A. J. J. (2007). High level production of beta carotene in *Saccharomyces cerevisiae* by successive transformation with carotenogenic genes from *Xanthophyllomyces dendrorhous. Applied and Environmental Microbiology, 73*, 4342–4350.

Villeneuve, P. (2007). Lipases in lipophilization reactions. *Biotechnology Advances, 25*, 515–536.

Wada, M., Sawada, K., Ogura, K., Shimono, Y., Hagiwara, T., Sugimoto, M., Onuki, A., & Yokota, A. (2016). Effects of phosphoenolpyruvate carboxylase desensitization on glutamic acid production of *Corynebacterium glutamicum* ATCC 13032. *Journal of Bioscience and Bioengineering, 121*, 172–177.

Wang, H. M. D., Chen, C. C., Huynh, P., & Chang, J. S. (2015). Exploring the potential of using algae in cosmetics. *Bioresource Technology, 184*, 355–362.

Willems, T., De Mol, M. L., De Bruycker, A., De Maeseneire, S. L., & Soetaert, W. K. (2020). Alkaloids from marine fungi: Promising antimicrobials. *Antibiotics, 9*, 340.

Wu, J., Liu, Y., Zhao, S., Sun, J., Jin, Z., & Zhang, D. (2019). Application of dynamic regulation to increase L-phenylalanine production in *Escherichia coli. Journal of Microbiology and Biotechnology, 29*, 923–932.

Wu, X., Zha, J., & Koffas, M. A. G. (2020). Microbial production of bioactive chemicals for human health. *Current Opinion in Food Science, 32*, 9–16.

Yang, J., Guo, L. (2014). Biosynthesis of β-carotene in engineered *E. coli* using the MEP and MVA pathways. *Micro Cell Fact, 13*, 160.

Yildiz, H., & Karatas, N. (2018). Microbial exopolysaccharides: Resources and bioactive properties. *Process Biochemistry, 72*, 41–46.

Yurekli, F., Yesilada, O., Yurekli, M., & Topcuoglu, S. F. (1999). Plant growth hormone production from olive oil and alcohol factory wastewater by white rot fungi. *World Journal of Microbiology and Biotechnology, 15*, 503–505.

Zhang, M., Zhang, Q., Zhang, Q., Cui, X., & Zhu, L. (2023). Promising antiparasitic natural and synthetic products from marine invertebrates and microorganisms. *Marine Drugs, 21*, 84.

Zheng, J., Zhu, H., Hong, K., Wang, Y., Liu, P., Wang, X., & Zhu, W. (2009). Novel cyclic hexapeptides from marine-derived fungus *Aspergillus sclerotiorum* PT06-1. *Organic Letters, 11*, 5262–5265.

Zhu, S., Wu, J., Du, G., Zhou, J., & Chen, J. (2014). Efficient synthesis of eriodictyol from L-tyrosine in *Escherichia coli. Applied and Environment Microbiology, 80*, 3072–3080.

The manufacturer's authorised representative in the EU is Springer Nature Customer Service Centre GmbH, Europaplatz 3, 69115 Heidelberg, Germany. If you have any concerns regarding our products, please contact ProductSafety@springernature.com

Printed and bound by CPI Group (UK) Ltd, Croydon, CR0 4YY
30/09/2026
02225516-0005